TRAFFIC-RELATED AIR POLLUTION

TRAFFIC-RELATED AIR POLLUTION

Edited by

HANEEN KHREIS

Center for Advancing Research in Transportation Emissions, Energy, and Health, Texas A&M Transportation Institute, TX, USA

MARK NIEUWENHUIJSEN

ISGlobal-CREAL, Barcelona, Spain

JOSIAS ZIETSMAN

Center for Advancing Research in Transportation Emissions, Energy, and Health, Texas A&M Transportation Institute, TX, USA

TARA RAMANI

Center for Advancing Research in Transportation Emissions, Energy, and Health, Texas A&M Transportation Institute, TX, USA

ELSEVIER

Elsevier
Radarweg 29, PO Box 211, 1000 AE Amsterdam, Netherlands
The Boulevard, Langford Lane, Kidlington, Oxford OX5 1GB, United Kingdom
50 Hampshire Street, 5th Floor, Cambridge, MA 02139, United States

Notices
Knowledge and best practice in this field are constantly changing. As new research and experience broaden our understanding, changes in research methods, professional practices, or medical treatment may become necessary.

Practitioners and researchers must always rely on their own experience and knowledge in evaluating and using any information, methods, compounds, or experiments described herein. In using such information or methods they should be mindful of their own safety and the safety of others, including parties for whom they have a professional responsibility.

To the fullest extent of the law, neither the Publisher nor the authors, contributors, or editors, assume any liability for any injury and/or damage to persons or property as a matter of products liability, negligence or otherwise, or from any use or operation of any methods, products, instructions, or ideas contained in the material herein.

Library of Congress Cataloging-in-Publication Data
A catalog record for this book is available from the Library of Congress

British Library Cataloguing-in-Publication Data
A catalogue record for this book is available from the British Library

ISBN: 978-0-12-818122-5

Publisher: Joe Hayton
Acquisitions Editor: Brian Romer
Editorial Project Manager: Aleksandra Packowska
Production Project Manager: Maria Bernard
Cover Designer: Christian J. Bilbow

Typeset by SPi Global, India

Contents

Contributors

Zorana Jovanovic Andersen
Section of Environmental Health, Department of Public Health, Faculty of Health and Medical Sciences, University of Copenhagen, Copenhagen, Denmark

Mohammad Hashem Askariyeh
Air Quality Scientist, Zachry Department of Civil and Environmental Engineering, Texas A&M University, College Station, TX, United States

Richard Baldauf
U.S. Environmental Protection Agency, Research Triangle Park, NC, United States

Sean D. Beevers
MRC-PHE Centre for Environment and Health; Environmental Research Group—King's College London, London, United Kingdom

Yoann Bernard
International Council on Clean Transportation, Berlin, Germany

Martin Birley
BirleyHIA, London, United Kingdom

Kanok Boriboonsomsin
College of Engineering—Center for Environmental Research and Technology, University of California at Riverside, Riverside, CA, United States

Doug Brugge
Department of Public Health Sciences, University of Connecticut, Farmington, CT, United States

Mark Burris
Texas A&M University, College Station, TX, United States

Christiana Demetriou
Department of Primary Care and Population Health, University of Nicosia Medical School, Nicosia, Cyprus

V. Dheeraj Alshetty
Department of Civil Engineering, Indian Institute of Technology Madras, Chennai, India

M. Diya
Department of Civil Engineering, Indian Institute of Technology Madras, Chennai, India

Mary Fox
Department of Health Policy and Management, Johns Hopkins Bloomberg School of Public Health, Baltimore, MD, United States

Christina H. Fuller
Department of Population Health Sciences, Georgia State University School of Public Health, Atlanta, GA, United States

Andrew Glazener
Center for Advancing Research in Transportation Emissions, Energy, and Health (CARTEEH), Texas A&M Transportation Institute (TTI), Bryan; Texas A&M | TTI Headquarters, Bryan; Langford College of Architecture, Texas A&M University, College Station; School of Architecture, University of Texas at Austin, Austin, TX, United States

Andrew Gouldson
School of Earth and Environment, University of Leeds, Leeds, United Kingdom

Peng Hao
College of Engineering—Center for Environmental Research and Technology, University of California at Riverside, Riverside, CA, United States

Sergio Ibarra-Espinosa
Departamento de Ciências Atmosféricas, Universidade de São Paulo, São Paulo, Brazil

Natalie Johnson
Department of Environmental and Occupational Health, Texas A&M University School of Public Health, College Station, TX, United States

Haneen Khreis
Center for Advancing Research in Transportation Emissions, Energy, and Health (CARTEEH), Texas A&M Transportation Institute (TTI), College Station, TX, United States; Barcelona Institute for Global Health (ISGlobal), Centre for Research in Environmental Epidemiology (CREAL), Barcelona; Universitat Pompeu Fabra (UPF), Barcelona; CIBER Epidemiología y Salud Pública (CIBERESP), Madrid, Spain

Kirsten Koehler
Department of Environmental Health and Engineering, Johns Hopkins Bloomberg School of Public Health, Baltimore, MD, United States

Juleen Lam
Department of Health Sciences, California State University East Bay, Hayward, CA, United States

Wen-Whai Li
Department of Civil Engineering, The Univeristy of Texas at El Paso, El Paso, TX, United States

Anthony D. May
University of Leeds, Leeds, United Kingdom

John Medeiros
Center for Advancing Research in Transportation Emissions, Energy and Health (CARTEEH), Texas A&M Transportation Institute (TTI), College Station, TX, United States

Jyothi S. Menon
Public Health Foundation of India, New Delhi, India

Mark R. Miller
University/BHF Centre for Cardiovascular Science, University of Edinburgh, Edinburgh, United Kingdom

Jennifer S. Mindell
Research Department of Epidemiology & Public Health, UCL (University College London), London, United Kingdom

Natalie Mueller
Barcelona Institute for Global Health (ISGlobal), Barcelona; Universitat Pompeu Fabra (UPF), Barcelona; CIBER Epidemiologia y Salud Pública (CIBERESP), Madrid, Spain

Mark J. Nieuwenhuijsen
Barcelona Institute for Global Health (ISGlobal), Centre for Research in Environmental Epidemiology (CREAL), Barcelona; Universitat Pompeu Fabra (UPF), Barcelona; CIBER Epidemiología y Salud Pública (CIBERESP), Madrid, Spain

Jennifer B. Raftis
University/BHF Centre for Cardiovascular Science, University of Edinburgh, Edinburgh, United Kingdom

Tara Ramani
Center for Advancing Research in Transportation Emissions, Energy, and Health (CARTEEH), Texas A&M Transportation Institute (TTI), College Station, TX, United States

April Willis Rodgers
Avatar Environmental and Technical Services, Marietta, GA; Samford University, Birmingham, AL, United States

Michael O. Rodgers
School of Civil and Environmental Engineering and School of Public Policy, University System of Georgia and Georgia Institute of Technology, Atlanta, GA, United States

David Rojas-Rueda
Barcelona Institute for Global Health (ISGlobal), Barcelona; Universitat Pompeu Fabra (UPF), Barcelona; CIBER Epidemiologia y Salud Pública (CIBERESP), Madrid, Spain; Department of Environmental and Radiological Health Sciences, Colorado State University, Fort Collins, CO, United States

Karl Ropkins
Institute for Transport Studies, University of Leeds, Leeds, United Kingdom

Kristen A. Sanchez
Center for Advancing Research in Transportation Emissions, Energy, and Health (CARTEEH), Texas A&M Transportation Institute (TTI); Texas A&M School of Public Health, College Station, TX, United States

Uwe Schlink
UFZ Helmholtz Centre for Environmental Research, Leipzig, Germany

S.M. Shiva Nagendra
Department of Civil Engineering, Indian Institute of Technology Madras, Chennai, India

Andrew Sudmant
School of Earth and Environment, University of Leeds, Leeds, United Kingdom

Shams Tanvir
College of Engineering—Center for Environmental Research and Technology, University of California at Riverside, Riverside, CA, United States

Suriya Vallamsundar
Associate Research Scientist, Texas A&M Transportation Institute, College Station, TX, United States

Ersilia Verlinghieri
Transport Studies Unit, School of Geography and the Environment, University of Oxford, Oxford, United Kingdom

Hanna M. Vesterinen
Program on Reproductive Health and the Environment, University of California, San Francisco, CA, United States

Paolo Vineis
MRC-PHE Centre for Environment and Health, Imperial College, London, United Kingdom

Martin L. Williams
MRC-PHE Centre for Environment and Health; Environmental Research Group—King's College London, London, United Kingdom

Tracey J. Woodruff
Program on Reproductive Health and the Environment, University of California, San Francisco, CA, United States

Yanzhi Ann Xu
Texas A&M Transportation Institute, Washington, DC, United States

Josias Zietsman
Center for Advancing Research in Transportation Emissions, Energy, and Health (CARTEEH), Texas A&M Transportation Institute (TTI), College Station, TX, United States

CHAPTER 1

Traffic-related air pollution: Emissions, human exposures, and health: An introduction

Haneen Khreis[a,b,c,d], Mark J. Nieuwenhuijsen[b,c,d], Josias Zietsman[a], and Tara Ramani[a]

[a]Center for Advancing Research in Transportation Emissions, Energy, and Health (CARTEEH), Texas A&M Transportation Institute (TTI), College Station, TX, United States
[b]Barcelona Institute for Global Health (ISGlobal), Centre for Research in Environmental Epidemiology (CREAL), Barcelona, Spain
[c]Universitat Pompeu Fabra (UPF), Barcelona, Spain
[d]CIBER Epidemiología y Salud Pública (CIBERESP), Madrid, Spain

Abbreviations

BC	black carbon
BoD	burden of disease
CARTEEH	Center for Advancing Research in Transportation Emissions, Energy, and Health.
CO	carbon monoxide
COPD	chronic obstructive pulmonary disease
EC	European Commission
EPA	US Environmental Protection Agency
HC	hydrocarbons
HIA	Health Impact Assessment
NO_2	nitrogen dioxide
NO_x	nitrogen oxides
O_3	ozone
PM	particulate matter
PM_{10}	particulate matter with diameter $< 10\,\mu m$
$PM_{2.5}$	particulate matter with diameter $< 2.5\,\mu m$
ROS	reactive oxygen species
SO_2	sulfur dioxide
SO_4	sulfates
TRAP	traffic-related air pollution
UFP	ultrafine particles
UK	United Kingdom
US	United States
WHO	World Health Organization

Traffic-Related Air Pollution
https://doi.org/10.1016/B978-0-12-818122-5.00001-6

Introduction

Traditionally, air pollution was recognized as an issue associated with domestic heating, coal burning, and industrial emissions (Vardoulakis et al., 2003). The most recent estimates by the Global Exposure Mortality Model reported over a doubling in the number of global deaths attributable to outdoor air pollution placing them at 8.9 million deaths in 2015 (95% confidence interval: 7.5–10.3); more than the number of deaths from cigarette smoking (Burnett et al., 2018). In the present postindustrial city scape, however, outdoor air pollution in many cities of the world (e.g., high-income countries) has become dominated by emissions attributable to road traffic, including dust, tailpipe, and non-tailpipe emissions of a wide variety of pollutants harmful to human health and well-being (Anderson, Favarato, & Atkinson, 2013; European Environment Agency, E.E.A, 2007). Nowadays, traffic-related air pollution (TRAP) represents a public health crisis, the extent and magnitude of which are large and keep growing as new knowledge and quantification methods become available.

TRAP refers to ambient air pollution resulting from the use of motorized vehicles such as heavy-duty and light-duty vehicles, buses, coaches, passenger cars, and motorcycles. TRAP is often also referred to as air pollution originating from on-road mobile sources. These vehicles emit a variety of air pollutants including but not limited to black carbon (BC), elemental carbon, carbon monoxide (CO), hydrocarbons (HC), nitrogen oxides (NO_x), nitrogen dioxide (NO_2), particulate matter with a diameter $<2.5\,\mu m$ ($PM_{2.5}$), particulate matter with a diameter $<10\,\mu m$ (PM_{10}), and particles with a diameter $<0.1\,\mu m$ which are referred to as ultrafine particles (UFP). These pollutants can be directly emitted through the vehicle exhaust, and are known as tailpipe emissions (Khreis, 2020). They can also be emitted through non-exhaust mechanisms such as evaporative emissions, the resuspension of dust, the wear of brakes and tires, and the abrasion of road surfaces, and are known as non-tailpipe emissions (Askariyeh et al., 2020; Khreis, 2020). The tailpipe emissions for key pollutants such as CO, NO_x, $PM_{2.5}$, and PM_{10}, and even particles number for UFP in Europe, are regulated (Khreis, 2020) and differ depending on the vehicle's fuel type (e.g., diesel versus petrol) and age. However, non-tailpipe emissions are not entirely regulated and their relative importance is growing with the expected widespread introduction of electric vehicles (Timmers & Achten, 2016). Vehicle emissions disperse into ambient air depending on multiple factors which are highly variable such as wind speed, wind direction and atmospheric stability, local and regional terrain, and background air pollution concentrations from other sources such as industry, agricultural emissions, and coal and wood burning (Khreis, 2020). The result of this dispersion is elevated concentrations of air pollutants, through primary emissions or through the formation of secondary pollutants. Humans are then exposed to these air pollutants in ambient air, or indoors through the infiltration of outdoor air pollutants. Human exposures and their inhaled doses which reach target organs or tissues are also determined by various dynamic factors such as mobility patterns, distance from the source, age, height, physical activity, respiratory rates, and transport mode choice (Khreis, 2020). Human exposures to TRAP can elicit a wide range of adverse health effects. The full chain of events covering traffic activity, vehicle emissions, the dispersion of these emissions, human exposures, and their ultimate health impacts as described in this chapter is depicted in Fig. 1.1.

The health effects of TRAP are very similar to the health effects of ambient air pollution in general. However, recent evidence shows that certain health effects, such as the onset of childhood asthma, are more strongly associated with the local (intra-urban) variation in air pollution, usually dominated by TRAP and not the regional variations in air pollution, usually dominated by other sources (Khreis, 2020). The differences in health effects from air pollution stemming from different sources may be explained by differential toxicity. Some of the factors which might drive this differential toxicity include the different chemical composition of the TRAP mixture as well as the different physical characteristics of the pollutants such as the size, surface area, and number of the smallest particles that are inhaled and

Fig. 1.1 The full chain of events linking TRAP to health impacts. *Source: Center for Advancing Research in Transportation Emissions, Energy and Health (CARTEEH), available from: https://www.carteeh.org/.*

reach target organs. These differences are being discussed more often, however, their relevance to the observed health effects are understudied in air pollution epidemiology and a common assumption is that all pollutants from the same species (e.g., all particles) are equally toxic (Khreis, 2020).

Recent studies, reviews, and meta-analyses found consistent associations between living near major roads and/or common traffic-related air pollutants and premature natural mortality from a wide variety of causes (Beelen et al., 2014; Brook et al., 2010; Cao et al., 2011; Hoek et al., 2002; Luo et al., 2019; Yang et al., 2013), supporting decades of research stemming from the seminal Harvard six cities study which demonstrated the association between ambient concentrations of sulfur dioxide (SO_2), particulate matter (PM), ozone (O_3), and sulfates (SO_4), and differences in the probability of survival (Dockery et al., 1993). Common traffic-related air pollutants and different TRAP metrics have also been associated with a wide spectrum of diseases, including, but not limited to, cardiovascular disease (Cesaroni et al., 2014; Link & Dockery, 2010; Mustafić et al., 2012; Stafoggia et al., 2014), lung cancer (Raaschou-Nielsen et al., 2013), diabetes (Eze et al., 2015; Liang et al., 2019), adverse birth outcomes (Fleischer et al., 2014; Pedersen et al., 2013; Sapkota et al., 2012), congenital anomalies (Vrijheid et al., 2010), pregnancy-induced hypertensive disorders and preeclampsia (Pedersen et al., 2014), and adverse respiratory outcomes, especially in susceptible populations like the children and the elderly (Andersen et al., 2011; Brunst et al., 2015; Gasana et al., 2012; Khreis et al., 2017a; Lindgren et al., 2009; MacIntyre et al., 2014; Nordling et al., 2008). This list of associated health effects continues to grow as more research becomes available and as new questions are being asked. The list now includes health effects which were not associated with TRAP a decade or so ago such as autism and child behavioral problems (Min & Min, 2017; Raz et al., 2015; Yolton et al., 2019), cognitive decline (Tonne et al., 2014; Weuve et al., 2012; Zhang, Chen, & Zhang, 2018), dementia and Alzheimer's disease (Carey et al., 2018; Chen et al., 2017; Oudin et al., 2016; Paul et al., 2019), obesity (Jerrett et al., 2014), and increased number of osteoporosis-related fracture hospital visits and decreased bone density (Prada et al., 2017). New adverse health effects associated with TRAP continue to emerge at a very rapid pace (Loxham, Davies, & Holgate, 2019), and the body of evidence has been strengthened substantially to demand urgent action (Khreis, 2020). Furthermore, many credible pathological mechanisms have been elucidated, lending biological plausibility to the epidemiological findings outlined earlier and strengthening the case for action (Block & Calderón-Garcidueñas, 2009; Costa et al.,

2017; Heusinkveld et al., 2016; Kelly & Fussell, 2017; Lodovici & Bigagli, 2011; Miller, Shaw, & Langrish, 2012). These mechanisms include airway remodeling, inflammation, oxidative stress, and a shift in immune function (Barthelemy et al., 2020; Thurston et al., 2020).

It is also important to note that the adverse health effects associated with TRAP, and air pollution in general, have been detected at relatively low pollutant concentrations, well below the European Commission's (EC), the United States Environmental Protection Agency's (EPA), and the more stringent World Health Organization's (WHO) air quality standards and guideline values (Beelen et al., 2014; Belanger et al., 2006; Castro et al., 2009; Chen & Omaye, 2001; Loxham et al., 2019; MacIntyre et al., 2014; Nishimura et al., 2013; Pedersen et al., 2013; Scoggins et al., 2004; Wei et al., 2019; World Health Organization, W.H.O., 2013). On the other hand, studies have also shown that relatively small reductions in ambient air pollution can result in beneficial health effects, reinforcing that no air pollution threshold for the associated adverse health effects can be established (Bayer-Oglesby et al., 2005; Currie, Ray, & Neidell, 2011). Currently, there is no known "safe" lower limit for exposure to air pollution, under which adverse health effects would not be observed. The science therefore seems to be outpacing the legislation and current air quality guideline values and standards, which many countries and regions are already failing to meet, do not reflect the latest evidence put forward by air pollution epidemiology (Hitchcock et al., 2014). Not only is the science outpacing the legislation, but some scientific evidence has yet no place in current legislation (Khreis, 2020). For example, pollutants like ammonia, BC, and UFPs have been shown to elicit adverse health effects and have been suggested to be abundant at the local scale due to traffic activity (Cape et al., 2004; Dennekamp et al., 2002; Durbin et al., 2001; Khreis et al., 2017a; Luben et al., 2017; Onat & Stakeeva, 2013; Perrino et al., 2002; Perrino, Catrambone, & Di Menno Di Bucchianico, 2003; Skjøth & Hertel, 2013; Tomlin, Sutton, & Tate, 2010). The levels of these pollutants in ambient air are not only unregulated and but are also not routinely measured or studied.

The public health implications of these health effects and their associations with TRAP are only becoming clearer as new burden of disease (BoD) and health impact assessments (HIA) provide means to their quantification. For example, in recent estimates, a quarter of new childhood asthma cases per year in an English urban area were attributed to TRAP (Khreis, de Hoogh, & Nieuwenhuijsen, 2018). In conservative global estimates, 184,000 deaths a year were attributable to TRAP (Bhalla et al.,

2014). Similarly, Lelieveld et al. estimated that land transportation-related air pollution is responsible for one-fifth of deaths from air pollution in the United Kingdom (UK), the United States (US), and Germany (Lelieveld et al., 2015). The public health burden of many of the health effects associated with TRAP has, however, not been quantified and therefore the full burden is yet to be fully elucidated. This is true for air pollution in general, but it is particularly true for TRAP, as specific source apportionment or full-chain BoD or HIA assessments are needed to come up with these estimates (Nieuwenhuijsen et al., 2017). Such exercises are challenging in their nature as they require many data inputs which are not routinely available, cross-disciplinary expertise, specific skills set, and need to be grounded in epidemiological observations and biological plausibility.

TRAP is not yesterday's problem. The importance and relevance of TRAP exposures continue to increase in a world that is witnessing rapid population growth and unprecedented urbanization (United Nations Department of Economic and Social Affairs, U.D., 2017). In many regions of the world, population growth and economic growth come hand in hand with an increase in the number of vehicles being manufactured, purchased, driven, and emitting more pollutants. The lack of public and active transportation culture and infrastructure, and the already deep-rooted cultural and economic dependence on motor vehicles being the primary mode of transportation which dominates transportation planning and policy (Khreis et al., 2016), exacerbate TRAP exposures and their health effects. The increased demand for travel and transportation activity continues to overpower emission regulations and advancements in vehicle and fuel technology (Metz et al., 2007). The adaptation of new technologies such as electric vehicles, which might alleviate TRAP, is rather slow and can be scrutinized due to the lack of progress with electricity decarbonization and non-tailpipe emissions mitigation (Khreis et al., 2019a). The rapid and unprecedented urbanization that the world is currently witnessing is also worrisome in this context. The world's urban population is projected to increase to 68% by 2050 (United Nations Department of Economic and Social Affairs, 2018). Cities and urban areas are hot spots for human exposure to TRAP, where traffic activity is not only higher, but also acts in close proximity to people increasing their harmful exposures and associated adverse health effects (Khreis, 2020; Khreis et al., 2017b). Of note, 92% of the world's population lives in cities where air pollution levels exceeds the WHO air quality guidelines (World Health Organization, W.H.O., 2016); which are arguably still too high to fully protect public health.

TRAP is particularly challenging to study and mitigate. TRAP is highly heterogeneous in space and in time. Concentrations of common vehicular pollutants are spatially misaligned and can significantly fluctuate even over no more than a few tens of meters (Bell, 2014; Briggs et al., 1997; Vardoulakis et al., 2003). The contribution of traffic to ambient air pollution levels is generally not well monitored or quantified and varies significantly between and within cities as it depends on numerous dynamic factors. These include the traffic flows, traffic speeds, congestion and start and stop driving, the vehicles' fleet mix including the age and maintenance of vehicles, fuel used, meteorology, local and regional terrain and characteristics of the built and natural environment such as the prevalence of street canyons, and availability of green spaces which influence pollutants' dispersion and deposition (Khreis, 2020). Also, some pollutants such as BC, NO_2, and urban UFPs are more specific markers of the TRAP mixture while others such as $PM_{2.5}$ and PM_{10} are shared across many other sources and their traffic contribution is less by definition.

Using different methods, some studies have reported on the contribution of traffic to ambient air pollution levels in different regions across the world. In Europe, for example, the traffic contribution to urban PM concentrations ranges from as little as 9% to up to 53% for PM_{10} and between 9% and 66% for $PM_{2.5}$ with an average of 39% and 43% at traffic sites, respectively (Sundvor et al., 2012). When observing for NO_2, this range can be much higher and can reach up to 80% (Sundvor et al., 2012). In low-income and developing countries, transportation activity typically accounts for significant overall shares of air pollution: ~67% of PM in Sao Paulo, Brazil (Josh Miller & Façanha, 2017), ~33% of all air pollutants throughout India (International Council on Clean Transportation, 2018), and up to 40% of air pollution in Chinese cities (World Resources Institute, 2018). These high percentages are likely due to the use of older and less efficient vehicles (WHO, 2018). Exposures to TRAP remain highly ubiquitous and of great relevance to human health as populations live, work, and play in close proximity to traffic activity. Notably, 24% of the population in Toronto (Canada), 41% of the population in New Delhi (India), 66% of the population in Beijing (China), 67% of the population in Paris (France), and 96% of the population in Barcelona (Spain) are potentially exposed to TRAP (Su et al., 2015).

Significant advances in TRAP and health research have been made including advancing the methods to assess traffic activity, vehicle emissions, air pollution, and human exposures. Furthermore, the associations between

TRAP and numerous health effects have been established in epidemiology and emerging health effects are continuously being studied. The strength of the overall body of evidence is assessed and the case for biological plausibility has been strengthened through toxicological and mechanistic studies and high-resolution and high-throughput technologies interrogating -omics. More stringent air quality guidelines have been developed and research is yet showing health risks occurring below these thresholds. BoD and HIA are being used more often to qualitatively asses and quantify the health burden attributable to TRAP and demonstrate the unequal distribution of that burden according to socioeconomic and sociodemographic factors. Policy options to mitigate TRAP and its adverse health effects are expanding and so are the studies quantifying the potential impacts and the cost-effectiveness of a wide range of policies. The potential health impacts of emerging technologies are being discussed and best practices to achieve TRAP reductions in addition to a multitude of goals, beyond air quality, are now documented. While many advances have been made in TRAP and health research and practice, the above emerging knowledge remains sporadic and there has not been a comprehensive synthesis of the health effects and public health impacts associated with TRAP exposures. Also, importantly, there are numerous new and existing methods and tools to quantify TRAP, its health impacts, and the impacts of potential policy and emerging technology options to mitigate its adverse health effects. However, these methods and tools are not mainstreamed in teaching, research, practice, or policymaking and the real-world impact of scientific and methodological advances is, therefore, very limited. The transportation sector is changing rapidly with new technologies and innovations such as autonomous and electric vehicles arising and being adopted across the world. The impacts of these technologies on TRAP levels and human health are also understudied but are critical to understand early on so that policy makers and planners can devise a proactive rather than a reactive approach to mitigate the potential adverse effects of these disruptors. Furthermore, the reduction in TRAP and the mitigation of its adverse health effects is only one objective in the policy decision-making realm. This objective, in fact, may be viewed as either a barrier to achieving other important objectives such as economic growth and efficiency, or as an enabler to achieving say climate change mitigation and equity. This overlap, and the lack of, is not fully understood as these issues are often discussed and studied in separate circles despite the ever-growing need for systemic and holistic approaches to tackle the multifaceted issues many cities struggle with, and to avoid unintended consequences (Cames & Helmers, 2013; Khreis et al., 2016, 2020). Urban traffic is responsible for many other adverse environmental exposures and associated adverse health effects that extend well beyond poor air quality (Khreis et al., 2016), and the relative importance of these different exposures varies significantly based on the local context and baseline environmental and health conditions (Khreis et al., 2019). The most effective and prudent mitigation efforts would consider this wider range of exposures and health effects which include motor vehicle crashes, physical inactivity, traffic-related noise, social exclusion, and greenhouse gases (Khreis, Glazener, et al., 2019). There is also little knowledge available on how integrated policy packages can address more than one of these issues and how effective these could be in achieving a multitude of goals such as increasing physical activity to promote public health and mitigating climate change (Glazener & Khreis, 2019; Gouldson et al., 2015; May, Khreis, & Mullen, 2018). Finally, of utmost importance is the environmental justice and inequity implications associated with TRAP, which are similar to the inequity concerns associated with other traffic-related exposure such as motor vehicle crashes, physical inactivity, traffic-related noise, and greenhouse gases (Khreis et al., 2016). Environmental justice is a historical and key issue which needs careful consideration when assessing the health impacts of transportation and planning or making policies to mitigate them. While there is no shortage of literature showing that exposures to TRAP (and to other traffic-related exposure), and therefore its associated adverse health effects, tend to be higher and more concentrated in lower socioeconomic locales and ethnic minority communities (Khreis et al., 2020; Lucas et al., 2019), there is a significant gap between the available scientific knowledge and real-world practice and policy decision-making. The most effective use of scarce resources may be made by better considering these susceptible populations in the planning process and addressing the range of intrinsic and extrinsic factors which makes them more exposed, and also more susceptible to their exposures (Khreis et al., 2016, 2019c). These factors include their living locations and conditions, their occupations, malnutrition and the lack of antioxidant intakes, exposure to stress, exposure to violence, genetics, and others. These factors can modify and often amplify the disproportionate adverse health effects of TRAP and other traffic-related exposures, contributing further to gross inequalities in health (Marmot, 2005). We have come a long way, but there are as yet critical knowledge gaps which need to be filled offering exciting research opportunities and a pathway to push and track progress toward the goal of clean air and protection of the public's health.

In this book, we focus on TRAP as one adverse and modifiable environmental exposure that significantly affects public health, especially in cities. Through the narrative of the book, we follow the full-chain framework linking traffic activity to its ultimate health impacts (Fig. 1.1), while considering the key issues surrounding this full chain. This book is structured as follows. Chapter 2 provides definitions for terminologies commonly used in describing the impacts of transportation operations on the air environment and human health, including describing the sources of air pollution in categories of origin (natural vs anthropogenic), mechanism (primary vs secondary), configuration (point, fugitive, vs mobile), and quantity (major vs area) of emissions. The chapter also provides a brief history and background for the evolution of air quality regulations along with trends of transportation emissions as well as air quality regulations and enforcement (Li, 2020). We then take the reader through every step between road traffic and its ultimate health impacts giving the reader the required knowledge, methods and tools to understand, assess and quantify road traffic, their associated emissions, air pollution, human exposures, health effects, and population-based health impacts. We overview methods and tools from both the monitoring and modeling worlds, in addition to qualitative and quantitative methods for assessing the health impacts of TRAP. Chapter 3 reviews the data and tools currently available for traffic monitoring and modeling in the context of transportation energy, emissions, air quality, and health analyses. The chapter outlines the existing data sources available for traditional traffic management and travel demand models and gives examples of novel data sources which can advance the field (Xu, 2020). Chapter 4 introduces conventional emissions monitoring and modeling practices used in the on-road transportation sector. The chapter critically describes both laboratory-based dynamometer and real-world emissions monitoring practices and identifies key events that helped to shape conventional practices. It also describes commonly used macro-, meso-, and microscale emissions modeling methods, and the changing nature of the relationship between monitoring and modeling activities and considers concerns regarding the effectiveness of existing practices (Ropkins, Ibarra-Espinosa, & Bernard, 2020). Chapter 5 goes on to describe the next element in the full-chain framework: air pollution monitoring and modeling methods, specifically in the context of TRAP, and overviews common metrics to compare modeled and monitored values and assess the validity of models. The chapter provides a high-level overview of TRAP in three subsections: monitoring, modeling, and model to monitor comparisons (Askariyeh, Khreis, & Vallamsundar, 2020). Chapter 6 takes the reader through the world of exposure assessment, a prerequisite to understanding and studying the health effects and impacts of TRAP and TRAP changes due to relevant policies. The chapter provides the reader with exposure definitions, discusses air pollution's impacts on health and associated legislation, which frames the spatial and temporal scales needed to correctly address human exposure. The chapter goes on to discuss exposure pathways and established and emerging exposure assessment methods covering the methods of fixed-site monitoring, land use regression, satellite remote sensing, atmospheric dispersion models, microenvironmental, and personal exposure models and personal monitoring (Beevers & Williams, 2020).

The following chapters move to the health research side. In Chapter 7, a brief history of air pollution epidemiology is provided alongside key examples of landmark studies which shaped the field. The differences between short-term and long-term air pollution studies are discussed, and the different study designs under each category are overviewed. The chapter also discusses the major strengths and challenges in air pollution epidemiology research, providing research recommendations and an outlook into the future of the field (Andersen, 2020). Chapter 8 provides basic definitions and methods used to synthesize epidemiological studies of air pollution effects through the use of systematic reviews and meta-analyses. The chapter discusses the utility of systematic reviews and meta-analyses in policy decision-making and BoD and HIA. The elements of systematic reviews are illustrated in this chapter through an evaluation of systematic reviews of TRAP and health. The strengths and limitations of systematic review methods used in this topic area are critically discussed in addition to how the strength of evidence as a whole can be assessed to increase its relevance and utility for evidence-based policy decision-making (Lam, Vesterinen, & Woodruff, 2020). Chapter 9 then introduces the reader to the wide range of health effects which have been associated with TRAP and discusses these effects in categories of established and emerging health effects. The chapter also presents special topics including cumulative risk assessment and environmental justice considerations in the context of TRAP exposures and human health (Fox, Koehler, & Johnson, 2020). Chapter 10 grounds the knowledge gained from air pollution epidemiology, systematic reviews, and meta-analyses in clinical and toxicological research and overviews a range of potential underlying mechanistic pathways which might be driving the observed effects. The chapter gives special attention to the effect of particles in TRAP on the cardiovascular system as an exemplar for the many mechanisms by which these pollutants can exert multiple effects on an organ system through a variety of underlying pathways (Miller & Raftis, 2020). Chapter 11 overviews the state-of-art research on exposure assessment using biomarkers which can increase the understanding of underlying mechanisms and enable the investigation of individual susceptibility, an area that is under researched in the field. The chapter introduces the exposome approach and demonstrates it as a promising approach to make sense of the multitude of signals emerging from -omics investigations and to strengthen causal inferences on the health effects of TRAP (Demetriou & Vineis, 2020).

The following chapters take the reader up from the biological and sample investigations of health effects and their underlying mechanisms through the public health impact implications at larger population levels; important for policy decision-making. Chapter 12 defines qualitative HIA, providing a brief history of its development and explaining its stages and the range of tools and methods used in HIA. The chapter clarifies the central role of assessment of inequalities in most forms of HIA, and how this is particularly pertinent for TRAP. The chapter discusses the strengths and limitations of qualitative HIAs and their use in policy decision-making (Mindell & Birley, 2020). Chapter 13 moves to quantitative HIA and BoD assessments as means to providing quantitative estimates of the current and/or expected health impacts which may be attributable to the exposure to TRAP, and the distribution of these health impacts. The strengths and limitations of quantitative HIA and BoD assessments are overviewed in addition to examples of key studies in this emerging field (Mueller, Nieuwenhuijsen, & Rojas-Rueda, 2020).

We then move on to the practice and policy decision-making realms, focusing on the real-world implications of this emerging science and overviewing the impacts of science on policy decision-making and the evidence on best practices as implemented in cities across the world. Chapter 14 discusses the development of air quality and public health policies within the context of contemporary societal needs and values. The chapter discusses the history of air pollution and how various cultures have developed public responses to respond to this dynamic issue over time (Rodgers & Rodgers, 2020). Chapter 15 follows to show the transportation policies which might be adopted to address the challenge of TRAP and its associated exposures and adverse health impacts. It considers how potential policies might best be identified and preferred policies selected. The chapter introduces the reader to the concept of and need for option generation as a formalized approach to selecting policy options and outlines the case for packaging of measures, and the use of packages to overcome constraints of individual policy measures (May, 2020). Chapter 16 extends beyond the isolated issue of poor air quality and overviews the best recent practices implemented in cities from across the world to both reduce TRAP and increase populations' physical activity, another major public health issue. The chapter overviews a wide range of measures and packages ranging from car-free policies, vehicle technologies, urban design interventions, green space provision, public transportation investment, and the integration thereof (Glazener & Khreis, 2020). Chapter 17 then focuses on one of these practices and explores the role of vegetation barriers and green space to mitigate TRAP and its adverse health effects. The chapter summarizes recommendations for roadside vegetation and green space characteristics that can lead to improved local air quality, and which can be used by planners and developers to achieve air quality and health benefits (Baldauf, 2020). Chapter 18 then delves into transportation investment appraisal and discusses why monetary values are needed for non-pecuniary items such as air quality and health. The chapter overviews different methods of quantifying the benefits of improved air quality and for calculating the cost-effectiveness of policies with a focus on one of the most commonly used methods in transportation investment appraisal: the benefit–cost analysis (Burris, 2020). Chapter 19 draws on a systematic review of core papers in the field of transportation studies and provides an overview of a broad range of potential benefits and negative impacts on health, congestion, employment, and poverty associated with three core categories of low carbon urban transportation policies which have significant air quality implications: land use changes, modal shift and public transportation improvements, and fleet improvement and transportation electrification. The chapter also overviews the barriers and the facilitators to policy implementation of low carbon urban transportation policies, in the context of mitigating traffic-related emissions, air pollution, and impacts on public health (Sudmant et al., 2020). Chapter 20 addresses the environmental justice issues related to TRAP exposures and associated health effects and provides real-world examples. The chapter explores the measurements of these linkages and makes recommendations on how to further our understanding of health inequity and how to support the inclusion of these implications into policy decision-making (Fuller & Brugge, 2020).

The next set of chapters overview the dynamic and changing nature of this field by looking into emerging transportation technologies, market solutions, new data sources, and the integration of emerging measurement and modeling technologies to advance our understanding of TRAP and its impacts. Chapter 21 explores connected and autonomous vehicles, on-demand mobility services, and zero-emission vehicles, as three key emerging technology categories that are profoundly changing the transportation sector and that have important implications for TRAP and its health effects. The chapter discusses the potential benefits and unintended consequences of the adoption of these technologies from an emissions, exposure, and health standpoints. The chapter also overviews implementation considerations specific to each of the three technologies for lowering emissions and mitigating the adverse associated health impacts (Tanvir, Hao, & Boriboonsomsin, 2020). Chapter 22 looks at market solutions that have been put forward to reduce personal exposure to TRAP, a major concern in cities of both the developed and developing worlds (Nagendra et al., 2020). Chapter 23 then introduces the reader to a new literature library which summarizes a vast literature on the elements of the full chain between traffic air pollution sources and associated health outcomes. The chapter describes the development, analysis, and benefits of the literature library, emphasizes the full-chain assessment of TRAP and health, describes the current state of literature regarding full-chain elements, making recommendations for future studies (Sanchez et al., 2020). Chapter 24 finally explores the current rise in the popularity of low-cost and mobile sensors, along with new approaches to modeling techniques. The chapter focuses on the increasing number of studies utilizing these technologies and combining them together in a single operation, examining how the use of these new methods is leading to improvements in the characterization of air pollution such as higher spatial and temporal resolution, measurements of street-level TRAP, source attribution/hotspot identification, and the assessment of personal exposure (Medeiros & Khreis, 2020). We conclude the book in Chapter 25 with an integrative summary of the information contained within and recommendations on how to better influence discourse and policy decision-making to account for the health effects of TRAP and their real societal cost.

This book is intended as a resource for educators, under- and postgraduate students, researchers, practitioners, advocacy groups, and policy makers interested in the area of transportation and health, especially the impact of transportation emissions and air pollution exposures on human health.

It is our hope that this book will provide the reader with a wide-ranging knowledge of the theory, methods, and tools available to assess and mitigate TRAP and its adverse health impacts. By the end of the book, the reader will appreciate that TRAP remains a public health crisis, and one that is gaining increased attention in both the transportation and public health spheres. The reader will become familiar with the existing and emerging methods and tools that can be used to assess TRAP and quantify its impacts for inclusion in practice and policy decision-making and the existing and emerging policy and technology options that can mitigate TRAP and its adverse health impacts but might have unintended consequences. A consistent message that emerges throughout the book is that when it comes to advancing knowledge and influencing policy decision-making, there is great value in mapping the holistic spectrum from road vehicles to their health impacts and a great value in cross-disciplinary research and practice.

Acknowledgments

Fig. 1.1 was created with funding from the Texas A & M Transportation Institute's Center for Advancing Research in Transportation Emissions, Energy, and Health (CARTEEH), a US Department of Transportation's University Transportation Center, College Station, TX. The grant number is 69A3551747128. More information about the Center for Advancing Research in Transportation Emissions, Energy, and Health is available at: https://www.carteeh.org/.

References

Andersen, Z. J. (2020). Air pollution epidemiology. In H. Khreis, et al. (Eds.), *Traffic-related air pollution*: Elsevier.

Andersen, Z. J., et al. (2011). Chronic obstructive pulmonary disease and long-term exposure to traffic-related air pollution: A cohort study. *American Journal of Respiratory and Critical Care Medicine, 183*(4), 455–461.

Anderson, H. R., Favarato, G., & Atkinson, R. W. (2013). Long-term exposure to air pollution and the incidence of asthma: Meta-analysis of cohort studies. *Air Quality, Atmosphere and Health, 6*(1), 47–56.

Askariyeh, M. H., Khreis, H., & Vallamsundar, S. (2020). Air pollution monitoring and modeling. In H. Khreis, et al. (Eds.), *Traffic-related air pollution*: Elsevier.

Askariyeh, M. H., et al. (2020). Near-road traffic-related air pollution: Resuspended PM2. 5 from highways and arterials. *International Journal of Environmental Research and Public Health, 17*(8), 2851.

Baldauf, R. (2020). Air pollution mitigation through vegetation barriers and green space. In H. Khreis, et al. (Eds.), *Traffic-related air pollution*. Elsevier.

Barthelemy, J., et al. (2020). New opportunities to mitigate the burden of disease caused by traffic related air pollution: Antioxidant-rich diets and supplements. *International Journal of Environmental Research and Public Health, 17*(2), 630.

Bayer-Oglesby, L., et al. (2005). Decline of ambient air pollution levels and improved respiratory health in Swiss children. *Environmental Health Perspectives, 113*, 1632–1637.

Beelen, R., et al. (2014). Effects of long-term exposure to air pollution on natural-cause mortality: An analysis of 22 European cohorts within the multicentre ESCAPE project. *The Lancet, 383*(9919), 785–795.

Beevers, S. D., & Williams, M. L. (2020). Traffic-related air pollution and exposure assessment. In H. Khreis, et al. (Eds.), *Traffic-related air pollution*: Elsevier.

Belanger, K., et al. (2006). Association of indoor nitrogen dioxide exposure with respiratory symptoms in children with asthma. *American Journal of Respiratory and Critical Care Medicine, 173*(3), 297–303.

Bell, L. M. (2014). *Critical issues of exposure assessment for human health studies of air pollution.* Available from: http://webcache.googleusercontent.com/search?q=cache:oGERjACtp1oJ:www.samsi.info/sites/default/files/BELL_Samsi_presentation.ppt+&cd=1&hl=en&ct=clnk&gl=uk.
Bhalla, K., et al. (2014). Transport for health: The global burden of disease from motorized road transport. In *Global road safety facility*. Washington, DC: Institute for Health Metrics and Evaluation and World Bank.
Block, M. L., & Calderón-Garcidueñas, L. (2009). Air pollution: Mechanisms of neuroinflammation and CNS disease. *Trends in Neurosciences, 32*(9), 506–516.
Briggs, D. J., et al. (1997). Mapping urban air pollution using GIS: A regression-based approach. *International Journal of Geographical Information Science, 11*(7), 699–718.
Brook, R. D., et al. (2010). Particulate matter air pollution and cardiovascular disease an update to the scientific statement from the American Heart Association. *Circulation, 121*(21), 2331–2378.
Brunst, K. J., et al. (2015). Timing and duration of traffic-related air pollution exposure and the risk for childhood wheeze and asthma. *American Journal of Respiratory and Critical Care Medicine, 192*(4), 421–427.
Burnett, R., et al. (2018). Global estimates of mortality associated with long-term exposure to outdoor fine particulate matter. *Proceedings of the National Academy of Sciences, 115*(38), 9592–9597.
Burris, M. (2020). Cost-effectiveness of projects and policies. In H. Khreis, et al. (Eds.), *Traffic-related air pollution*: Elsevier.
Cames, M., & Helmers, E. (2013). Critical evaluation of the European diesel car boom-global comparison, environmental effects and various national strategies. *Environmental Sciences Europe, 25*(1), 1–22.
Cao, J., et al. (2011). Association between long-term exposure to outdoor air pollution and mortality in China: A cohort study. *Journal of Hazardous Materials, 186*(2–3), 1594–1600.
Cape, J., et al. (2004). Concentrations of ammonia and nitrogen dioxide at roadside verges, and their contribution to nitrogen deposition. *Environmental Pollution, 132*(3), 469–478.
Carey, I. M., et al. (2018). Are noise and air pollution related to the incidence of dementia? A cohort study in London, England. *BMJ Open, 8*(9), e022404.
Castro, H. A., et al. (2009). Effect of air pollution on lung function in schoolchildren in Rio de Janeiro, Brazil. *Revista de Saúde Pública, 43*(1), 26–34.
Cesaroni, G., et al. (2014). Long term exposure to ambient air pollution and incidence of acute coronary events: Prospective cohort study and meta-analysis in 11 European cohorts from the ESCAPE project. *BMJ, 348*, f7412.
Chen, L., & Omaye, S.T. (2001). Air pollution and health effects in Northern Nevada. *Reviews on Environmental Health, 16*(2), 133–149.
Chen, H., et al. (2017). Living near major roads and the incidence of dementia, Parkinson's disease, and multiple sclerosis: A population-based cohort study. *The Lancet, 389*(10070), 718–726.
Costa, L. G., et al. (2017). Neurotoxicity of traffic-related air pollution. *Neurotoxicology, 59*, 133–139.
Currie, J., Ray, S. H., & Neidell, M. (2011). Quasi-experimental studies suggest that lowering air pollution levels benefits infants' and children's health. *Health Affairs, 30*(12), 2391–2399.
Demetriou, C., & Vineis, P. (2020). Biomarkers and OMICS of health effects associated with traffic-related air pollution. In H. Khreis, et al. (Eds.), *Traffic-related air pollution*: Elsevier.
Dennekamp, M., et al. (2002). Exposure to ultrafine particles and PM2. 5 in different micro-environments. *Annals of Occupational Hygiene, 46*(suppl 1), 412–414.
Dockery, D. W., et al. (1993). An association between air pollution and mortality in six US cities. *New England Journal of Medicine, 329*(24), 1753–1759.

Durbin, T. D., et al. (2001). Emissions of ammonia from light duty vehicles. In *10th international emission inventory conference.* United States Environmental Protection Agency: Denver.

European Environment Agency, E.E.A. (2007). *Transport contribution to air quality.* 18 January 2014 [cited 2014 6 January 2016]; Available from: http://www.eea.europa.eu/data-and-maps/indicators/transport-contribution-to-air-quality-3.

Eze, I. C., et al. (2015). Association between ambient air pollution and diabetes mellitus in Europe and North America: Systematic review and meta-analysis. *Environmental Health Perspectives (Online), 123*(5), 381.

Fleischer, N. L., et al. (2014). Outdoor air pollution, preterm birth, and low birth weight: Analysis of the world health organization global survey on maternal and perinatal health. *Environmental Health Perspectives, 122*(4), 425–430.

Fox, M., Koehler, K., & Johnson, N. (2020). Established and emerging effects of traffic-related air pollution. In H. Khreis, et al. (Eds.), *Traffic-related air pollution*: Elsevier.

Fuller, C. H., & Brugge, D. (2020). Environmental justice. In H. Khreis, et al. (Eds.), *Traffic-related air pollution*: Elsevier.

Gasana, J., et al. (2012). Motor vehicle air pollution and asthma in children: A meta-analysis. *Environmental Research, 117*, 36–45.

Glazener, A., & Khreis, H. (2019). Transforming our cities: Best practices towards clean air and active transportation. *Current environmental health reports, 6*(1), 22–37.

Glazener, A., & Khreis, H. (2020). Best practices for air quality and active transportation. In H. Khreis, et al. (Eds.), *Traffic-related air pollution*: Elsevier.

Gouldson, A., et al. (2015). *Accelerating low-carbon development in the world's cities. Contributing paper for seizing the global opportunity: Partnerships for better growth and a better climate.* London and Washington, DC: New Climate Economy. Available at: http://newclimateeconomy.report/misc/working-papers.

Heusinkveld, H. J., et al. (2016). Neurodegenerative and neurological disorders by small inhaled particles. *Neurotoxicology, 56*, 94–106.

Hitchcock, G., et al. (2014). *Air quality and road transport: Impacts and solution.* The Royal Automobile Club Foundation for Motoring Ltd.

Hoek, G., et al. (2002). Association between mortality and indicators of traffic-related air pollution in the Netherlands: A cohort study. *The Lancet, 360*(9341), 1203–1209.

International Council on Clean Transportation. (2018). *India* Available from https://theicct.org/india.

Jerrett, M., et al. (2014). Traffic-related air pollution and obesity formation in children: A longitudinal, multilevel analysis. *Environmental Health, 13*(1), 49.

Josh Miller, L. D., & Façanha, C. (2017). *Vehicle and fuel standards can dramatically reduce air pollution in Brazil, São Paulo's air quality data demonstrate.* Available from: https://theicct.org/blogs/staff/vehicle-and-fuel-stds-can-dramatically-reduce-air-pollution-brazil.

Kelly, F. J., & Fussell, J. C. (2017). Role of oxidative stress in cardiovascular disease outcomes following exposure to ambient air pollution. *Free Radical Biology and Medicine, 110*, 345–367.

Khreis, H. (2020). Traffic, air pollution, and health. In *Advances in transportation and health* (pp. 59–104): Elsevier.

Khreis, H., de Hoogh, K., & Nieuwenhuijsen, M. (2018). Full-chain health impact assessment of traffic-related air pollution and childhood asthma. *Environment International, 114*, 365–375. Available online 27 March 2018.

Khreis, H., Glazener, A., Ramani, T., Zietsman, J., Nieuwenhuijsen, M. J., Mindell, J. S., et al. (2019). *Transportation and Health: A Conceptual Model and Literature Review.* T.C.f.A.R.i.T.E. College Station, Energy, and Health. Available at http://www.carteeh.org/wp-content/uploads/2019/04/14-Pathways-Project-Brief_Final-version_24April2019.pdf.

Khreis, H., et al. (2016). The health impacts of traffic-related exposures in urban areas: Understanding real effects, underlying driving forces and co-producing future directions. *Journal of Transport & Health, 3*, 249–267.

Khreis, H., et al. (2017a). Exposure to traffic-related air pollution and risk of development of childhood asthma: A systematic review and meta-analysis. *Environment International, 100*, 1–31.

Khreis, H., et al. (2017b). Commentary: How to create healthy environments in cities. *Epidemiology, 28*(1), 60–62.

Khreis, H., et al. (2019a). Transport policy measures for climate change as drivers for health in cities. In *Integrating human health into urban and transport planning* (pp. 583–608): Springer.

Khreis, H., et al. (2019c). Traffic-related air pollution and the local chronic burden of childhood asthma in Bradford, UK. *International Journal of Transportation Science and Technology, 8*(2), 116–128.

Khreis, H., et al. (2020). *Traffic-related air pollution: Emissions, human exposures, and health*. Vol. In Preparation: Elsevier.

Lam, J., Vesterinen, H., & Woodruff, T. (2020). Systematic reviews and meta-analyses of air pollution epidemiological studies. In H. Khreis, et al. (Eds.), *Traffic-related air pollution*: Elsevier.

Lelieveld, J., et al. (2015). The contribution of outdoor air pollution sources to premature mortality on a global scale. *Nature, 525*(7569), 367–371.

Li, W.-W. (2020). Air pollution, air quality, vehicle emissions, and environmental regulations. In H. Khreis, et al. (Eds.), *Traffic-related air pollution*: Elsevier.

Liang, F., et al. (2019). Long-term exposure to ambient fine particulate matter and incidence of diabetes in China: A cohort study. *Environment International, 126*, 568–575.

Lindgren, A., et al. (2009). Traffic-related air pollution associated with prevalence of asthma and COPD/chronic bronchitis. A cross-sectional study in Southern Sweden. *International Journal of Health Geographics, 8*(1), 2.

Link, M. S., & Dockery, D. W. (2010). Air pollution and the triggering of cardiac arrhythmias. *Current Opinion in Cardiology, 25*(1), 16.

Lodovici, M., & Bigagli, E. (2011). Oxidative stress and air pollution exposure. *Journal of toxicology, 2011*, 487074.

Loxham, M., Davies, D. E., & Holgate, S. T. (2019). The health effects of fine particulate air pollution. *British Medical Journal Publishing Group*.

Luben, T. J., et al. (2017). A systematic review of cardiovascular emergency department visits, hospital admissions and mortality associated with ambient black carbon. *Environment International, 107*, 154–162.

Lucas, K., et al. (2019). *Measuring transport equity* (1st ed.).

Luo, L., et al. (2019). Time series analysis of ambient air pollution effects on dynamic stroke mortality. *The International Journal of Health Planning and Management, 35*, 79–103.

MacIntyre, E. A., et al. (2014). Air pollution and respiratory infections during early childhood: An analysis of 10 European birth cohorts within the ESCAPE project. *Environmental Health Perspectives, 122*(1), 107–113.

Marmot, M. (2005). Social determinants of health inequalities. *The Lancet, 365*(9464), 1099–1104.

May, A. D. (2020). Policy option generation and selection. In H. Khreis, et al. (Eds.), *Traffic-related air pollution*: Elsevier.

May, A. D., Khreis, H., & Mullen, C. (2018). Option generation for policy measures and packages: An assessment of the KonSULT knowledgebase. *Case studies on transport policy, 6*(3), 311–318.

Medeiros, J., & Khreis, H. (2020). How emerging technology and its integrations is advancing our understanding of urban and traffic related air pollution. In H. Khreis, et al. (Eds.), *Traffic-related air pollution*: Elsevier.

Metz, B., et al. (2007). *Contribution of working group III to the fourth assessment report of the intergovernmental panel on climate change*.

Miller, M. R., & Raftis, J. B. (2020). Evidence from toxicological and mechanistic studies. In H. Khreis, et al. (Eds.), *Traffic-related air pollution*: Elsevier.

Miller, M. R., Shaw, C. A., & Langrish, J. P. (2012). From particles to patients: Oxidative stress and the cardiovascular effects of air pollution. *Future Cardiology*, *8*(4), 577–602.

Min, J.-y., & Min, K.-b. (2017). Exposure to ambient PM10 and NO2 and the incidence of attention-deficit hyperactivity disorder in childhood. *Environment International*, *99*, 221–227.

Mindell, J. S., & Birley, M. H. (2020). Qualitative health impact assessment. In H. Khreis, et al. (Eds.), *Traffic-related air pollution*: Elsevier.

Mueller, N., Nieuwenhuijsen, M. J., & Rojas-Rueda, D. (2020). Quantitative health impact and burden of disease assessment of traffic-related air pollution. In H. Khreis, et al. (Eds.), *Traffic-related air pollution*: Elseveir.

Mustafić, H., et al. (2012). Main air pollutants and myocardial infarction: A systematic review and meta-analysis. *JAMA*, *307*(7), 713–721.

Nagendra, S. S., et al. (2020). Traffic-related air pollution, human exposure and commercially available market solutions: Perspectives from the developing nation context. In H. Khreis, et al. (Eds.), *Traffic-related air pollution*: Elsevier.

Nieuwenhuijsen, M. J., et al. (2017). Participatory quantitative health impact assessment of urban and transport planning in cities: A review and research needs. *Environment International*, *103*, 61–72.

Nishimura, K. K., et al. (2013). Early-life air pollution and asthma risk in minority children. The GALA II and SAGE II studies. *American Journal of Respiratory and Critical Care Medicine*, *188*(3), 309–318.

Nordling, E., et al. (2008). Traffic-related air pollution and childhood respiratory symptoms, function and allergies. *Epidemiology*, *19*(3), 401–408.

Onat, B., & Stakeeva, B. (2013). Personal exposure of commuters in public transport to PM2.5 and fine particle counts. *Atmospheric Pollution Research*, *4*, 329–335.

Oudin, A., et al. (2016). Traffic-related air pollution and dementia incidence in northern Sweden: A longitudinal study. *Environmental Health Perspectives*, *124*(3), 306–312.

Paul, K. C., et al. (2019). Ambient air pollution, noise, and late-life cognitive decline and dementia risk. *Annual Review of Public Health*, *40*, 203–220.

Pedersen, M., et al. (2013). Ambient air pollution and low birthweight: A European cohort study (ESCAPE). *The Lancet Respiratory Medicine*, *1*(9), 695–704.

Pedersen, M., et al. (2014). Ambient air pollution and pregnancy-induced hypertensive disorders: A systematic review and meta-analysis. *Hypertension*, *64*(3), 494–500.

Perrino, C., Catrambone, M., & Di Menno Di Bucchianico, A. (2003). Gaseous ammonia from traffic emissions in the urban area of Rome. *Advances in air pollution series*. (pp. 601–609).

Perrino, C., et al. (2002). Gaseous ammonia in the urban area of Rome, Italy and its relationship with traffic emissions. *Atmospheric Environment*, *36*(34), 5385–5394.

Prada, D., et al. (2017). Association of air particulate pollution with bone loss over time and bone fracture risk: Analysis of data from two independent studies. *The Lancet Planetary Health*, *1*(8), e337–e347.

Raaschou-Nielsen, O., et al. (2013). Air pollution and lung cancer incidence in 17 European cohorts: Prospective analyses from the European study of cohorts for air pollution effects (ESCAPE). *The Lancet Oncology*, *14*(9), 813–822.

Raz, R., et al. (2015). Autism spectrum disorder and particulate matter air pollution before, during, and after pregnancy: A nested case–control analysis within the nurses' health study II cohort. *Environmental Health Perspectives*, *123*(3), 264–270.

Rodgers, M. O., & Rodgers, A. W. (2020). Impacts of traffic-related air pollution on policy and decision making. In H. Khreis, et al. (Eds.), *Traffic-related air pollution*: Elsevier.

Ropkins, K., Ibarra-Espinosa, S., & Bernard, Y. (2020). Vehicle emissions measurement and modeling. In H. Khreis, et al. (Eds.), *Traffic-related air pollution*: Elsevier.

Sanchez, K. A., et al. (2020). The state of the literature on traffic-related emissions, air pollution, human exposures, and health. In H. Khreis, et al. (Eds.), *Traffic-related air pollution*: Elsevier.

Sapkota, A., et al. (2012). Exposure to particulate matter and adverse birth outcomes: A comprehensive review and meta-analysis. *Air Quality, Atmosphere and Health, 5*(4), 369–381.

Scoggins, A., et al. (2004). Spatial analysis of annual air pollution exposure and mortality. *Science of the Total Environment, 321*(1), 71–85.

Skjøth, C. A., & Hertel, O. (2013). Ammonia emissions in Europe. In *Urban air quality in Europe* (pp. 141–163): Springer.

Stafoggia, M., et al. (2014). Long-term exposure to ambient air pollution and incidence of cerebrovascular events: Results from 11 European cohorts within the ESCAPE project. *Environmental Health Perspectives, 122*(9), 919–925.

Su, J. G., et al. (2015). Populations potentially exposed to traffic-related air pollution in seven world cities. *Environment International, 78*, 82–89.

Sudmant, A., et al. (2020). The social, environmental, health and economic impacts of low carbon transport policy: A review of the evidence. In H. Khreis, et al. (Eds.), *Traffic-related air pollution*: Elsevier.

Sundvor, I., et al. (2012). *Road traffic's contribution to air quality in European cities, ETC/ACM technical paper 2012/14*. [cited 2016 17th April 2016]; Available from: http://acm.eionet.europa.eu/reports/docs/ETCACM_TP_2012_14_traffic_contribution_city_aq.pdf.

Tanvir, S., Hao, P., & Boriboonsomsin, K. (2020). Emerging transportation technologies and implications for traffic-related emissions, air pollution exposure, and health. In H. Khreis, et al. (Eds.), *Traffic-related air pollution*: Elsevier.

Thurston, G. D., et al. (2020). Outdoor air pollution and new-onset airway disease. An official American Thoracic Society workshop report. *Annals of the American Thoracic Society, 17*(4), 387–398.

Timmers, V. R., & Achten, P. A. (2016). Non-exhaust PM emissions from electric vehicles. *Atmospheric Environment, 134*, 10–17.

Tomlin, A., Sutton, A., & Tate, J. (2010). The effect of commuter route choice on particulate exposure of an urban cyclist. In *The 2010 annual UK review meeting on outdoor and indoor air pollution research meeting*: Department of Health: Cranfield University.

Tonne, C., et al. (2014). Traffic-related air pollution in relation to cognitive function in older adults. *Epidemiology (Cambridge, Mass.), 25*(5), 674.

United Nations Department of Economic and Social Affairs. (2018). *2018 revision of world urbanization prospects*. United Nations Department of economic and social affairs.

United Nations Department of Economic and Social Affairs, U.D. (2017). *World population prospects, the 2017 revision, volume I: Comprehensive tables*. New York: United Nations Department of Economic & Social Affairs.

Vardoulakis, S., et al. (2003). Modelling air quality in street canyons: A review. *Atmospheric Environment, 37*(2), 155–182.

Vrijheid, M., et al. (2010). Ambient air pollution and risk of congenital anomalies: A systematic review and meta-analysis. *Environmental Health Perspectives, 119*(5), 598–606.

Wei, Y., et al. (2019). Short term exposure to fine particulate matter and hospital admission risks and costs in the medicare population: Time stratified, case crossover study. *BMJ, 367*.

Weuve, J., et al. (2012). Exposure to particulate air pollution and cognitive decline in older women. *Archives of Internal Medicine, 172*(3), 219–227.

WHO. (2018). *Health and sustainable development. Air pollution.* [cited 12.10.2018]; Available from:http://www.who.int/sustainable-development/transport/health-risks/air-pollution/en/.

World Health Organization, W.H.O. (2013). *Health effects of particulate matter. Policy implications for countries in Eastern Europe, Caucasus and Central Asia.* World Health Organization.

World Health Organization, W.H.O. (2016). *Ambient air pollution: A global assessment of exposure and burden of disease.* World Health Organization.

World Resources Institute. (2018). *Reducing China's urban air pollution with transport solutions at Beijing workshop.* Available from https://wrirosscities.org/news/reducing-china%E2%80%99s-urban-air-pollution-transport-solutions-beijing-workshop.

Xu, Y. A. (2020). Traffic monitoring and modeling for energy, air quality, and health. In H. Khreis, et al. (Eds.), *Traffic-related air pollution.* Elsevier.

Yang, Y., et al. (2013). The association between ambient air pollution and daily mortality in Beijing after the 2008 olympics: A time series study. *PLoS One, 8*(10), e76759.

Yolton, K., et al. (2019). Lifetime exposure to traffic-related air pollution and symptoms of depression and anxiety at age 12 years. *Environmental Research, 173*, 199–206.

Zhang, X., Chen, X., & Zhang, X. (2018). The impact of exposure to air pollution on cognitive performance. *Proceedings of the National Academy of Sciences, 115*(37), 9193–9197.

CHAPTER 2

Air pollution, air quality, vehicle emissions, and environmental regulations

Wen-Whai Li
Department of Civil Engineering, The Univeristy of Texas at El Paso, El Paso, TX, United States

Abbreviations

C_{mass}	mass concentration of an air pollutant, $\mu g/m^3$
C_{volume}	volumetric concentration of an air pollutant, m^3/m^3
CAAA	clean air act amendments
CAP	compliance assurance program
CFR	Code of Federal Regulations
CO	carbon monoxide
CO_2	carbon dioxide
EC	elemental carbon
EPA	Environmental Protection Agency
HC	hydrocarbons
HS	hydrated sulfate
IARC	International Agency for Research on Cancer
K	absolute temperature, K
MPO	Metropolitan Planning Organization
MTBE	methyl tertiary-butyl ether
MW	molecular weight, g/mol
NAAQS	National Ambient Air Quality Standards
NESHAP	National Emission Standards for Hazardous Air Pollutants
NH_3	ammonia
NO	nitric oxide
NHTSA	National Highway Traffic Safety Administration
NO_2	nitrogen dioxide
NO_x	nitrogen oxides, $NO_x = NO + NO_2$
O_3	ozone
OC	organic carbon
p	standard atmospheric pressure, $1.01 \times 10^5\,N\,m^2$
ppmv	parts per million by volume, 10^{-6}
ppbv	parts per billion by volume, 10^{-9}
pptv	parts per trillion by volume, 10^{-12}
PM	particulate matter
$PM_{2.5}$	particulate matter of $<2.5\,\mu m$ in aerodynamic diameter
PM_{10}	particulate matter of $<10\,\mu m$ in aerodynamic diameter

Traffic-Related Air Pollution
https://doi.org/10.1016/B978-0-12-818122-5.00002-8

R	Universal gas constant, 8.31 N m/K mol
SIP	State Implementation Plan
TRAP	traffic-related air pollution
VOCs	volatile organic compounds
WHO	World Health Organization

Air becomes inseparable from a man's life the moment he leaves his mother's womb. Every living being would cease in a few minutes without air while without water or food life could continue for at least a few days. The availability and quality of air are of paramount importance to human life, yet how to protect the quality of the air we breathe has not been addressed as adequately. The quality of air (air quality) affects the perception of the air we breathe. Brisk air with low temperature and humidity gives our olfactory epithelium a more pleasant sensory stimulus than humid, musky, warm air, which may not have any effect on human health. On the other hand, sweet and delicate odors, artificial aromas of pine tree, fruits, candies, and flowers, or incenses emanated from man-made materials such as smells of a new car interior [various volatile organic compounds (VOCs)], carpets (Caprolactam), scented candles, deodorants, air fresheners (phthalates), fresh paint, potpourri, among many others, present false perception of cleanness while exposure to these chemical-containing fumes may have adverse effects on human health. As a matter of fact, the air we breathe is indeed filled with harmful and even carcinogenic substances. Once these substances are released into the atmosphere, they disperse rapidly, and are almost impossible to remove except by naturally occurring processes. Most of the man-made substances will have long residence time and persistent toxicity in the atmosphere due to lack of natural removal mechanisms. The International Agency for Research on Cancer (IARC) has classified outdoor air pollution as a cancer-causing agent and "outdoor air pollution is not only a major environmental risk to health in general, it is the most important environmental cancer killer due to the large number of people exposed" (World Health Organization (WHO), 2013).

As commonly defined in the literature, air pollution is a situation in which airborne substances, resulting from anthropogenic activities or natural occurrences, are present at concentrations sufficiently high above their normal ambient levels to produce measurable adverse health effects on humans or animals or measurable damaging effects on vegetation, materials, or the environment (e.g., Refs. Ayra, 1997; de Nevers, 2017). While the word "pollution" is distinct from "contamination," the two words are frequently used with no distinction in air quality engineering. Indeed, "contamination" refers to a situation where foreign substances are injected into an environment that initially does not contain any of these substances. The foreign substance can be anything (particles, VOCs, nutrients, toxins, inert gases, etc.) which poses no to high risk to human health or the environment. A common example for the distinction between pollution and contamination can be found on the label of a can of nuts which states that the product may be contaminated with peanuts because the process line may be previously used for processing peanuts. Another example can be seen in a clean room used for the fabrication of high tech chipsets where the room is considered contaminated if a few particles accidently enter the room, which by no means present any harmful effect on human health or the environment. In this book, we use the two words interchangeably for pollution without further explanation.

Sources of air pollution

Air pollution may be the consequence of natural processes that include volcanic eruptions, plant and animal decomposition, hydrocarbon emissions from vegetation, pollen, mold, and spore releases, background ozone formation, wind-blown dust, and dust storms, among many other mechanisms. Some of these events such as volcanic eruption, forest fires, or dust storms may generate enormous amount of pollutants; however, the occurrence of these events are considered rare to low, the locations of the events and the areas that are likely to be impacted are removed from major population centers, or the amounts generated by the processes such as background ozone formation and hydrocarbon emissions from vegetation are relatively insignificant when comparing the same substances that are generated by anthropogenic sources. Therefore, control and impact of these processes are typically not discussed by air pollution professionals.

Anthropogenic air pollution is the main concern of the modern society. Anthropogenic air pollution is generally more persistent and toxic, with profound effects. It could cause significant adverse health effects in human beings and irreversible damage to the environment. Anthropogenic air pollution can be easily observed as smokestack emissions from industrial facilities, trapped emissions in routinely occurring urban temperature inversion layers, and hazes in national parks. Anthropogenic air pollutants can be emitted directly from a source or formed indirectly among airborne substances in the atmosphere chemically through chemical and/or photochemical reactions or physically through nucleation, accumulation, agglomeration, phase change, or decomposition. Air pollutants emitted directly from a source are called primary pollutants whereas those formed indirectly in the atmosphere are called secondary pollutants. The sources of air pollution are also classified into three categories according to the nature of the emission sources. A "point source" refers to a source that emits from a stationary, identifiable location such as a smokestack or a leak point in a container vessel. A "mobile source" refers to sources that are powered by combustion engines and can move on-road or off-road. Any other source that cannot be classified as point source or mobile source is called an "area source," which in general represents sources with multiple unidentifiable locations or groups of sources with small amounts and/or unsteady-state emissions. Fugitive emissions from industry-related operations, small businesses, and household activities (heating, cooking, cleaning, gardening) that are difficult to quantify and locate fall into the category of area sources. In the United States, for regulatory compliance enforcement, those point sources which emit or have the potential to emit 10 tons per year or more of a hazardous air pollutant or 25 tons per year or more of a combination of hazardous air pollutants are called "major sources," as defined in Section 112 of the 1997 Clean Air Act Amendments (CAAA). Likewise, an "area source" is defined as any stationary source that is not a major source, i.e., emitting < 10 tons per year or more of a hazardous air pollutant or 25 tons per year or more of a combination of hazardous air pollutants.

Pollutant concentration and units

The level of air pollution for a substance is quantified either as mass concentration (C_{mass}) or as volumetric concentration (C_{vol}). The mass concentration for an air pollutant is expressed as

$$C_{mass} = \frac{Mass_{Pol}}{Volume_{Air}}\left[\frac{\mu g_{Pol}}{m^3_{Air}}\right]$$

where $Mass_{Pol}$ is the total mass of a pollutant in a known volume of air, $Volume_{Air}$.

Mass concentration can be easily converted to volumetric concentration, which is defined as the ratio of the volume occupied by a pollutant to a unit volume of air in which the pollutant resides by dividing the mass concentration by the density of the pollutant (ρ_{Pol}):

$$C_{vol} = C_{mass} \times \left(\frac{1}{\rho_{Pol}}\right)$$

$$\left[\frac{\mu g_{Pol}}{m^3_{Air}} \times \left(\frac{m^3_{Pol}}{\mu g_{Pol}}\right)\right] = \left[\frac{m^3_{Pol}}{m^3_{Air}}\right] = \left[\frac{m^3_{Pol}}{m^3_{Air}} \times 10^6 \times 10^{-6}\right] = \left[\frac{m^3_{Pol}}{m^3_{Air}} \times 10^6 \times \text{ppmv}\right]$$

For airborne particulate matter (PM), the density is approximately 1 g/cm^3 (called the standard density of PM), therefore the volumetric concentration for 1 μg/m^3 of PM mass concentration would be equivalent to 1×10^{-6} ppmv or 1 pptv.

$$C_{vol} = 1\frac{\mu g_{PM}}{m^3_{Air}} \times \frac{1}{1\frac{g}{cm^3}} \times \left(10^{-6}\frac{g}{\mu g}\right) \times \left(10^{-6}\frac{m^3}{cm^3}\right) = 1 \times 10^{-12}\frac{m^3_{PM}}{m^3_{Air}}$$
$$= 1 \times 10^{-6}\,\text{ppmv} = 1 \times 10^{-3}\,\text{ppbv} = 1\text{pptv}$$

where

1 ppmv $= 10^{-6}$ by volume;

1 ppbv $= 10^{-9}$ by volume;

1 pptv $= 10^{-12}$ by volume.

It is understandable from the above expression that a volumetric expression of PM pollution would yield a very small number which may inadvertently convey a false impression to the public that the magnitude of the pollutant concentration is of such a minute quantity that the level is of no significance to human health or the environment. Therefore, mass concentration is preferred for the expression of PM pollution.

For gaseous pollutants, the density of gas is depicted by the gas law such that

$$pV = nRT \rightarrow pV = \frac{m}{MW}RT \rightarrow p \cdot MW = \frac{m}{V}RT \rightarrow \rho = \frac{p \cdot MW}{RT}$$

where

p = atmospheric pressure;

V = volume of the gaseous pollutant;

R = universal gas constant;

T = atmospheric temperature;

m = mass of the pollutant;

MW = molecular weight of the pollutant;

ρ = density of the gas.

Under the standard condition of $p = 1$ atm and $T = 293$ K, the volumetric concentration for a gaseous pollutant can be expressed as

$$C_{vol} = C_{mass} \cdot \left(\frac{RT}{p \cdot MW}\right) = C_{mass}\left(\frac{\mu g}{m^3}\right) \cdot \left(\frac{8.31\frac{N \times m}{K \times mol} \cdot 293K}{1.01 \cdot 10^5 \frac{N}{m^2} \cdot MW\left(\frac{g}{mol}\right)}\right) \cdot 10^{-6}\frac{g}{\mu g}$$
$$= C_{mass} \cdot \left(2.41 \cdot 10^{-8}\right)\frac{1}{MW} = 0.0241\frac{1}{MW}C_{mass}\,(\text{ppmv})$$

where

C_{vol} = volumetric concentration for a gaseous species, ppmv;

C_{mass} = mass concentration for a gaseous species, $\mu g/m^3$;

R = Universal gas constant, 8.31 N m/K mol;

p = atmospheric pressure, $1.01 \times 10^5\,N/m^2$;

MW = molecular weight of a gaseous species; g/mol.

For example, the equivalent volumetric concentration for $1\,\mu g/m^3$ of NO_2 in an environment under 1 atm and at 293 K would be 5.24×10^{-4} ppmv or 0.524 ppbv.

History of air pollution control regulations in the United States

Air pollution control by regulations has been widely adopted globally. Industrialized countries were quick to recognize the urgency of taking actions against air pollution in the early 20th century. The United Kingdom began to regulate air emissions as early as 1906 (Kuklinska, Wolska, & Namiesnik, 2015). However, the United States has been the world leader in taking systemic regulatory actions against air pollution since 1955. Many countries have since developed their own air quality control acts by adopting or modifying some landmark US regulations. As of this date, air quality regulations are well developed in many countries and may differ from one another. This section dose not attempt to discuss the difference, effectiveness, and/or appropriateness among the various air quality guidelines and standards created by different nations. Rather, we focus on several milestones in the development of air quality legislations in the United States as an illustration of the evolution of many similar air quality standards in the world.

Air quality in the United States deteriorated dramatically in the first half of the 20th century due to significant population growth, accelerated industrial outputs, and increased consumption of fossil fuels. As a result, an Air Pollution Control Act was promulgated in 1955 which paved the way for a series of air pollution reform acts later. Prior to 1955, only a limited number of air ordinances were installed in various municipalities, primarily cities with high industrial activities. For instance, Chicago and Cincinnati first established smoke ordinances in 1881 and Philadelphia passed an ordinance limiting the amount of smoke in flues, chimneys, and open spaces in 1904 (U.S. Environmental Protection Agency (U.S. EPA), 2018). In 1947,

California authorized the creation of Air Pollution Control Districts in every county of the state.

There have been several amendments made to The Air Pollution Control Act of 1955. The Clean Air Act of 1963 was the first federal legislation regarding air pollution control. It established a federal program within the US Public Health Service and authorized research into techniques for monitoring and controlling air pollution. In 1967, the Air Quality Act of 1967 was passed to expand federal government's activities in air pollution control. In accordance to this law, enforcement proceedings were initiated in areas subject to interstate air pollution transport. As part of these proceedings, the federal government for the first time conducted extensive ambient air monitoring studies and stationary source inspections (U.S. Environmental Protection Agency (U.S. EPA), 2018). In 1969, another amendment was made to the act. This amendment further expanded the research on low-emissions, fuels, and automobiles (Forswall & Higgins, 2005).

Public awareness on the state of the environment intensified in the 1960s and reached a new peak in the 1970s when the nation went through a tumultuous era of anti-war, liberalism, civil and women's rights, energy crisis, and environmental movements. The 1970 Clean Air Act was enacted and completely replaced the 1967 Air Quality Act. This new act also created the US Environmental Protection Agency (EPA), as advised by President Nixon's Advisory Council on Executive Organization (EPA, 1970). It identified six common air pollutants of concern, called *criteria pollutants*, and required the EPA to develop technology-based National Ambient Air Quality Standards (NAAQS) to protect public health and welfare. Table 2.1 lists the six criteria pollutants identified in 1970: sulfur oxide (as sulfur dioxide), particulate matter, carbon monoxide, photochemical oxidants, hydrocarbons, and nitrogen dioxide, and their respective primary and secondary National Ambient Air Quality Standards. The primary and secondary NAAQSs were designed, respectively, to provide adequate margin of safety to protect public health and public welfare (including visibility, animals, crops, vegetation, and structures).

Among the many amendments and revisions made after 1970, the 1990 Clean Air Act Amendments (1990 CAAA) was the most significant amendment to the Clean Air Act. The 1990 amendments granted significantly more authority to the federal government than any prior air quality legislation. It set more deadlines for compliance which posed burdens on the

Table 2.1 The 1971 National Ambient Air Quality Standards.

Pollutant	Primary secondary	Averaging time		NAAQS	Form/enforcement
Carbon monoxide (CO)	Primary and secondary	8 h	10 mg/m^3	9 ppm	Not to be exceeded more than once per year
		1 h	40 mg/m^3	35 ppm	
Hydrocarbons	Primary and secondary	Max 3 h (6–9AM)	160 μg/m^3	0.08 ppm	Not to be exceeded more than once per year
Nitrogen dioxide (NO_2)	Primary and secondary	1 year	100 μg/m^3	0.05 ppm	Not to be exceeded
Photochemical oxidants (corrected for NO_x and SO_2)	Primary and secondary	1 h	160 μg/m^3	0.08 ppm	Not to be exceeded more than once per year
	Primary	1 year	75 μg/m^3		As a guide for state implementation
Particulate matter (TSP)	Secondary	1 year	60 μg/m^3		As a guide for state implementation
	Primary	24 h	260 μg/m^3		Not to be exceeded more than once per year
	Secondary	24 h	150 μg/m^3		
Sulfur oxides (measured as SO_2)	Primary	1 year	80 μg/m^3	0.03 ppm	
		24 h	365 μg/m^3	0.14 ppm	Not to be exceeded more than once per year
		3 h	–	–	
	Secondary	1 year	60 μg/m^3	0.02 ppm	
		24 h	260 μg/m^3	0.1 ppm	Not to be exceeded more than once per year
		3 h	1300 μg/m^3	0.5 ppm	

regulatory agencies to enforce the regulations and on the industries to reduce emissions. In all, 11 "Titles" were listed in the 1990 CAAA and many subjects were addressed. The subjects covered by different "Titles" vary in a wide range including attainment and maintenance of NAAQS, mobile sources emissions, toxic air pollutants emissions, acid deposition control, air permits, stratospheric ozone and global climate, regulation enforcement, air quality research and development, air monitoring, visibility improvement at national parks, and unemployment and training programs for workers affected by the 1990 CAAA.

In brief, air pollution control regulations in the United States have continued to evolve with amendments, updates, and revisions since 1970, as new findings of air pollutants on human health became available and new developments in control and monitoring technologies became applicable and feasible. For instance, ozone (O_3) was designated to replace the total photochemical oxidants as one of the criteria pollutants in 1979 and an annual fourth highest daily maximum 8-h O_3 average of 0.08 ppmv was installed to replace the not-to-be-exceeded 1-h standard of 0.12 ppmv in 1997. EPA routinely conducts scientific reviews through the Clean Air Scientific Advisory Committee and makes new recommendations for public comments prior to issuing the final rules. Table 2.2 lists the evolution of the NAAQS for ozone whereas Table 2.3 presents the evolution of the NAAQS for PM, and two criteria pollutants that have gained increased attention due to their ubiquity and severe adverse health effects. The current NAAQS in the United States is listed in Table 2.4.

Table 2.2 Historical development of NAAQS for ozone.

Year	Final rule	Pollutant	Averaging time	NAAQS (ppm)	Primary/secondary
1971	36 FR 8186	TPC	1 h	0.08 ppm	Primary and secondary
1979	44 FR 8202	O_3	1 h	0.12 ppm	Primary and secondary
1997	62 FR 38856	O_3	8 h	0.08 ppm	Primary and secondary
2008	73 FR 16483	O_3	8 h	0.075 ppm	Primary and secondary
2015	80 FR 65292	O_3	8 h	0.070 ppm	Primary and secondary

Table 2.3 Historical development of NAAQS for particulate matter[a] (PM).

Year	Final rule	Pollutant	Averaging time	NAAQS ($\mu g/m^3$)	Primary/secondary
1971	36 FR 8186	TSP	24 h	260	Primary
				150	Secondary
			Annual	75	Primary
				60	Secondary
1987	52 FR 24634	PM_{10}	24 h	150	Primary and secondary
			Annual	50	Primary and Secondary
1997	52 FR 38652	$PM_{2.5}$	24 h	65	Primary and Secondary
			Annual	15	Primary and secondary
		PM_{10}	24 h	150	Primary and secondary
			Annual	50	Primary and secondary
2006	71 FR 61144	$PM_{2.5}$	24 h	35	Primary and secondary
			Annual	15	Primary and secondary
		PM_{10}	24 h	150	Primary and secondary
2012	78 FR 3085	$PM_{2.5}$	24 h	35	Primary
			Annual	12	Primary
				15	Secondary
		PM_{10}	24 h	150	Primary and secondary

[a] *TSP*, Total suspended particulate (or PM_{35}); *PM10*, particulate matter with aerodynamic diameter of <10 μm; *PM2.5*, particulate matter with aerodynamic diameter of <2.5 μm; *PM35*, total suspended particles or particulate matter with aerodynamic diameter of <35 μm.

Pollutant emission trends in the United States

Air quality in the United States has improved since the enactment of the 1990 CAA. Fig. 2.1 shows the emission trends for the criteria pollutants (except lead) and two major precursors (VOC and NH_3) for ozone. Lead was added to the list of NAAQS in 1978 while hydrocarbons were delisted (Federal Register 1978, 43FR46246, 1978). In 1979, ozone replaced photochemical oxidants as one of the six criteria air pollutants (44 CFR 8202). The lead standard was set at $1.5\,\mu g/m^3$ on the basis that children are most sensitive to lead exposure and that a blood level of $>30\,\mu g$ of lead per 1 dL of blood (μg/dL) was associated with the impairment of theme synthesis in cells indicated by elevated erythrocyte protoporphyrin (EP), which EPA regards as adverse to the health of chronically exposed children. Ambient lead concentrations in the United States have reduced dramatically since 1980 when the US EPA gradually phased out lead in gasoline starting in the 1970s. Levels of lead in the air have decreased by 99% between 1980 and 2017 (U.S. Environmental Protection Agency (U.S. EPA), 2019a).

Carbon monoxide (CO) is a product of combustion and combustion of hydrocarbons is the primary source of CO emissions. Incomplete combustion,

Table 2.4 The 2018 National Ambient Air Quality Standards.

Pollutant		Primary/secondary	Averaging time	NAAQS	Form/enforcement
Carbon monoxide (CO)		Primary	8 h	9 ppm	Not to be exceeded more than once per year
			1 h	35 ppm	
Lead (Pb)		Primary and secondary	Rolling 3 month	0.15 $\mu g/m^3$ [a]	Not to be exceeded
Nitrogen dioxide (NO_2)		Primary	1 h	100 ppb	98th percentile of 1-h daily maximum concentrations, averaged over 3 years
		Primary and secondary	1 year	53 ppb[b]	Annual mean
Ozone (O_3)		Primary and secondary	8 h	0.070 ppm[c]	Annual fourth-highest daily maximum 8-h concentration, averaged
Particulate matter (PM)	$PM_{2.5}$	Primary	1 year	12.0 $\mu g/m^3$	annual mean, averaged over 3 years
		Secondary	1 year	15.0 $\mu g/m^3$	annual mean, averaged over 3 years
		Primary and secondary	24 h	35 $\mu g/m^3$	98th percentile, averaged over 3 years
	PM_{10}	Primary and secondary	24 h	150 $\mu g/m^3$	Not to be exceeded more than once per year on
Sulfur dioxide (SO_2)		Primary	1 h	75 ppb[d]	99th percentile of 1-h daily maximum concentrations, averaged over 3 years
		Secondary	3 h	0.5 ppm	Not to be exceeded more than once per year

[a]In areas designated nonattainment for the Pb standards prior to the promulgation of the current (2008) standards, and for which implementation plans to attain or maintain the current (2008) standards have not been submitted and approved, the previous standards (1.5 $\mu g/m^3$ as a calendar quarter average) also remain in effect.
[b]The level of the annual NO_2 standard is 0.053 ppm. It is shown here in terms of ppb for the purposes of clearer comparison to the 1-h standard level.
[c]Final rule signed October 1, 2015, and effective December 28, 2015. The previous (2008) O_3 standards additionally remain in effect in some areas. Revocation of the previous (2008) O_3 standards and transitioning to the current (2015) standards will be addressed in the implementation rule for the current standards.
[d]The previous SO_2 standards (0.14 ppm 24-h and 0.03 ppm annual) will additionally remain in effect in certain areas: (1) any area for which it is not yet 1 year since the effective date of designation under the current (2010) standards, and (2) any area for which an implementation plan providing for attainment of the current (2010) standard has not been submitted and approved and which is designated nonattainment under the previous SO_2 standards or is not meeting the requirements of a SIP call under the previous SO2 standards [40 CFR 50.4(3)]. A SIP call is an EPA action requiring a state to resubmit all or part of its State Implementation Plan to demonstrate attainment of the required NAAQS.
Source: US EPA, https://www.epa.gov/criteria-air-pollutants/naaqs-table#1.

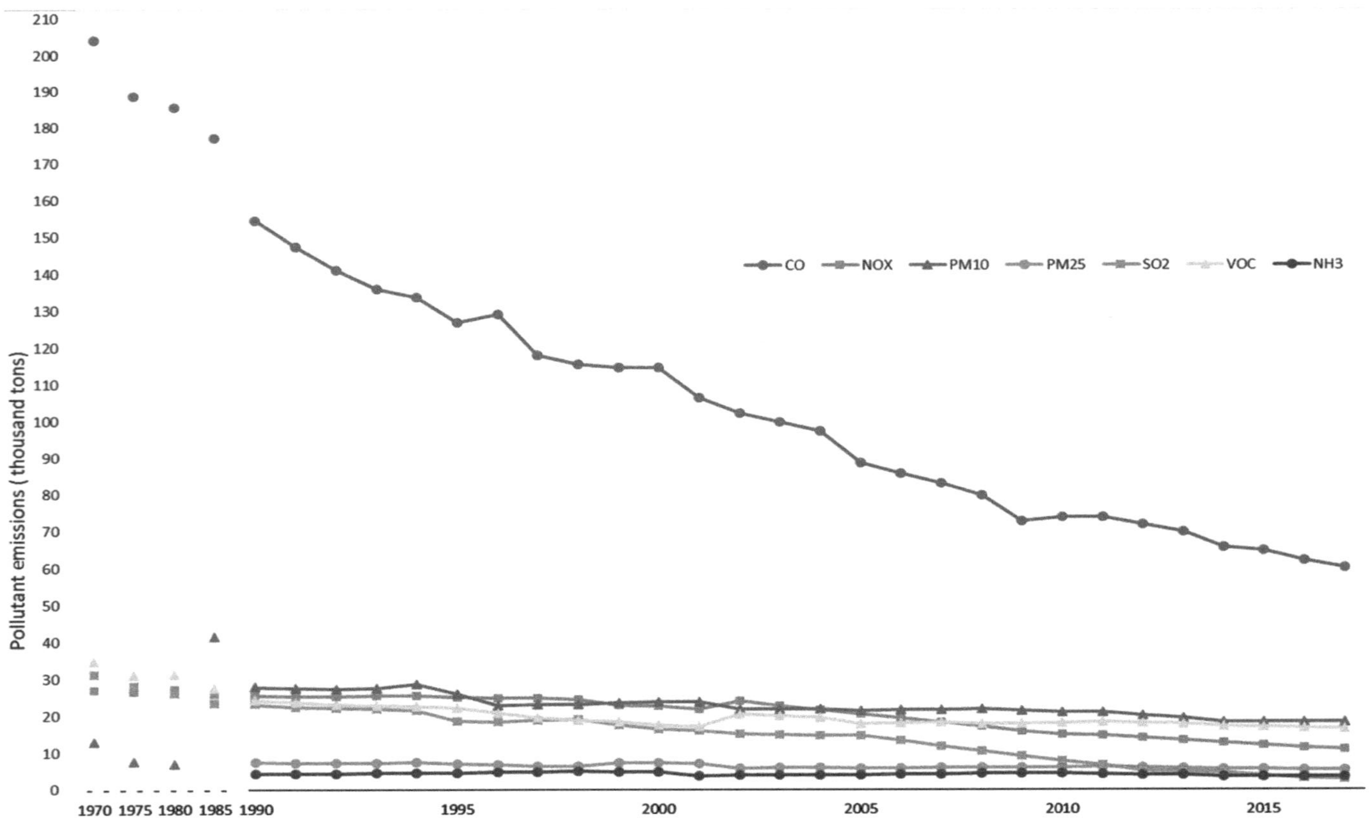

Fig. 2.1 National emission trends in the United States, 1970–2017. *Data source: US EPA, https://www.epa.gov/air-emissions-inventories/air-pollutant-emissions-trends-data.*

in particular, would result in higher levels of CO emissions. CO emissions in the United States reached the peak before 1970 and emissions have since significantly reduced due to more stringent vehicle emission regulations, improved engine performance and fuel economy, better roadways design and traffic conditions, and the introduction of oxygenates in automobile gasoline, although one of the additives, methyl tertiary-butyl ether (MTBE), was later found to have the propensity to contaminate groundwater and subsequently the use as gasoline additive was banned by many states and phased out by EPA (U.S. Environmental Protection Agency (U.S. EPA), 2008). In the United States, the national CO emission rate has decreased continuously from 204 million tons to 60 million tons albeit the number of vehicles has increased from 108 million in 1970 to 272 million in 2018, and from 1.12 trillion to 3.2 trillion miles for the total vehicle miles traveled (U.S. Department of Transportation (U.S.DOT), 2019). Indeed, vehicle emission-related CO concentrations have decreased to levels of little concern. A recent study has shown that the CO levels observed at 60 near-road monitors are relatively insignificant compared to their respective NAAQS (De Winter, Brown, Seagram, & Landsberg, 2018).

Nitrogen oxides (NO_x) are a group of gases consisting of nitric oxide (NO) and nitrogen dioxide (NO_2). NO is a precursor of NO_2 through oxidization after being released from a combustion source and an active compound in the nitrogen dioxide photolytic cycle for ozone formation. NO_2 is not only an irritant gas which could cause inflammation in human's respiratory track but also a precursor to the formation of tropospheric ozone, another criteria air pollutant. NO_2 is released into the atmosphere primarily from combustion sources such as fossil fuel powered vehicles, equipment, and power plants. NO and NO_2 are referenced together as NO_x for their convertibility in the atmosphere. Transportation (47.4%), power plant (19.7%), industrial (15.3%), agriculture and forestry (9.8%), and residential and commercial activities (7.1%) account for $>$99% of the total NO_x emissions in the United States. Fig. 2.1 shows the trends in NO_x emissions between 1970 and 2017. Total NO_x emissions in the United States decreased by 57% from 24,955 Gg (1 Gg = 1000 metric tons) in 1970 to 10,777 Gg in 2017. Similar trends are observed in Fig. 2.1 for SO_2 which decreased by 91% from 31,218 to 2815 Gg in the same period, and by 53% from 34,659 to 16,232 Gg from 1970 to 2017 for VOC. Slower reductions (18% and 29%) were observed for NH_3 and $PM_{2.5}$, which decreased from 4320 to 3563 Gg and from 7520 to 5340 Gg between 1990 and 2017, respectively. The only exception is PM_{10} whose emissions actually increased from 13,022 to 18,152 Gg between 1970 and 2017.

Transportation emissions in the United States

Among the many sources of pollution, pollution caused by transportation-related activities is of particular importance to humans due to its ubiquity, close proximity to humans, increased usage, less controlled emissions, and toxic effects on human health and the environment. Transportation-related emission sources include vehicles, engines, and motorized equipment that produce exhaust and evaporative emissions on-road or off-road. "On-road" emissions refer to emissions from vehicles traveling on roadways including freeways, arterial, secondary, and surface roads. These sources are primarily associated with discussions on traffic-related air pollution (TRAP). Off-road emissions refer to emissions from vehicle or machinery/equipment powered by internal combustion engines such as equipment used for construction, agriculture, recreation, and many other purposes. Within these two broad categories, on-road and off-road sources can be further distinguished by size, weight, use, and/or horsepower.

While the trends of all emissions (except PM_{10}) in the United States have decreased since 1970, transportation-related emissions have not decreased in the same pace as other pollutants due to the increases in the total number of vehicles and the total number of vehicle miles traveled. Roadway infrastructures and on-road vehicles in the United States increased dramatically in the early 20th century. The public road and street mileage in the United States has reached 3.5 million in 1960 but slowly grown to 4.1 million in 2017 while the number of registered vehicles grew rapidly from 74 million in 1960 to 269 million in 2016 (U.S. Department of Transportation (U.S.DOT), 2018). Nationwide, energy-efficient, low-emission vehicles (including gasoline hybrid-electric, plug-in hybrid-electric and electric vehicles) have steadily increased from a sale of merely 17 vehicles in 1999 to 370,000 in 2017 while the average age of household vehicles has increased from 8.4 years in 1995 to 11.6 years in 2016 and from 11.6 to 17.3 years for trucks. It is worth noting that while the growth of total number of roads has been slow and steady, the vehicle miles traveled in the United States has grown significantly from 858 million in 1960 to 6.2 billion in 2016, an increase of 7.2 fold compared to only a fraction of 17% increase in the total miles of roads (U.S. Department of Transportation (U.S.DOT), 2018). Air pollutant emissions from the activities associated with the operations of these vehicles have been considered a source of major pollution. In 2017, the CO_2 emissions from transportation exceeded all other major sources and became the largest source in the country (Fig. 2.2).

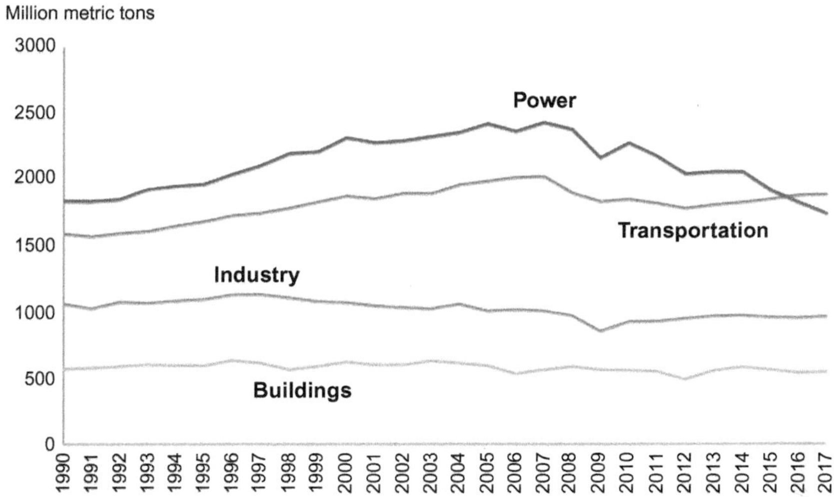

Fig. 2.2 CO_2 emission trends for the largest major sources in the United States.

Characterization of transportation-related emissions

Air pollutants emanated from traffic-related activities associated with the operations of transportation vessels (or the traffic-related air pollutants) have the most profound impact on human health because of (i) the toxicities associated with a vast number of chemicals; (ii) the quantity of pollutants emitted; and (iii) the relatively close proximity between the source and the population. While the health impacts and burdens of disease associated with exposures to traffic-related air pollutants are discussed in other chapters of this book, this section characterizes the various transportation emissions by their activities. Emissions from transportation-related activities are generally classified into evaporative emissions, tailpipe emissions, and nonroad emissions. Emissions from other transportation vessels such as aircraft, locomotives, waterborne vessels, and military equipment are not the focus of this chapter and therefore are briefly discussed under the nonroadway emissions in this section.

Evaporative emissions

Volatile organic compounds (VOCs) can evaporate from multiple points under the hood and during engine operations. Fuel losses during engine resting and running include liquid and vapor leaks in the fuel system, diurnal emissions (or tank breathing loss) due to venting of vapors from the fuel tank as ambient temperature rises over the course of the day, and hot soak emissions that occur due to residual engine heat at the end of a trip after the engine has been shut down. Analyses of gasoline samples from California for both summer and winter indicated significant differences in liquid fuel and vapor chemical composition due to intentional seasonal adjustments (Gentner et al., 2013). The California Emissions Factor Model estimates that 30% of gasoline-related VOCs emissions are fuel vapor, that is, nontailpipe emissions account for around 30% of gasoline-related emissions in urban regions. Similar weight % for the most abundant VOCs in the motor vehicle emissions were reported during a Houston tunnel air pollution study in Texas (McGaughey et al., 2003). Table 2.5 lists the chemical composition of gasoline and diesel along with the calculated composition of nontailpipe, i.e., headspace vapor, and gasoline emissions (Genter, Harley, Miller, & Goldstein, 2009). Besides gasoline fuel, many vehicles are powered by diesel fuel. Diesel fuel contains low-volatility hydrocarbons and as a result evaporative contributions of diesel fuel were found to be negligible compared to exhaust emissions (Hilpert, Adria-Mora, Ni, Rule, & Nachman, 2015).

Tailpipe emissions

Vehicle exhaust (or tail pipe) emission is a major concern of air pollution. Tail pipe emissions are dominated by complete and incomplete combustion products such as CO, CO_2, VOCs or HCs, NO_x, and PM. CO, CO_2, NO_x, and PM emissions are well reported in the literature and strictly regulated by federal and state air pollution control agencies whereas VOCs are less reviewed in the literature, even though they are generally more toxic than the criteria pollutants. The speciation of VOC emitted from motor vehicles reflects the chemical composition of the fuel consumed, with additional species formed and emitted as products of incomplete combustion. Isopentane, methane, MTBE, ethylene, toluene, 2,2,4-trimethylpentane, acetylene, 2-methylpentane, m- and p-xylene, and isobutylene are the primary compounds in the motor vehicle exhausts that account for approximately 50% of the emissions (Araizaga, Mancilla, & Mendoza, 2013; McGaughey et al., 2003). In urban areas where gasoline accounts for a range of 73%–90% of total on-road fuel use, gasoline sources are responsible for 69%–96% of reactive organic carbon emissions and 79%–97% of organic precursors to ozone from motor vehicles in the United States. Isopentane is a major constituent of the headspace vapors accounting for about 40% of the VOC while it accounts for about 10% of the composition of the base gasoline (Gentner et al., 2013). These compounds are responsible for the majority of ozone formation due to organic emissions from on-road motor vehicles.

Table 2.5 Most abundant compounds in gasoline and headspace vapors in percentage and range.[a]

Compound	Headspace vapor		Liquid gasoline	
	Summer	Winter	Summer	Winter
Isopentane	31.7 (30.9–32.6)	24.2 (23.6–24.8)	6.9 (6.7–7.1)	7.4 (7.2–7.5)
n-Butane	4.7 (4.4–5.1)	24.2 (24.1–24.3)	0.5 (0.5–0.5)	3.4 (3.4–3.4)
2-Methylpentane	6.5 (6.3–6.7)	4.0 (3.6–4.3)	3.9 (3.8–3.9)	3.3 (3.0–3.6)
3-Methylpentane	3.5 (3.4–3.7)	2.5 (2.3–2.6)	2.4 (2.3–2.4)	2.3 (2.2–2.5)
Ethanol	7.5 (7.5–7.6)	5.4 (5.4–5.5)	6.0 (6.0–6.1)	6.3 (6.2–6.4)
n-Pentane	8.3 (8.2–8.3)	7.8 (7.1–8.4)	2.4 (2.4–2.4)	3.2 (2.9–3.5)
Toluene	2.0 (1.9–2.1)	1.3 (1.2–1.4)	8.2 (7.7–8.6)	7.6 (7.0–8.3)
m-Xylene	0.3 (0.3–0.3)	0.2 (0.2–0.2)	3.9 (3.9–4.0)	4.1 (4.0–4.3)
2,2,4-Trimethylpentane	3.3 (3.2–3.4)	1.3 (1.0–1.7)	6.3 (6.0–6.7)	3.5 (2.6–4.5)
Isobutane	0.7 (0.7–0.7)	6.2 (4.9–7.5)	0.05 (0.05–0.05)	0.6 (0.5–0.7)

[a]Liquid gasoline composition acquired from CARB and is based on 20 liquid gasoline samples collected each season and aggregated into two mixtures for detailed hydrocarbon analysis (ranges are over the two aggregates). Some individual liquid samples may have been affected by weathering in the fuel tank.
Table copied from Genter, D.R., Harley, R.A., Miller, A.A., Goldstein, A.H. (2009) Diurnal and seasonal variability of gasoline-related volatile organic compound emissions in Riverside, California. Environmental Science & Technology 43(12), 4247–4252.

Nonfuel-related emissions

Nonfuel-related emissions associated with transportation activities refer to PM emissions of resuspended road dust, tire and brake pad wear, clutch wear, road surface deterioration, and entrainment of roadside materials due to vehicle-induced turbulence. Among nonfuel sources, brake wear can be a significant PM contributor, particularly within areas with high traffic density and braking frequency. In urban environments, brake wear can contribute up to 55% by mass to total nonexhaust traffic-related PM_{10} emissions and up to 21% by mass to total traffic-related PM_{10} emissions; the respective contribution is lower on freeway due to lower braking frequency (Grigoratos & Martini, 2015). The composition of nonfuel-related emissions depends strongly on the activities and materials involved in the emission processes. Road dust are likely to contain naturally occurring elements such as silicon, aluminum, potassium, calcium, and iron whereas brake and clutch wears are likely to emit titanium, iron, copper, barium, magnesium, antimony, zinc, and other metals. Organic compounds could be part of the nonfuel emissions. Polyalkylene glycol ethers (56.9%) and n-alkanoic acids (34.3%) are the most abundant species while n-alkanes, PAHs, and substituted PAHs are also detected in trace concentrations.

Nonroadway emissions

Nonroadway emissions are those emissions resulting from the use of nonroad engines. Nonroad engines are used in an extremely wide range of applications, each involving great differences in operating characteristics, engine technology, and market dynamics. EPA has adopted emission standards for all types of nonroad engines, equipment, and vehicles including emissions from aircraft, locomotives, waterborne vessels, and heavy equipment. For instance, marine diesel engine at various load conditions emits significant amounts of elemental carbon (EC), organic carbon (OC), hydrated sulfate (HS), ash, and sulfate (Petzold et al., 2010). Aircraft emits PM of various sizes and heavy metals (Boyle, 1996). Soot from jet engines operating at low- and high-power levels may contain varied percentages of elements associated with lube oil—namely the metals from the organometallic additives (Wal, Bryg, & Huang, 2016).

Transportation air quality regulations in the United States

Transportation emissions have been regulated since the enactment of the 1970 Clean Air Act when EPA was authorized to regulate air pollution from transportation facilities. As a result, the new passenger vehicles are 98%–99% cleaner when comparing the current tailpipe emissions to that of the 1960s (U.S. Environmental Protection Agency (U.S. EPA), 2019b). A major achievement of the transportation regulation is the implementation of a complete ban of leaded gas in 1995. EPA started a "phasedown" program in 1973 to bring the levels of lead down at all refineries to an average of leaded and unleaded gasoline of 0.5 g/gal by 1982. EPA first changed the standard to 1.10 g/gal in 1982 for leaded gallon and eliminated the "average" provision between unleaded and leaded gasoline. EPA later, in 1985, revised the standard to 0.1 g/gal, a 90% reduction from the 1982 standard of 1.10 g/gal, beginning on January 1, 1986. Lead was completely banned as a fuel additive in the United States in 1996 (Newell & Rogers, 2003). Fig. 2.2 shows a significant reduction in blood lead level in children aged 1–5 years from 1976 to 2008 since EPA started to regulate lead concentration in gasoline and other lead-containing consumer products such as lead paints, plumbing, food cans, etc.

The US EPA continues to regulate transportation emissions throughout the years. The timeline of EPA's major accomplishments in regulating transportation air quality and vehicle emissions can be found at EPA's webpage (U.S. Environmental Protection Agency (U.S. EPA), 2019b). Most noticeably, the final national emission standards for hazardous air pollutants (NESHAP) for engine test cells/stands were promulgated in May 2003 and new fuel and vehicle standards (Tier 3 standards) that require the oil companies to produce gasoline of lower sulfur content of gasoline and vehicle manufacturers to improve emission control technologies to reduce harmful tailpipe pollution were enacted in 2014 (40CFR Part79). EPA began to regulate greenhouse gas emissions from vehicles in 2009 when it concluded that the current and projected concentrations of the six key well-mixed greenhouse gases in the atmosphere threaten the public health and welfare of current and future generations. In 2011, EPA and the Department of Transportation's National Highway Traffic Safety Administration (NHTSA) announced the first-ever regulations to reduce greenhouse gas emissions and improve fuel efficiency of heavy-duty trucks and buses. In 2015, EPA and NHTSA proposed model years 2018–2027 greenhouse gas emissions and fuel economy standards for medium- and heavy-duty vehicles. A graphic timeline of the US federal vehicle emission regulatory compliance and control program was prepared by the International Council on Clean Transportation (He & Jin, 2017) and reproduced in Fig. 2.3.

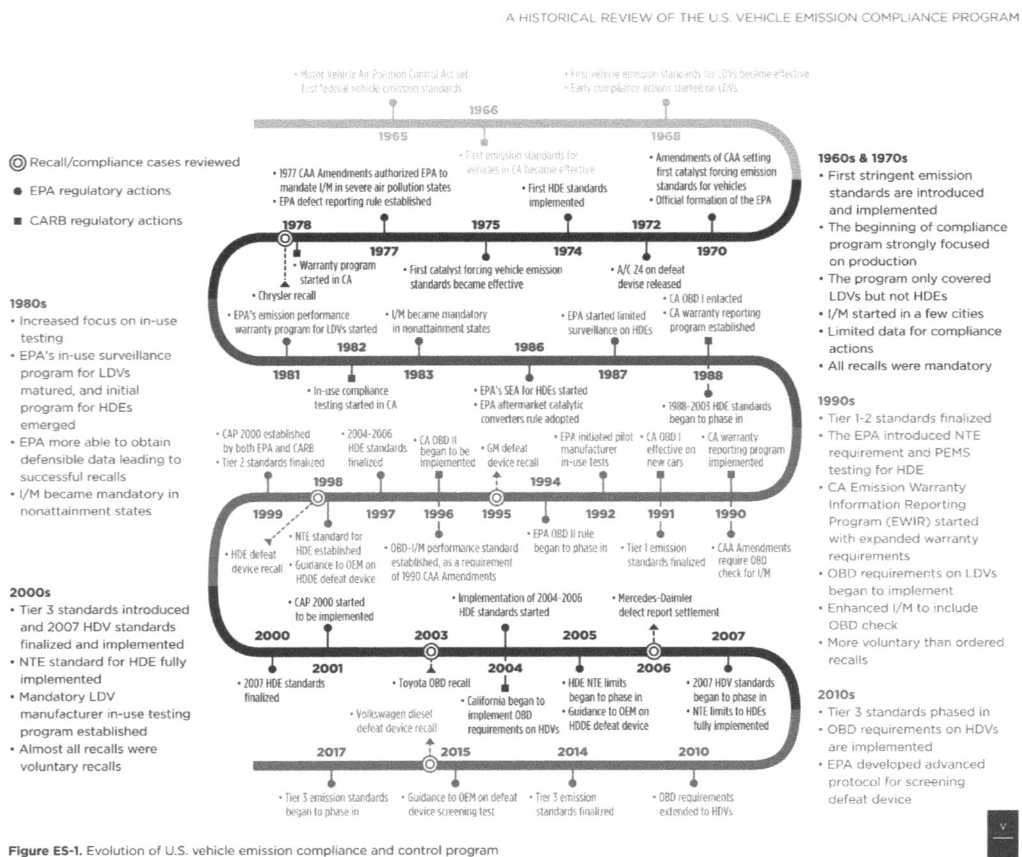

Fig. 2.3 Evolution of US vehicle emission compliance and control program (He & Jin, 2017).

Transportation conformity

Transportation conformity is defined in Section 176(c)(4) of the 1990 CAAA as conformity for new or existing highway or transit projects to a state implementation plan's (SIP's) purpose of eliminating or reducing the severity and number of violations of the NAAQS and achieving expeditious attainment of such standards. The intent is to prevent the air quality impacts of federal actions from causing or contributing to a violation of the NAAQS or interfering with the purpose of a SIP. The EPA has established criteria and procedures for determining if transportation activities conform to the SIP. In general, transportation conformity regulations in the United States act as a means to ensure that metropolitan planning organizations and state departments of transportation consider transportation emissions as part of their transportation planning activities in air quality nonattainment areas, i.e., areas that do not meet the NAAQS.

Transportation conformity is implemented in three phases: (i) applicability analysis; (ii) conformity determination; and (iii) review process. Basically, the applicability analysis requires actions for those activities that cause emissions in designated nonattainment and maintenance areas. If an application analysis shows a need for the action, then a conformity determination must be performed. In the conformity determination, it must (i) identify the total direct and indirect emissions associated with the activity; (ii) demonstrate that these emissions would not exceed the SIP emissions; (iii) obtain statements from state and local metropolitan planning organization (MPO) that the emissions are included in the SIP and the area's transportation plan; (iv) fully offset the emissions in the same nonattainment or maintenance area; and (v) demonstrate, by conducting air quality modeling, that the emissions will not cause or contribute to new violations of the standards, or increase the frequency or severity of any existing violations of the standards. Final revisions to the General Conformity Regulations were published on March 24, 2010 (40 CFR Parts 51 and 93).

Transportation regulations enforcement

The EPA regulates transportation emissions through regulating not only fuel sulfur contents but also individual passenger, commercial, and non-road vehicles. The regulations and programs affecting individual vehicle are standards for exhaust and evaporative emissions, control of hazardous air pollutants and air toxics, national low-emission vehicle program, CAP 2000 (Compliance Assurance Program), onboard refueling vapor recovery, and inspection and maintenance programs. Compliance of these regulations is a major challenge to gasoline refiners, automobile manufactures, and the public. Refiners and fuel importers have the primary responsibility of compliance with the motor vehicle fuels standards. Parties in the fuel distribution system are responsible for ensuring that motor vehicle fuel is not contaminated and is used in the proper locations and times. Vehicle or engine manufacturers are responsible for compliance with the vehicle and engine emissions standards. Private vehicle owners are responsible for compliance of vehicle emission standards through state-implemented inspection and maintenance programs.

Tempering of emissions tests is not limited to private citizens, especially with people who do not abide by the law, but also to major auto manufacturers. In 2018, the German auto manufacturers Volkswagen and Audi (Dorenkamp et al., 2018) was charged in the United States for defrauding the United States by working collaboratively in designing, testing, implementing, and improving software they knew to cheat the US testing process by making it appear as if the emissions standards were met (referred to as the *Diesel Gate scandal*). Whether it is intentional or due to serious technology deficiencies, violation of emission standards by auto manufacturers is not unique to Volkswagen. Emission control devices on certain passenger vehicles manufactured by Fiat Chrysler (U.S. Department of Justice, 2019), Daimler (Reuters, 2018), and GM (Fenner et al., 2018) were found to shut off completely during driving. In other cases, vehicles may be recalled for excessive emissions in violation of EPA's emission standards (e.g., Reuters, 2019). These companies failed to comply with the emissions standards have since agreed to pay up to $25 billion in fine to settle the charges.

Transportation emissions and near-road communities

Air pollution is a complex issue which affects every living being in the world. New data from the WHO shows that 9 out of 10 people breathe air containing high levels of pollutants. The WHO estimates that around 7 million people die every year from exposure to $PM_{2.5}$ in polluted air that penetrate deep into the lungs and cardiovascular system, causing diseases including stroke, heart disease, lung cancer, chronic obstructive pulmonary diseases, and respiratory infections, including pneumonia (Prüss-Ustün, Wolf, Corvalán, Bos, & Neira, 2016). Ambient air pollution alone caused some 4.2 million deaths in 2016, while household air pollution from cooking with polluting fuels caused an estimated 3.8 million deaths in the same period. In the United States, approximately 20% of mortality could be attributed to air pollution exposure (Jerrett, 2015; Lee et al., 2017). Air pollution deaths cost global economy US $225 billion in lost labor income in 2013 (The World Bank Group, 2016). It costs the global economy >$5 trillion annually in welfare costs, with the most devastating damage occurring in the developing world.

Air pollution is not only a public health but also a social and economic inequality issue. Globally, > 90% of air pollution-related deaths occur in low- and middle-income countries, mainly in Asia and Africa. In the developed countries, people living in underserved, low-income neighborhoods are likely to be exposed to more severe air pollution in the same city. Residents of underserved communities (low-income minority communities, in particular) are more likely to be exposed to excessive levels of air pollution. The US demographics have shown that 68% of African Americans live within 30 miles of a coal-fired power plant and 66% of Latinos live in areas that do not meet the federal government's air quality standards. Highly polluting industries are likely to locate their facilities in less affluent and less regulated areas. Individuals with higher education and income are likely to be more aware or better informed of the causes and impacts of air pollution and have the financial means to move away from poor air quality areas than people who don't. As we all know, public policy is heavily influenced by interested parties and industries are said to be willing to invest more fund to support less restrictive environmental regulations than that to invest on pollution control equipment. All these result in a "pollution inequality," as concluded in a recently published study that non-Hispanic whites experience a "pollution advantage" vs. Blacks and Hispanics (Tessum et al., 2019). On average, non-Hispanic whites experience ~17% less $PM_{2.5}$ exposures than Black and Hispanics. This disparity reflects a "pollution burden" of 56% and 63% excess exposure by Black and Hispanics, respectively, relative to the exposure caused by their consumption (or economic affluence).

Among the many sources of pollution, pollution caused by transportation-related activities is of particular importance to near-road communities due to its close proximity to the sources and toxicity of the pollutants. Traffic emissions have a substantial impact on indoor and outdoor exposures in addition to personal exposures resulting in substantial detrimental health effects (Janssen, Van Vliet, Aarts, Harssema, & Brunekreef, 2001; Spira-Cohen, Chen, Kendall, Lall, & Thurston, 2011). A 2011 national household survey (American Housing Survey (AHS), 2015) showed that 16.88 million households in the United States lived within 1/2 block (~100 m) from a four-or-more-lane highway, railroad, or airport. This implies that approximately 43.5 million people were exposed to traffic-related emissions in 2011, using an average people per household of 2.58 for that year. The numbers are consistent with a widely quoted statistic of 22 million total housing units and 45 million of population living near traffic facilities (U.S. Environmental Protection Agency (U.S. EPA), 2010; Weinstock, Watkins, Wayland, & Baldauf, 2013). Emerging evidence suggests that living in close proximity to traffic is particularly harmful to children. Between 2005 and 2006, it was estimated that approximately 3.2 million students attended schools located within 100 m of a major roadway and an additional 3.2 million students attended schools located 100–250 m from major roadways, as reported by Kingsley and his colleagues (Kingsley et al., 2014). School children living 30–300 m from a major roadway had increased arterial stiffness (Iannuzi, Verga, & Renis, 2010), increased carotid intima-media thickness (Armijos et al., 2015), decreased academic performance (Gilliland et al., 2001), increased absenteeism (Chen, Jenison, Yang, & Omaye, 2000), and increased clinical asthma symptoms (Wendt, Symanski, Stock, Chan, & Du, 2014). A multitude number of cross-sectional studies have also been conducted to study traffic-related air pollutants and respiratory health, behavioral problems, and physical activities children living near busy highways. It is well documented in the literature that TRAPs pose adverse effects on children respiratory health (Barone-Adesi et al., 2015; Gehring et al., 2013; Ierodiakonou et al., 2016), behavioral problems (Forns, Dadvand, Foraster, & Alvarez-Pedrerol, 2016), and physical (Lovinsky-Desir et al., 2016).

Conclusion

Traffic-related air pollutants are known to contribute to poor air quality in urban areas and result in damages to the environment and human health. This chapter provides general information on air quality, air pollution, sources of emissions, history of air pollution control and regulations, traffic-related air pollutants, and characterization of transportation emissions. It also updates the readers with recent developments in air quality standards, emission trends, transportation emission regulations and enforcements, and impacts of traffic emissions on near-road communities. It provides the readers with basic background knowledge to further explore emerging issues associated with traffic-related air pollution discussed in the later chapters of this book.

References

American Housing Survey (AHS). (2015). *AHS 2013 national summary tables*. U.S. Department of Housing and Urban Development. https://www.census.gov/programs-surveys/ahs/data/2013/ahs-2013-summary-tables/national-summary-report-and-tables---ahs-2013.html. (Accessed 1 July 2017).

Araizaga, A. E., Mancilla, Y., & Mendoza, A. (2013). Volatile organic compound emissions from light-duty vehicles in Monterrey, Mexico: A tunnel study. *International Journal of Environmental Research*, 7(2), 277–292.

Armijos, R. X., Weigel, M. M., Myers, O. B., Li, W.-W., Racines, M., & Berwick, M. (2015). Residential exposure to urban traffic is associated with increased carotid intima-media thickness in children. *Journal of Environmental and Public Health*, 1–11.

Ayra, S. P. (1997). *Air pollution meteorology and dispersion*. New York, New York: Oxford University Press.

Barone-Adesi, F., Dent, J. E., Dajnak, D., Beevers, S., Anderson, H. R., Kelly, F. J., et al. (2015). Long term exposure to primary traffic pollutants and lung function in children: Cross-sectional study and meta-analysis. *PLoS One*, *10*(11), e0142565.

Boyle, K. A. (1996). Evaluating particulate emissions from jet engines: Analysis of chemical and physical characteristics and potential impacts on coastal environments and human health. *Transportation Research Record*, *1517*(1), 1–9. https://doi.org/10.3141/1517-01.

Chen, L., Jenison, B. L., Yang, W., & Omaye, S. T. (2000). Elementary school absenteeism and air pollution. *Inhalation Toxicology*, *12*(11), 997–1016.

de Nevers, N. (2017). *Air pollution control engineering* (3rd ed.). Long Grove, Illinois: Waveland Press.

De Winter, J. L., Brown, S. G., Seagram, A. F., & Landsberg, K. (2018). A national-scale review of air pollutant concentrations measured in the U.S. near-road monitoring network during 2014 and 2015. *Atmospheric Environment*, *183*, 94–105.

Dorenkamp, R., et al. (2018). *U.S. District Court Eastern District of Michigan Southern Division, United State of America*. https://www.justice.gov/opa/press-release/file/1059821/download. (Accessed 18 March 2019).

EPA. (1970). *Ash Council Memo, Federal Organization for Environmental Protection, Executive Office of the President, President's Advisory Council on Executive Organization*. Washington, DC: EPA.

Fenner, A., et al. (2018). *U.S. District Court Eastern District of Michigan Northern Division*. v. General Motor LLC et al. https://www.hbsslaw.com/uploads/case_downloads/silverado_sierra/02.20.18_hagens_berman_gm_duramax_order.pdf. (Accessed 14 March 2019).

Forns, J., Dadvand, P., Foraster, M., & Alvarez-Pedrerol, M. (2016). Traffic-related air pollution, noise at school, and behavioral problems in Barcelona schoolchildren: A cross-sectional study. *Environmental Health Perspectives*, *124*(4), 529–535.

Forswall, C. D., & Higgins, K. E. (2005). *Clean Air Act implementation in Houston: An historical perspective 1970-2005*. Rice University Environmental and Energy Systems Institute Shell Center for Sustainability.

Gehring, U., Gruzieva, O., Aguis, R. M., Beelen, R., Custovic, A., Cyrys, J., et al. (2013). Air pollution exposure and lung function in children: The ESCAPE project. *Environmental Health Perspectives*, *121*(11–12), 1357.

Genter, D. R., Harley, R. A., Miller, A. A., & Goldstein, A. H. (2009). Diurnal and seasonal variability of gasoline-related volatile organic compound emissions in Riverside, California. *Environmental Science & Technology*, *43*(12), 4247–4252.

Gentner, D. R., Worton, D. R., Isaacman, G., Davis, L. C., Dallmann, T. R., Wood, E. C., et al. (2013). Chemical composition of gas-phase organic carbon emissions from motor vehicles and implications for ozone production. *Environmental Science & Technology*, *47*(20), 11837–11848. https://doi.org/10.1021/es401470e.

Gilliland, F. D., Berhane, K., Rappaport, E. B., Thomas, D. C., Avol, E., Gauderman, W. J., et al. (2001). The effects of ambient air pollution on school absenteeism due to respiratory illnesses. *Epidemiology, 12*, 43–54.

Grigoratos, T., & Martini, G. (2015). Brake wear particle emissions: A review. *Environmental Science and Pollution Research*, (22), 2491–2504. https://doi.org/10.1007/s11356-014-3696-8.

He, H., & Jin, L. (2017). *White paper: A historical review of the U.S. vehicle emission compliance program and emission recall cases*. Washington DC: The International Council on Clean Transportation. https://www.theicct.org/sites/default/files/publications/EPA-Compliance-and-Recall_ICCT_White-Paper_12042017_vF.pdf. (Accessed 10 March 2019).

Hilpert, M., Adria-Mora, B., Ni, J., Rule, A., & Nachman, K. (2015). Hydrocarbon release during fuel storage and transfer at gas stations: Environmental and health effects. *Current Environmental Health Reports, 2*, 412–422.

Iannuzi, A., Verga, M. C., & Renis, M. (2010). Air pollution and carotid arterial stiffness in children. *Cardiology in the Young, 20*(2), 186–190.

Ierodiakonou, D., Zanobetti, A., Coull, B. A., Melly, S., Postma, D. S., Boezen, H. M., et al. (2016). Ambient air pollution, lung function, and airway responsiveness in asthmatic children. *The Journal of Allergy and Clinical Immunology, 137*(2), 390–399.

Janssen, N. C. H., Van Vliet, P. H. N., Aarts, F., Harssema, H., & Brunekreef, B. (2001). Assessment of exposure to traffic related air pollution of children attending schools near motorways. *Atmospheric Environment, 35*, 3875–3884.

Jerrett, M. (2015). Atmospheric science: The death toll from air-pollution sources. *Nature, 525*(7569), 330–331.

Kingsley, S. L., Eliot, M., Carlson, L., Finn, J., MacIntosh, D. L., Suh, H. H., et al. (2014). Proximity of US schools to major roadways: A nationwide assessment. *Journal of Exposure Science & Environmental Epidemiology, 24*(3), 253–259.

Kuklinska, K., Wolska, L., & Namiesnik, J. (2015). Air quality policy in the U.S. and the EU—A review. *Atmospheric Pollution Research, 20*(1), 129–137.

Lee, E. Y., Lin, J., Noth, E. M., Hammond, K., Nadeau, K., Eisen, E. A., et al. (2017). Traffic-related air pollution and telomere length in children and adolescents living in Fresno, CA: A pilot study. *Journal of Occupational and Environmental Medicine, 59*(5), 446–452.

Lovinsky-Desir, S., Jung, K. H., Rundle, A. G., Hoepner, L. A., Bautista, J. B., Perera, F. P., et al. (2016). Physical activity, black carbon exposure and airway inflammation in an urban adolescent cohort. *Environmental Research, 151*, 756–762.

McGaughey, G. R., Desai, N. R., Allen, D., Seila, R. L., Lonneman, W. A., Fraser, M. P., et al. (2003). Analysis of motor vehicle emissions in a Houston tunnel during the Texas air quality study. In *Presented at NARSTO workshop on innovative methods for emission-inventory development and evaluation, Austin, TX, October 14–17.*

Newell, R. G., & Rogers, K. (2003). The U.S. experience with the phasedown of lead in gasoline. In *Resources for the Future*. Discussion Paper, Washington DC.

Petzold, A., Weingartner, E., Hasselbach, J., Lauer, P., Kurok, C., & Fleischer, F. (2010). Physical properties, chemical composition, and cloud forming potential of particulate emissions from a marine diesel engine at various load conditions. *Environmental Science & Technology, 44*(10), 3800–3805. https://doi.org/10.1021/es903681z.

Prüss-Ustün, A., Wolf, J., Corvalán, C., Bos, R., & Neira, M. (2016). *Preventing disease through healthy environments: A global assessment of the burden of disease from environmental risks*. Geneva, Switzerland: WHO Press, World Health Organization.

Reuters. (2018). *Germany threatens Daimler with 3.75 billion euro fine over emissions-Spiegel*. https://www.reuters.com/article/us-germany-emissions-daimler/germany-threatens-daimler-with-3-75-billion-euro-fine-over-emissions-spiegel-idUSKCN1IX4MT. (Accessed 15 March 2019).

Reuters. (2019). *Fiat Chrysler recalling 965,000 U.S., Canadian vehicles over excess emissions.* https://www.reuters.com/article/us-fiat-chrysler-emissions-exclusive/fiat-chrysler-to-recall-862520-u-s-vehicles-over-excess-emissions-idUSKBN1QU1WM. (Accessed 14 March 2019).

Spira-Cohen, A., Chen, L. C., Kendall, M., Lall, R., & Thurston, G. D. (2011). Personal exposures to traffic-related air pollution and acute respiratory health among Bronx school children with asthma. *Environmental Health Perspectives*, *119*(4), 559–565.

Tessum, C. W., Apte, J. S., Goodkind, A. L., Muller, N. Z., Mullins, K. A., Paolella, D. A., et al. (2019). Inequity in consumption of goods and services adds to racial–ethnic disparities in air pollution exposure. *Proceedings of the National Academy of Sciences of the United States of America*. https://doi.org/10.1073/pnas.1818859116. Accessed 15 March 2019.

The World Bank Group. (2016). *The cost of air pollution: Strengthening the economic case for action.* Washington, DC: Institute for Health Metrics and Evaluation, World Bank. https://openknowledge.worldbank.org/handle/10986/25013. (Accessed 15 March 2019).

U.S. Department of Justice. (2019). *In Civil Settlements with the United States and California, Fiat Chrysler will Resolve Allegations of Cheating on Federal and State Vehicle Emission Tests.* https://www.justice.gov/opa/pr/civil-settlements-united-states-and-california-fiat-chrysler-will-resolve-allegations. (Accessed 14 March 2019).

U.S. Department of Transportation (U.S.DOT). (2018). *Highway statistics 2016.* Washington DC: Office of Policy and Governmental Affairs. https://www.fhwa.dot.gov/policyinformation/statistics/2016/. (Accessed 1 March 2019).

U.S. Department of Transportation (U.S.DOT). (2019). *Traffic volume trends.* Washington DC: Office of Policy and Governmental Affairs. https://www.fhwa.dot.gov/policyinformation/travel_monitoring/tvt.cfm. (Accessed 1 March 2019).

U.S. Environmental Protection Agency (U.S. EPA). (2008). *Regulatory determinations support document for selected contaminants from the second drinking water contaminant candidate list (CCL2).* EPA-OGWDW.

U.S. Environmental Protection Agency (U.S. EPA). (2010). Primary National Ambient Air Quality Standards for nitrogen dioxide—Final rule. *Federal Register*, *75*, 6482.

U.S. Environmental Protection Agency (U.S. EPA). (2018). *History of air pollution.* Durham, NC: U.S. EPA Research. https://www.epa.gov/air-research/history-air-pollution#then. (Accessed 11 December 2018).

U.S. Environmental Protection Agency (U.S. EPA). (2019a). *National trends in lead levels.* Durham, NC: U.S. EPA Research. https://www.epa.gov/air-trends/lead-trends. (Accessed 1 March 2019).

U.S. Environmental Protection Agency (U.S. EPA). (2019b). https://www.epa.gov/transportation-air-pollution-and-climate-change/timeline-major-accomplishments-transportation-air. (Accessed 1 March 2019).

Wal, R. L., Bryg, V. M., & Huang, C. (2016). Chemistry characterization of jet aircraft engine particulate matter by XPS: Results from APEX III. *Atmospheric Environment*, *140*, 623–629. https://doi.org/10.1016/j.atmosenv.2016.05.039.

Weinstock, L., Watkins, N., Wayland, R., & Baldauf, R. (2013). EPA's emerging near-road ambient monitoring network: A progress report. *Environmental Magazine*, 7, 6–10.

Wendt, J. K., Symanski, E., Stock, T. H., Chan, W., & Du, X. L. (2014). Association of short-term increase in ambient air pollution and timing of initial asthma diagnosis among Medicaid-enrolled children in a metropolitan area. *Environmental Research*, *131*, 50–58.

World Health Organization (WHO). (2013). *Outdoor air pollution cause cancer.* https://www.cancer.org/latest-news/world-health-organization-outdoor-air-pollution-causes-cancer.html. (Accessed 1 March 2003).

CHAPTER 3

Traffic monitoring and modeling for energy, air quality, and health

Yanzhi Ann Xu
Texas A&M Transportation Institute, Washington, DC, United States

List of abbreviations

AB	activity based
ARPA-E	Advanced Research Projects Agency-Energy
CDR	call detail record
DEM	digital elevation model
DOE	Department of Energy
DTA	dynamic traffic assignment
EIA	Energy Information Agency
EPA	Environmental Protection Agency
FHWA	Federal Highway Administration
GIS	geographic information system
GPS	global positioning system
GTFS	general transit feed specification
NEI	National Emissions Inventory
NTD	National Transit Database
OBD	on-board diagnostic
O-D	origin-destination
SAE	Society of Automotive Engineers
TNC	transportation network companies
TRANSNET	Traveler Response Architecture Using Novel Signaling for Network Efficiency in Transportation
TSDC	Transportation Secure Data Center
U.S.	United States

Introduction

Traffic monitoring and modeling are the cornerstones of the field of transportation systems engineering. The vast amount of traffic monitoring data and the deep expertise in traffic modeling are beyond what a single chapter can cover. Instead, this chapter focuses only on the existing and potential

Traffic-Related Air Pollution
https://doi.org/10.1016/B978-0-12-818122-5.00003-X

application of traffic monitoring data and traffic modeling efforts to research questions related to energy, air quality, and health.

Traditionally, detailed traffic monitoring data, travel demand models, and simulation models are mostly used for applications related to transportation planning, traffic operations and safety, such as congestion relief and incident response. In the United States, there has also been an established practice linking traffic data and models with emission models for analyses to meet the transportation conformity regulations (US Environmental Protection Agency (EPA), 2012). Due to regulatory requirements, these modeling schemes are rather rigid and take a long time to run, and therefore not suitable for scenario analysis. Further, these efforts are scattered, because a consistent national approach requires too much data for standard tools to handle.

Beyond regulatory requirements, traffic data and models have been used in academia to study policy implications. The conventional approach to model the air quality impacts from policies such as congestion reduction measures require a long chain, including land use, travel behavior, traffic flow, and vehicle emission models. At each step, the models require large amounts of data and embody intrinsic uncertainty. As researchers propagate data through the modeling chain, the levels of uncertainty compound, often resulting in little explanatory power to inform policy making (Battelle and Texas A&M Transportation Institute, 2015). Nevertheless, significant progress has taken place at the component level of the modeling chain, as one will see in the remaining chapter. Hopefully, existing efforts can be organized and leveraged in such a way to analyze the full-chain impacts from traffic to air quality to health.

The rest of this chapter will first review the types of conventional and emerging traffic monitoring data available for energy, air quality, and health research. The author will then give an overview on modeling efforts integrating traffic with energy and air quality. The reviews of data and models will culminate into a discussion on gaps and opportunities, followed by concluding remarks.

Traffic monitoring data

Traffic-related data can be broadly categorized into three groups:

1. Traffic operations: data that indicates the patterns of roadways, including counts, density, speed, and volume;
2. Travel activity: data that indicates the patterns of travelers, including trip frequency, origin, destination, and travel time; and

3. Vehicle operations: data that indicates the patterns of movements, often referring to the second-by-second speed-acceleration profiles of trips or trip segments.

The three categories generally apply to all modes of travel, including automobile, transit, walking, bicycling, and other alternative modes. However, due to the dominance of automobiles in trip making in the United States, data related to vehicular travel is far more abundant than data related to other modes. As such, the review below first focuses on vehicular travel in the three categories, and then groups all other modes into one section. The review will also discuss data availability on transportation infrastructure because infrastructure elements (e.g., roads, bridges, sidewalks, parking facilities, etc.) constitute the supply that meets travel demand.

This review of traffic monitoring data focuses on people's movement rather than goods movement. This is because, in traditional traffic monitoring data discussions, people movement, especially across surface modes, often receives preferential emphasis, due to a large number of constituents, i.e., the traveling public, and the proportion of vehicles that are dedicated to people movement. However, in energy, air quality, and health discussions, it would be amiss to overlook goods movement and non-surface modes, due to their disproportionate contribution to energy consumption and pollutant emissions. For example, air travel constitutes about 0.1% of all vehicle miles in the United States in 2017, according to the U.S. Bureau of Transportation Statistics (BTS) (2018), but jet fuel supplies 12% of all transportation energy in the same year, according to the U.S. Energy Information Administration (EIA) (2018). Similar patterns exist regarding air pollutants that are harmful to human health. According to the U.S. EPA's 2014 National Emissions Inventory (NEI) (U.S. EPA, 2014) (most recent year of NEI at the time of writing this chapter), commercial marine vessels (tugboats, barges, ocean-going vessels, etc.) contributed 15% of total mobile source NO_x emissions, compared to only 28% from all non-diesel light-duty vehicles (mostly used for people movement). The 2014 NEI also reported that heavy-duty on-road vehicles, mostly used for goods movement, contributed to 30% of mobile source black carbon emissions, compared to only 8% from light-duty on-road vehicles, mostly used for people movement. Such statistics underscore the need to thoroughly consider goods movement and non-surface modes for energy, air quality, and health-related transportation studies. Readers are encouraged to refer to a review of freight data by Mani and Prozzi (2004).

Traffic operations data

Traffic operations data have long been collected for system performance. These include spot traffic data and incident data. Traditionally, spot traffic data, including count, speed, volume, and occupancy, is collected by state and local agencies through a network of detectors. As technologies evolve, the detectors have advanced from permanent and temporary loop detectors to acoustic, radar, and video detection systems. The Federal Highway Administration provides a comprehensive overview of the detection methods, theory, and attributes of traffic operations data in its Traffic Monitoring Guide (FHWA, 2016).

Travel activity data

Travel activity data related to travelers' trip making patterns. Some travel activity data are more closely aligned with research around travel behavior, such as trip frequency, mode of travel, and origin-destination (OD) pairs. Other measures, such as travel time, are more closely related to traffic operations and are sometimes grouped as such. The rationale for grouping travel time as travel activity data, along with other trip-making attributes, is that travel time data is not easily collected using point-based detectors. Instead, travel activity data, in general, requires tracking individual travelers.

Household travel surveys

Household travel surveys have long served as the primary source of travel activity data. Household travel surveys administered at the national, state, and regional levels have been the backbone of transportation statistics and travel demand models. In studies of smaller scales or special purposes, researchers also tend to resort to surveys to gain insight into travel activity.

Global positioning system data

In recent years, tracking devices equipped with a global positioning system (GPS) have grown popular to supplement or even replace traditional travel surveys. As the quality of GPS data continues to improve and the cost of deploying GPS devices continues to decrease, GPS data has become increasingly available. Accordingly, a wealth of research has been conducted to automate GPS data processing. Algorithms have been developed to automatically identify the trip purpose (Krause & Zhang, 2018) and mode (Yang, Stewart, Tang, Xie, & Li, 2018). Together, the advances in hardware and software have made GPS data ubiquitous. Just about a decade ago, GPS devices are mostly used to equip vehicles, and such data is often referred to as probe vehicle data. More recently, GPS data is often available from mobile phones, and a discussion on GPS data quickly expands into a discussion on mobile phone data.

Mobile phone data

Mobile phone data, like GPS data, is passively collected. However, mobile phone datasets have distinct characteristics compared to GPS datasets. Mobile phone data contains location information at much coarser temporal resolution than GPS data. As reviewed by Chen, Ma, Susilo, Liu, and Wang (2016), there are two types of mobile phone data: call detail record (CDR) data and sightings data. These two data types differ in temporal resolution, spatial resolution, and user interactions. Sightings data has higher temporal and spatial resolutions but is not based on user interactions as the case for CDR data. Due to its relatively low temporal and spatial resolutions, the most intuitive application of mobile phone data is to infer trip rates and OD pairs in studying mobility patterns. However, recent advances have demonstrated inferred travel times (Toole et al., 2015) and modes (Wang, He, & Leung, 2018).

Participatory sensing network data

A more novel data source is participatory sensing networks (PSN) (Burke et al., 2006). The proliferation of smartphones has given rise to such networks, where passively collected sensor data (e.g., GPS and cellular network data) is a couple with proactive user-provided data, such as from Waze and FourSquare. Silva, Vaz de Melo, Almeida, and Loureiro (2013) provide a comprehensive overview of the opportunities and challenges of using such data in mobility studies. Another emerging data source is transportation providers, such as taxi fleets and transportation network companies (TNCs). Recently, Uber started releasing anonymized O-D and travel time data via Uber Movement (movement.uber.com).

Due to the disaggregate nature of travel activity data, data privacy and security are of paramount concern. As such, the archival and sharing of detailed activity data requires significant investment in IT infrastructure, especially as it relates to security. Additionally, access to such data often requires a lengthy approval process. A good example of data archival and sharing of travel activity data mostly resulting from travel surveys is the National Renewable Energy Laboratory's Transportation Secure Data Center (TSDC) (Gonder, Burton, & Murakami, 2015). When travel activity data is released publicly, as with Uber Movement, the granularity of the data is greatly reduced to protect privacy. As such, methodological innovations are required to utilize these new data streams to augment transportation research.

Vehicle operations data

Transportation research related to energy, air quality, and health requires more granular data than travel activity data. The energy consumption and emissions patterns are driven by the vehicle use patterns, also known as duty cycles. This is especially relevant to vehicle emissions because different pollutants have very different emission processes in engine operations. Additionally, air quality issues occur at both the regional level, such as with ozone formation and the local level (e.g., near road and neighborhood), such as particulate matter and carbon monoxide. As such, energy, air quality, and health studies often require vehicle operations data with high temporal and spatial resolutions.

In the real-world setting, vehicle trajectory data comes mostly from probe vehicles instrumented with tracking devices. One of the most common tracking methods is GPS, which provides second-by-second location information from which speed and acceleration information can be derived. Even though there are data quality issues associated with GPS, such as wander at low speeds (Xu & Guensler, 2015), the technology and associated filtering and processing methods have matured substantially in the past decade, which GPS has become an affordable and reliable source for vehicle trajectory data.

A second data source for vehicle trajectory data is the onboard diagnostic (OBD) system. Most modern vehicles are equipped with OBD systems for self-diagnostic and reporting. The Society of Automotive Engineers (SAE) has developed the OBD-II standard (SAE, 2001) in the United States to enable regulation of vehicle emissions. Unlike GPS, OBD data does not contain spatial information. OBD speed is measured from wheel rotations. OBD data is especially popular in energy and emissions research because it contains vehicle operations information such as throttle position and manifold absolute pressure (SAE, 2017), and therefore gives researchers direct insights to real-time fuel consumption rate and emission characteristics. OBD data related to engine parameters is especially valuable for heavy-duty vehicles, whose fuel consumption rates vary widely even given the same speed/acceleration profile. With the advances in communication technologies and miniaturization of sensors, OBD II readers have been increasingly combined with GPS (Wahlström, Skog, Nordström, & Händel, 2018) and

WiFi (Malekian, Moloisane, Nair, Maharaj, & Chude-Okonkwo, 2017) for telematics systems.

Fused data sets from multiple data streams, including GPS and OBD, among other sensor systems, have become increasingly available. Many of these data sets were originally intended for safety and operations research but are nonetheless invaluable for energy and emissions research. One such example is the Naturalistic Driving Study (NDS) data (Campbell, 2012). Using the NDS data, researchers have derived drive cycles for energy and emissions research (Liu, Ivanco, & Filipi, 2015).

A complete departure from instrumented vehicles to obtain vehicle movement trajectories is through video processing. The proliferation of traffic cameras and advances in video processing capabilities have made it possible to extract vehicle movement to study vehicle emissions (Suh et al., 2017). Compared to instrumented vehicles, vehicle trajectories automatically generated from videos provide a much larger cross section of vehicles on the roadway and thus can be particularly useful in the analysis of sensitive locations such as at intersections or highway ramps.

Alternative modes

Compared to data on vehicular traffic, data on alternative transportation modes are much scarcer and more scattered. Regarding transit, the National Transit Database (NTD) has served as a foundational data source in the United States. The NTD has been employed in understanding energy consumption patterns in urban rail systems (Gbologah, Xu, Rodgers, & Guensler, 2014; Wang & Rakha, 2017). More recently, new data sources such as smart card data have been used in travel behavior research (Kusakabe & Asakura, 2014). On a limited but growing basis, microtransit service providers are also sharing their data with public agencies (He, Chow, & Nourinejad, 2017). Regarding pedestrian and bicycle data, Schneider, Patten, and Toole (2005) analyzed data in 29 state and local agencies in the United States, including manual counts, automated counts, general and targeted surveys, inventories, and spatial analyses. Since then, technology changes have enabled more data sources. Notably, data from bike-sharing services have proven to be useful in understanding cycling behavior. For example, O'Brien, Cheshire, and Batty (2014) mined bike-sharing data to compare usage patterns across the world. Similar data sources are emerging from novel modes such as ride-sharing, microtransit, and micro-mobility options such as e-scooters, but the data availability varies significantly across service providers, and the application of such data in transportation modeling is still in its infancy.

Long-distance travel is often analyzed and modeled separately from urban transportation networks. The data sources, such as aviation, rail, and intercity bus, are also siloed from vehicular traffic data, collected, and archived by different departments from roadway agencies. From a travel demand and traffic management point of view, such a separation is justifiable as an efficient way of organization. However, from points of view that are outside of urban transportation management, intermodal comparisons and integration become relevant policy and implementation questions. For example, Liu, Xu, Stockwell, Rodgers, and Guensler (2016) compared aviation, intercity buses, and driving for long-distance travel regarding life-cycle energy consumption and found that the modes would trade off in energy efficiency depending on the distance traveled—air is less efficient compared to surface modes only for shorter trips. Such context-sensitive comparisons will prove essential with the anticipation of automated vehicles serving as "mobile hotels" for travelers. Therefore, it is sensible that researchers should expand their scope of data discovery and analysis in energy, air quality, and health to consider all modes.

Infrastructure

Data regarding the transportation infrastructure has been an integral part of transportation modeling. For example, the transportation network serves as a fundamental input in transportation modeling. To construct the network at a high-enough resolution, with the requisite geometric attributes, such as a number of lanes and presence of shoulders, and operational attributes such as speed limits and traffic signal timing plans, can be one of the most time-consuming tasks in a modeling effort. The open data movement is starting to lift this constraint, however. For example, many modeling projects are building projects on the foundation of OpenStreetMap (www.openstreetmap.org). U.S. EPA's Smart Location Database (https://www.epa.gov/smartgrowth/smart-location-mapping) provides additional land use and urban design data that are relevant to transportation modeling.

In addition, in the context of energy, emissions, and health, there are specific infrastructure elements that have not traditionally received enough emphasis on data collection, processing, and archival. Regarding energy consumption, emissions, and air quality, these data elements include road grade and charging and fueling infrastructure. Regarding active transportation, exposure while engaging in alternative transportation modes and health, these data elements include the presence and quality of sidewalks, bike facilities, and transit supply.

Road grade

Road grade information is traditionally very difficult to obtain. Recently, there have been methodological advances that show the promise of using the Digital Elevation Model (DEM) to derive road grade at satisfactory accuracy levels for emissions estimation (Liu, Li, Rodgers, & Guensler, 2018; Wood, Burton, Duran, & Gonder, 2014). These methods need to be further scaled to provide consistent road grade data across the United States.

Charging infrastructure

Charging infrastructure is a key piece of information in understanding the potential widespread deployment of electric vehicles, and the related impacts on energy consumption and air quality. Fortunately, the US Department of Energy provides central and open access to charging station locations in the United States and Canada through the Alternative Fuels Data Center (USDOE, 2019).

Data related to infrastructure for alternative modes

In the United States, due to the dominance of automobiles, data on infrastructure for pedestrian, bicycling, and transit is scarce. Recent research has demonstrated innovative ways to collect such data, but in general, such data only exists in a scattered, fractured fashion. For example, Li et al. (2018) showcased a semiautomated method to generate GIS-based sidewalk networks. Smartphone apps have been utilized to collect data on sidewalk conditions (Erraguntla, Delen, Agrawal, Madanagopal, & Mayer, 2017). Still, complete sidewalk networks, including both sides of roadways, crossings, and curb cuts, are difficult to find. Similar challenges exist for bicycle infrastructure. Schoner and Levinson (2014) evaluated the network structure of bicycle infrastructure in 74 US cities, but the authors had to collect maps from individual cities, and such data is still not centrally available for further research by the broader community. Data on transit supply, including routes and schedules, is becoming easier to obtain, thanks to the General Transit Feed Specification (GTFS). However, data coverage and quality are inconsistent across transit agencies.

In general, open data initiatives such as Open Street Map and open data standards such as GTFS are quickly changing the landscape of transportation data and are being aligned with policy objectives and practical problems (Goldstein, 2013). Therefore, the traffic monitoring and modeling community should actively seek to contribute to and leverage the open data movement.

Modeling advances

Zmud et al., (2018) categorized three types of modeling systems in transportation planning: trip-based models, activity-based (AB) models and dynamic traffic assignment (DTA) models, and strategic models. Table 3.1 summarizes the differences in these modeling systems by the level of aggregation in their components. Trip-based models and strategic models have existed for a long time and are still the state of practice in most places in the United States and around the world. In comparison, disaggregate modeling systems that combine AB and DTA models have only recently become available due to increased computational capacity. This trend is evident in recent large-scale transportation network modeling programs funded by the US DOE. This chapter will review these newest efforts that were explicitly designed for energy modeling and can generally be applied for emissions modeling in air quality analyses.

From 2015 to 2018, the Advanced Research Projects Agency-Energy (ARPA-E) funded five teams across the United States in the Traveler Response Architecture using Novel Signaling for Network Efficiency in Transportation (TRANSNET) program (ARPA-E, 2015a). The motivation of the program was to "develop mechanisms for individual travelers that both signal and guide them toward the improvement of the energy efficiency of the transportation network in multimodal urban areas" (ARPA-E, 2014). To explore the efficacy of the signal and guidance mechanisms, a key component of this program is a system model—high-fidelity regional transportation simulation model that can both reflect the impacts on energy consumption from changes in individual travelers (e.g., mode shifts and departure time changes) and in the transportation network (e.g., incidents and work zones). As such, a TRANSNET system model must meet five requirements:

1. Integrated—from travel demand, traffic assignment, to energy consumption.
2. Dynamic—iterative, equilibrium-seeking algorithms based on user-experienced travel costs (Chiu et al., 2011).

Table 3.1 Level of aggregation in modeling systems.

	Travel demand	Transportation supply
Trip-based models	Aggregate	Disaggregate
AB and DTA models	Disaggregate	Disaggregate
Strategic models	Disaggregate	Aggregate

3. (Near) real-time—fast runtime.
4. Regional—reflecting an entire network of more than three million inhabitants, as specified by the TRANSNET funding opportunity announcement (FOA).
5. Disaggregate—able to model individual travelers.

At the time of the program announcement, components of such a modeling system have already existed. However, several aforementioned requirements remained as unresolved conflicts against one another. For example, dynamic models are intrinsically computationally intensive and therefore are less advantageous in solving large-scale problems compared to static models (Chiu et al., 2011). Moreover, activity-based travel demand models were already widely adopted among metropolitan planning organizations (MPOs), but there were almost always linked with static models instead of dynamic traffic models (Zmud et al., 2018). Therefore, the TRANSNET program posed to the transportation research community a challenge that was at the time seemed very difficult to tackle. Moreover, since it was geared toward energy assessments and thus can be linked to emissions modeling, it is particularly relevant to this book. Fortunately, research teams were able to leverage both their existing expertise and parallel advances in computation to each come up with unique solutions. A brief overview of each of the TRANSNET[a] modeling efforts is given below. The review follows the general definitions of macroscopic, mesoscopic, and microscopic simulation models set forth in the Federal Highway Administration's (FHWA) Traffic Analysis Tools Program (Cambridge Systematics, Inc, 2014).

Georgia tech research corporation (GTRC)—Network performance monitoring and distributed simulation

The GT team's overview document (Guensler et al., 2018) to their Github code repository provides an excellent introduction to the architecture of their modeling approach. The team's transportation network simulation model incorporates the activity-based travel demand model used by the Atlanta Regional Commission (ARC), the DIRECT DTA model (Alnawaiseh, Abdelghany, & Hassan, 2014), and a Vissim microsimulation model coded for a subarea for the metropolitan Atlanta region. The centerpiece of the

[a]In full disclosure, the author served as Program Director for TRANSNET with in-depth knowledge of the projects, but the descriptions in this chapter are syntheses of publicly available information, derived from published literature and/or confined to the system model components, which were required by the FOA to be developed under open software standards and are publicly available through Github.

GT team's modeling system is the Space-Time Memory (STM), a data architecture implemented in MongoDB that can integrate model outputs and real-world data irrespective of the source. The team used Atlanta, Georgia as a case study. The team's energy model includes MOVES-Matrix, a large matrix constructed by running U.S. EPA's MOVES model millions of times in a high-performance computing (HPC) environment (Guensler et al., 2017) for tailpipe emissions, and the Fuel and Emissions Calculator (FEC) (Xu, Gbologah, Liu, Rodgers, & Guensler, 2015), a life-cycle assessment model to estimate upstream energy consumption and emissions especially for hybrid and electric vehicles (EV).

Massachusetts Institute of Technology (MIT)—Sustainable travel incentives with prediction, optimization, and personalization (TRIPOD)

MIT's Tripod system uses SimMobility for transportation network modeling and TripEnergy for energy modeling (ARPA-E, 2015b). SimMobility is an activity- and agent-based modeling software that includes three levels (De Lima, Danaf, Akkinepally, De Azevedo, & Ben-Akiva, 2018): long-term, which handles land use and vehicle ownership, midterm, which handles daily travel patterns, and short-term, which is a microscopic traffic simulator. TripEnergy was run concurrently with SimMobility midterm to estimate energy consumption and CO_2 emissions (Needell & Trancik, 2018). MIT's modeling system was applied to demonstrate the potential impact of automated mobility-on-demand services on mass transit (Basu et al., 2018). The team applied the modeling system to the greater Boston area in Massachusetts under TRANSNET award.

National Renewable Energy Laboratory (NREL)—The connected traveler: A framework to reduce energy use in transportation

The NREL team did not develop a system model for the transportation network under TRANSNET funding. Rather, the team used a mobility platform developed by Metropia, Inc., which utilizes ABM with DTA as part of its urban analytics capabilities (Metropia, Inc., 2017). The NREL team used Future Automotive Systems Technology Simulator (FASTSim) to model vehicle energy consumption (Duvall & Young, 2018). FASTSim is a vehicle powertrain model to compare vehicle efficiency, performance, cost, and battery life (Brooker et al., 2015). The NREL team used Austin, Texas as its case study metropolitan area.

Palo Alto Research Center (PARC)—Collaborative optimization and planning for transportation energy reduction (COPTER)

The PARC team uses INTEGRATION as its transportation network model. INTEGRATION is a microscopic dynamic traffic assignment and simulation software (Van Aerde & Yagar, 1988) with built-in modules to estimate fuel consumption and emissions (Rakha, Ahn, & Trani, 2004). INTEGRATION includes two main components—traffic assignment and microscopic traffic simulation (Rakha, Ahn, & Moran, 2012). The traffic assignment component first represents travel demand as O-D pairs, then dynamically assigns the demand onto the roadway network based on traffic flow theory. The traffic simulation component estimates vehicle location every deci-second based on the route assignment from the first component. The simulation model employs a vehicle dynamics model to ensure realistic speed-acceleration representations in the model (Rakha & Lucic, 2002), making the modeled results suitable for fuel consumption and emissions analysis. The PARC team used Los Angeles, California as its case study location.

University of Maryland—Traveler information and incentive technology

The University of Maryland (UMD) team developed a modeling system for the Baltimore-Washington metro area, spanning Washington, DC, northern Virginia, and parts of Maryland. UMD's transportation network model comprises of a custom-built travel demand component based on behavioral user-equilibrium theory (ARPA-E, 2018) and DTALite for traffic assignment. DTALite is a mesoscopic, lightweight DTA to allow rapid dynamic traffic analysis (Zhou & Taylor, 2014). DTALite is an open-source software package. For energy and emissions modeling, the UMD team adopted fuel and emission rates derived from the U.S. EPA's MOVES model (Zhou et al., 2015). The modeling system developed by the UMD team heavily leveraged computational advances in cloud and high-performance computing to achieve real-time trip and energy use prediction (ARPA-E, 2018).

Discussion

As the statistician, Box (1976) famously declared, "all models are wrong." The above review has shown that the recent modeling efforts all show a unique combination of strengths and weaknesses. Some favored simplified simulation routines to achieve faster run time. Others excelled in a microscopic simulation that can reflect individual travel and driving behaviors.

Some teams focused on energy modeling that is based on high-fidelity vehicle powertrain dynamics. Others featured comprehensive characterizations of vehicle emissions that can be extended to air quality and health studies. The teams all had to decide on some trade-offs given resource constraints and implementation feasibility. Consequently, instead of developing an "all-purpose" model, it is perhaps more prudent for research sponsors and model developers to collectively design modeling systems that will answer specific questions.

Beyond the TRANSNET program, other ongoing modeling efforts show promise. For example, US DOE has also been funding the SMART Mobility program (Sarkar & Ward, 2016), under which POLARIS (Auld et al., 2016) and BEAM (Sheppard, Waraich, Campbell, Pozdnukhov, & Gopal, 2017) are examples of large-scale agent-based transportation network simulation models for energy analysis. Beyond energy and emissions modeling, off-the-shelf modeling systems from traffic to air quality and health are more difficult to find, due to the long modeling chain. For example, Lefebvre et al. (2013) presented an integrated model chain to answer traffic- and health-related policy questions and the model chain consists of six stand-alone model components.

It is hopeful, however, that integrated traffic, energy, air quality, and health models will become available, given two general trends in the modeling community. First, computational speed and storage capacity continue to increase, making it possible not only to implement complex algorithms but also to store multidisciplinary data that have traditionally been siloed. Second, open-source standards continue to gain momentum. All the models reviewed in this chapter (except for DynusT, which was not developed under the TRANSNET program) are modular and publicly available, making it possible for the broader community to build upon the most recent advances. For example, modelers that intend to build integrated traffic, air quality, and health model are likely to benefit from the integration that has already taken place in traffic and energy modeling, instead of starting from scratch.

Gaps and opportunities

There is no doubt that the field of traffic monitoring and modeling is changing rapidly—the types and amount of data are ballooning, the scale and complexity of the questions to be answered are increasing, and the mathematical methods to analyze the data to answer these questions are becoming ever more sophisticated. It is in the context of such complexity that one can benefit from a look at the broader landscape, not only as technical experts but also as entrepreneurs ready to tackle the new challenges brought about in the urban era. Therefore, the following discussion takes on a business angle to analyze the gaps and opportunities in traffic monitoring and modeling. The discussion first analyzes the demand, or the "market pull" for solutions—who are the "clients" the research community is serving, and what types of questions are to be addressed. Then the supply, or the "technology push" is summarized as general trends. Finally, gaps and opportunities emerge logically as the match and the lack thereof between demand and supply.

Market pull

Traditionally, traffic monitoring systems and predictive models in the United States have served state and local governments for traffic operations and to meet regulatory requirements. More recently, however, traffic data and models are being used to serve a much broader audience for applications that are much beyond the data and models' original intent. The widening of scope goes both in the direction of a global scale and all the way to individuals.

Global challenges

The need for models that can answer global-scale questions stems from high-level trends such as the rapid urbanization process, the urgency of climate change mitigation and adaptation, and the United Nations' 2030 Agenda for Sustainable Development. At this scale, the primary consumers of data and models are policy makers. Because of these challenges are large scale by nature, and the options policy makers have to weight are numerous, models and the data that supply these models must be scalable and allow for rapid assessment.

Communities

Parallel to the market requirement to address global-scale challenges, communities have been advocating for policies that will equitably address their unique local challenges. A most recent example is California's Assembly Bill (AB) 617, which resulted in the California Air Resources Board (CARB) establishing the Community Air Protection Program (CARB, 2018). Such trends dictate that transportation data and models need to serve communities in their advocacy for air quality and health improvements, and support policy makers to analyze and communicate the local impacts of transportation projects and policies.

Individuals

Employing data and developing models to serve individual travelers is perhaps the most fundamental paradigm shift for transportation researchers. Because transportation research is mostly funded by public entities, transportation researchers are accustomed to applying their work to policymaking. With the technological disruptions such as connected and automated vehicles and personal communications currently underway or looming large, however, there is great potential in using data and models to drive technology adoption and behavioral change. In this context, the consumers of traffic data and models are individuals, and the questions the data and models must address span across the entire spectrum of decisions related to travel, from vehicle ownership, to travel mode, to departure time. This shift in the audience requires a much higher resolution and instantaneous feedback mechanism.

Technology push

Recent scientific and technological advances have afforded researchers and developers with a suite of tools that can address the market pull discussed above. The convergence of progress in the following four areas holds great promise for the field of traffic monitoring and modeling.

Computational power

The vast increase in computational power has been at the heart of transportation models' ever-expanding capabilities. The advances in computing hardware, including the speed of microprocessors and the cost of storage devices, mean that models can process more data and handle more calculations. More recently, the rise of cloud computing and high-performance computing has both stemmed from the need and driven the development of parallel computing. The change of computational resources from personal computers to on-premise servers to computing clusters or clouds signals a shift of data and modeling architecture that can take advantage of the ever-increasing computing power.

Sensors and communication technologies

Wireless sensors and communication technologies are enabling the "Internet of Things." In-vehicle and in-road sensors, along with increased communications capacity and edge computing, provide entirely new ways to access, collect, and process data, and operate the transportation system. The advances in sensors and communications technologies will likely drive progress in traffic monitoring and models in two ways—more data will become available, and edge computing will be able to run models in a distributed fashion and communicate the results with or without humans in the loop.

Data

The increasing availability of existing data and the decreasing cost to collect more data beacon a need for change in the way data are perceived. On the one hand, "data is the new oil," as the common refrain goes. Entities, including the White House, are viewing data as a "strategic asset" (President's Management Council and the Executive Office of the President, 2018). Such emphasis has elevated the governance and accountability of data and can, therefore, benefit the research community. However, taken in a narrow sense, treating data as an asset can have the unintended consequence of data protectionism. On the other hand, unlike oil, which can be extracted profitably from individual wells, data is much less useful when siloed and scattered. Much progress in the field of transportation modeling is due to the standardization and sharing of data, as observed in the development of GTFS.

Algorithms

Along with the growth in data and computational resources, algorithms, and methods for analysis and modeling have also progressed. Such progress stems from two sources. First, the theories related to travel behavior and traffic flow have matured over the past couple of decades. Activity-based travel demand models and dynamic traffic assignment models have become widely adopted, albeit separately (Zmud et al., 2018). The integration of the two types of models to form an agent-based model chain from travel demand to traffic simulations has become the natural next step has the field moves forward. Second, machine learning techniques that were developed independently of the transportation field have been demonstrated to be effective in analyzing transportation data and explaining traffic phenomena. For example, Pereira, Rodrigues, Polisciuc, and Ben-Akiva (2015) applied latent Dirichlet allocation, a hierarchical Bayesian model originally developed for text analysis (Blei, Ng, & Jordan, 2003) to analyze transit smartcard data to model overcrowding. Machine learning will prove especially useful in implementing online modeling systems that can be continuously updated with data feeds provided by sensors and communication devices that are becoming ubiquitous.

Market/product match

With the broadening of the audience asking for more innovative solutions and the impetus of technology moving the field forward, the field of transportation monitoring and modeling is facing tremendous opportunity to fundamentally change the landscape of mobility for the betterment of the environment and public health.

To achieve this goal, all stakeholders have a role to play. Innovators must understand the market and their needs and roles and provide solutions that match their requirements. Decision-makers in government, industry, and philanthropic organizations need to emphasize asking the right questions, the answers to which are actionable and have a sizable and equitable impact. The public's increasing engagement and empowerment, aided by the proliferation of social media and smartphones, are the ultimate driving force for innovation. Ultimately, traffic data and models, and the energy, air quality, and health impacts they estimate, concern every citizen in their choices ranging from home location to trip planning.

It is in the context of such a broad range of stakeholders that a collaborative and iterative approach should be emphasized. A transportation modeler has long been stereotyped to be an introvert hunching over his or her computer all day, crunching numbers. A funder would issue a grant or contract, hear about the progress only a handful times throughout, and receive a final report or a model at the end of the project, at which time not much room is left to adjust. The public is often at the mercy of such research processes, having little to no input into the data collection, model setup, or the application of the research product. To challenge the status quo, the transportation monitoring and modeling community should embrace the open-source movement—foster a culture of codesign and co-modeling with the end users, encourage crowdsourcing, sharing, and standardization of data, and develop models in the open so that the development efforts are transparent, and results are reproducible. Researchers should strive to iterate with funders and end users to obtain frequent feedback. Funders need to design research policies and contracting mechanisms to encourage or even mandate an open, iterative, and agile development approach. Iteration is the only way to achieve market/product fit.

The call for collaborative iteration culminates with an emphasis on scale and speed. The traffic monitoring and modeling field have come to a point where solutions must scale, or they will fade. The scale is both horizontal and vertical. Horizontal scaling in this sense refers to the fact that the data and modeling systems must cover more locations and populations. Vertical scaling refers to the integration across an entire modeling chain that can explain and predict impacts from travel to health. Such scales must also be achieved with unprecedented speed, necessitated by both demand and supply. First, on the supply side, the field is ripe for massive scaling. Many of the scientific advances underlying the modeling efforts have undergone three decades of development. Time has come to coalesce independent scientific endeavors to create innovative solutions. Second, on the demand side, both the urgency of solving urban and global challenges such as mobility and climate change and the rapid changes in technology deployment and adoption in transportation require that the modeling community to respond. Therefore, the field as a whole need to move forward with a sense of urgency that prioritizes speed and scale.

Conclusions

This chapter has reviewed the current status of traffic monitoring data and transportation models in the context of energy, air quality, and health impacts. The review has evidenced the fact that there are abundant data and modeling effort, but mostly for other purposes than energy, air quality, and health. A few integrated solutions have started coalescing regarding energy impacts, but similar efforts are lacking to extend the modeling chain all the way to air quality and health, due to the increased complexity and uncertainty. The following list summarizes the key takeaways from this chapter:

- The state of practice in transportation, air quality, and health modeling remains in silos.
- In each discipline, well-researched and validated models already exist.
- Integrated models for transportation and energy are emerging.
- Recent software and hardware advances make it possible to further extend the modeling chain to air quality and health.
- Integrated modeling efforts should be scalable across multiple locations, taking advantage of new data sources.

The research community in the traffic monitoring and modeling field is called upon for scalable solutions to address unique issues in energy, air quality, and health. Decision-makers and the public must press the research community for answers to questions that matter. Together, all stakeholders need collaborative and coordinated efforts to build on recent advances in software and hardware technologies to bring multidisciplinary expertise required by the scale and complexity of the challenges.

References

Advanced Research Projects Agency-Energy (ARPA-E). (2014). *Traveler response architecture using novel signaling for network efficiency in transportation (TRANSNET) funding opportunity announcement*. Funding Opportunity No. DE-FOA-0001199. https://arpa-e-foa.energy.gov/FileContent.aspx?FileID=9fdd2747-283e-4c45-afeb-3e2b90da867a (Accessed March 2019).

Alnawaiseh, A., Abdelghany, K., & Hassan, A. (2014). Rollback approach for demand consistency checking of real-time traffic network state estimation models. *Transportation Research Record*, *2467*(1), 30–39. https://doi.org/10.3141/2467-04.

ARPA-E. (2015a). *TRANSNET: Traveler response architecture using novel signaling for network efficiency in transportation*. https://arpa-e.energy.gov/?q=arpa-e-programs/transnet (Accessed March 2019).

ARPA-E. (2015b). *Sustainable travel incentives with prediction, optimization and personalization (tripod)*. https://arpa-e.energy.gov/?q=slick-sheet-project/sustainable-travel-incentives-prediction-optimization-and-personalization-tripod (Accessed March 2019).

ARPA-E. (2018). Incentivizing individual travel behavior changes for system-level energy savings. In Y. Xu (Ed.), Vol. III. *ARPA-E impacts: A sampling of project outcomes*. https://arpa-e.energy.gov/?q=impact-sheet/university-maryland-transnet (Accessed March 2019).

Auld, J., Hope, M., Ley, H., Sokolov, V., Xu, B., & Zhang, K. (2016). POLARIS: Agent-based modeling framework development and implementation for integrated travel demand and network and operations simulations. *Transportation Research Part C: Emerging Technologies*, *64*, 101–116. https://doi.org/10.1016/j.trc.2015.07.017.

Basu, R., Araldo, A., Akkinepally, A. P., Nahmias Biran, B. H., Basak, K., Seshadri, R., et al. (2018). Automated mobility-on-demand vs. mass transit: A multi-modal activity-driven agent-based simulation approach. *Transportation Research Record*, https://doi.org/10.1177/0361198118758630.

Battelle and Texas A&M Transportation Institute. (2015). *MAP-21 air quality and congestion mitigation measure outcomes assessment study*. Report No. FHWA-HEP-15-008. Federal Highway Administration.

Blei, D. M., Ng, A. Y., & Jordan, M. I. (2003). Latent dirichlet allocation. *Journal of Machine Learning Research*, *3*(Jan), 993–1022.

Box, G. (1976). Science and statistics. *Journal of the American Statistical Association*, *71*(356), 791–799.

Brooker, A., Gonder, J., Wang, L., Wood, E., Lopp, S., & Ramroth, L. (2015). *FASTSim: A model to estimate vehicle efficiency, cost, and performance*. SAE technical paper 2015-01-0973. https://doi.org/10.4271/2015-01-0973.

Burke, J., Estrin, D., Hansen, M., Parker, A., Ramanathan, N., Reddy, S., et al. (2006). *Participatory sensing*. In: *Workshop on WorldSensor-Web (WSW'06), Boulder, Colorado*. (pp. 117–134).

California Air Resources Board. (2018). *Community air protection program blueprint & recommended communities factsheet*. https://ww2.arb.ca.gov/sites/default/files/2018-08/final_draft_community_air_protection_blueprint_august_2018_fact_sheet.pdf (Accessed March 2019).

Cambridge Systematics, Inc. (2014). *Traffic analysis toolbox. Volume 1: Traffic analysis tools primer* Publication No. FHWA-HRT-04-038. https://ops.fhwa.dot.gov/trafficanalysistools/tat_vol1/sect4.htm (Accessed March 2019).

Campbell, K. (2012). The SHRP2 naturalistic driving study. *TR News*, *282*, 30–36.

Chen, C., Ma, J., Susilo, Y., Liu, Y., & Wang, M. (2016). The promises of big data and small data for travel behavior (aka human mobility) analysis. *Transportation Research Part C: Emerging Technologies*, *68*, 285–299. https://doi.org/10.1016/j.trc.2016.04.005.

Chiu, Y.-C., Bottom, J., Mahut, M., Paz, A., Balakrishna, R., Waller, T., et al. (2011). *TRB transportation research E-circular E-C153: Dynamic traffic assignment: A primer*. https://doi.org/10.17226/22872.

De Lima, I. V., Danaf, M., Akkinepally, A., De Azevedo, C. L., & Ben-Akiva, M. (2018). Modeling framework and implementation of activity- and agent-based Simulation: An application to the Greater Boston Area. *Transportation Research Record*, https://doi.org/10.1177/0361198118798970.

Duvall, A. L., & Young, S. E. (2018). *The connected traveler*. In: *Poster presentation at the ARPA-E TRANSNET Program Review, 27–29 June, Santa Clara, California*. https://www.osti.gov/biblio/1471176-connected-traveler (Accessed March 2019).

Erraguntla, M., Delen, D., Agrawal, R. K., Madanagopal, K., & Mayer, R. (2017). Mobile-based sidewalk inventory app for smart communities, health, and safety. *Suburban Sustainability*, *5*(1). Article 3. http://scholarcommons.usf.edu/subsust/vol5/iss1/3 (Accessed March 2019).

FHWA. (2016). *Traffic monitoring guide*. https://www.fhwa.dot.gov/policyinformation/tmguide/ (Accessed December 2019).

Gbologah, F. E., Xu, Y., Rodgers, M. O., & Guensler, R. (2014). Demonstrating a bottom-up framework for evaluating energy and emissions performance of electric rail transit options. *Transportation Research Record*, *2428*(1), 10–17. https://doi.org/10.3141/2428-02.

Goldstein, B. (2013). Preface. In B. Goldstein & L. Dyson (Eds.), *Beyond transparency*. San Francisco, CA: Code for America Press. codeforamerica.org/beyond-transparency (Accessed March 2019).

Gonder, J., Burton, E., & Murakami, E. (2015). Archiving data from new survey technologies: Enabling research with high-precision data while preserving participant privacy. *Transportation Research Procedia*, *11*, 85–97. https://doi.org/10.1016/j.trpro.2015.12.008.

Guensler, R., Liu, H., Xu, Y., Akanser, A., Kim, D., Hunter, M. P., et al. (2017). Energy consumption and emissions modeling of individual vehicles. *Transportation Research Record*, *2627*(1), 93–102. https://doi.org/10.3141/2627-11.

Guensler, R., Rodgers, M. O., Hunter, M., Guin, A., Abdelghany, K., Laval, J., et al. (2018). *Network performance monitoring and distributed simulation to improve transportation energy efficiency: TRANSNET-Atlanta*. Project overview of Github repository. https://github.com/arpa-e-transnet/gtstm (Accessed March 2019).

He, Y., Chow, J. Y. J., & Nourinejad, M. (2017). *A privacy design problem for sharing transport service tour data*. In: *2017 IEEE 20th international conference on intelligent transportation systems (ITSC), Yokohama* (pp. 1–6). https://doi.org/10.1109/ITSC.2017.8317692.

Krause, C. M., & Zhang, L. (2018). Short-term travel behavior prediction with GPS, land use, and point of interest data. *Transportation Research Part B: Methodological*, https://doi.org/10.1016/j.trb.2018.06.012. In press.

Kusakabe, T., & Asakura, Y. (2014). Behavioural data mining of transit smart card data: A data fusion approach. *Transportation Research Part C: Emerging Technologies*, *46*, 179–191. https://doi.org/10.1016/j.trc.2014.05.012.

Lefebvre, W., Degrawe, B., Beckx, C., Vanhulsel, M., Kochan, B., Bellemans, T., et al. (2013). Presentation and evaluation of an integrated model chain to respond to traffic-and health-related policy questions. *Environmental Modelling & Software*, *40*, 160–170. https://doi.org/10.1016/j.envsoft.2012.09.003.

Li, H., Cebe, J., Khoeini, S., Xu, Y., Dyess, C., & Guensler, R. (2018). A semi-automated method to generate GIS-based sidewalk networks for asset management and pedestrian accessibility assessment. *Transportation Research Record*, *2672*(44), 1–9. https://doi.org/10.1177/0361198118757981.

Liu, Z., Ivanco, A., & Filipi, Z. (2015). Naturalistic drive cycle synthesis for pickup trucks. *Journal of Safety Research*, *54*, 109–115. https://doi.org/10.1016/j.jsr.2015.06.005.

Liu, H., Li, H., Rodgers, M. O., & Guensler, R. (2018). Development of road grade data using the United States geological survey digital elevation model. *Transportation Research Part C: Emerging Technologies*, *92*, 243–257. https://doi.org/10.1016/j.trc.2018.05.004.

Liu, H., Xu, Y., Stockwell, N., Rodgers, M. O., & Guensler, R. (2016). A comparative life-cycle energy and emissions analysis for intercity passenger transportation in the U.S. by aviation, intercity bus, and automobile. *Transportation Research Part D: Transport and Environment*, *48*, 267–283. https://doi.org/10.1016/j.trd.2016.08.027.

Malekian, N., Moloisane, R., Nair, L., Maharaj, B. T., & Chude-Okonkwo, U. A. K. (2017). Design and implementation of a wireless OBD II fleet management system. *IEEE Sensors Journal, 17*(4), 1154–1164. https://doi.org/10.1109/JSEN.2016.2631542.

Mani, A., & Prozzi, J. (2004). In *State-of-the-practice in freight data: A review of available freight data in the US (no. 0-4713-P2). University of Texas at Austin. Center for Transportation Research.* Report no. 0-4713-P2. https://rosap.ntl.bts.gov/view/dot/6410 (Accessed March 2019).

Metropia, Inc. (2017). *Urban analytics.* http://www.metropia.com/urban-analytics-overview (Accessed March 2019).

Needell, Z. A., & Trancik, J. E. (2018). Efficiently simulating personal vehicle energy consumption in Mesoscopic transport models. *Transportation Research Record, 2672*(25), 163–173. https://doi.org/10.1177/0361198118798244.

O'Brien, O., Cheshire, J., & Batty, M. (2014). Mining bicycle sharing data for generating insights into sustainable transport systems. *Journal of Transport Geography, 34*, 262–273. https://doi.org/10.1016/j.jtrangeo.2013.06.007.

Pereira, F. C., Rodrigues, F., Polisciuc, E., & Ben-Akiva, M. (2015). Why so many people? Explaining nonhabitual transport overcrowding with internet data. *IEEE Transactions on Intelligent Transportation Systems, 16*(3), 1370–1379. https://doi.org/10.1109/TITS.2014.2368119.

President's Management Council and the Executive Office of the President. (2018). *President's management agenda.* https://www.performance.gov/PMA/Presidents_Management_Agenda.pdf (Accessed March 2019).

Rakha, H. A., Ahn, K., & Moran, K. (2012). Integration framework for modeling eco-routing strategies: Logic and preliminary results. *International Journal of Transportation Science and Technology, 1*(3), 259–274. https://doi.org/10.1260/2046-0430.1.3.259.

Rakha, H., Ahn, K., & Trani, A. (2004). Development ofVT-micro model for estimating hot stabilized light duty vehicle and truck emissions. *Transportation Research Part D: Transport and Environment, 9*(1), 49–74. https://doi.org/10.1016/S1361-9209(03)00054-3.

Rakha, H., & Lucic, I. (2002). Variable power vehicle dynamics model for estimating truck accelerations. *Journal of Transportation Engineering, 128*(5), 412–419. https://doi.org/10.1061/(ASCE)0733-947X(2002)128:5(412).

SAE. (2001). *OBD II scan tool equivalent to ISO/DIS 15031-4, SAE Standard J1978 199203.*

SAE. (2017). *E/E diagnostic test modes, SAE Standard J1979 201702.*

Sarkar, R., & Ward, J. (2016). DOE SMART mobility: Systems and modeling for accelerated research in transportation. In G. Meyer & S. Beiker (Eds.), *Road vehicle automation 3. Lecture notes in mobility*. Cham: Springer. https://doi.org/10.1007/978-3-319-40503-2_4.

Schneider, R. J., Patten, R. S., & Toole, J. L. (2005). Case study analysis of pedestrian and bicycle data collection in U.S. communities. *Transportation Research Record, 1939*(1), 77–90. https://doi.org/10.1177/0361198105193900110.

Schoner, J. E., & Levinson, D. M. (2014). The missing link: Bicycle infrastructure networks and ridership in 74 US cities. *Transportation, 41*, 1187. https://doi.org/10.1007/s11116-014-9538-1.

Sheppard, C., Waraich, R., Campbell, A., Pozdnukhov, A., & Gopal, A. R. (2017). *Modeling plug-in electric vehicle charging demand with BEAM. LBNL-2001018.* https://eta.lbl.gov/sites/default/files/publications/lbnl-2001018.pdf (Accessed March 2019).

Silva, T. H., Vaz de Melo, P. O. S., Almeida, J. M., & Loureiro, A. A. F. (2013). *Challenges and opportunities on the large scale study of city dynamics using participatory sensing.* In: *2013 IEEE symposium on computers and communications (ISCC), 000528-000534.* IEEE. https://doi.org/10.1109/ISCC.2013.6755000.

Suh, W., Henclewood, D., Guin, A., Guensler, R., Hunter, M., & Fujimoto, R. (2017). Dynamic data driven transportation systems. *Multimedia Tools and Applications, 76*(23), 25253–25269. https://doi.org/10.1007/s11042-016-4318-x.

Toole, J. L., Colak, S., Sturt, B., Alexander, L. P., Evsukoff, A., & González, M. C. (2015). The path most traveled: Travel demand estimation using big data resources. *Transportation Research Part C: Emerging Technologies*, *58*, 162–177. https://doi.org/10.1016/j.trc.2015.04.022.

U.S. Bureau of Transportation Statistics. (2018). *National Transportation Statistics 2018 4th quarter, Table 1–35*. https://www.bts.gov/content/us-vehicle-miles (Accessed March 2019).

U.S. Department of Energy. (2019). *Electric vehicle charging station locations*. https://afdc.energy.gov/fuels/electricity_locations.html#/find/nearest?fuel=ELEC (Accessed March 2019).

U.S. Energy Information Administration. (2018). *Monthly energy review, Tables 2.5, 3.8c, and 10.2b*. Preliminary data. https://www.eia.gov/energyexplained/?page=us_energy_transportation (Accessed March 2019).

U.S. Environmental Protection Agency (EPA). (2012). *Transportation conformity regulations, 42 pp, 288 K, EPA-420-B-12-013*. https://nepis.epa.gov/Exe/ZyPDF.cgi/P100E7CS.PDF?Dockey=P100E7CS.PDF (Accessed March 2019).

U.S. EPA. (2014). *Profile of version 1 of the 2014 National Emissions Inventory*. U.S. EPA 2014 NEI Version 1.0, April 2017. https://www.epa.gov/sites/production/files/2017-04/documents/2014neiv1_profile_final_april182017.pdf (Accessed March 2019).

Van Aerde, M., & Yagar, S. (1988). Dynamic integrated freeway/traffic signal networks: A routing-based modelling approach. *Transportation Research Part A: General*, *22*(6), 445–453. https://doi.org/10.1016/0191-2607(88)90048-9.

Wahlström, J., Skog, I., Nordström, R. L., & Händel, P. (2018). Fusion of OBD and GNSS measurements of speed. *IEEE Transactions on Instrumentation and Measurement*, *67*(7), 1659–1667. https://doi.org/10.1109/TIM.2018.2803998.

Wang, Z., He, S. Y., & Leung, Y. (2018). Applying mobile phone data to travel behaviour research: A literature review. *Travel Behaviour and Society*, *11*, 141–155. https://doi.org/10.1016/j.tbs.2017.02.005.

Wang, J., & Rakha, H. A. (2017). Electric train energy consumption modeling. *Applied Energy*, *193*, 346–355. https://doi.org/10.1016/j.apenergy.2017.02.058.

Wood, E., Burton, E., Duran, A., & Gonder, J. (2014). *Appending high-resolution elevation data to GPS speed traces for vehicle energy modeling and simulation*. National Renewable Energy Laboratory. NREL/TP-5400-61109.

Xu, Y., Gbologah, F., Liu, H., Rodgers, M. O., & Guensler, R. (2015). Assessment of alternative fuel and powertrain transit bus options using real-world operations data: Life-cycle fuel and emissions modeling. *Applied Energy*, *154*, 143–159. https://doi.org/10.1016/j.apenergy.2015.04.112.

Xu, Y., & Guensler, R. (2015). *Capturing personal modality styles using multiday GPS data—Findings from the San Francisco Bay Area*. In: *Multiday GPS travel behavior data for travel analysis, Report No. FHWA-HEP-015-026, Federal Highway Administration. March*.

Yang, X., Stewart, K., Tang, L., Xie, Z., & Li, Q. (2018). A review of GPS trajectories classification based on transportation mode. *Sensors (Basel, Switzerland)*, *18*(11), 3741. https://doi.org/10.3390/s18113741.

Zhou, X., Tanvir, S., Lei, H., Taylor, J., Liu, B., Rouphail, N. M., et al. (2015). Integrating a simplified emission estimation model and mesoscopic dynamic traffic simulator to efficiently evaluate emission impacts of traffic management strategies. *Transportation Research Part D: Transport and Environment*, *37*, 123–136. https://doi.org/10.1016/j.trd.2015.04.013.

Zhou, X., & Taylor, J. (2014). DTALite: A queue-based mesoscopic traffic simulator for fast model evaluation and calibration. *Cogent Engineering*, *1*(1), https://doi.org/10.1080/23311916.2014.961345. Filippo Pratico (Reviewing Editor).

Zmud, J., Williams, T., Outwater, M., Bradley, M., Kalra, N., & Row, S. (2018). *Updating regional transportation planning and modeling tools to address impacts of connected and automated vehicles. Volume 2: Guidance*. NCHRP research report 896. Washington, DC: The National Academies Press. https://doi.org/10.17226/25332.

CHAPTER 4

Vehicle emissions measurement and modeling

Karl Ropkins[a], Sergio Ibarra-Espinosa[b], and Yoann Bernard[c]
[a]Institute for Transport Studies, University of Leeds, Leeds, United Kingdom
[b]Departamento de Ciências Atmosféricas, Universidade de São Paulo, São Paulo, Brazil
[c]International Council on Clean Transportation, Berlin, Germany

Abbreviations

ARB	(Californian) Air Resources Board
CADC	common Artemis drive cycle
CH_4	methane
CO	carbon monoxide
CO_2	carbon dioxide
CVS	constant volume sampler
EDAR	emission detection and reporting
EEA	European Environment Agency
EPA	(US) Environmental Protection Agency
FEAT	fuel efficiency automobile test
FID	flame ionization detection
GM	General Motors
GPS	global positioning system
HBEFA	handbook of emission factors for road transport
I&M	inspection and maintenance
IVE	international vehicle emissions
NDIR	nondispersive infrared
NEDC	New European Drive Cycle
NH_3	ammonia
NO	nitric oxide
NO_2	nitrogen dioxide
NO_x	oxides of nitrogen or nitrogen oxides
OBD	onboard diagnostic
OHMS	on-road heavy-duty measurement system
PEAQS	portable emissions acquisition system
PEMS	portable emissions measurement system
PFD	proportional (or partial) flow dilutors
PM	particulate mass
PN	particulate number
PTI	periodic technical inspection
RDE	real driving emissions

Traffic-Related Air Pollution
https://doi.org/10.1016/B978-0-12-818122-5.00004-1

ROE	real-time on-road emission
RSD	remote sensing device
THC	total hydrocarbons
TRAP	traffic-related air pollution
VEIN	vehicular emissions inventory
VERSS	vehicle emissions remote sensing system
VSP	vehicle specific power
VW	Volkswagen
WHO	World Health Organization
WLTP	worldwide harmonized light-vehicle test procedure

Abbreviations used in Formulae

$(CO_2 \rightarrow \mathit{Fuel})_{Equivalent}$	the fuel consumption equivalency term, mass amount of fuel consumed per unit volume of CO_2 emitted
$[n]$	the measured concentration of a species, *n*, in exhaust gas
ACTIVITY	the measure of vehicle activity, distance travelled, etc.
EF_n	the emission rate factor
mwt_n	the molecular weight of emitted species, *n*
nMASS	the mass emissions of emitted species, *n*
nRATIO	the ratio of the measured concentrations of emitted species of interest, *n*, and carbon dioxide
Q_{EX}	the exhaust flow rate
x	the normalization factor, e.g., unit time, distance traveled, or fuel consumed for g/s, g/km, g/kg, etc. emission rates

Introduction

Few of us would claim to be unaware that vehicle emissions, and in particular those from on-road traffic, are a significant source of airborne pollution. In almost all of the most polluted areas, the local vehicle fleet is the major source of many of the pollutants adversely affecting both local air quality and international climate change (see, e.g., Seinfeld & Pandis, 2016). Over 90% of us globally are living in places where air pollution exceeded World Health Organization Air Quality Guideline Levels (WHO, 2018a, 2018b). There is also significant evidence regarding the damage traffic-related air pollution (or TRAP) is doing to us and our environment, and the associated costs of not taking rapid and effective action to reduce impacts (Jacob & Winner, 2009; Lelieveld, Evans, Fnais, Giannadaki, & Pozzer, 2015; Stern & Stern, 2007). Action on vehicle emissions is already a key element of administrative strategies for the delivery of improved air quality at local, national, and international levels. However, growing vehicle fleets and rapidly evolving vehicle and emissions management technologies, as well as our own advancing monitoring and modeling capabilities, all make this a moving target. With this in mind, this chapter provides a contextualized discussion of monitoring and modeling practices used as part of conventional vehicle emissions management activities.

This chapter provides an introduction to this broad and complex topic. It begins by describing the key principles of conventional vehicle emissions monitoring methods and provides an overview of some of the events that helped shape current practices. The chapter highlights some key concerns regarding the effectiveness of our ongoing efforts to deliver reduced emissions vehicle fleets. It also provides a similar overview of conventional emissions modeling practices and discusses the changing nature of the relationship between emissions monitoring and modeling activities. It acknowledges limitations in current practices and considers options for the delivery of better, more results-focused vehicle emissions monitoring and modeling practices in future. Finally, it provides recommendations regarding further reading for those interested in further studies.

In the context of this chapter, discussion of vehicle emissions monitoring and modeling focuses of activities specifically intended to characterize and quantify the emissions levels of individual vehicles or vehicles of a particular class for regulatory compliance, fleet inventorying, or land-use planning. Although many of measurement and modeling strategies discussed have been adopted in other sectors (off-road, air, and water transport), attention and examples are provided for on-road vehicles reflecting the focus of most emissions monitoring and modeling activity to date.

Vehicle emissions monitoring

When discussing current vehicle emissions monitoring practices, it is useful to consider the events that helped to shape their development. It is also important to acknowledge the huge step change in both our understanding of the by-products of combustion and our ability to manage their environmental impacts over the same timescales.

Roadside stop-and-inspection testing

Some of the earliest vehicle monitoring data were gathered using stop-and-inspection-based vehicle emissions testing. As the name suggests, these tests involved a vehicle being stopped and measurements made at the roadside to determine emissions levels. The procedures and constraints are broadly similar to other roadside testing activities, e.g., driver sobriety and vehicle roadworthiness testing: The test needs to be quick and easy to administer because the tester will be disrupting journeys of potentially innocent drivers and the test procedure highly standardized and the tester trained to ensure consistency. If the test is integrated into a vehicle management activity such as an Inspection-and-Maintenance program where high emissions could result in legal action, e.g., a fine for the driver, a notice to repair vehicle or confiscation of the vehicle, test procedure, and instrument calibration also need to be fully documented in case findings are legally challenged.

Warren Springs, then the UK Government's national environmental research laboratory, developed some of the earliest vehicle stop-and-inspection procedures in the 1960s. At the time regulators were most concerned by visible exhaust emissions, and a number of optical systems were investigated as potential smoke meters. Although exhaust monitoring technologies are now obviously more sophisticated and the range of emissions tested is much wider, such stop-and-inspection style testing procedures are still widely used today as part of Inspection and Maintenance (I&M) and Periodic Technical Inspection (PTI) programs in many countries (see, e.g., Posada, Yang, & Muncrief, 2015; Burtscher, Lutz, & Mayer, 2019). There are, however, significant limitations with regard to the representativeness of the approach due to the short duration of these procedures and differences in engine management strategies under the low-engine-load and low-acceleration conditions of these tests and the higher engine-load, higher acceleration activities that are the major source of real-world vehicle emissions.

Dynamometer and drive cycle testing

At about the same time, the Californian Environmental Protection Agency was developing early laboratory-based vehicle emissions testing procedures. In these dynamometers were used to simulate vehicle operation. Two types of systems were developed: *Chassis dynamometers* where the test vehicle was driven on a rolling road which was fitted with a dynamometer to apply load to the wheels of the cars; and *engine dynamometers* where engines were tested in isolation and the dynamometer typically used to apply load to the prop shaft. These approaches (describe in further detail in Franco et al., 2013) have a number of clear advantages over roadside testing, but most notably: (1) being laboratory-based, they could readily incorporate sophisticated exhaust gas conditioning and sample transfer methods, making them amenable to use with a wide range of measurement methods; and (2) being dynamometer-based, the applied load could be varied to simulate different driving activities within a journey and once a combination of activities was defined, it could be readily repeated with high precision and accuracy to produce a highly standardized test procedure known as a drive cycle. Logging the concentrations of vehicle emissions in the exhaust, exhaust gas flow rates, and vehicle activity using the vehicle's own onboard diagnostics (OBDs) or feedbacks from the dynamometer system, emission rates can be determined in the general form

$$nMASS = \frac{([n] \times mwt_n \times Q_{EX})}{x}$$

where $nMASS$ is the mass emissions of emitted species, here n; $[n]$ is the measured concentration in the exhaust gas; mwt_n is the molecular weight of n; Q_{EX} is the exhaust flow rate; and x is the normalization factor, e.g., unit time, distance traveled, or fuel consumed for g/s, g/km, g/kg fuel emissions, respectively.

Attention focused on the main combustion productions, e.g., carbon dioxide (CO_2), carbon monoxide (CO), nitric oxide (NO), oxides of nitrogen (NO_x), hydrocarbons, and particulate matter. Even the earliest dynamometer facilities were typically able to simultaneously measure other species, making them highly flexible tools for a wide range of research and development applications. Fig. 4.1 includes a schematic of a chassis dynamometer and Table 4.1 provides summary information on common measurement methods.

These and other early monitoring activities provided policy makers with the evidence needed to call for emissions control, but it was the dynamometer procedures, or more specifically the drive cycle-based testing methods that provided the framework for early enforcement programs. Associated legislation defined achievement of emissions standards (limits not to exceeded) as a prerequisite of approval to sell vehicles, and the high reproducibility of dynamometer and drive cycle testing provided the "level playing field" that competing vehicle manufacturers expected. The approach was widely adopted and in the early years, many administrative authorities developed dedicated drive cycles for the vehicle fleets they had jurisdiction over (see, e.g., Andre, 1996, for further discussion of early drive cycle procedures).

These emissions standards were then incrementally tightened over several generations of legislation (such as the TIER standards in the US and EURO standards in Europe) to provide vehicle manufacturers with the

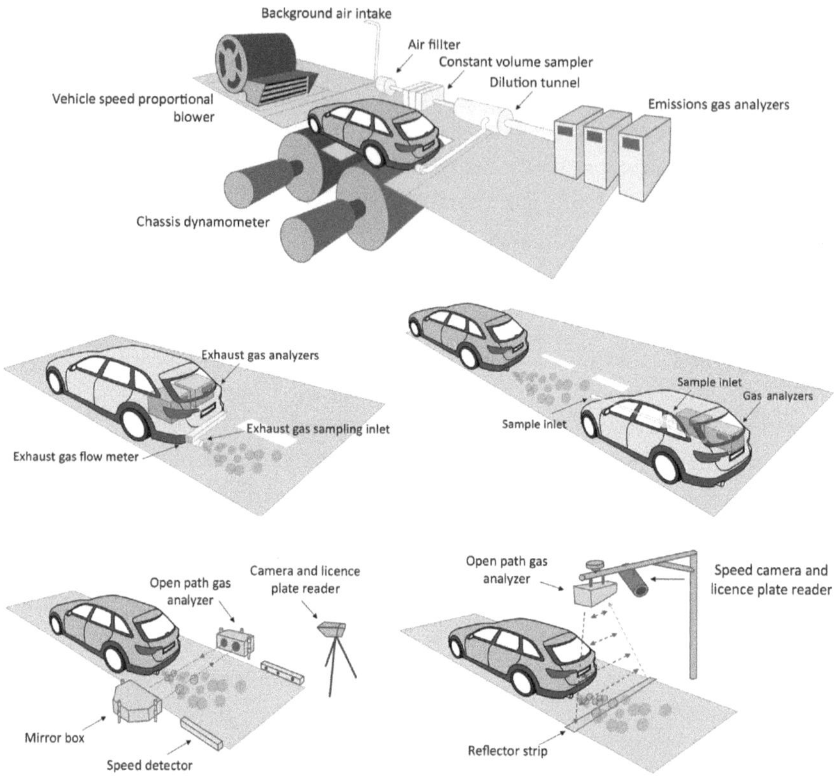

Fig. 4.1 Vehicle emissions measurement system schematics for *(top)* chassis dynamometer, *(middle left)* portable emissions measurement system, *(middle right)* vehicle chaser, and *(bottom left)* across-road and *(bottom right)* down-facing vehicle emissions remote sensing systems.

incentive to develop emission abatement technologies at what policy makers considered challenging but economically viable rates. This process, in turn, fostered both investment and growth in the sector: new markets for analytical instrument makers, and new consultant, technical and work-floor expertise requirements for personnel in the industry, government, and academia. Information from research and development activities also fed back into regulatory practices, refining the technologies used to monitor and manage vehicle emissions and extending the range of species regulated as new pollutants of concern were identified. So, unsurprisingly, dynamometer technologies developed rapidly during this period. Early fixed dilution sample transfer systems were replaced by constant volume samplers (CVSs) and proportional or partial flow dilutors (PFDs), designed to regulate the volume of sample passed to analyzers, further extending the range and number

Table 4.1 Typical monitoring methods.

Analyte	Common measurement method[a]	Analytical performance[b]			Further information[d]
		Measurement range[c]	Accuracy (%)	Precision (%)	
Carbon dioxide (CO_2)	Nondispersive infrared (NDIR)	0.5–5000 to 200–200,000 ppm	±5–10	±1–2	Wong and Anderson (2012), Dinh, Choi, Son, and Kim (2016)
Carbon monoxide (CO)	Nondispersive infrared (NDIR)	0.1–1000 to 200–200,000 ppm	±5–10	±1–2	Wong and Anderson (2012), Dinh et al. (2016)
Oxides of nitrogen (NO, NO_2, NO_x)	Chemiluminescence	0.1–10 to 100–10,000 ppm	±10–20[e]	±2–5[e]	Glover (1975), Klingenberg (2012)
Total hydrocarbons (THC)	Flame ionization detection (FID)	0.1–100 to 100–60,000 ppmC	±5–10	±2–5	Nakamura, Kihara, Adachi, Nakamura, and Ishida (2003), Klingenberg (2012)
	Nondispersive infrared (NDIR)	2–5000 ppmC	±5–10	±2–5	
Methane (CH_4)	Dual FID	0.1–100 to 100–60,000 ppmC	±5–10	±2–5	Klingenberg (2012)
Particulate mass (PM)	Gravimetric filter	10–10,000 μg	±5	±5–10	Kittelson, Arnold, and Watts (1999), Giechaskiel et al. (2014)
	Microbalance	50–10,000 $\mu g\,m^{-3}$	±15–30	±10–20	
	Optical (opacity or light scattering)				
Particulate number (PN)	Condensation counter[f]	10^6–10^{14} # m^{-3}	±20	±10–20	Kittelson (1998), Giechaskiel et al. (2014)

[a] Please note, measurement methods reported here are commonly used methods, and these are recommended methods for many regulatory data collection activities, they are not the only methods used to monitoring these analytes.
[b] In many cases analytical performance is a function of individual instrument design, and values here should be considered generally indicative of instruments used for vehicle emissions measurement rather than specific instruments.
[c] In many cases, different instruments are recommended for different vehicle emissions measurement applications, e.g., diesel and gasoline engines or light- and heavy-duty vehicles, so two ranges are given as indicative of instrument ranges for cleaner and dirtier applications.
[d] Suggested references for those seeking more information on these measurement methods only, other instrument texts should provide similar discussions.
[e] Accuracy and precision for NO only.
[f] Measured as number of particles (#).

Based on Siegel, R. D. (1972). A review of instrumentation, test, and sampling procedures. Journal of the Air Pollution Control Association 22(11), 845–853, Ropkins, K., Beebe, J., Li, H., Daham, B., Tate, J., Bell, M., Andrews, G. (2009). Real-world vehicle exhaust emissions monitoring: Review and critical discussion. Critical Reviews in Environmental Science and Technology 39(2), 79–152, Franco, V., Kousoulidou, M., Muntean, M., Ntziachristos, L., Hausberger, S., Dilara, P. (2013). Road vehicle emission factors development: A review. Atmospheric Environment 70, 84–97, Giechaskiel, B., Maricq, M., Ntziachristos, L., Dardiotis, C., Wang, X., Axmann, H., Bergmann, A., Schindler, W. (2014). Review of motor vehicle particulate emissions sampling and measurement: From smoke and filter mass to particle number. Journal of Aerosol Science 67, 48–86, and references in these reviews.

of species that dynamometer facilities could simultaneously measure (see, e.g., Kittelson et al., 1999; Giechaskiel et al., 2014). Although less widely adopted, other vehicles/dynamometer interfaces were also developed, e.g., the heavy-duty portable chassis dynamometer developed by West Virginia University that significantly reduced facility infrastructure by replacing the rolling road with an eddy suppressor/flywheel combination that could be connected directly to the vehicle wheels (Bata et al., 1991).

Analysis of both vehicle emissions and air quality data suggests that the incremental tightening of engine emissions standards provided a very effective means for reducing emissions in the earliest generations of associated regulations when large emission benefits were more easily achieved with less costly abatement technologies (Carslaw, Beevers, Tate, Westmoreland, & Williams, 2011; Hooftman, Messagie, Van Mierlo, & Coosemans, 2018; Vestreng et al., 2009). However, as regulatory standards became more stringent, the cost and complexity of the abatements required increased. While early standards could be achieved with passive catalyst-based approaches alone, later generation standards required combustion and exhaust conditions monitoring, active management and/or responsive strategies such as tight air to fuel ratio control, exhaust gas recirculation, selective catalytic reduction, NO_x after-treatment and particulate filtering or combinations of these. Associated increases in manufacturing costs and fuel consumption penalties made abatement technologies unpopular with both vehicle builders and owners. A combination of diminishing returns and unforeseen circumstance also began to undermine the effectiveness of this fleet management strategy.

Firstly, the more aggressive later generation emission abatement methods were typically behavior/response strategies that were mapped onto engine operation. The use of these in combination with the highly standardized testing procedures led to later generations of vehicles that were optimized for the driving activities most like those used in regulatory testing. By comparison, driving activities outside this range, never encountered on the test beds and often termed "off-cycle" by the research community of the time, were effectively penalty free, creating a significant divergence between regulatory and on-road emissions (Weiss, Bonnel, Hummel, Provenza, & Manfredi, 2011).

Secondly, the definition of specific regulatory metrics created regulatory rather than environmental targets for abatement technology developers. For example, the development of a particulate emissions standard based on mass-calibrated measurement of opacity resulted in early generation emission reduction strategies that reduced the concentrations of the larger particulate emissions that these methods were most sensitive to rather than particulate emissions more generally. Consequently, the next generation of vehicles emitted much finer, much less visible and more subtly harmful particulates (Kittelson, 1998), and this, in turn, triggered the need to review and rethink particulate-related regulatory monitoring (Giechaskiel et al., 2014, 2018).

But also, perhaps most concerning, the increasing cost of legitimate solutions meant other less ethical options became more attractive. Many of these were not strictly illegal, but are probably best considered "gaming the system," e.g., strategies that produced saving on test beds that were not transferrable to the real world, such as removing passenger seats to reduce vehicle weight and taping doors and windows to reduce drag (Kadijk et al., 2012; Stewart, Hope-Morley, Mock, & Tietge, 2015). No doubt some simply saw these as performance target to be met, and their actions no different from a professional tennis player perfecting a winning shot. While others, knowing that competitors were doing such things and getting away with them, maybe felt they had no choice but to do similar to remain competitive. However, such erosion of standards is always a downward spiral, perpetuating larger and larger malpractices. Here, perhaps the most contemptible of these were the use of "Golden Vehicles," preproduction vehicles deliberately built differently from those coming off standard production lines, in test procedures and "Defeat Devices," onboard systems designed to detect emissions testing drive cycle conditions and deliberately use more aggressive emission abatement strategies as "cycle beating" strategy. While other strategies were maybe to a degree defendable, these were very much deliberate and cynical attempts to circumvent emission regulations (Archer & Director, 2015).

Real-world emissions measurement

Concerns about the widening gap between test bed and on-road vehicle emissions led to the development of a wide range of measurement strategies intended to provide real-world measurements of vehicle emissions. These included a broad spectrum of methods, from onboard systems that provided high resolution information about the emissions of a single vehicle to fixed site systems that provided briefer emissions measurements but for the majority of vehicles in the passing fleet (see, e.g., Ropkins et al., 2009; Franco et al., 2013). Although these systems provided valuable insights, they lacked the commercial market created by mandatory regulatory testing programs.

As a result, early generations of these monitoring technologies were not as rapidly or as widely adopted as dynamometer-based technologies and their use was initially limited to much smaller-scale research-based applications. The order and rate of growth of interest in the different real-world vehicle emissions measurement methods were driven by the creation of business models for significant investment in the different technologies.

Portable emissions measurement systems

Onboard vehicle emission systems, known as portable emissions measurement (or sometimes monitoring) systems (or PEMS), were arguably the first to be successfully commercialized. PEMS are instrument platforms that are installed on or inside a vehicle and typically include a series of individual analyzers that measure both the concentrations of emitted species in gases and the flow rate of gases in the exhaust of that vehicle (Fig. 4.1). Other onboard monitors were also used to log vehicle position and speed, engine activity and, sometimes, driver activity. PEMS are typically operated at 1–10 Hz sampling rates (1–10 measurements per second) and generate high temporal resolution time-series datasets of emissions measurements for the test vehicle they are installed on (Weiss et al., 2011). Emissions are typically measured and reported in a similar form to dynamometer outputs.

The many analytical challenges of building instruments that can be installed and operated on vehicles in real-world on-road situations, e.g., size, weight, and power constraints of both sampling handling and monitoring systems, makes the accuracies and precisions achieved in the much larger scale and much more controllable conditions of dynamometer facilities challenging. Likewise, the inherently variable nature of real-world conditions, e.g., changing weather and other driver behavior even on the most highly defined routes, means that tests can never be rerun with anything approaching the same level of reproducibility. However, the obvious tradeoff for this increased measurement uncertainty was that data could be collected over a much broader range of conditions and that the measurements obtained more directly reflect the actual on-road performance of the test vehicle. Also, many of the species monitored using dynamometer facilities can be monitored with PEMS using highly similar measurement strategies. Consequently, PEMS have been used in a wide range of comparative studies, driving different vehicles, or vehicles using different fuels or employing different engine technology on defined routes, which is arguably the closest analog to dynamometer/drive cycle test under real-world conditions.

But they can also be deployed in a much more flexible fashion to quickly investigate "atypical" behavior in further detail when it is encountered.

Therefore, it is unsurprising that PEMS found its niche in the fight against Defeat Devices.

Here, it is important to note that "Dieselgate," the 2014 Defeat Device Scandal that focused on Volkswagen (VW), was by no means the first or only time that vehicle manufacturers had been caught using such technologies (Bernard, German, Kentroti, & Muncrief, 2019). Between 1973 and 1998 Chrysler, Fiat, Ford, Honda, General Motors (GM), Toyota, and VW were all caught using highly questionable emission abatement system strategies. In early cases, the distinction between an intentional deception and valid engine protection strategies made prosecution uncertain and kept fines low, but regulators, most notably the US Environmental Protection Agency (EPA), worked hard to tighten legislation and identify deliberate malpractice. So, when in 1998, a number of heavy-duty truck manufacturers (Caterpillar, Cummins Engine, Detroit Diesel, Mack Trucks, Navistar International, Renault Véhicules Industriels, and Volvo Trucks) were caught using a timer-based Defeat Device that switched abatement controls off under highway driving conditions after times longer than those in US test procedures, the EPA were able to impose what were the largest environmental action fines of their day, just over $80 million.

Early PEMS systems were used to identify unusually high on-road emissions indicative of the presence of a Defeat Device, but they lacked the accuracy required for legal action. So, the methods of the day were similar to breathalyzer/blood test drink-driving regulations. The first test (PEMS) identified likely cases and provided the precedence to call for other more intrusive but rigorous testing. Similarly, when the EPA looked to add an on-road component to their related vehicle regulation procedures, it was a PEMS method they endorsed (Feist, Sharp, & Spears, 2009). Given the regulatory nature of this application and the understandably conservative nature of those developing regulatory codes of practices, attention naturally focused on the highest possible accuracy and favored the larger more complex instruments that the more established instrument manufacturers were reverse engineering from instruments previously developed for other larger sectors.

So, while fines from these early defeat device convictions generated both the need for such systems and significant investment, they perhaps did not foster the high risk/return opportunities that innovators typically look for, and early commercial PEMS were relatively large, heavy, and cumbersome instruments.

In situ or passing vehicle emissions measurement methods

Research interest in better measurement of vehicle fleet emissions also led to the development of alternative monitoring strategies (Huang et al., 2018; Ropkins et al., 2009). As the names suggest, many of these deployed monitors at fixed locations to provide measures of the emissions of vehicles as they passed that monitor. By comparison to PEMS, these approaches provided a highly cost-effective means of measuring the emissions of a large proportion of the local vehicle fleet and arguably a better estimate of average fleet emissions. However, there are several trade-offs to be aware of: (1) even with the fastest analyzers, only a very small number of measurements are possible per passing vehicle, so information about individual vehicles is often limited and not readily extrapolated to rest of the vehicle journey, (2) monitoring at a fixed location (or a small number of fixed locations in larger studies) only tells you about emissions under driving conditions at similar locations, and the type and range of locations that can be used are often limited by the measurement strategy, and (3) being remote measurements, they typically do not provide direct measures like exhaust flow rate, which restricts reported emissions to measurements based on concentration alone. Results are commonly reported as a ratio of concentrations of other species and CO_2 and extrapolated from these to emissions on a "per fuel consumed" basis, in the general form:

$$nRATIO = \frac{[n]}{[CO_2]}; nMASS = nRATIO \times \frac{mwt_n}{(CO_2 \rightarrow Fuel)_{Equivalent}}$$

where *nRATIO* is the ratio of the measured concentrations of emitted species of interest, here $[n]$, and carbon dioxide, $[CO_2]$; *nMASS* is the emissions of *n*, typically reported on a "grams of *n* emitted per kg of fuel co*nsumption" basis,* mwt_n is the molecular weight of *n*; and $(CO_2 \rightarrow Fuel)_{Equivalent}$ is fuel consumption equivalency terms, the mass amount of fuel consumed per unit volume of CO_2 emitted. In more sophisticated approaches, the equivalency term is expanded to include all carbon and hydrogen-containing species measured to provide a more accurate estimate of fuel consumption.

Methods have also been proposed to estimate other emissions rates, but they have not been as widely used to date, and these are unlikely to be as accurate as emissions rate measurements determined using direct measures of other parameters like exhaust flow.

In the late 1960s and 1970s, tunnel-based methods were arguably the first of these approaches to be systematically used (El-Fadel & Hashisho, 2001).

These typically used the difference between airborne concentrations of emitted species inside and outside the tunnel, e.g., from outtake and intake points in tunnel ventilation systems, as a measure of local emissions. Regressing these against simultaneous in-tunnel vehicle counts, they provided a measure of typical emissions per vehicle, or per vehicle of different types (e.g., cars, buses, and trucks) if counts were similarly disaggregated. Although widely used in the years that followed their development, they could not provide the "per individual vehicle" resolution of other in situ techniques, and today they are much less commonly used except in special cases, e.g., where the tunnel itself is identified as a source of air quality concern.

Across-road remote sensing is arguably the most widely used of these other techniques (Borken-Kleefeld & Dallmann, 2018; Zhang, Stedman, Bishop, Guenther, & Beaton, 1995). These use open-path optical instruments that incorporate separate light sources and analyzers and measure the absorption of light by species in ambient air between the two. They have been used in various environmental applications since the 1960s, but were first successfully used to measure the emissions of individual passing vehicles by researchers at the University of Denver working in collaboration with Ford and the Californian Air Resources Board (ARB) in the 1980s. Their instrument, known as Fuel Efficiency Automobile Test or FEAT, used an across-road configuration with the source on one side of the road and analyzer on the other. This configuration or a mirror-box variation, with the mirror-box on the other side of road, reflect light from the source to the analyzer position adjacent to or incorporated with the same units as the source (Fig. 4.1), has been adopted by almost all Vehicle Emissions Remote Sensing System (VERSS) manufacturers since then. The instruments are typically used in combination with speed traps, cameras, and automatic license plate recognition software to measure emissions, speed and acceleration, and record vehicle registrations. The latter can also be used to obtain further information about observed vehicles, such as vehicle type, engine size, emission standard, fuel, and age, from national archives.

In the 1990s and early 2000s, VERSSs were first identified as potential tools for future regulatory practices, and the sector received some early investment. It provided early indications of the importance of high emitters, a relatively small proportion of vehicles that appear to emit significantly more than most other similar vehicles (Bishop, Stedman, & Ashbaugh, 1996). However, uncertainty regarding the representativeness of measurements of individual vehicles hindered regulatory uptake and ongoing investment was modest until very recently.

In the 2010s, a down-facing VERSS known as Emission Detection and Reporting (EDAR) was developed. This used a laser-light source and analyzer unit mounted over the road and a reflect strip on (or later in) the road surface to provide a similar open-path measurement of passing vehicle emissions (Ropkins et al., 2017; Fig. 4.1). While the earlier across-road VERSSs typically used a single across-road light beam, making emissions measurements very sensitive to exhaust height, this newer system used a "curtain" beam rapidly scanned back and forth across the road, and this in combination with its down-facing configuration makes it much less sensitive to exhaust configuration.

However, all VERSSs, across-road and down-facing alike, remain subject to concerns about the representativeness of measurement obtained for individual vehicles. They also measure emissions in the air after a passing vehicle, so require a small gap between vehicles, significantly reducing their capture rates in the congested and stop-start driving conditions that tend to be prevalent in most urban pollution hot spots. In addition, being open-path, they are limited to use with optical (light absorption, obstruction, or scattering) measures that tend to be less sensitive to the finer particulates emitted by more recent vehicle exhaust systems.

At about the same time that EDAR was being developed, other researchers and instrument manufacturers were also looking for other in situ emissions monitoring options. Many of these use analyzers or a suite of analyzers deployed at roadside (or near-road) locations run at high data collection rates to log the evolution of passing emission plumes. Emissions are then separated from similar species in background ambient air by extracting the plume peak or feature from the time series measured as the emitting vehicle passes by. As with VERSSs, there is no direct measure of parameters like exhaust flow and measurements tend to be reported as CO_2 ratios. Their sampling points are unavoidably further from vehicle exhausts than VERSS (which is extrapolating to a point just after the exhaust), and plume peaks seen using these methods tend to be more diffuse, which means that sensitivity can be lower, the overlap between emissions plumes from different vehicles can be more commonplace and capture rates can be lower than for VERSS. As a result, active sampling methods, rapidly drawing large volumes of air through the analyzers, are often used to counter this issue. Here, two of the earliest of these active sampling methods, On-road Heavy-duty Measurement System (OHMS) and Portable Emissions AcQuisition System (PEAQS), illustrate the common sampling strategies. OHMS, which was developed by the University of Denver and Texas A&M University

(Bishop et al., 2015), uses an open tent or similar structure that the tested vehicle can drive through. The roof of the tent acts as a funnel, concentrating emissions at its apex where they are sampled and from there rapidly pumped to the analyzer(s). PEAQS, which was developed by ARB (Smith et al., 2018), and other similar systems (e.g., Dallmann, Harley, & Kirchstetter, 2011) use an arguably cruder approach, very high-volume sampling from either a high-up point above road if passing vehicles have high exhausts or near-road-surface if passing vehicles have lower exhausts. Although both are arguably more intrusive than VERSS, being active sampling procedures, they are not limited to use with only optical measurement methods, which means they are arguably better suited to measurement parameters like particle number that generally associates with fine particulate emissions.

Other emissions measurement methods

Other emission measurement methods typically sit in an analytical "no man's land": less often deployed either because they provided less data-per-vehicle or less direct measurements than PEMS or because they provided significantly lower fleet-coverage than VERSS. However, while admittedly no single method provides the optimal combination of vehicle and fleet information to become a stand-alone emission measurement solution, they have the potential to enhance effort in particular applications.

Trailers and roof and rear-mounted racks have all been widely used to increase instrumentation capacity in PEMS studies. They have been used to install extra or larger analyzers that would not fit in the test vehicle or power supplies such as generators for high power demand instruments that might be suitable for in-cabin deployment. However, in recent years they have become more common options for conventional PEMS deployment because they are quick to transfer from one vehicle to another, so provide a highly convenient alternative for those working on high-throughput testing activities.

Chaser vehicles are vehicles fitted with onboard monitors with sample inlets positioned and configured to measure the emissions of other vehicles, most commonly the vehicle they are following (Kolb et al., 2004; Zavala et al., 2006; Fig. 4.1). Although much of the instrumentation is very similar to that used by PEMS, the measurement strategy is perhaps best considered the mobile equivalent of in situ VERSS and active sampling approaches because, as with VERSS, there is no direct measure of parameters like exhaust flow, and, as a result, quantification strategies are typically similar to those used with the passing-vehicle monitoring methods. However, their mobility means that like conventional PEMS, car chasers can be used to monitor a vehicle's emission over a range of operating conditions, e.g., tracking chased/monitored vehicle speed and acceleration, typically inferred from car-chaser speed and chased vehicle distance. As monitoring equipment is installed in the chaser vehicle, not the vehicle under investigation, the approach provides a much less intrusive option for third-party vehicle testing than PEMS, and as the amount of data collected per-vehicle is decided by the chase vehicle driver, the method is ideally suited to the gathering of the larger volume of data on the emissions behavior of vehicles initially observed to be behaving atypically. This makes the approach particularly attractive to those looking for evidence of behavior linked with the presence of Defeat Devices and other forms of vehicle emissions systems tampering (Bernard, German, & Muncrief, 2019).

Vehicle emissions modeling

Current vehicle emission modeling practices, like current monitoring practices, have rapidly evolved in recent decades in response to concerns about the environmental impact of the vehicle emissions, and they will in all likelihood evolve more rapidly in future, as computer power and access to big data increase. While this process has been influenced by the events previously described, contemporary scientific understanding, and technology and, as in any field, funding availability and the lower overheads of research and development work in the modeling sector allowed earlier and more speculative work. Policy makers were also arguably quicker to take a more proactive role in shaping modeling capabilities. As a result, there was smoother, less evidence-driven progression in the emission modeling sector.

One of policy makers' earliest requirements was for fleet-level emissions information, most notably the emissions inventories that made environmental impact assessment possible. However, the emissions information of the time, e.g., the emission rates determined in early on-road and dynamometer testing, were not readily extrapolated to the larger fleet. Therefore, the earliest vehicle emissions modeling activities were typically "hole-filling" exercises, focused on the characterization of vehicle fleets and vehicle activity patterns and designed to bridge the knowledge gap between available emissions measurements and the information needed to estimate fleet emissions. These early inventories were used in combination with source apportionment and dispersion modeling outputs to provide early predictions of how much airborne pollution came from traffic-related sources and where it was going when it was emitted from vehicles.

The limitations of using emissions data from regulatory dynamometer/drive cycle testing as the basis for emissions modeling activities were soon to become very apparent. The widening gap between the emissions performance of vehicles on the test beds and in the real-world led to the development of a generation of emissions models that overpredicted the emissions reductions associated with later emission abatement technologies. This effect was particularly pronounced in countries where real-world emissions research was limited, and there was little local evidence of the scale of problem associated with assuming similar test bed and on-road emissions performance. Consequently, some of the earliest calls for better real-world emissions information were from the modelers wanting better data to build their models with and from the policy makers wanting to use these models to help them develop next-generation emission policies. This changed the data gathering dynamic and repositioned the modelers as the clients in many of the funded measurement studies of the years that followed. This allowed them to define the scope of monitoring activities, incorporate novel monitoring methods and investigate more nonregulatory emission species such as non-exhaust emissions, and this, in turn, provided them with the information needed to significantly refine their emissions models. It also provided some of the early (but albeit modest) investment that the more entrepreneurial instrument developers needed to sustain them through the fallow years prior to more widespread interest.

Over the same period, the need for local as well as national vehicle emissions management became increasingly evident, as did the full economic costs of dedicated monitoring programs. This led many local authorities, bus and lorry fleet operators, and those working in the commercial sector that were being asked to consider for the first time TRAP to actively lobby for alternative inventorying and assessment options. Here, specialist software, models developed using best available data and best scientific understanding, designed to provide an expert assessment on a highly specific topic, and packaged for guided use by trained but nonexpert personnel, were seen as highly cost-effective tools. This, in turn, led to the rapid development of a wide range of emissions modeling packages, some authority-led and often freely distributed, but also many more developed commercially within the private sector.

General principles of emissions modeling

All vehicle emissions models, regardless of complexity or intended purpose, are built on the base principle that emissions can be estimated in the form

$$nMASS = EF_n \times ACTIVITY$$

where $nMASS$ is the mass emissions of emitted species, here n; EF_n is an emission rate factor and $ACTIVITY$ is an associated measure of the vehicle activity.

In the simplest of these models, emissions rate and activity terms can be single numeric values, but more commonly one or both are categorical terms or mathematics functions designed to model the real-world emissions or behavior of in-use vehicles. Categorical emission functions are effectively lookup tables of emission factors for different modes of operation of a vehicle, e.g., the classical modal structure includes emission factors for idling, accelerating, cruising and decelerating vehicles, and is used in combination with measures of distance traveled in each of these modes. Functions accept one or more activity terms, e.g., speed, acceleration, and road slope, and use a mathematical model to predict associated emissions. These can be used to model individual vehicles or engines, e.g., in early research and development work as part of the commercial production of a new car, or nested and summed to provide a prediction of emissions from a larger fleet

$$nMASS_{fleet} = \sum^{i=I} EF_{n,i} \times ACTIVITY_i \times COUNT_i$$

where $nMASS_{fleet}$ is the fleet total mass emissions of emitted species, here n; $EF_{n,i}$ is an emission rate factor for species and vehicle class i; $ACTIVITY_i$ is the related measure of activity for vehicle class i; $COUNT_i$ is the number of vehicles of class I in the fleet; and $nMASS_{fleet}$ is calculated as the sum of emissions for all classes of vehicle (e.g., $i=1$ to I) considered by the fleet emissions model.

Vehicle emissions models can be grouped in numerous ways, e.g., according to spatial resolution, temporal resolution, the model application, the type of inputs and the resolution they are required at, the modeling strategy and assumptions used to develop the models, the type of outputs and the resolution they are reported at. Generally, there are no strict divisions and all combinations have been at least investigated at some time, but typically, and arguably for practical purposes, these model groupings tend to associate and the smaller-scale models tend to accept more inputs, apply more sophisticated modeling strategies, and provide more detailed but shorter timescale predictions (see, e.g., Faris, Rakha, Kafafy, Idres, & Elmoselhy, 2011; Linton, Grant-Muller, & Gale, 2015). Therefore, we group these here according to the spatial scale for discussion. We also use the common terminology, macro-, meso-, and microscale for the larger-, medium-, and smaller-scale vehicle emissions models, but highlight that these terms are not rigorously defined and that except at the extremes those reading further on this topic should expect to find some variation in the point at which different model developers have drawn the divide between, e.g., macro- and meso- and meso- and microscale models.

Macroscale vehicle emissions models

Macroscale vehicle emissions models are widely used to generate large-scale emission inventories, and they are most commonly used by national and regional authorities to inform decision-making processes when developing vehicle emissions management plans. The earliest tended to be "top-down" models that used highly aggregated input data, e.g., fuel sales for a region, local fleet composition summary information from vehicle registration authorities, and emissions factors from national archives, to provide relatively crude total emissions reports and summaries. In many cases, "top-down" emissions models have since been superseded by "bottom-up" models that use much more disaggregated or local-level data, e.g., vehicle counts and average speeds by street (or road network link), to generate higher resolution emissions maps which many authorities now routinely publish. As part of this move, many associated data gathering and modeling activities have been passed to local authorities and other transport network management service providers, and many national governments have developed their own macroscale vehicle emissions models and codes of practice to standardize outputs. Full "bottom-up" modeling is not always possible, and some "top-down" information often has to be used in some regional models, and such guided modeling practices play an important part in ensuring consistency of modeling practices across regions.

Examples of such models include MOBILE (USEPA, 2002) and later MOVES (Koupal, Cumberworth, Michaels, Beardsley, & Brzezinski, 2003) which were developed by the EPA for use in the US, EMFAC which was developed by ARB for use in California (CARB, 2002), COPERT which was developed by the European Environment Agency (EEA) for use in Europe (Ntziachristos, Gkatzoflias, Kouridis, & Samaras, 2009), EMBEV which was developed for the Chinese Ministry of Ecology and Environment (Wu et al., 2012), the Handbook of Emission factors (HBEFA) model follows the approach of a traffic situation with emission factors from the PHEM model (Andre et al., 2009), and the International Vehicle Emissions (IVE) model which has been more widely used in Asia, Africa, and Australia (Davis, Lents, Osses, Nikkila, & Barth, 2005). Although vehicle speed has never been considered particularly robust input for an emission model, most model developers pragmatically adopted speed-based strategies for most of these macroscale models because of the relatively high availability of speed data from, e.g., automatic traffic counters, routine on-road surveying, and traffic network models. Here, one notable exception is the EPA MOVES. MOVES replaced the earlier speed-based MOBILE model and uses a categorical emission model based on binned measurements of speed and vehicle specific power (or VSP), a measure of work done by a vehicle. Other inputs have also been investigated, e.g., traffic engineering variables such as traffic flow, traffic descriptors such as "stop-stop" and "free-flow" and driving modes descriptors such "idle," "accelerating," "cruising," and "decelerating" in models such as ARTEMIS (Andre et al., 2009), VERSIT+ (Smit, Smokers, & Rabé, 2007), PHEM (Zallinger, Tate, & Hausberger, 2008), and CMEM (Scora & Barth, 2006) although these are less readily implemented at regional scales, and their use in macroscale models is not yet commonplace. Most of the more established macroscale emissions models incorporate correction factors to account for other influences on emissions, e.g., driving conditions or amount of congestion, average air temperature and humidity, vehicle age, or mileage.

The performance of macroscale emissions models is not readily evaluated because "true" emissions are not strictly known at these levels. However, those that have considered this question in detail have often noted that more sophisticated models do not always guarantee greater predictive power (see, e.g., Smit, Ntziachristos, & Boulter, 2010).

Microscale vehicle emissions models

Microscale vehicle emissions models are widely used to model emissions trends at smaller scales, e.g., typically from a road network link, street or junction or a small network sections of several links. They are extensively used to assess a wide range of local traffic network management schemes, e.g., road network expansion, traffic calming and flow management, bus and cycle route prioritization, and as a real-time input for Intelligent Transport Systems. They are also used in eco-driving and other vehicle management/

routing tools, and in supporting regional development and other construction work. They are widely used both predictively for future-casting, scenario testing, and as part of planning and approvals stages of preliminary work, and as apportionment tools in post-implementation scheme appraisals. Many vehicle emissions models used in larger-scale applications have also been used at one time or another in microscale models also can be applied at the microscale, e.g., MOVES, CMEM and PHEM. As with macroscale models there were a number that have been identified as endorsed, recommended or preferred methods by national, local, and network management authorities, e.g., EMIT in the UK (CERC, 2019), COPERT Street Level in Europe (Yudego et al., 2018) and PΔP in Australia (Smit, Ellison, & Greaves, 2014). However, the wide range of potential applications and end users have also generated commercial interest, and private sector microscale emissions models include, amongst others, NetSim, VERSIT, VeTESS, and VT-Micro (Algers et al., 1997; Linton et al., 2015). In addition, many software developers have also developed specialist emission modeling modules for already established traffic micro-simulation models, e.g., AIMSUN, CORSIM, DRACULA, INTEGRATION, PARAMICS SUMO, andVISSIM. The latter are often highly attractive options for transport network managers, who are already likely to be collecting the inputs for traffic micro-simulation models and often have in-house expert users.

The smaller scales of these models allow greater complexity. Inputs are typically higher resolution traffic information, e.g., vehicle speed profiles from "floating" car and dedicated Global Positioning System (GPS) activity studies or high-resolution micro-simulation traffic flow modeling. Unlike macroscale emissions models, which all tend to have similar structures and be used in model chains with other discrete models (e.g., the traffic simulation, emissions models, dispersion model, health impacts chains described by Viaene et al., 2016), there is more variation in the scope of emissions models, and therefore the inputs required, at these smaller scales. For example, some microscale emissions models require information about road and building geometry and weather conditions, and provide predictions of local air quality contributions but are perhaps more correctly considered hybrid emission and dispersion models.

The simplest microscale emissions models are effectively higher-resolution versions of the simplest macroscale models: accepting higher spatial and/or temporal frequency average speed or vehicle activity inputs, and applying similar speed-based emission models to provide more detailed but smaller scale inventories or maps. However, the majority of current generation microscale emissions models are what are known as *instantaneous models*, emissions models that estimate emissions at the same resolution as the supplied inputs. The instantaneous modeling strategy can be similar, e.g., categorical or function, and can be strictly instantaneous, using only the inputs from a single time interval, or in more sophisticated time-period-based models, using the inputs from several intervals to better map emissions onto the combinations of vehicle activities that generated them. At this resolution the modeling strategy can also be much more refined, e.g., taking into account different driving conditions, likely gear position and vehicle properties such as mass and drag that are less easily modeled with more aggregated data. The obvious trade-off here is the potential to introduce bias when scaling up from a set of predicted emissions profiles, based on tracked vehicles, to the larger, and mostly untracked, fleet. As a result, a range of scaling and modeling strategies have been proposed to handle this step.

While strong evidence regarding their performance is limited and there is still much debate regarding best practices on their evaluation, microscale emissions models, particularly those that employ instantaneous modeling and robust fleet scaling strategies, are widely regarded as effective tools for assessing the emissions impacts of a wide range of traffic and fleet management activities (Nocera, Basso, & Cavallaro, 2017).

Mesoscale vehicle emissions models

Mesoscale vehicle emissions models are employed at geographical scales in between micro and macroscale vehicle emissions models. Several regulatory authorities have attempted to more rigorously define terminologies, but the divisions micro/meso and meso/macro remain debatable. However, mesoscale models tend to be implemented at the "neighborhood" to "district" levels. Modeling strategies also reflect this positioning with many of the methods used at macro and microlevels, being used at mesolevels.

Here, it is also worth highlighting one notable trend: Advances in computing power, particularly large-scale computation on personal computers and online parallel process services, and data availability, e.g., accessible GPS vehicle activity archives, mean that we can use increasing higher-resolution modeling strategies on increasing larger scales. So, what was yesterday only routinely possible on a microscale will most likely be a meso or macroscale option tomorrow. This is seen in the increasing larger-scale and higher-resolution use of GPS data vehicle activity data in emission modeling activities. For instance, vehicle activity models derived from the analysis of large-scale (100 million records plus) local fleet GPS datasets have already been used as inputs for bottom-up emission inventory models, first with data from one vehicle class in, e.g., Singapore-MIT Alliance for Research & Technology studies in Singapore (Nyhan et al., 2016), and more recently with GPS data from different vehicle classes all uniquely handled in, e.g., Vehicular Emissions INventory (VEIN) studies in Brazil (Ibarra-Espinosa et al., 2018; Ibarra-Espinosa, Ynoue, Giannotti, Ropkins, & de Freitas, 2019). There still remains some work to be done before GPS data can be robustly employed in larger-scale emission inventorying work because many current-generation emission factors archives are not strictly suitable for use with higher resolution or instantaneous vehicle activity data (Ntziachristos & Samaras, 2016), but Wu, Chang, Wang, Hang, and Zhang (2019) very recently reported on the Real-time On-road Emission (ROE) v1.0 that reads and models vehicle activity data directly from livestreamed internet traffic information services broadcasts in China, demonstrating the very real achievability of such efforts.

Current and future challenges

The 2015 "Dieselgate" Defeat Device Scandal was neither the first scandal of its type or the only challenge we face in the vehicle emissions management sector. That said, it was the subject of significant media and public attention, and as in the years following the Heavy-Duty Diesel Vehicle Defeat Device Scandal of the 1990s, it did provide a catalyst for change, focusing efforts to review and revise monitoring and modeling practices.

Work had already begun to address some of the challenges we face prior to Dieselgate. Many aspects of regulatory dynamometer test procedures were under review and new drive cycles, the Worldwide Harmonized Light-vehicle Test Procedures (WLTPs), considered more representative of real-world driving, had been proposed as replacements for conventional tests such as the New European Drive Cycle (NEDC). Similarly, new particle number monitoring standards have been developed to address concerns about the finer particulate emissions of modern vehicles, but this time are intended to run alongside rather than replace existing particulate mass-based monitoring standards (Giechaskiel et al., 2018). While Dieselgate did not trigger these work programs, it did help to acceleration their adoption. It was also the incentive for a number of other long called for revisions to practices. Prior, confirmatory real-world PEMS testing had previously only been required as part of Heavy-Duty Vehicle approval procedures in the US. Post-Dieselgate, authorities in the US extended similar on-road testing to the light-duty vehicle fleet, and elsewhere, where on-road testing had previously not been part of regulatory practices, authorities were quick to adopt similar procedures. For example, in Europe, China, Japan, and India new PEMS-based Real Driving Emissions (RDE) procedures have or are being introduced into legislation (Mock & German, 2015; Posada, Bernard, Yang, & Bandivadekar, 2018).

However, there is still much to do (Fig. 4.2) and here we highlight some of the key challenges ahead and opportunities to shape next-generation vehicle emissions monitoring and modeling practices.

New monitoring technologies and increasing data collection capabilities have already had an impact on the way we work. Global Positioning System (GPS)-based vehicle activity data is already collected for and used in a number of applications, most notably tracking services for logistics, theft-prevention, and insurance. But emissions modelers are using it at increasing large scales as a source of vehicle activity data. We are now, once

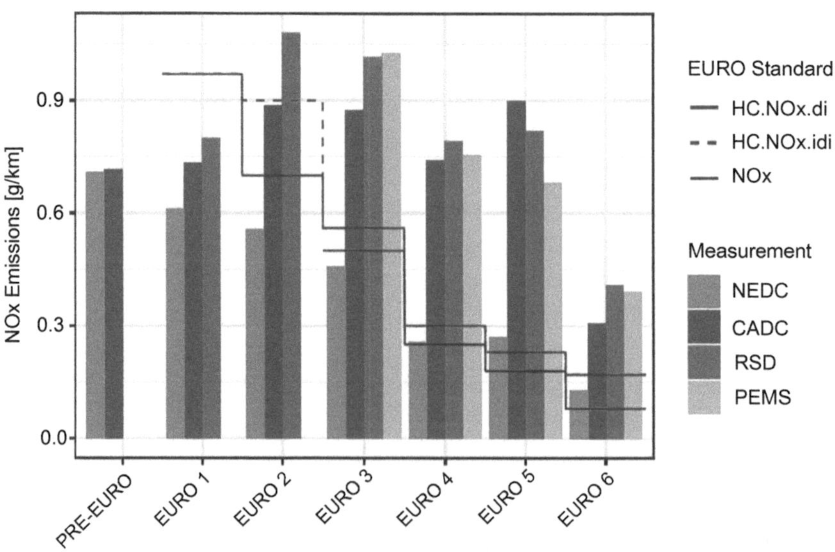

Fig. 4.2 Typical diesel passenger car NO_x emissions as report by Dynamometer-based testing using the NEDC, CADC, RSD, and PEMS. Data sources and handling: NEDC Dynamometer-based measurement using New European Drive Cycle and CADC Dynamometer-based measurement using Common Artemis Drive Cycle (Hooftman et al., 2018, and references therein); RSD (mean average of values from Carslaw & Rhys-Tyler, 2013, and Chen & Borken-Kleefeld, 2014, both calculated as g/km emissions assuming COPERT5 CO_2 emissions, and Bernard, Tietge, German, & Muncrief, 2018); and PEMS (mean average of values from Weiss et al., 2012, Franco, Sánchez, German, & Mock, 2014, and O'Driscoll et al., 2016).

again, looking at how we reduce the size, weight, cost, and energy demand of PEMS so we can deploy the technique more widely in future, and how we can extend the monitoring capabilities of the onboard diagnostic (OBD) sensors already incorporated into the exhaust management systems of modern vehicles to provide more direct emissions data for vehicles in routine use. As such bodies of work evolve and align, we are rapidly moving to a situation where we may all be individually accountable for our environmental burden. We can currently estimate emissions with reasonable accuracy, but in future, we will most likely know them. Many see these data-streams as "triple win" resources: lower-cost, better fleet coverage, and most importantly, providing more timely evidence on the fleet-level performance of emissions reduction strategies by comparison to conventional data gathering and modeling activities. However, such transitions are never smooth, and they will most likely change the data collection (emissions monitoring)/data modeler (emissions modeling) dynamic once again. In future, there is a very real possibility that vehicle data-linked service providers, rather than regulators, vehicle manufacturers, or model developers, will be the major clients of work shaping the data gathering efforts.

GPS and OBD-based data-streams provide us with the opportunity to comprehensively measure the emissions of future vehicle fleets, but complete reliance on these would make us even more vulnerable to malicious efforts to circumvent emission standards. Therefore, independent regulatory monitoring will need to continue, and dynamometer and standardized drive cycle-based test procedures will most likely remain the cornerstone of legislative action. However, we are already acknowledging the need for more structured data gathering activities, especially if we want to start to actively target the dirtiest vehicles in our fleets. This means we need to be looking again at all the real-world emissions monitoring methods and how they can be best used in combination to inform and assess our vehicle approval procedures and emissions reduction programs because none is realistically a "stand-alone" solution (Table 4.2). We also need to be thinking harder about the combinations needed to quickly and most comprehensively assess new vehicle engine and emission abatement technologies, fuels, and retrofit systems for existing fleets because we need to be able to better identify and champion the options that will deliver real emissions savings in a more timely fashion. Even a fleet of electric vehicles will produce emissions, e.g., tire and brake-wear particulate emissions, (Timmers & Achten, 2016), and we need to be ready to monitor them before rather than after they become a problem.

Table 4.2 Vehicle emissions measurements methods, summary information.

Measurement				Approximate measurement rate	
Method	**Strategy**	**Species**[a]	**Report units**	**Vehicles per day**	**per vehicle**
Laboratory					
Dynamometer	Chassis or engine dynamometer facility simulation of on-road vehicle operation	CO_2, CO, NO, NO_x, NO_2, N_2O, THCs, CH_4, NH_3, PM, PN (plus)	g/s, g/km, g/kg fuel	1–20[b]	10,000 (plus)
Real-world					
PEMS	Onboard monitors directly measuring exhaust flow and emissions concentrations of that vehicle	CO_2, CO, NO_x (or NO, NO_2), THCs, PM (or PN)	g/s, g/km, g/kg fuel	1–2	10,000 (plus)
Vehicle chaser	Onboard monitors on one vehicle measuring emissions of other followed vehicles	CO_2[c], CO, NO_x (or NO, NO_2), THCs, PM (or PN)	CO_2 ratio, others estimated[c]	5–25	500–1000
In situ	Roadside deployed (often active sampling) monitors measuring passing vehicle emissions	CO_2[c], CO, NO_x (or NO, NO_2), THCs, PM (or PN)	CO_2 ratio, others estimated[c]	1000 (plus?)[d]	1 (plus?)
VERSS	Across-road or down-facing open-path monitoring measuring passing vehicle emissions	CO_2[c], CO, NO_x (or NO, NO_2), THCs, PM (proxy)	CO_2 ratio, others estimated[c]	2000–8000	1

PEMS, portable emissions measurement system; *VERSS*, vehicle emissions remote sensing system.
[a] Please note: This is species that systems typically measure rather than a comprehensive list of all species they can (or have been reported to) measure.
[b] Higher "Vehicles per day" rates on specialist "roll-on/roll-off" facilities.
[c] CO_2 measurement used to determine ratio, other reported emissions measurements are typically estimated from this, e.g., most commonly reported g/kg fuel emissions are typically estimated using carbon balance.
[d] The "plus?" indicates "or possibly more."

We know that engine and emission abatement system performance typically deteriorate over the operational lifetime of a vehicle. We know that the level and effectiveness of vehicle maintenance, the quality of the driving environment and the driving style of the individual driver can also affect the deterioration rate. But we also know that not all vehicles are equal, and some for whatever reason (bad design, inferior build, operational damage, poor maintenance, and deliberate tampering) emit more than others and believe in many cases that the "high emitter" components of our vehicle fleets are responsible for the majority of real-world emissions (Bishop et al., 1996). We had some idea of the scale of this problem, but are only really recently starting to gather robust evidence on this from the work of those that had the foresight to implement long-term VERSS studies (Borken-Kleefeld & Chen, 2015; Chen & Borken-Kleefeld, 2016) and which pollutants are the most likely to be higher emitter, rather than fleet-wide, issues. We also know that while many countries adopted similar "up-front" vehicle approval-based emissions control strategies, countries like the US that adopted complementary on-road surveillance and Inspection and Maintenance (I&M) programs were more effective in the delivery of their emissions reduction programs than those that did not (Posada et al., 2015). Therefore, we need to be looking at these programs, which by their nature are unavoidably costly and identifying best options for future fleet surveillance activities. Here, remote sensing is arguably one of the most cost-effective strategies but uncertainties (e.g., the overlap between the occasional high emissions of an otherwise clean vehicle and the lower level emissions of a bad vehicle) currently hinder its more widespread use as a fleet screening tool.

Intervention is arguably the area within the vehicle emissions regulatory paradigm (Fig. 4.3) where least progress has been made to date, but it is also the area where our efforts could provide the largest returns: A bad vehicle removed from the fleet or effectively repaired and returned to fleet could potentially provide large and timely emissions benefits. We know something of the scale of the emission costs of broken and faulty engine and emission abatement system parts (Huang et al., 2019; Ntziachristos, Papadimitriou, Ligterink, & Hausberger, 2016), but questions remain about our ability to effectively integrate emissions management into the frontline point-of-service vehicle repair activities of commercial garages and vehicle fleet maintenance departments, where, understandably, the focus is getting broken vehicles running and back on the road as quickly as possible. Here, there is a need for low-cost expert-system monitors that provide diagnostic information, not just a pass or fail measure of emissions, but a

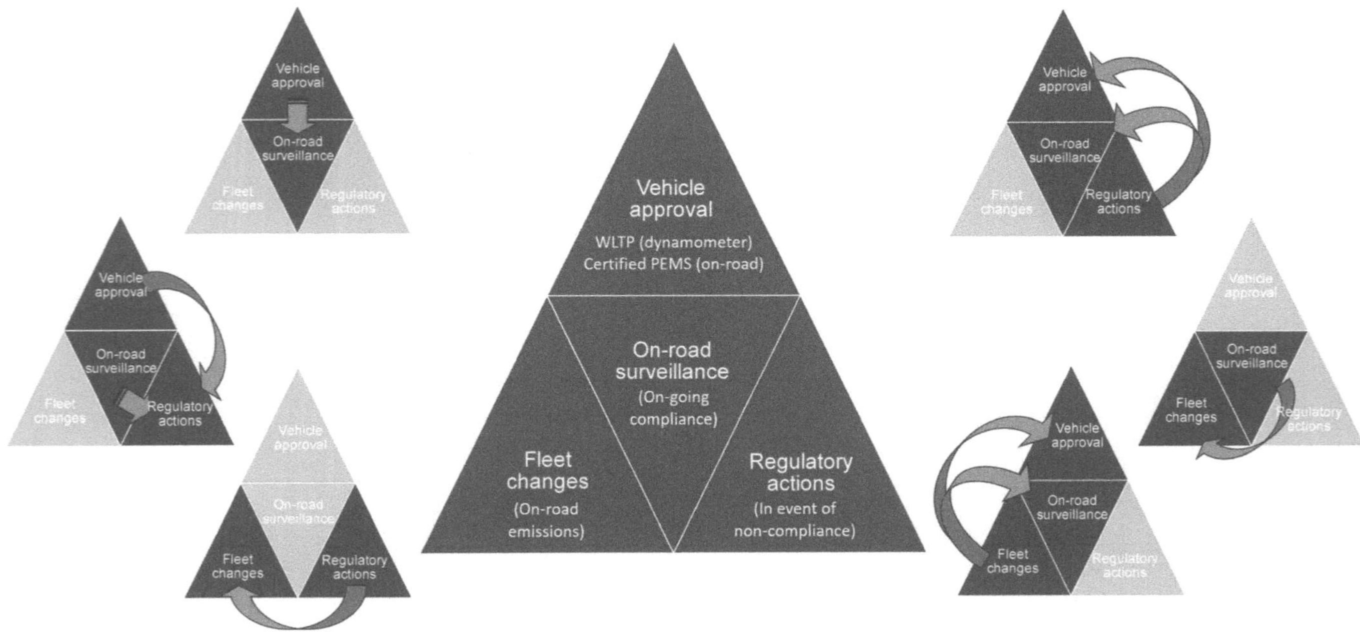

Fig. 4.3 The vehicle emissions regulatory paradigm.

fault identification and, ideally, associated guidance of "best practice" for emission-friendly repair. Because, as with Defeat Devices, deliberate malpractice is involved, vehicle tampering investigations are likely to be the earliest focus for intervention. However, even this is likely to be sensitive work because new regulations are never popular, particularly if they are seen to be removing existing freedoms or to target most heavily those least able to bear the costs.

Summary and conclusions

This chapter describes conventional vehicle emissions monitoring practices, focusing on how these evolved alongside our understanding of the environmental impacts of our vehicle fleets:

- From the earliest roadside methods developed to provide initial evidence on the levels of vehicle emissions.
- Through the development of dynamometer-and-drive-cycle-based procedures that provided the reproducible regulatory framework needed for the introduction of emissions standards legislation and vehicle emissions reduction policies.
- To the development of real-world monitoring procedures including portable emissions measurement systems (PEMS) and vehicle emissions remote sensing systems (VERSS), designed to address more recent concerns about the widening gap between test bed and on-road vehicle emissions, and potential abuses of testing procedures by some vehicle manufacturers.

This chapter also describes the parallel development of vehicle emissions modeling practices, focusing on the changing role of the emissions modelers:

- The earliest emission modeling activities focused on the estimation of vehicle fleet sizes and activity patterns, bridging the knowledge gap between emissions measurements and the information needed by policy makers to estimate fleet-level emissions.
- These activities led to the development of large (national and regional)-scale or macroscale emission models that were widely adopted by national and local government bodies, and smaller-scale (*meso* and micro) emission models that other organizations were encouraged to use as a practical alternative to more costly monitoring programs as part of environmental impact assessment activities.

- The need for better models and models for species less readily monitored, e.g., non-exhaust emissions, meant that modelers moved over time from users of existing monitoring data to the clients informing and shaping the scope of future monitoring activities.

Finally, looking forward, the chapter also discusses some important limitations of current practices and options for more effective future practices:

- Novel emission sensing technologies, increasing computing power and lower-cost data storage all allow us to continually refine our vehicle monitoring and modeling capabilities; new sensors increase the scale and scope of onboarding monitoring, newer roadside monitoring provides an ever more credible option for routine passing-vehicle fleet surveillance, and rapidly evolving data sources such as Global Positioning System (GPS) and OnBoard Diagnostics (OBD) provide us with ever more accurate measures of vehicle fleet emissions.
- But looking forward, our biggest challenge remains intervention, the necessary targeting of higher emitting technologies and vehicles within our fleets, because, although this may seem the most obvious and direct means of delivering the largest and most timely emissions benefits, it may not always make to the popular policy unless carefully implemented.

Suggested further reading

Any of the references cited in the chapter could be of potential interest to those wanting to undertake further studies of the topics. However, the following are particularly recommended as earlier texts: Jacob and Winner (2009) regarding air pollution and climate change; Ropkins et al. (2009) or Franco et al. (2013) regarding vehicle emissions monitoring; Faris et al. (2011) regarding emissions modeling; and Vestreng et al. (2009) regarding vehicle emissions control strategies.

Acknowledgments

We gratefully acknowledge the valuable comments and insights of many reviewers and apologize to those whose significant contributions to the sector could not be documented more fully in a work of this length.

References

Algers, S., Bernauer, E., Boero, M., Breheret, L., Di Taranto, C., Dougherty, M., et al. (1997). *Review of micro-simulation models*. Review report of the SMARTEST project.

Andre, M. (1996). *Driving cycles development: Characterization of the methods (no. 961112)*. SAE Technical Paper.

Andre, M., Keller, M., Sjödin, Å., Gadrat, M., Mc Crae, I., & Dilara, P. (2009, April). The Artemis European tools for estimating the transport pollutant emissions. In *Proceedings of the 18th international emission inventories conference* (pp. 1–10).

Archer, G., & Director, C.V. (2015). *Mind the gap 2015: Closing the chasm between test and real-world car CO2 emissions*. Brussels: Transport & Environment.

Bata, R., Clark, N., Gautam, M., Howell, A., Long, T., Loth, J., et al. (1991). The first transportable heavy-duty vehicle emissions testing laboratory. *SAE Transactions*, 433–440.

Bernard, Y., German, J., Kentroti, A., & Muncrief, R. (2019). *Catching defeat devices: How systematic vehicle testing can determine the presence of suspicious emissions control strategies.* ICCT white paper. https://theicct.org/publications/detecting-defeat-devices-201906. (Accessed August 20, 2019).

Bernard, Y., German, J., & Muncrief, R. (2019). *Worldwide use of remote sensing to measure motor vehicle emissions.* ICCT white paper. https://theicct.org/publications/worldwide-use-remote-sensing-measure-motor-vehicle-emissions. (Accessed August 10, 2019).

Bernard, Y., Tietge, U., German, J., & Muncrief, R. (2018). *Determination of real-world emissions from passenger vehicles using remote sensing data.* ICCT report. https://theicct.org/publications/real-world-emissions-using-remote-sensing-data (Accessed August 10, 2019).

Bishop, G. A., Hottor-Raguindin, R., Stedman, D. H., McClintock, P., Theobald, E., Johnson, J. D., et al. (2015). On-road heavy-duty vehicle emissions monitoring system. *Environmental Science & Technology, 49*(3), 1639–1645.

Bishop, G. A., Stedman, D. H., & Ashbaugh, L. (1996). Motor vehicle emissions variability. *Journal of the Air & Waste Management Association, 46*(7), 667–675.

Borken-Kleefeld, J., & Chen, Y. (2015). New emission deterioration rates for gasoline cars—Results from long-term measurements. *Atmospheric Environment, 101*, 58–64.

Borken-Kleefeld, J., & Dallmann, T. (2018). *Remote sensing of motor vehicle exhaust emissions.* ICCT white paper. https://theicct.org/sites/default/files/publications/Remote-sensing-emissions_ICCT-White-Paper_01022018_vF_updated.pdf (Accessed August 10, 2019).

Burtscher, H., Lutz, T., & Mayer, A. (2019). A new periodic technical inspection for particle emissions of vehicles. *Emission Control Science and Technology, 5*, 1–9.

CARB (California Air Resources Board). (2002). EMFAC 2002 on-road emissions inventory model. In *Presented by mobile source analysis branch, planning and technical support division, El Monte.* California Air Resources Board.

Carslaw, D. C., Beevers, S. D., Tate, J. E., Westmoreland, E. J., & Williams, M. L. (2011). Recent evidence concerning higher NOx emissions from passenger cars and light duty vehicles. *Atmospheric Environment, 45*(39), 7053–7063.

Carslaw, D. C., & Rhys-Tyler, G. (2013). New insights from comprehensive on-road measurements of NOx, NO2 and NH3 from vehicle emission remote sensing in London, UK. *Atmospheric Environment, 81*, 339–347.

CERC (Cambridge Environmental Research Consultants). (2019). *EMIT, Comprehensive emissions inventory toolkit.* https://www.cerc.co.uk/environmental-software/EMIT-tool/data.html. (Accessed August 10, 2019).

Chen, Y., & Borken-Kleefeld, J. (2014). Real-driving emissions from cars and light commercial vehicles—Results from 13 years remote sensing at Zurich/CH. *Atmospheric Environment, 88*, 157–164.

Chen, Y., & Borken-Kleefeld, J. (2016). NOx emissions from diesel passenger cars worsen with age. *Environmental Science & Technology, 50*(7), 3327–3332.

Dallmann, T. R., Harley, R. A., & Kirchstetter, T. W. (2011). Effects of diesel particle filter retrofits and accelerated fleet turnover on drayage truck emissions at the Port of Oakland. *Environmental Science & Technology, 45*(24), 10773–10779.

Davis, N., Lents, J., Osses, M., Nikkila, N., & Barth, M. (2005). Development and application of an international vehicle emissions model. *Transportation Research Record, 1939*(1), 156–165.

Dinh, T.V., Choi, I.Y., Son, Y. S., & Kim, J. C. (2016). A review on non-dispersive infrared gas sensors: Improvement of sensor detection limit and interference correction. *Sensors and Actuators B: Chemical, 231*, 529–538.

El-Fadel, M., & Hashisho, Z. (2001). Vehicular emissions in roadway tunnels: A critical review. *Critical Reviews in Environmental Science and Technology, 31*(2), 125–174.

Faris, W. F., Rakha, H. A., Kafafy, R. I., Idres, M., & Elmoselhy, S. (2011). Vehicle fuel consumption and emission modelling: An in-depth literature review. *International Journal of Vehicle Systems Modelling and Testing, 6*(3–4), 318–395.

Feist, M. D., Sharp, C. A., & Spears, M. W. (2009). Determination of PEMS measurement allowances for gaseous emissions regulated under the heavy-duty diesel engine in-use testing program: Part 1-project overview and PEMS evaluation procedures. *SAE International Journal of Fuels and Lubricants, 2*(1), 435–454.

Franco, V., Kousoulidou, M., Muntean, M., Ntziachristos, L., Hausberger, S., & Dilara, P. (2013). Road vehicle emission factors development: A review. *Atmospheric Environment, 70*, 84–97.

Franco, V., Sánchez, F. P., German, J., & Mock, P. (2014). Real-world exhaust emissions from modern diesel cars. *Communications, 49*(30), 847129-102.

Giechaskiel, B., Lahde, T., Suarez-Bertoa, R., Clairotte, M., Grigoratos, T., Zardini, A., et al. (2018). Particle number measurements in the European legislation and future JRC activities. *Combustion Engines, 57*, 3–16.

Giechaskiel, B., Maricq, M., Ntziachristos, L., Dardiotis, C., Wang, X., Axmann, H., et al. (2014). Review of motor vehicle particulate emissions sampling and measurement: From smoke and filter mass to particle number. *Journal of Aerosol Science, 67*, 48–86.

Glover, J. H. (1975). Chemiluminescence in gas analysis and flame-emission spectrometry. A review. *Analyst, 100*(1192), 449–464.

Hooftman, N., Messagie, M., Van Mierlo, J., & Coosemans, T. (2018). A review of the European passenger car regulations—Real driving emissions vs local air quality. *Renewable and Sustainable Energy Reviews, 86*, 1–21.

Huang, Y., Ng, E. C., Yam, Y. S., Lee, C. K., Surawski, N. C., Mok, W. C., et al. (2019). Impact of potential engine malfunctions on fuel consumption and gaseous emissions of a Euro VI diesel truck. *Energy Conversion and Management, 184*, 521–529.

Huang, Y., Organ, B., Zhou, J. L., Surawski, N. C., Hong, G., Chan, E. F., et al. (2018). Remote sensing of on-road vehicle emissions: Mechanism, applications and a case study from Hong Kong. *Atmospheric Environment, 182*, 58–74.

Ibarra-Espinosa, S., Ynoue, R., Giannotti, M., Ropkins, K., & de Freitas, E. D. (2019). Generating traffic flow and speed regional model data using internet GPS vehicle records. *MethodsX, 6*, 2065–2075. https://doi.org/10.1016/j.mex.2019.08.018.

Ibarra-Espinosa, S., Ynoue, R., O'Sullivan, S., Pebesma, E., Andrade, M. D. F., & Osses, M. (2018). VEIN v0.2.2: An R package for bottom–up vehicular emissions inventories. *Geoscientific Model Development, 11*, 2209–2229. https://doi.org/10.5194/gmd-11-2209-2018.

Jacob, D. J., & Winner, D. A. (2009). Effect of climate change on air quality. *Atmospheric Environment, 43*(1), 51–63.

Kadijk, G., Verbeek, R. P., Smokers, R. T. M., Spreen, J. S., Patuleia, A. F., & van Ras, M. G. (2012). *Supporting analysis regarding test procedure flexibilities and technology deployment for review of the light duty vehicle CO_2 regulations.* European Commission report (DG CLIMA). Accessed from TNO archive: http://publications.tno.nl/publication/34622247/QesCIL/kadijk-2012-supporting.pdf. (Accessed August 10, 2019).

Kittelson, D. B. (1998). Engines and nanoparticles: A review. *Journal of Aerosol Science, 29*(5), 575–588.

Kittelson, D. B., Arnold, M., & Watts, W. F. (1999). *Review of diesel particulate matter sampling methods*. Minneapolis, Minnesota, USA: University of Minnesota, Department of Mechanical Engineering, Center for Diesel Research.

Klingenberg, H. (2012). *Automobile exhaust emission testing: Measurement of regulated and unregulated exhaust gas components, exhaust emission tests*. Springer Science & Business Media.

Kolb, C. E., Herndon, S. C., McManus, J. B., Shorter, J. H., Zahniser, M. S., Nelson, D. D., et al. (2004). Mobile laboratory with rapid response instruments for real-time measurements of urban and regional trace gas and particulate distributions and emission source characteristics. *Environmental Science & Technology*, *38*(21), 5694–5703.

Koupal, J., Cumberworth, M., Michaels, H., Beardsley, M., & Brzezinski, D. (2003). Design and implementation of MOVES: EPA's new generation mobile source emission model. *Ann Arbor*, *1001*(48), 105.

Lelieveld, J., Evans, J. S., Fnais, M., Giannadaki, D., & Pozzer, A. (2015). The contribution of outdoor air pollution sources to premature mortality on a global scale. *Nature*, *525*(7569), 367.

Linton, C., Grant-Muller, S., & Gale, W. F. (2015). Approaches and techniques for modelling CO_2 emissions from road transport. *Transport Reviews*, *35*(4), 533–553.

Mock, P., & German, J. (2015). The future of vehicle emissions testing and compliance. *Communications*, *49*(30), 847129-102.

Nakamura, H., Kihara, N., Adachi, M., Nakamura, S., & Ishida, K. (2003). Development of hydrocarbon analyzer using heated-NDIR method and its application to on-board mass emission measurement system. *JSAE Review*, *24*(2), 127–133.

Nocera, S., Basso, M., & Cavallaro, F. (2017). Micro and macro modelling approaches for the evaluation of the carbon impacts of transportation. *Transportation Research Procedia*, *24*, 146–154.

Ntziachristos, L., Gkatzoflias, D., Kouridis, C., & Samaras, Z. (2009). COPERT: A European road transport emission inventory model. In *Information technologies in environmental engineering* (pp. 491–504). Berlin, Heidelberg: Springer.

Ntziachristos, L., Papadimitriou, G., Ligterink, N., & Hausberger, S. (2016). Implications of diesel emissions control failures to emission factors and road transport NOx evolution. *Atmospheric Environment*, *141*, 542–551.

Ntziachristos, L., & Samaras, Z. (2016). Passenger cars, light commercial trucks, heavy-duty vehicles including buses and motor cycles category. In 2016. *EMEP/EEA emission inventory guidebook*.

Nyhan, M., Sobolevsky, S., Kang, C., Robinson, P., Corti, A., Szell, M., et al. (2016). Predicting vehicular emissions in high spatial resolution using pervasively measured transportation data and microscopic emissions model. *Atmospheric Environment*, *140*, 352–363.

O'Driscoll, R., ApSimon, H. M., Oxley, T., Molden, N., Stettler, M. E., & Thiyagarajah, A. (2016). A portable emissions measurement system (PEMS) study of NOx and primary NO2 emissions from Euro 6 diesel passenger cars and comparison with COPERT emission factors. *Atmospheric Environment*, *145*, 81–91.

Posada, F., Bernard, Y., Yang, L., & Bandivadekar, A. (2018). *Technical considerations for implementing a real driving emissions test for passenger vehicles in India*. https://theicct.org/publications/real-driving-emissions-test-pv-india. (Accessed August 10, 2019).

Posada, F., Yang, Z., & Muncrief, R. (2015). *Review of current practices and new developments in heavy-duty vehicle inspection and maintenance programs.* https://theicct.org/publications/review-current-practices-and-new-developments-heavy-duty-vehicle-inspection-and. (Accessed August 10, 2019).

Ropkins, K., Beebe, J., Li, H., Daham, B., Tate, J., Bell, M., et al. (2009). Real-world vehicle exhaust emissions monitoring: Review and critical discussion. *Critical Reviews in Environmental Science and Technology*, *39*(2), 79–152.

Ropkins, K., DeFries, T. H., Pope, F., Green, D. C., Kemper, J., Kishan, S., et al. (2017). Evaluation of EDAR vehicle emissions remote sensing technology. *Science of the Total Environment, 609*, 1464–1474.

Scora, G., & Barth, M. (2006). *Comprehensive modal emissions model (CMEM), version 3.01. User guide* (p. 1070). Riverside: Centre for Environmental Research and Technology, University of California.

Seinfeld, J. H., & Pandis, S. N. (2016). *Atmospheric chemistry and physics: From air pollution to climate change*. John Wiley & Sons.

Smit, R., Ellison, A. B., & Greaves, S. (2014). The use of GPS and high resolution vehicle emission modeling to assess environmental benefits of driver behaviour change programs. In *ARRB Conference, 26th, 2014, Sydney, New South Wales, Australia*.

Smit, R., Ntziachristos, L., & Boulter, P. (2010). Validation of road vehicle and traffic emission models—A review and meta-analysis. *Atmospheric Environment, 44*(25), 2943–2953.

Smit, R., Smokers, R., & Rabé, E. (2007). A new modelling approach for road traffic emissions: VERSIT +. *Transportation Research Part D: Transport and Environment, 12*(6), 414–422.

Smith, J. D., Ham, W., Burnitzki, M., Downey, S., Howard, C., Quiros, D., et al. (2018). Quantification of HD in-use vehicles using the portable emissions AcQuisition system (PEAQS). In *28th coordinated research council (CRC) real world emissions workshop, California, UK, 18 March 2018*. https://www.nctcog.org/nctcg/media/Transportation/Present/JeremySmith_CARBCRC_Presentation.pdf. (Accessed August 10, 2019).

Stern, N., & Stern, N. H. (2007). *The economics of climate change: The stern review*. Cambridge University press.

Stewart, A., Hope-Morley, A., Mock, P., & Tietge, U. (2015). *Quantifying the impact of real-world driving on total CO2 emissions from UK cars and vans.* ElementEnergy and ICCT report. https://www.theccc.org.uk/publication/impact-of-real-world-driving-emissions/ (Accessed November 10, 2019).

Timmers, V. R., & Achten, P. A. (2016). Non-exhaust PM emissions from EVs. *Atmospheric Environment, 134*, 10–17.

USEPA (United States Environmental Protection Agency). (2002). *User's guide to MOBILE6.1 and MOBILE6.2: Mobile source emission factor model EPA420-R-02-028*. United States Environmental Protection Agency.

Vestreng, V., Ntziachristos, L., Semb, A., Reis, S., Isaksen, I. S., & Tarrasón, L. (2009). Evolution of NOx emissions in Europe with focus on road transport control measures. *Atmospheric Chemistry and Physics, 9*(4), 1503–1520.

Viaene, P., Belis, C. A., Blond, N., Bouland, C., Juda-Rezler, K., Karvosenoja, N., et al. (2016). Air quality integrated assessment modelling in the context of EU policy: A way forward. *Environmental Science & Policy, 65*, 22–28.

Weiss, M., Bonnel, P., Hummel, R., Provenza, A., & Manfredi, U. (2011). On-road emissions of light-duty vehicles in Europe. *Environmental Science & Technology, 45*(19), 8575–8581. [or associated JRC report Weiss, M., Bonnel, P., Hummel, R., Manfredi, U., Colombo, R., Lanappe, G., Le Lijour, P. and Sculati, M. (2011). Analyzing on-road emissions of light-duty vehicles with portable emission measurement systems (PEMS). JRC scientific and technical reports, EUR, 24697].

Weiss, M., Bonnel, P., Kühlwein, J., Provenza, A., Lambrecht, U., Alessandrini, S., et al. (2012). Will Euro 6 reduce the NOx emissions of new diesel cars?—Insights from on-road tests with portable emissions measurement systems (PEMS). *Atmospheric Environment, 62*, 657–665.

WHO (World Health Organization). (2018a). *Burden of disease from the joint effects of household and ambient air pollution for 2018*. https://www.who.int/airpollution/data/cities/en/ (Accessed August 10, 2019).

WHO (World Health Organization). (2018b). *Ambient (outdoor) air quality and health: Key facts*. https://www.who.int/en/news-room/fact-sheets/detail/ambient-(outdoor)-air-quality-and-health. (Accessed August 10, 2019).

Wong, J. Y., & Anderson, R. L. (2012). *Non-dispersive infrared gas measurement.* USA: International Frequency Sensor Association Publishing.

Wu, L., Chang, M., Wang, X., Hang, J., & Zhang, J. (2019). Development of a real-time on-road emission (ROE v1.0) model for street-scale air quality modeling based on dynamic traffic big data. *Geoscientific Model Development Discussion*, https://doi.org/10.5194/gmd-2019-74. in review.

Wu, Y., Zhang, S. J., Li, M. L., Ge, Y. S., Shu, J. W., Zhou, Y., et al. (2012). The challenge to NOx emission control for heavy-duty diesel vehicles in China. *Atmospheric Chemistry and Physics, 12*(19), 9365–9379.

Yudego, E. A., Candás, J. L., Álvarez, E.Á., López, M. J., García, L., & Fernández-Pacheco, V. M. (2018). Computational tools for analysing air pollutants dispersion: A comparative review. Multidisciplinary Digital Publishing Institute. *Proceedings, 2*(23), 1408–1412.

Zallinger, M., Tate, J., & Hausberger, S. (2008). *An instantaneous emission model for the passenger car fleet.* In: *16th international transport and air pollution Congress Technical University Graz.*

Zavala, M., Herndon, S. C., Slott, R. S., Dunlea, E. J., Marr, L. C., Shorter, J. H., et al. (2006). Characterization of on-road vehicle emissions in the Mexico City metropolitan area using a mobile laboratory in chase and fleet average measurement modes during the MCMA-2003 field campaign. *Atmospheric Chemistry and Physics, 6*(12), 5129–5142.

Zhang, Y., Stedman, D. H., Bishop, G. A., Guenther, P. L., & Beaton, S. P. (1995). Worldwide on-road vehicle exhaust emissions study by remote sensing. *Environmental Science & Technology, 29*(9), 2286–2294.

CHAPTER 5

Air pollution monitoring and modeling

Mohammad Hashem Askariyeh[a], Haneen Khreis[b], and Suriya Vallamsundar[b]
[a]Air Quality Scientist, Zachry Department of Civil and Environmental Engineering, Texas A&M University, College Station, TX, United States
[b]Associate Research Scientist, Texas A&M Transportation Institute, College Station, TX, United States

Monitoring

Introduction

Numerous studies have shown associations between exposure to traffic-related air pollution (TRAP) and a variety of environmental (both localized and regional impacts) and adverse health effects (short- and long-term effects) (Alotaibi et al., 2019; Bowatte et al., 2017; HEI, 2010; WHO, 2005; Zamora et al., 2018; Zhang & Batterman, 2013). To assess TRAP, and its environmental and health impacts, TRAP levels need to be determined. Levels of TRAP are either directly measured or when measurements are unfeasible or unavailable, levels are typically modeled using a variety of methods. A key factor to consider with respect to TRAP is the spatial resolution or the variation in concentration levels with distance from the roadway edge. This variation is high and depends on many factors such as the pollutant, time-of-day, season, and meteorology (Askariyeh, Vallamsundar, & Farzaneh, 2018; Batterman, Cook, & Justin, 2015). Motor vehicle pollutants tend to peak within a short distance from the source and rapidly decline with increasing distance from the source. The peaking effect is observed within a distance of 500 ft. (approximately 150 m) before the levels decrease to background concentrations at about 2000 ft. (approximately 600 m) from the roadway (Askariyeh et al., 2018; Askariyeh, Zietsman, & Autenrieth, 2020; Khan, Ketzel, Kakosimos, Sørensen, & Jensen, 2018).

TRAP is ubiquitous and thus a concerning exposure for human health. In the United States, for example, more than 40 million people reside within a few hundred feet from a major roadway (U.S. Department of Housing and Urban Development, 2016), which may be an indicator of a much larger population exposed to TRAP throughout the world. Vehicular emission sources are also estimated to contribute up to 80%, 66%, and 53% of nitrogen dioxide (NO_2), $PM_{2.5}$, and PM_{10}, respectively, in Europe (EEA, 2012).

Traffic-Related Air Pollution
https://doi.org/10.1016/B978-0-12-818122-5.00005-3

Studies have shown that populations residing close to roads are more likely to experience adverse health effects such as respiratory and cardiovascular diseases compared to people who live 500 m or further away. For example, researchers from Copenhagen Municipality studied transportation-related air pollution and showed how the reduction of urban areas' NO_2 concentrations to a rural level increases life expectancy (Brønnum-Hansen et al., 2018). A study on a group of children and their parents in Cincinnati demonstrated a significant association between childhood exposure to traffic-related elemental carbon and child-reported depression and anxiety (Yolton et al., 2019). A recent study on pregnant women, as a susceptible population, revealed an association between exposure to TRAP and blood pressure (Liang et al., 2019). Different studies around the world have shown a consistent association between childhood exposure to TRAP and risk of asthma development (Khreis et al., 2017). Other adverse health effects of exposure to TRAP include but are not limited to childhood obesity (Bloemsma et al., 2019), cardiac anomalies (Girguis et al., 2016), lung function decrements, and respiratory health problems (Cakmak, Hebbern, Cakmak, & Vanos, 2016), and these have been discussed elsewhere in this book (Chapters 7–9).

Traffic-related air pollutants

Transportation sector's activity results in the emission of different air pollutants including exhaust and nonexhaust pollutants, in addition to facilitating the formation of secondary pollutants in ambient air. For research and regulatory studies, TRAP may be classified into two broad classes of (a) regulated air pollutants and (b) other indicators (Baldauf et al., 2009).

Regulated air pollutants (criteria pollutants)

Among a variety of air pollutants, a limited number are often times selected and regulated as *criteria* pollutants to monitor and ensure standards are met. In this part, air pollutants that the World Health Organization (WHO) and the US National Ambient Air Standard (NAAQS) consider as the criteria pollutants are briefly discussed with a focus on the transportation sector.

Carbon monoxide (CO). Vehicles running on fossil fuels are major sources of CO in urban areas. Light-duty gasoline vehicles among vehicular emission sources significantly contribute to transportation-related CO (Askariyeh & Arhami, 2013). The importance of CO from a monitoring point of view is in its adverse health outcomes (Thom, 2011) and its nonreactivity in the near-road environment. Control measures have been successful in the reduction of transportation-related CO emissions in the recent decades (Eisinger, Dougherty, Chang, Kear, and Morgan, 2011), and other pollutants became more specific markers of the TRAP mixture.

Oxides of nitrogen (NO_x). Fossil fuel combustion produces NO_x in all vehicle engines, but diesel vehicles are major contributors to overall NO_x emissions. Vehicles emit NO_x in both forms of nitrogen oxide (NO) and nitrogen dioxide (NO_2). NO forms the majority of primary emitted NO_x; however, emitted NO reacts with ozone (O_3) and produces secondary NO_2. With respect to health effects, NO_x is associated with short- and long-term adverse health effects (Latza, Gerdes, Fau-Baur, and Baur, 2009).

Sulfur dioxide (SO_2). Vehicles emit SO_2, and other compounds containing sulfur, but they are not significant contributors to ambient air sulfur pollution (Baldauf et al., 2009). The main reason for the small contribution is the low sulfur content of fuels used to power vehicles and the diesel vehicles' after-treatment devices. Unlike road mobile sources, SO_2 emissions are higher from marine vessels, locomotives, and air crafts due to the high sulfur content in the fuel used to power these modes.

Ozone (O_3). O_3 is a highly reactive secondary gas formed as a result of chemical reactions among transportation-related pollutants, in the presence of sunlight. Tropospheric (ground-level) O_3 is primarily the result of a complex set of photochemical reactions, in the presence of sunlight, that involve volatile organic compounds (VOCs) and NO_x. These pollutants are referred to as the ozone precursors. High ozone concentration episodes generally occur during summer time. The transportation sector contributes to ambient O_3 concentrations since it emits both NO_x and VOCs in urban areas. However, a lower level of O_3 concentration in the proximity of the transportation network is observed. The main reason for this is the consumption of O_3 in the presence of an elevated level of NO, which is expected in the near-road microenvironment, and NO_2 formation reaction. As such, O_3 studies involve many other variables including but not limited to the presence of other air pollutants in target microenvironment (Baldauf et al., 2009; Vallero, 2014).

Lead (Pb). Transportation contributes to lead levels due to vehicle fuel and lubricating oils. Over the last decades, lead emissions from tailpipe exhaust have significantly reduced due to the introduction of unleaded fuel. Lead can also be emitted from brake wear, tire wear, and resuspension of deposited lead on the road surface. Different modes of transportation other than road transportation can emit a significant amount of lead depending on their fuel type (Masiol & Harrison, 2014). Currently, aircrafts are the largest source of lead pollution in the United States (Carr et al., 2011).

Particulate matter (PM). Vehicular activity results in the emission, formation, and dispersion of PM through fuel combustion, brake wear, tire wear, resuspension of deposited dust, and formation of secondary pollutants. Particular matters are generally classified based on their size such as PM_{10}, $PM_{2.5}$, PM_1, and $PM_{0.1}$, which are defined as particles with an aerodynamic diameter smaller than 10, 2.5, 1, and 0.1 μm, respectively. The tailpipe exhaust contribution to the near-road PM mass concentration is not significant since combustion mechanisms mainly result in ultrafine PM ($PM_{0.1}$), but its contribution to the near-road PM number concentration is significant (Baldauf et al., 2009). Brake wear, tire wear, and resuspended dust contribute significantly to the near-road traffic-related coarser particles (generally PM_{10}) and increase the near-road PM mass concentration. Traffic-related PM can be involved in complex photochemical reactions and form secondary aerosols, which depend on ambient air conditions, the presence of other pollutants, and residence time.

Other traffic-related air pollution indicators

Transportation sources are associated with the emission of some hazardous air pollutants (air toxics), which are not included in the *criteria* pollutant list. The most important group of these pollutants is VOCs. VOCs are defined as "any compound of carbon, excluding carbon monoxide, carbon dioxide, carbonic acid, metallic carbides or carbonates, and ammonium carbonate, which participates in atmospheric photochemical reactions" (Code of Federal Regulations (CFR), 2008). VOCs are classified into three groups of very volatile, volatile, and semivolatile compounds. VOCs are associated with adverse health effects (WHO, 2005) and also contribute to the formation of photochemical smog (Zeman, 2012). The different compounds of VOCs released from vehicular sources in the near-road environment are benzene, ethylbenzene, toluene, xylene, and styrene. Aldehydes (e.g., formaldehyde and acetaldehyde) are also among unregulated traffic-related air pollutions with adverse health effects (Baldauf et al., 2009).

There are some surrogates that can be considered and measured as indicators of TRAP. Fossil fuel combustion results in ultrafine particles, which do not have significant mass but form an extremely high number of particles (by count). As such, particle number can be used as a marker for transportation sector activity in the near-road environment. Combustion of carbonaceous material also results in black carbon (BC) formation. Mobile emission sources contribute significantly to BC emissions (Briggs & Long, 2016), and as such, BC is also considered as a good marker for TRAP in the near-road environment. PM number (count) concentration and BC have been used as the indicators of TRAP in numerous TRAP and health research studies (Lee et al., 2018; Lee, Stenstrom, & Zhu, 2015; Li, Lee, Zhou, Liu, & Zhu, 2017; Ranasinghe et al., 2019; Yu, Shu, Lin, & Zhu, 2018). The length traveled by vehicle fleet (vehicle miles traveled (VMT) or vehicle kilometers traveled (VKT)) is another common indicator of traffic activity, which can be used to evaluate different traffic scenarios in transportation engineering (Renne & Tolford, 2018; Rentziou, Gkritza, & Souleyrette, 2012; Zhang, Guhathakurta, & Khalil, 2018) or be used as a surrogate for exposure. Greenhouse gases (GHGs) and carbon dioxide equivalent (CO_2e) are two main and general TRAP indicators that are widely being used to evaluate high-level policies and also TRAP due to different transportation scenarios in urban areas (Figliozzi, Saenz, & Faulin, 2020; Kissinger & Reznik, 2019; Niedertscheider, Haas, & Görg, 2018).

Air pollution measurement

The main objective of ambient air monitoring is to obtain an understanding of air pollutant concentrations at a particular time and location and their changes overtime or location. Measuring air pollutant concentrations in target areas at all times is not possible. As such, measurement of contaminants captured in a limited volume of ambient air which can be collected in a canister or a bag (sample) is a widely used strategy to monitor air pollutant concentrations. Measured air pollutants in a limited time and at a particular location are then usually assigned to a much larger area and used for exposure assessment. Air pollution measurements can be conducted either directly or indirectly. Measurement of the concentrations of particles and gases are examples of direct measurements. Light scattering as an indicator of particles number is an example of an indirect measurement. There are other ways of ambient air pollution measurements like continuous ambient air monitoring and periodic sampling with portable devices. Ambient air pollutants can be measured in the gas or vapor phase or in form of particles (Vallero, 2014). Based on the type of pollutant, air pollution monitoring methods are generally classified into two main categories of gases and vapors monitoring and particulate monitoring, which are discussed in the following sections.

Gases and vapors measurement

Most air pollution sampling systems are comprised of a combination of devices to move a specific volume of ambient air to a collection medium without any changes over a specific time period. The medium can be a solid sorbent, liquid, an evacuated flask, or a cryogenic trap (Vallero, 2014). For gaseous pollutants, sampling bubbler systems can be used to hydrate or sorb target species. Then, the sealed tube, that contains air sample, will be analyzed in the laboratory using methods like liquid or gas chromatography to measure target gases in the sample. Considering chemical reactivity of gases, Tedlar bags, or stainless-steel canisters coated with inert substances are generally used to keep air pollutants before gas chromatography (GC) or mass spectrometry (MS).

The in situ continuous sampling of gaseous pollutants is becoming more prevalent. This sampling consists of collecting the ambient air flow from the monitoring equipment, which is kept in a shelter. Then, a small fraction of the inlet flow is extracted automatically for analysis. This method uses real-time GC and MS to measure a variety of toxic air pollutants and VOCs. This procedure is also being used for continuous emission monitoring (CEM) for regulatory purposes like compliance assurance monitoring (CAM). There are some open-path air pollution monitoring techniques based on advanced technologies, including the tunable diode laser (TDL), differential absorption lidar (DIAL), and Fourier transform infrared (FTIR).

TDL is a spectroscopy technique used for measuring the concentration of pollutants in the gaseous state such as water vapor or methane. In addition to concentration, the technique can also measure other parameters related to the temperature, humidity, pressure, and velocity of the gases. The setup for the technique consists of a tunable diode laser light source, transmitting and receiving optics, detectors, and an optically accessible medium. Advantages of the technique include very low detection limits, stable calibration, continuous in situ measurements, and reduced interference from other gases (Vallero, 2014).

DIAL uses light detection and ranging (LIDAR) to measure a variety of gaseous pollutants in the atmosphere such as O_3 and CH_4 (North, Pyle, & Zhang, 2015). In this technique, the difference between the energy of pulsed laser radiation and backscattered radiation (absorbed by the gas, the pollutant of concern) is used to measure the gas concentration. This technique can be used to provide a three-dimensional (3D) profile of air pollutant concentrations measured in urban areas. The distance for gases measurement is variable, and ranges between 1 and 3 km (North et al., 2015; Vallero, 2014).

The measurement techniques which use infrared are generally based on the relationship between energy and gas molecules. The gas molecular vibrations determine how absorbing light promotes the molecule's energy state (Marshall, Chaffin, Hammaker, & Fateley, 1994). Many different gaseous pollutants have distinct energy absorbance (Schütze et al., 2013). The main advantage of FTIR techniques is its capability of simultaneous measurement of many air pollutants (Vallero, 2014).

Particulate matter measurement

Measurement of particulate matter (PM) with different aerodynamic diameters is generally performed by separating the particles from a known volume of air (sample) using filtration. The filtration is conducted through four mechanisms depending on the particle size (a) diffusion which is based on the particle's random and Brownian motion, and cause filtration of extremely small particles with diameter smaller than 0.1 μm, (b) interception which cause filtration of particles with diameter between 0.1 and 1 μm due to their contact with filter medium while they are in the airstream, (c) inertial impaction which cause filtration of particles with diameter greater than 1 μm due to inertia, and (d) electrostatics as the effect of electrical interaction between particles and filter surface which can capture a wide range of sizes. Particles mass is measured as the difference between filter weight before and after filtration (Chapman, 2010). Filters' material, quality, pore size, and strand diameter vary widely for different purposes.

The sampling conducted using filtration is done over a period of time and eventually provides the average concentration (e.g., 24-h average) and as such cannot capture hourly variations (unless the filter is changed every hour). As an improvement over passive sampling, optical methods are used to perform continuous PM measurement. Therefore, beta attenuation technology and tapered element oscillating microbalance (TEOM) technology are used. In beta attenuation technology, the absorbed beta rays which pass through the collected particles in a fibrous filter tap are used to predict particles mass. The rational of this method lies in the fact that the mass of PM in the sample should be proportional to the difference between the beta count after sampling and the baseline beta count (Macias & Husar, 1976). The TEOM monitoring technology utilizes a gravimetric instrument and an exchangeable filter cartridge. This technology gets ambient air through the filter mounted onto the end of a tapered tube. The tapered tube and its filter maintain an oscillation, which its frequency varies by the mass of particles in inlet flow. The near-real-time concentration of particles can be measured by tracking the oscillation frequency (Patashnick & Rupprecht, 1991; Schwab, Felton, Rattigan, & Demerjian, 2006). There are two concerns about continuous PM monitoring using TEOM technology. First, these devices generally cannot capture extremely small particles (diameters smaller than 0.1 μm), hence they miss a portion of particulate matter. Second, TEOM technology may not be able to capture all semivolatile organic compounds in the inlet flow (Northam, 2017; Tortajada-Genaro & Borrás, 2011).

Monitoring in tracer studies

Transportation activity causes the emission of numerous air pollutants. These same air pollutants also exist in the ambient air (background concentration) as the result of activities of sectors other than transportation. As such, air pollutants measured in ambient air, even in the proximity of vehicular emission sources (e.g., near-roadways), represent the result of emission from different sources including the transportation sector. Moreover, the air pollutants emitted from traffic involve in different chemical reactions and can be consumed or produced (secondary pollutants) as the results of these reactions.

Tracer studies are conducted in order to focus on the dispersion of TRAP particularly. In tracer studies, a known amount of an inert gas (e.g., SF6: sulfur hexafluoride) is emitted from vehicles and its concentration is measured at different distances from the road. Considering the baseline zero concentration of the tracer gas in ambient air and its nonreactivity, monitored concentrations solely represent what is directly emitted from vehicles during the experiment. Hence, obtained results (monitored tracer gas) well illustrate the effect of the dispersion process on mass transport from vehicular emission sources to the surrounding area. A comprehensive tracer study with a focus on TRAP was performed by General Motors (GM) in 1975 (Cadle, Chock, Monson, & Heuss, 1977). In this study, eight trucks emitted SF6 at a known rate among 352 light-duty vehicles driving at a constant speed of 80 km/h at GM Proving Ground. The SF6 concentrations were measured at 30-min time periods using gas chromatography (Cadle et al., 1977). This study has been the most comprehensive tracer study with a focus on traffic-related air pollution since the tracer gas was measured at different heights from ground level and different distances from the roadway at both upwind and downwind. SF6 was also used for "Caltrans Highway 99" (Benson, 1989) and "Idaho Falls" (Finn et al., 2010) tracer studies to focus on TRAP dispersion. The results obtained from tracer studies have been widely used to discuss the effect of meteorological variables (e.g., wind speed and direction) and vehicle-induced turbulence (VIT) on near-road traffic-related air pollution (Askariyeh, Kota, Vallamsundar, Zietsman, & Ying, 2017).

Modeling

Introduction

Ambient monitoring stations are commonly established to measure regional pollutant concentration levels and also cannot trace concentrations back to their sources. Their locations are often based on regulatory purposes, rather than exposure or human health considerations. The spatial coverage of ambient monitoring networks is limited to several miles between locations. This spatial coverage would fail to capture the highly fluctuating spatial patterns of TRAP, which is highly heterogeneous in space, with air pollutant concentrations that are spatially misaligned. Indeed, TRAP significantly fluctuates even over no more than a few tens of meters (Briggs et al., 1997; Vardoulakis, Fisher, Pericleous, & Gonzalez-Flesca, 2003).

To better capture some of that variability and to extend coverage beyond the locations of ambient monitoring stations, air quality models are often used in exposure assessment and health effect studies. Air quality models represent mathematical methods and numerical techniques in order to simulate the dispersion and transport of pollutants in the atmosphere. These models also overcome the limitation in ambient monitoring by tracking concentrations back to the specific sources causing them and can provide highly resolved temporal and spatial variations in the concentrations. Moreover, air quality models are helpful for cases, where monitoring data are not available (e.g., evaluation of future scenarios) or where data collection is expensive. Air quality modeling utilizes a series of input data sets to represent different physical and chemical processes in order to predict the concentration of a particular air pollutant, in a specific time and location, and under certain assumptions (Seinfeld & Pandis, 2016; Thunis et al., 2016).

There are two main equivalent approaches in order to investigate a fluid motion, which have been extensively used to develop air quality models. These two approaches are Lagrangian and Eulerian (Leelőssy, Mona, Mészáros, Lagzi, & Havasi, 2016). In the Lagrangian approach, the emitted pollutant is followed along with its trajectory in the air. It is like when we follow the trajectory of a boat when moving down the river. As such, the mass conservation equation is governed to predict air pollutant concentrations at target times and locations (Leelossy et al., 2014). The analytical solution of the mass conservation equation to predict concentrations under steady-state conditions with the assumption of Gaussian distribution of air pollutants forms the basis of a series of *air quality dispersion models* extensively used for TRAP studies (Leelossy et al., 2014). These models are known for their relative simplicity and their limitations particularly in dealing with chemical processes. In the Eulerian approach, the focus is on a specific location in space as time passes. It is like when we focus on one point in a river and watch water flows. As such, air pollutant concentrations at target locations overtime are predicted for unsteady-state conditions (Barth et al., 2011). Through this way, air pollutant concentration changes due to mass transport and chemical reactions can be simulated in *atmospheric chemical transport models*. These models divide the target space under investigation into grid cells in both horizontal and vertical dimensions. Concentration changes in each grid cell are predicted as a function of physical and chemical processes as well as mass exchange with surrounding cells over the time. The mathematical expressions used for prediction of concentrations in each cell can be directly used to represent different chemical reactions (Seinfeld & Pandis, 2016). *Computational fluid dynamics* (CFD) *models* are capable of utilizing both Eulerian and Lagrangian approaches. Some sophisticated CFD models govern numerical methods in order to simulate air flow around the obstacles and predict air pollutant concentrations (Herring & Huq, 2018). There are also some mathematical techniques to use speciation monitoring data along with physical and chemical characteristics of emitted pollutants (from different emission sources) in order to estimate the contribution of each source to the monitored air pollutant. These mathematical and statistical methods provide the basis of *receptor models*.

There are different classifications for air quality models based on different methods, assumptions, and required input sets. This section provides a short introduction and a quick review of air quality models in four main categories of dispersion, atmospheric chemical transport, CFD, and receptor modeling as well as hybrid modeling, which can be used for transportation air quality, exposure, and health effects analysis.

Dispersion modeling

Dispersion modeling is based on a steady-state formulation of the dispersion problem with a focus on advection and diffusion (Leelossy et al., 2014). Gaussian air dispersion models represent the dispersion of pollutant in the vertical, horizontal, and lateral direction to follow a Gaussian distribution. The Gaussian dispersion models predict primary air pollutants, which are emitted to a homogenous body of the air for receptors located at a radius of a few tens of kilometers from the emission sources. The changes in air pollution concentrations due to chemical reactions cannot be taken into consideration using these models (Leelossy et al., 2014). The geometry of emission sources and target points for air pollution prediction (receptors) and emission rates as well as processed meteorological variables are required to conduct dispersion modeling and predict hourly concentrations. The Gaussian air dispersion modeling is widely used in urban-scale air quality and exposure modeling as well as project-level regulatory analysis and decision-making. The Atmospheric Dispersion Modeling System (ADMS) (CERC, 2016) and the American Meteorological Society/Environmental Protection Agency Regulatory Model (AERMOD) (U.S. EPA, 2019a) are two of well-known air dispersion models for research and regulatory purposes. There are numerous studies on the evaluation of traffic-related air pollution using air dispersion models (Askariyeh et al., 2018; Gibson, Kundu, & Satish, 2013; Heist et al., 2013; Khan et al., 2018; Misra, Roorda, & MacLean, 2013). One of the advantages of air dispersion models is their capability of predicting population's exposure when moving across different microenvironments. This capability has been utilized to predict dynamic exposure of population to TRAP using activity-based travel demand modeling (Tayarani & Rowangould, 2020) and population's specific location monitoring data (Askariyeh, Vallamsundar, Zietsman, & Ramani, 2019). These studies showed that TRAP exposure estimation increases by about 50% when taking population mobility into account (Askariyeh et al., 2019; Tayarani & Rowangould, 2020).

Atmospheric chemical transport modeling

Atmospheric chemical transport models predict air pollutant concentrations while considering the transformation and chemical reactions of the pollutants. These models simulate the pollutant dispersion by dividing the atmosphere into one-dimensional (1D), two-dimensional (2D), or 3D grid cells. The simplest definition of grid cells is the 1D one, which makes the entire target atmosphere as one box for modeling. It should be noted that the Lagrangian approach can be used to predict concentrations in a box model. However, Eulerian approach-based methods are widely used to predict concentrations in the 3D computational domain (grid cells) covering the effect of emission sources, meteorological variables, mixing of pollutants among layers, and chemical reactions. Typically, these grid cells are a few kilometers wide and their thickness increases with the distance above the ground (cells near the ground are typically 30m thick, which increases to a few kilometers with height above the ground).

The entire computational domain can vary from a few meters to thousands of kilometers to cover local, regional, and global air quality modeling. In order to consider uneven terrain (e.g., mountains), the terrain-following coordinate can be used instead of flat terrain while defining the 3D computational domain (Byun & Schere, 2006; Seinfeld & Pandis, 2016).

The advantage of chemical transport models is their capability of dealing with chemical reactions, which makes them complicated (Byun & Schere, 2006). These models have been used to predict air pollutants for grids with dimension of 4, 12, 36 km, and larger (Jiang & Yoo, 2018; Schaap et al., 2015). Atmospheric chemical transport modeling uses differential equations, which require initial conditions and boundary conditions. Initial conditions are concentrations of pollutant species involved in chemical reactions at the beginning of modeling in each grid cell. Since not enough measurements are generally available for all of the species in all of the grid cells, extrapolation of available measurements can be used to obtain the initial concentrations for all grid cells. Considering the effect of initial conditions on predicted concentrations, atmospheric simulations need to be run first for enough time to obtain a stable modeling system dominated by emission sources before getting to modeling the target time period. This time is called a "startup" period (Seinfeld & Pandis, 2016). Boundary conditions are pollutant species concentrations at both ends of each direction (three directions for 3D modeling) throughout the model simulation (Seinfeld & Pandis, 2016). Ideally, to avoid uncertainties in model estimates, the initial and boundary conditions should be determined from field observations. However, such estimates are not always available especially for analyses limited to a smaller domain and limited time period. Methods developed in the literature (Seinfeld, 1991) to minimize the impact of the initial and boundary conditions are related to (1) using the output from a larger-scale model simulation, (2) isolating simulation domain from other sources, and (3) using objective or interpolation techniques to ambient monitoring data.

Atmospheric chemical transport models are increasingly used for source apportionment of air pollutants and exposure assessments as well as policy evaluation. Community Multiscale Air Quality (CMAQ) (U.S. EPA, 2020), Comprehensive Air Quality Model with Extensions (CAMx) (CAMx, 2018), and Goddard Earth Observing System Atmospheric Chemistry (GEOS-Chem) model (GEOS-Chem, 2020) are some of the currently popular models for chemical transport modeling. These models have been used to predict TRAP and its associated adverse health effects across regions as broad as a continent. For example, Barrett et al. (2015) used chemical transport modeling to estimate premature mortality due to violation of standards and excessive NO_x emissions by a group of light-duty diesel vehicles in the

United States. In another study, a global chemical transport study revealed that $PM_{2.5}$ and ozone emission associated with the transportation sector resulted in "an estimated 7.8 million years of life lost" around the world in 2015 (Anenberg, Miller, Henze, Minjares, & Achakulwisut, 2019). This study and many others put emphasis on the global importance of TRAP adverse health effects (Chang et al., 2017; Kheirbek, Haney, Douglas, Ito, & Matte, 2016). The advantage of chemical transport models for TRAP studies is their capability of tracking the contribution of every subset of emission sources to every air pollutant (Anenberg et al., 2019; Barrett et al., 2015), which helps both researchers and policy makers better understand the influential parameters and define best mitigation efforts for emissions reduction (Chemel et al., 2014).

Computational fluid dynamics

The air quality models discussed in the previous sections predict air pollutant concentrations over open terrain, while in reality there are many structures and complexities including the near-road barriers and geometry of street canyons and buildings in urban areas (Di Sabatino, Buccolieri, Pulvirenti, & Britter, 2007). The complex geometry surrounding the roadways should be taken into consideration when predicting TRAP and in exposure assessment. The air dispersion models and chemical mass transport models cannot reflect the effect of these structures around the roadways on their own right. The CFD method overcomes this drawback by incorporating the impacts of surrounding obstacles on fluid flow by imposing constraints on the flow pattern (Herring & Huq, 2018; Schiffman, McLaughlin, Katul, & Nagle, 2005).

The CFD modeling, as a branch of fluid mechanics, governs numerical methods to solve partial differential equations representing conservation laws of mass, energy, and momentum, and can predict air flow qualitatively and quantitatively (Herring & Huq, 2018; Leelossy et al., 2014). The model predicts air pollutant concentrations considering the effect of obstacles and structures located around the emission source on air flow. The CFD modeling can be resource intensive, depending on the study area and the extent of analyses. However, some models incorporate CFD-like solutions (Pardyjak & Brown, 2003) to take the effect of surrounding structures into account when predicting indoor and outdoor TRAP, but perform much faster (Brown, Williams, Nelson, & Werley, 2016; Misra et al., 2013). CFD modeling can be conducted using researchers' developed modules like the Comprehensive Turbulent Aerosol Dynamics and Gas Chemistry (CTAG)

(Tong, Baldauf, Isakov, Deshmukh, & Zhang, 2016), commercial software like Ansys Fluent (Ansys, 2020), or open access sources like OpenFOAM (OpenCFD Ltd, 2020), QUIC (LANL, 2011), and SimScale (SimScale, 2020).

There are numerous studies with the application of CFD modeling in TRAP research including the evaluation of the effect of surrounding buildings (Boppana, Wise, Ooi, Zhmayev, & Poh, 2019), roadside barriers (Steffens, Wang, & Zhang, 2012; Tong et al., 2016), and roadway configuration (Steffens et al., 2014; Wang & Zhang, 2009) on near-road TRAP in urban areas (Neofytou et al., 2008). The CFD modeling is also being used for microenvironmental air pollution prediction and exposure assessment (Dong et al., 2017; Tong, Yang, Hopke, & Zhang, 2017). The CFD modeling is capable of predicting in-cabin exposure to TRAP (Li et al., 2017; Li, Lee, Liu, & Zhu, 2015). As such, in-cabin self-pollution of vehicles is proven, which varies by the location of tailpipe, exhaust flow rate, and vehicle speed (Li et al., 2015). In addition to directly modeling pollutant concentrations, the CFD modeling results can be used as inputs for dispersion modeling. The wind field simulated using CFD modeling is incorporated into dispersion modeling to predict traffic-related air pollutant levels in street canyons in an urban area (Fu et al., 2017). Obtained results show that the taller the building and the narrower the roadways are, the higher pedestrian exposure to TRAP is in urban areas (Fu et al., 2017).

Receptor modeling

The fundamental difference between previous air quality models and receptor models is the approach followed in estimating pollutant concentrations. The abovementioned air quality models follow a "bottom-up" approach, where they mainly use the source emission characteristics and meteorological variables as well as modeling domain specifications to estimate the source contributed pollutant concentrations. Source-specific concentrations are combined with the background concentrations representing other sources to get an estimate of the overall pollutant concentration levels. On the other hand, receptor models follow a "top-down" approach, where pollutant concentrations are collected from ambient monitors, along with the speciation information. The pollutant signatures are analyzed in terms of their chemical and physical characteristics to trace the pollutants back to their sources, and quantify the contribution from each source (Hopke, 2016; Park, Sullivan, Kang, Ying, & Spiegelman, 2020). Hence, they need complete information about source profiles and chemical speciation. Inputs required for receptor modeling vary depending on the model. Important inputs for receptor models include information about the source-specific emission factors, source emission profiles, and chemical speciation, fate and transport properties of the pollutants. Positive matrix factorization (PMF), chemical mass balance (CMB) models, and EPA UNMIX are some of the currently available models for receptor modeling (Hopke, 2016). Receptor modeling has been widely used in TRAP studies since 1965 (Hopke, 2016). For example, Zhang et al. used PMF and concluded that mobile sources have the highest contribution to polycyclic aromatic hydrocarbons (PAH) in Dresden, Germany (Zhang, Li, Zhang, Ding, & Hua, 2019). The study of temporal and spatial variation of traffic-related $PM_{2.5}$ using receptor modeling in two near-road sites in Toronto revealed significant hourly variations and contribution of motor vehicles to $PM_{2.5}$ (Jeong et al., 2019). This study showed that exhaust and nonexhaust TRAP can contribute to up to half of total $PM_{2.5}$ in the morning peak hours. Researchers used receptor modeling and determined that the contribution of motor vehicle exhaust was 25% of total nonmethane hydrocarbons (TNMHC) in El Paso, Texas (Park et al., 2020).

Hybrid modeling

As discussed, each air quality model has advantages and limitations. To overcome the shortcoming of each model, two or more air quality models may be combined together. This utilization of any combination of different models is called hybrid modeling. The classical example of coupling air quality models is using a Eulerian model (e.g., chemical transport models) to predict regional air pollutant concentrations along with a dispersion model and a Lagrangian model (e.g., dispersion models) to predict concentrations due to a local set of emission sources. The main advantage of the hybrid approach is to benefit from capabilities and overcome limitations of both models. In a recent study, Parvez and Wagstrom utilized R-Line and CAMx to predict near-road NO_x and $PM_{2.5}$ concentrations (Parvez & Wagstrom, 2019). In this study, they used dispersion modeling to predict local air pollutants at a fine resolution (40 m × 40 m) and chemical transport model for regional nonroad air pollutants at a coarse resolution (12 km × 12 km). Obtained results show improved prediction using hybrid model compared to what a regional model (CAMx) provides. Evaluation of different traffic scenarios using hybrid modeling with CMAQ and AERMOD puts emphasis on nonlinear relation between emission rates and TRAP. In this regard, a 20% and

30% reduction in vehicular emissions yield 4.3% and 6.6% in total predicted $PM_{2.5}$ concentrations (Lai, Ma, Chen, Hsiao, & Pan, 2019). Obtained results highlight the importance and necessity of comprehensive studies to support air quality control strategies rather than just focusing on emission reductions (Lai et al., 2019). In coupling air dispersion and chemical transport models to use in environmental health studies, attention must be paid to avoid double counting of transportation emissions due to inclusion of on-road emissions in both models (Isakov, Lobdell, Palma, Rosenbaum, & Ozkaynak, 2009). A comprehensive evaluation of air quality models shows that the hybrid modeling is vital to account for all physical and chemical mechanisms in the evaluation of the effect of near-road green infrastructure in reducing TRAP and related exposures (Tiwari et al., 2019). Fleet passing the roadways causes a turbulence in the air, which is called vehicle-induced turbulence (VIT) and needs to be addressed in TRAP modeling. A hybrid CFD and dispersion modeling shows a significant improvement in dispersion modeling performance (Sahlodin, Sotudeh-Gharebagh, & Zhu, 2007). Currently, improved versions of Gaussian air dispersion models utilize dispersion parameters to reflect the VIT effect on near-road TRAP modeling (U.S. EPA, 2019b).

Model to monitor comparisons

While air quality models have several advantages, there is an essential need to validate these models by comparing their estimates with real-world air quality observations. In order to compare predictions from a model with observational measurements, several statistical performance measures can be used. These measures generally indicate three qualities in predicted values compared to observations: accuracy, precision, and bias. Accuracy refers to how predicted values are close to observations quantified by error as the difference between predicted and observed values. Precision refers to how predicted results are reproducible and close to each other for comparable observations. Bias shows inaccuracy of predicted values in a systematic way (Steidl & Carriere, 2010). In other words, bias shows how a model may tend to shift predicted values in one direction from observations and it is also called the systematic error. Fig. 5.1 schematically illustrates the concept of accuracy, precision, and bias.

Table 5.1 lists some of the most commonly used performance measures in the literature for validation of model predictions (Belis et al., 2020; Bravo, Fuentes, Zhang, Burr, & Bell, 2012; Gibson et al., 2013; Heist et al., 2013).

In addition to the abovementioned statistical measures, visual evaluation of the model to monitor can be highly useful to evaluate the overall

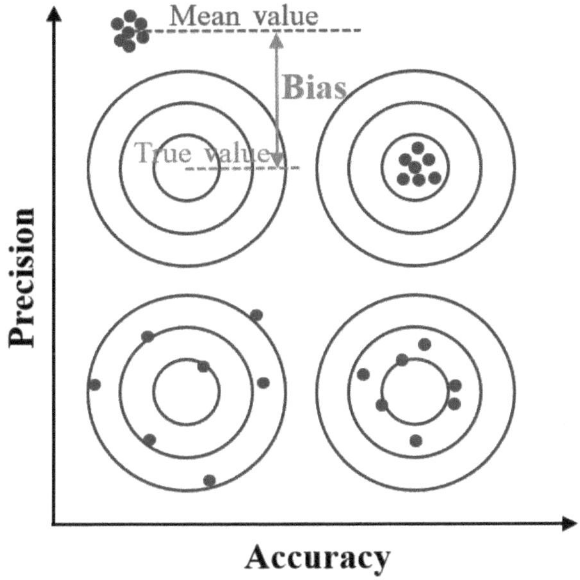

Fig. 5.1 Accuracy, precision, and bias (Steidl & Carriere, 2010; Stilling, 2009).

Table 5.1 Performance measures for model-to-monitor comparison (Askariyeh et al., 2017; Heist et al., 2013; Seinfeld & Pandis, 2016).

Performance measures	Equation/definition
Normalized mean square error (NMSE)	$\text{NMSE} = \dfrac{\overline{(C_P - C_O)^2}}{\overline{C_P C_O}}$
The linear Pearson correlation coefficient (R)	$R = \dfrac{\overline{(C_P - \overline{C_P})(C_O - \overline{C_O})}}{\sigma_{C_P}\sigma_{C_O}}$
Fractional bias (FB)	$\text{FB} = 2\dfrac{(\overline{C_P} - \overline{C_O})}{(\overline{C_P} + \overline{C_O})}$
Fraction of predictions within a factor of two of observations (FAC2)	$0.5 \le \dfrac{C_P}{C_O} \le 2.0$
The mean normalized bias (MNB)	$\text{MNB} = \overline{\left(\dfrac{C_P}{C_O} - 1\right)}$
The mean bias (MB)	$\text{MB} = \overline{(C_P - C_O)}$
The mean absolute normalized gross error (MANGE)	$\text{MANGE} = \overline{\left\lvert \dfrac{C_P}{C_O} - 1 \right\rvert}$
The mean error (ME)	$\text{ME} = \overline{\lvert C_P - C_O \rvert}$

C_P, predicted concentrations; C_O, observed concentrations; σ, standard deviation.

performance and identify any correlations between influential parameters like time, space, and air pollutants concentration (Askariyeh et al., 2017; Heist et al., 2013). Examples of visual evaluation include comparing the predicted and observed paired concentrations with factor of 2 and 1:1 line, residual plots and time series of modeled and measured concentrations, etc. As discussed earlier, air pollutants monitored in ambient air represent emissions of different sources including transportation sector. Hence, predicted TRAP, using air quality modeling, cannot be compared with ambient air pollutants directly, unless supplemented with background concentrations. One solution to overcome this obstacle is to compare predicted TRAP and observed tracer gases. Evaluation of TRAP modeling using tracer gas data is done using normalized mean square error (NMSE), the linear Pearson correlation coefficient (R), fractional bias (FB), and fraction of predictions within a factor of two of observations (FAC2) (Askariyeh et al., 2017; Heist et al., 2013). Another solution is to use atmospheric chemical transport models and predict air pollutant concentrations resulting from the emissions of all sources and then compare with ambient measurements. As such, four statistical measures of the mean bias (MB), the mean normalized bias (MNB), the mean absolute normalized gross error (MANGE), and the mean error (ME) are listed as the quantitative measures routinely used for purposes like validation of ozone prediction (Seinfeld & Pandis, 2016).

Summary and conclusion

The association between exposure to TRAP and a variety of environmental and adverse health effects has been investigated and established in numerous studies. TRAP levels and their temporal and spatial variations need to be determined in order to evaluate TRAP's associated environmental and adverse health effects. TRAP can be classified into two broad classes of regulated air pollutants and other indicators. CO, NO_x, SO_2, O_3, Pb, and PM are regulated air pollutants; however, TRAP now has a lower level of contribution to CO, SO_2, and Pb due to technology advancements and tighter environmental standards in the recent decades. There are also some other pollutants like air toxics and VOCs as well as surrogates like particle numbers, and BC which are being used in transportation air quality analysis. To better capture temporal and spatial variability of TRAP and to extend coverage beyond the locations of ambient monitoring stations, air quality models are often used in exposure assessment and health effects analyses. Moreover, air quality models are helpful for cases that monitoring data are not available (e.g., evaluation of future scenarios) or data collection is expensive. The air quality dispersion models are known for their simplicity and are widely being used for TRAP assessment in urban areas. The atmospheric chemical transport models benefit from high-performance computing systems to simulate complex chemical reactions over large domains as broad as one or more continent. To simulate air flow and TRAP dispersion in the presence of specific structures (e.g., near-road sound barrier walls, roadside buildings, and bridges, etc.), computational fluid dynamics models are being utilized. The availability of monitoring for a wide variety of air pollutants and different species allows utilizing mathematical techniques and statistical analyses to determine contribution of different emission sources to air pollution levels with a top-down approach using receptor modeling. In the recent years, air quality modeling using a combination of two or more different models, which is known as hybrid modeling, has become more prevalent in order to capture the advantages of both models in one study. There is an essential need to validate models by comparing their estimates with real-world air quality observations, along a range of metrics. There are several statistical performance measures, which can be used to compare predictions from a model to observational measurements. These measures are generally defined based on three qualities of accuracy, precision, and bias. Collection of more accurate data with a higher level of temporal and spatial variation of TRAP, making progress in modeling TRAP with a higher level of detail, exploring the combination of different models in addition to models and measurements, as well as evaluating populations' exposures, accounting for mobility patterns, are all ongoing efforts to obtain a better understanding of TRAP and its environmental and health.

References

Alotaibi, R., Bechle, M., Marshall, J. D., Ramani, T., Zietsman, J., Nieuwenhuijsen, M. J., et al. (2019). Traffic related air pollution and the burden of childhood asthma in the contiguous United States in 2000 and 2010. *Environment International*, *127*, 858–867.

Anenberg, S. C., Miller, J., Henze, D. K., Minjares, R., & Achakulwisut, P. (2019). The global burden of transportation tailpipe emissions on air pollution-related mortality in 2010 and 2015. *Environmental Research Letters*, *14*(9), 094012.

Ansys. (2020). *Ansys fluent*. Accessed at: https://www.ansys.com/products/fluids/ansys-fluent.

Askariyeh, M. H., & Arhami, M. (2013). Projecting emission reductions from prospective mobile sources policies by road link-based modelling. *International Journal of Environment and Pollution*, *53*(1–2), 87–106.

Askariyeh, M. H., Kota, S. H., Vallamsundar, S., Zietsman, J., & Ying, Q. (2017). AERMOD for near-road pollutant dispersion: Evaluation of model performance with different emission source representations and low wind options. *Transportation Research Part D: Transport and Environment*, ***57***(Suppl. C), 392–402.

Askariyeh, M. H., Vallamsundar, S., & Farzaneh, R. (2018). Investigating the impact of meteorological conditions on near-road pollutant dispersion between daytime and nighttime periods. *Transportation Research Record*, *2672*(25), 99–110.

Askariyeh, M. H., Vallamsundar, S., Zietsman, J., & Ramani, T. (2019). Assessment of traffic-related air pollution: Case study of pregnant women in south texas. *International Journal of Environmental Research and Public Health*, *16*(13).

Askariyeh, M. H., Zietsman, J., & Autenrieth, R. (2020). Traffic contribution to PM2.5 increment in the near-road environment. *Atmospheric Environment*, 224.

Baldauf, R., Watkins, N., Heist, D., Bailey, C., Rowley, P., & Shores, R. (2009). Near-road air quality monitoring: Factors affecting network design and interpretation of data. *Air Quality, Atmosphere & Health*, *2*(1), 1–9.

Barrett, S. R. H., Speth, R. L., Eastham, S. D., Dedoussi, I. C., Ashok, A., Malina, R., et al. (2015). Impact of the Volkswagen emissions control defeat device on US public health. *Environmental Research Letters*, *10*(11), 114005.

Barth, T. J., Griebel, M., Keyes, D. E., Nieminen, R. M., Roose, D., & Schlick, T. (2011). *Lecture notes in computational science and engineering*. Springer.

Batterman, S., Cook, R., & Justin, T. (2015). Temporal variation of traffic on highways and the development of accurate temporal allocation factors for air pollution analyses. *Atmospheric Environment*, *107*, 351–363.

Belis, C. A., Pernigotti, D., Pirovano, G., Favez, O., Jaffrezo, J. L., Kuenen, J., et al. (2020). Evaluation of receptor and chemical transport models for PM10 source apportionment. *Atmospheric Environment: X*, *5*, 100053.

Benson, P. E. (1989). *CALINE4—A dispersion model for predicting air pollutant concentrations near roadways*. FHWA/CA/TL/-84/15. Accessed at: https://www.weblakes.com/products/calroads/resources/docs/CALINE4.pdf.

Bloemsma, L. D., Wijga, A. H., Klompmaker, J. O., Janssen, N. A. H., Smit, H. A., Koppelman, G. H., et al. (2019). The associations of air pollution, traffic noise and green space with overweight throughout childhood: The PIAMA birth cohort study. *Environmental Research*, *169*, 348–356.

Boppana, V. B. L., Wise, D. J., Ooi, C. C., Zhmayev, E., & Poh, H. J. (2019). CFD assessment on particulate matter filters performance in urban areas. *Sustainable Cities and Society*, *46*, 101376.

Bowatte, G., Erbas, B., Lodge, C. J., Knibbs, L. D., Gurrin, L. C., Marks, G. B., et al. (2017). Traffic-related air pollution exposure over a 5-year period is associated with increased risk of asthma and poor lung function in middle age. *European Respiratory Journal*, *50*(4), 1602357.

Bravo, M. A., Fuentes, M., Zhang, Y., Burr, M. J., & Bell, M. L. (2012). Comparison of exposure estimation methods for air pollutants: Ambient monitoring data and regional air quality simulation. *Environmental Research*, *116*, 1–10.

Briggs, D. J., Collins, S., Elliott, P., Fischer, P., Kingham, S., Lebret, E., et al. (1997). Mapping urban air pollution using GIS: A regression-based approach. *International Journal of Geographical Information Science*, *11*(7), 699–718.

Briggs, N. L., & Long, C. M. (2016). Critical review of black carbon and elemental carbon source apportionment in Europe and the United States. *Atmospheric Environment*, *144*, 409–427.

Brønnum-Hansen, H., Bender, A. M., Andersen, Z. J., Sørensen, J., Bønløkke, J. H., Boshuizen, H., et al. (2018). Assessment of impact of traffic-related air pollution on morbidity and mortality in Copenhagen municipality and the health gain of reduced exposure. *Environment International*, *121*, 973–980.

Brown, M. J., Williams, M. D., Nelson, M. A., & Werley, K. A. (2016). QUIC transport and dispersion modeling of vehicle emissions in cities for better public health assessments. *Environmental health insights*, *9*(Suppl. 1), 55–65.

Byun, D., & Schere, K. L. (2006). Review of the governing equations, computational algorithms, and other components of the models-3 community multiscale air quality (CMAQ) modeling system. *Applied Mechanics Reviews*, *59*(2), 51–77.

Cadle, S. H., Chock, D. P., Monson, P. R., & Heuss, J. M. (1977). General motors sulfate dispersion experiment: Experimental procedures and results. *Journal of the Air Pollution Control Association*, *27*(1), 33–38.

Cakmak, S., Hebbern, C., Cakmak, J. D., & Vanos, J. (2016). The modifying effect of socioeconomic status on the relationship between traffic, air pollution and respiratory health in elementary schoolchildren. *Journal of Environmental Management*, *177*, 1–8.

CAMx. (2018). *Comprehensive air quality model with extensions.* Accessed at: http://www.camx.com/.

Carr, E., Lee, M., Marin, K., Holder, C., Hoyer, M., Pedde, M., et al. (2011). Development and evaluation of an air quality modeling approach to assess near-field impacts of lead emissions from piston-engine aircraft operating on leaded aviation gasoline. *Atmospheric Environment*, *45*(32), 5795–5804.

CERC. (2016). *Atmospheric dispersion modeling system (ADMS) 5 user guide: Atmospheric dispersion modeling system V5.2.* Accessed at: https://www.cerc.co.uk/environmental-software/assets/data/doc_userguides/CERC_ADMS_5_2_User_Guide.pdf.

Code of Federal Regulations (CFR). (2008). *Title 40: Chapter 1, Subchapter C, Part 51, Subpart F, 51100.* Accessed at: https://www.ecfr.gov/cgi-bin/text-idx?SID=387a1dbfd-3f26c488ff218f51bc0dc3f&mc=true&tpl=/ecfrbrowse/Title40/40CIsubchapC.tpl.

Chang, S. Y., Vizuete, W., Serre, M., Vennam, L. P., Omary, M., Isakov, V., et al. (2017). Finely resolved on-road PM2.5 and estimated premature mortality in Central North Carolina. *Risk Analysis*, *37*(12).

Chapman, R. A. (2010). *Applications of nonwovens in technical textiles* (p. 224). Woodhead.

Chemel, C., Fisher, B. E. A., Kong, X., Francis, X. V., Sokhi, R. S., Good, N., et al. (2014). Application of chemical transport model CMAQ to policy decisions regarding PM2.5 in the UK. *Atmospheric Environment*, *82*, 410–417.

Di Sabatino, S., Buccolieri, R., Pulvirenti, B., & Britter, R. (2007). Chapter 1.1 Application and validation of FLUENT flow and dispersion modelling within complex geometries. In C. Borrego, & E. Renner (Eds.), *Developments in environmental science* (pp. 3–11). Elsevier.

Dong, L., Zuo, H., Hu, L., Yang, B., Li, L., & Wu, L. (2017). Simulation of heavy gas dispersion in a large indoor space using CFD model. *Journal of Loss Prevention in the Process Industries*, *46*, 1–12.

EEA. (2012). *Road traffic's contribution to air quality in European cities.* The European Environment Agency. Accessed at: https://www.eionet.europa.eu/etcs/etc-atni/products/etc-atni-reports/etcacm_tp_2012_14_traffic_contribution_city_aq.

Eisinger, D., Dougherty, K., Chang, D., Kear, T., & Morgan, P. (2011). *A reevaluation of carbon monoxide: Past trends, future concentrations, and implications for conformity "hot-spot" policies* (pp. 1096–2247). (Print).

Figliozzi, M., Saenz, J., & Faulin, J. (2020). Minimization of urban freight distribution lifecycle CO2e emissions: Results from an optimization model and a real-world case study. *Transport Policy*, *86*, 60–68.

Finn, D., Clawson, K. L., Carter, R. G., Rich, J. D., Eckman, R. M., Perry, S. G., et al. (2010). Tracer studies to characterize the effects of roadside noise barriers on near-road pollutant dispersion under varying atmospheric stability conditions. *Atmospheric Environment*, *44*(2), 204–214.

Fu, X., Liu, J., Ban-Weiss, G. A., Zhang, J., Huang, X., Ouyang, B., et al. (2017). Effects of canyon geometry on the distribution of traffic-related air pollution in a large urban area: Implications of a multi-canyon air pollution dispersion model. *Atmospheric Environment*, *165*, 111–121.

GEOS-Chem. (2020). *Model info.* Accessed. at: http://acmg.seas.harvard.edu/geos/index.html.

Gibson, M. D., Kundu, S., & Satish, M. (2013). Dispersion model evaluation of PM2.5, NO_x and SO_2 from point and major line sources in Nova Scotia, Canada using AERMOD Gaussian plume air dispersion model. *Atmospheric Pollution Research*, *4*(2), 157–167.

Girguis, M. S., Strickland, M. J., Hu, X., Liu, Y., Bartell, S. M., & Vieira, V. M. (2016). Maternal exposure to traffic-related air pollution and birth defects in Massachusetts. *Environmental Research, 146*, 1–9.

HEI. (2010). *Traffic-related air pollution: A critical review of the literature on emissions, exposure, and health effects, in Special Report 17*. Boston, MA: Health Effects Institute Panel on the Health Effects of Traffic-Related Air Pollution. Accessed at: https://www.healtheffects.org/publication/traffic-related-air-pollution-critical-review-literature-emissions-exposure-and-health.

Heist, D., Isakov, V., Perry, S., Snyder, M., Venkatram, A., Hood, C., et al. (2013). Estimating near-road pollutant dispersion: A model inter-comparison. *Transportation Research Part D-Transport and Environment, 25*, 93–105.

Herring, S., & Huq, P. (2018). A review of methodology for evaluating the performance of atmospheric transport and dispersion models and suggested protocol for providing more informative results. *Fluids, 3*(1).

Hopke, P. K. (2016). Review of receptor modeling methods for source apportionment. *Journal of the Air & Waste Management Association, 66*(3), 237–259.

Isakov, V., Lobdell, D., Palma, T., Rosenbaum, A., & Ozkaynak, H. (2009). Combining regional- and local-scale air quality models with exposure models for use in environmental health studies. *Journal of the Air & Waste Management Association, 59*(4), 461–472.

Jeong, C.-H., Wang, J. M., Hilker, N., Debosz, J., Sofowote, U., Su, Y., et al. (2019). Temporal and spatial variability of traffic-related PM2.5 sources: Comparison of exhaust and non-exhaust emissions. *Atmospheric Environment, 198*, 55–69.

Jiang, X., & Yoo, E.-h. (2018). The importance of spatial resolutions of community multiscale air quality (CMAQ) models on health impact assessment. *Science of the Total Environment, 627*, 1528–1543.

Khan, J., Ketzel, M., Kakosimos, K., Sørensen, M., & Jensen, S. S. (2018). Road traffic air and noise pollution exposure assessment—A review of tools and techniques. *Science of the Total Environment, 634*, 661–676.

Kheirbek, I., Haney, J., Douglas, S., Ito, K., & Matte, T. (2016). The contribution of motor vehicle emissions to ambient fine particulate matter public health impacts in New York City: A health burden assessment. *Environmental Health, 15*(1), 89.

Khreis, H., Kelly, C., Tate, J., Parslow, R., Lucas, K., & Nieuwenhuijsen, M. (2017). *Exposure to traffic-related air pollution and risk of development of childhood asthma: A systematic review and meta-analysis*. (1873–6750 (Electronic)).

Kissinger, M., & Reznik, A. (2019). Detailed urban analysis of commute-related GHG emissions to guide urban mitigation measures. *Environmental Impact Assessment Review, 76*, 26–35.

Lai, H., Ma, H., Chen, C., Hsiao, M., & Pan, B. (2019). Design and application of a hybrid assessment of air quality models for the source apportionment of PM2.5. *Atmospheric Environment, 212*, 116–127.

LANL. (2011). *QUIC: Fast building-aware atmospheric dispersion modeling*. Accessed at: https://www.lanl.gov/projects/quic/.

Latza, U., Gerdes, S., & Bauer, X. (2009). Effects of nitrogen dioxide on human health: Systematic review of experimental and epidemiological studies conducted between 2002 and 2006. *International Journal of Hygiene and Environmental Health, 212*(3), 271–287. https://doi.org/10.1016/j.ijheh.2008.06.003.

Lee, E. S., Ranasinghe, D. R., Ahangar, F. E., Amini, S., Mara, S., Choi, W., et al. (2018). Field evaluation of vegetation and noise barriers for mitigation of near-freeway air pollution under variable wind conditions. *Atmospheric Environment, 175*, 92–99.

Lee, E. S., Stenstrom, M. K., & Zhu, Y. (2015). Ultrafine particle infiltration into passenger vehicles. Part I: Experimental evidence. *Transportation Research Part D: Transport and Environment, 38*, 156–165.

Leelossy, Á., Molnár, F., Izsák, F., Havasi, Á., Lagzi, I., & Mészáros, R. (2014). Dispersion modeling of air pollutants in the atmosphere: A review. *Central European Journal of Geosciences, 6*(3), 257–278.

Leelőssy, Á., Mona, T., Mészáros, R., Lagzi, I., & Havasi, Á. (2016). Eulerian and lagrangian approaches for modelling of air quality. In *Mathematical problems in meteorological modelling*. Cham: Springer International Publishing.

Li, F., Lee, E. S., Liu, J., & Zhu, Y. (2015). Predicting self-pollution inside school buses using a CFD and multi-zone coupled model. *Atmospheric Environment, 107*, 16–23.

Li, F., Lee, E. S., Zhou, B., Liu, J., & Zhu, Y. (2017). Effects of the window openings on the micro-environmental condition in a school bus. *Atmospheric Environment, 167*, 434–443.

Liang, D., Ladva, C. N., Golan, R., Yu, T., Walker, D. I., Sarnat, S. E., et al. (2019). Perturbations of the arginine metabolome following exposures to traffic-related air pollution in a panel of commuters with and without asthma. *Environment International, 127*, 503–513.

Macias, E. S., & Husar, R. B. (1976). Atmospheric particulate mass measurement with beta attenuation mass monitor. *Environmental Science & Technology, 10*(9), 904–907.

Marshall, T. L., Chaffin, C. T., Hammaker, R. M., & Fateley, W. G. (1994). An introduction to open-path FT-IR atmospheric monitoring. *Environmental Science & Technology, 28*(5), 224A–232A.

Masiol, M., & Harrison, R. M. (2014). Aircraft engine exhaust emissions and other airport-related contributions to ambient air pollution: A review. *Atmospheric Environment, 95*, 409–455.

Misra, A., Roorda, M. J., & MacLean, H. L. (2013). An integrated modelling approach to estimate urban traffic emissions. *Atmospheric Environment, 73*, 81–91.

Neofytou, P., Haakana, M., Venetsanos, A., Kousa, A., Bartzis, J., & Kukkonen, J. (2008). Computational fluid dynamics modelling of the pollution dispersion and comparison with measurements in a street canyon in Helsinki. *Environmental Modeling & Assessment, 13*(3), 439–448.

Niedertscheider, M., Haas, W., & Görg, C. (2018). Austrian climate policies and GHG-emissions since 1990: What is the role of climate policy integration? *Environmental Science & Policy, 81*, 10–17.

North, G. R., Pyle, J., & Zhang, F. (2015). *Encyclopedia of atmospheric sciences* (2nd ed., p. 2998). Elsevier.

Northam, A. (2017). *Development and evaluation of a model to correct tapered. element oscillating microbalance (TEOM) readings of PM2.5 in Chullora, Sydney*. School of Earth & Environmental Science, University of Wollongong. Accessed at: https://ro.uow.edu.au/cgi/viewcontent.cgi?article=1153&context=thsci.

OpenCFD Ltd. (2020). *OpenFOAM*. Accessed at: https://www.openfoam.com/.

Pardyjak, E. R., & Brown, M. J. (2003). *QUIC URB v1.1 theory and users guide*. Los Alamos, NM: Los Alamos National Laboratory. Accessed at: https://www.lanl.gov/projects/quic/open_files/QUICURB_UsersGuide.pdf.

Park, E. S., Sullivan, D. W., Kang, D. H., Ying, Q., & Spiegelman, C. H. (2020). Assessment of mobile source contributions in El Paso by PMF receptor modeling coupled with wind direction analysis. *Science of the Total Environment, 720*, 137527.

Parvez, F., & Wagstrom, K. (2019). A hybrid modeling framework to estimate pollutant concentrations and exposures in near road environments. *Science of the Total Environment, 663*, 144–153.

Patashnick, H., & Rupprecht, E. G. (1991). Continuous PM-10 measurements using the tapered element oscillating microbalance. *Journal of the Air & Waste Management Association, 41*(8), 1079–1083.

Ranasinghe, D., Lee, E. S., Zhu, Y., Frausto-Vicencio, I., Choi, W., Sun, W., et al. (2019). Effectiveness of vegetation and sound wall-vegetation combination barriers on pollution dispersion from freeways under early morning conditions. *Science of the Total Environment, 658*, 1549–1558.

Renne, J. L., & Tolford, T. (2018). A planning tool for evaluating vehicles miles travelled and traffic safety forecasts of growth management scenarios: A case study of Baton Rouge and New Orleans. *Transportation Research Part D: Transport and Environment, 59*, 237–245.

Rentziou, A., Gkritza, K., & Souleyrette, R. R. (2012). VMT, energy consumption, and GHG emissions forecasting for passenger transportation. *Transportation Research Part A: Policy and Practice*, *46*(3), 487–500.

Sahlodin, A. M., Sotudeh-Gharebagh, R., & Zhu, Y. (2007). Modeling of dispersion near roadways based on the vehicle-induced turbulence concept. *Atmospheric Environment*, *41*(1), 92–102.

Schaap, M., Cuvelier, C., Hendriks, C., Bessagnet, B., Baldasano, J. M., Colette, A., et al. (2015). Performance of European chemistry transport models as function of horizontal resolution. *Atmospheric Environment*, *112*, 90–105.

Schiffman, S. S., McLaughlin, B., Katul, G. G., & Nagle, H. T. (2005). Eulerian-Lagrangian model for predicting odor dispersion using instrumental and human measurements. *Sensors and Actuators B: Chemical*, *106*(1), 122–127.

Schütze, C., Lau, S., Reiche, N., Sauer, U., Borsdorf, H., & Dietrich, P. (2013). Ground-based remote sensing with open-path Fourier-transform infrared (OP-FTIR) spectroscopy for large-scale monitoring of greenhouse gases. *Energy Procedia*, *37*, 4276–4282.

Schwab, J. J., Felton, H. D., Rattigan, O. V., & Demerjian, K. L. (2006). New York state urban and rural measurements of continuous PM2.5 mass by FDMS, TEOM, and BAM. *Journal of the Air & Waste Management Association*, *56*(4), 372–383.

Seinfeld, J. (1991). *Rethinking the ozone problem in urban and regional air pollution. 524.* Washington, DC: National Academies Press.

Seinfeld, J., & Pandis, S. (2016). *Atmospheric chemistry and physics: From air pollution to climate change* (3rd ed.). New Jersey: John Wiley & Sons.

SimScale. (2020). *Simulatio software: SimScale.* Accessed at: https://www.simscale.com/.

Steffens, J. T., Heist, D. K., Perry, S. G., Isakov, V., Baldauf, R. W., & Zhang, K. M. (2014). Effects of roadway configurations on near-road air quality and the implications on roadway designs. *Atmospheric Environment*, *94*, 74–85.

Steffens, J. T., Wang, Y. J., & Zhang, K. M. (2012). Exploration of effects of a vegetation barrier on particle size distributions in a near-road environment. *Atmospheric Environment*, *50*, 120–128.

Steidl, B., & Carriere, Y. (2010). *Applied biostatistics*. Accessed at: https://cals.arizona.edu/classes/rnr613/.

Stilling, M. (2009). Polyethylene wear analysis. In *Health Sciences*. Denmark: University of Aarhus.

Tayarani, M., & Rowangould, G. (2020). Estimating exposure to fine particulate matter emissions from vehicle traffic: Exposure misclassification and daily activity patterns in a large, sprawling region. *Environmental Research*, *182*, 108999.

Thom, S. R. (2011). Carbon monoxide transport and actions in blood and tissues. *Comprehensive Physiology*, *1*(1), 421–446. https://doi.org/10.1002/cphy.c091005.

Thunis, P., Miranda, A., Baldasano, J. M., Blond, N., Douros, J., Graff, A., et al. (2016). Overview of current regional and local scale air quality modelling practices: Assessment and planning tools in the EU. *Environmental Science & Policy*, *65*, 13–21.

Tiwari, A., Kumar, P., Baldauf, R., Zhang, K. M., Pilla, F., Di Sabatino, S., et al. (2019). Considerations for evaluating green infrastructure impacts in microscale and macroscale air pollution dispersion models. *Science of the Total Environment*, *672*, 410–426.

Tong, Z. M., Baldauf, R. W., Isakov, V., Deshmukh, P., & Zhang, K. M. (2016). Roadside vegetation barrier designs to mitigate near-road air pollution impacts. *Science of the Total Environment*, *541*, 920–927.

Tong, Z., Yang, B., Hopke, P. K., & Zhang, K. M. (2017). Microenvironmental air quality impact of a commercial-scale biomass heating system. *Environmental Pollution*, *220*, 1112–1120.

Tortajada-Genaro, L.-A., & Borrás, E. (2011). Temperature effect of tapered element oscillating microbalance (TEOM) system measuring semi-volatile organic particulate matter. *Journal of Environmental Monitoring*, *13*(4), 1017–1026.

U.S. Department of Housing and Urban Development. (2016). *AHS 2013 national summary tables V1.2.* American Housing Survey (AHS). Accessed at: https://www.census.gov/programs-surveys/ahs/data/2013/ahs-2013-summary-tables/national-summary-report-and-tables---ahs-2013.html.

U.S. EPA. (2019a). *User's guide for the AMS/EPA regulatory model (AERMOD).* North Carolina: Office of Air Quality Planning and Standards: Research Triangle Park. EPA-454/B-19-027. Accessed at: https://www3.epa.gov/ttn/scram/models/aermod/aermod_userguide.pdf.

U.S. EPA. (2019b). *AERMOD model formulation and evaluation.* North Carolina: Office of Air Quality Planning and Standards, Air Quality Assessment Division: Research Triangle Park. Accessed at: https://www3.epa.gov/ttn/scram/models/aermod/aermod_mfed.pdf.

U.S. EPA. (2020). *CMAQ: The community multiscale air quality modeling system.* Accessed at: https://www.epa.gov/cmaq.

Vallero, D. (2014). *Fundamentals of air pollution* (5th ed.). Elsevier.

Vardoulakis, S., Fisher, B. E. A., Pericleous, K., & Gonzalez-Flesca, N. (2003). Modelling air quality in street canyons: A review. *Atmospheric Environment, 37*(2), 155–182.

Wang, Y. J., & Zhang, K. M. (2009). Modeling near-road air quality using a computational fluid dynamics Model, CFD-VIT-RIT. *Environmental Science & Technology, 43*(20), 7778–7783.

WHO. (2005). *Health effects of transport-related air pollution.* Copenhagen, Denmark. 205.

Yolton, K., Khoury, J. C., Burkle, J., LeMasters, G., Cecil, K., & Ryan, P. (2019). Lifetime exposure to traffic-related air pollution and symptoms of depression and anxiety at age 12 years. *Environmental Research, 173*, 199–206.

Yu, N., Shu, S., Lin, Y., & Zhu, Y. (2018). Assessing and reducing fine and ultrafine particles inside Los Angeles taxis. *Atmospheric Environment, 181*, 155–163.

Zamora, M. L., Pulczinski, J. C., Johnson, N., Garcia-Hernandez, R., Rule, A., Carrillo, G., et al. (2018). Maternal exposure to PM2.5 in south Texas, a pilot study. *Science of the Total Environment, 628–629*, 1497–1507.

Zeman, F. (2012). *Metropolitan sustainability: Understanding and improving the urban environment.* Woodhead.

Zhang, K., & Batterman, S. (2013). Air pollution and health risks due to vehicle traffic. *The Science of the Total Environment, 450–451*, 307–316.

Zhang, W., Guhathakurta, S., & Khalil, E. B. (2018). The impact of private autonomous vehicles on vehicle ownership and unoccupied VMT generation. *Transportation Research Part C: Emerging Technologies, 90*, 156–165.

Zhang, J., Li, R., Zhang, X., Ding, C., & Hua, P. (2019). Traffic contribution to polycyclic aromatic hydrocarbons in road dust: A source apportionment analysis under different antecedent dry-weather periods. *Science of the Total Environment, 658*, 996–1005.

CHAPTER 6

Traffic-related air pollution and exposure assessment

Sean D. Beevers[a,b] and Martin L. Williams[a,b]
[a]MRC-PHE Centre for Environment and Health—King's College London, London, United Kingdom
[b]Environmental Research Group—King's College London, London, United Kingdom

Abbreviations

AOD	aerosol optical depth
BC	black carbon
CAMx	comprehensive air quality model with extensions
cc	cubic centimeter
Chimere	a chemical transport model
CMAQ	community multiscale air quality modeling system
CO	carbon monoxide
CTM	chemical transport model
EGR	exhaust gas recirculation
EMEP	European Monitoring and Evaluation Programme (a Cooperative Programme under CLRTAP)
EU	European Union
Euro 1/6/I/VI	vehicle emissions standards
GBD	global burden of disease
HRAPIE	health risks of air pollution in Europe
ICCT	International Council on Clean Transport
LNT	low NO_x traps
Lotos-Euros	a chemical transport model
LUR	land use regression
$\mu g\,m^{-3}$	micrograms per cubic metre of air
NO_x	nitrogen oxides ($NO + NO_2$)
NO_2	nitrogen dioxide
OA	organic aerosol
O_3	ozone
PEMS	portable emissions measurement system
PM	particulate matter
PM_{10}	airborne particulate matter passing a sampling inlet with a 50% efficiency cut-off at 10-μm aerodynamic diameter and which transmits particles of below this size
$PM_{2.5}$	airborne particulate matter passing a sampling inlet with a 50% efficiency cutoff at 2.5-μm aerodynamic diameter and which transmits particles of below this size
P_{num}	particle number
POA	primary organic aerosol

Traffic-Related Air Pollution
https://doi.org/10.1016/B978-0-12-818122-5.00006-5

PSS	photo stationary state
RDE	real driving emissions
REVIHAAP	review of evidence on health aspects of air pollution
SCR	selective catalytic reduction
SHEDS	Stochastic Human Exposure and Dose Simulation
SOA	secondary organic aerosol
UFP	ultrafine particles
VOC	volatile organic compound (nonmethane)
WHO	World Health Organization
WLTP	Worldwide Harmonized Light-Vehicle Test Procedure

Introduction

There are numerous anthropogenic sources of air pollution from energy, manufacturing and production industries, nonindustrial combustion, such as heating and cooking, extraction and distribution of fossil fuels, use of solvents, nonroad mobile machinery, aircraft and ships, waste treatment and disposal, and agriculture, as well as biogenic sources, such as forestry, and natural sources such as windblown dust and sea salt. However, whilst in this chapter we discuss the air pollutants that exist as a consequence of these sources, we focus our attention on one of the most important, road transport. Later in the "Exposure Assessment Methods," we discuss the transition between the health evidence and exposure to air pollution, which was originally based on a small number of ambient city measurements, assuming each city population had the same exposure, to currently, the use of sophisticated modeling methods, satellite measurements, and small personal sensors.

Exposure definition Exposure, according to the Oxford English Dictionary, is "The state of having no protection from something harmful." In this chapter, exposure refers to the concentration of toxic gases and particles in the air we breathe, and where account is also taken of breathing rate, this is termed dose. In the context of air pollution impacts, exposure has often been defined as the time-weighted average over mean concentrations in the various microenvironments experienced by people. This averaging removes the detail of shorter time periods of high concentrations so that exposure is better defined as the time series of instantaneous or very short time period concentrations as an individual moves through a polluted space:

$$E = \left(C_1, C_2, C_3, \ldots, C_n\right)$$

where the $C_{1\ldots n}$ in practical applications are not instantaneous values but are the concentrations over the shortest averaging time available to the measurement method or the modeling system. This broader definition then allows any subsequent averaging to be carried out.

In air pollution research there are a number of pollutants with strong evidence of human health effects and that also have an important contribution from traffic sources. These include $PM_{2.5}$ mass concentrations, ozone (O_3), carbon monoxide (CO), nitrogen dioxide (NO_2) (the latter two being good indicators of city traffic emissions), and PM_{10} mass concentrations (which includes $PM_{2.5}$ plus other larger particles from abrasion sources, including metals). In health research, there is also interest in which component of PM is responsible for the health impacts of air pollution; particle nitrate, particle sulfate, primary and secondary organic aerosol (POA/SOA), black carbon (BC), and particle number (P_{num}) which is a measure of ultrafine particles with a diameter equal or less than 0.1 μm. Recently, the WHO, Review of evidence on health aspects of air pollution (REVIHAAP) report (World Health Organisation, 2013), reviewed the growing health effects evidence, concluding that: "a considerable amount of new scientific information on the adverse effects on the health of particulate matter, ozone, and nitrogen dioxide, observed at levels commonly present in Europe, has been published in recent years" as well as supporting the possibility of effects below the current WHO guidelines (World Health Organisation, 2005). Furthermore, the WHO, health risks of air pollution in Europe (HRAPIE) report (Henschel & Chan, 2013) looked at both sources and pollutants of most concern, concluding that road traffic, space heating and shipping were important sources and that fine and ultrafine particles and associated metals were of concern. Finally, both BC and P_{num} have a considerable contribution from diesel vehicle exhaust, which is considered to be carcinogenic (International Agency for Research on Cancer, 2012).

Environmental standards are established to avoid harmful effects on human health and the environment and reflect both the health evidence, and the political framework in which they exist (European Union Air Quality Standards (AQS), n.d.; United States Environmental Protection Agency, National Ambient Air Quality Standards (NAAQS), n.d.). The health effects, be they acute or chronic, determine the averaging time of the environmental standard and typically for O_3 concentrations are given as a maximum daily 8 hours average, for PM_{10} and $PM_{2.5}$ mass as both daily and annual averages, for NO_2 as hourly and annual averages and for CO as 8 hours and 1 hourly averages. Known carcinogens are given as annual averages. In setting standards, the health effects evidence is in virtually all cases based on fixed point monitored concentrations and these values are used as surrogates for personal exposure.

So, whilst the pollutants and the temporal scales over which air pollution should be considered are reasonably well defined, the spatial scale over which exposure should be considered is less clear. There are, for example, some details with regard to the spatial scale over which you need to comply with European Union (EU) limit values. In the case of annual average NO_2, the EU limit value is to be met throughout the territory of the EU Member State via monitoring and/or modeling at fixed point locations. There are attempts in the EU legislation to bridge the divide between fixed point concentrations and true personal exposures. There are areas exempt from compliance where no people have access, and where there is no habitation. Similarly, concentrations must be assessed "where the highest concentrations occur to which the population is likely to be directly or indirectly exposed for a period which is significant in relation to the averaging period of the limit value(s)" and "in other areas....which are representative of the exposure of the general population" (European Union Air Quality Standards (AQS), n.d.).

However, the most appropriate spatial scale for human exposure is still unclear and is to a large degree determined by the magnitude, granularity and release conditions of the emissions sources, the atmospheric lifetime of the pollutant species and the role played by atmospheric chemistry. For example, the within city spatial contrast of average air pollution in bounded by on the one hand: particle number concentrations close to traffic (and airports), which at roadsides are typically 10^5 per cubic centimeter (cc), but can fall by two orders of magnitude within tens of meters of the road, and on the other hand secondary particulate pollutants that are created from precursor emissions hundreds of kilometers upwind over time periods of many hours and more. The secondary pollutants include particle nitrate, sulfate and SOA and have spatial gradients measured in tens or even hundreds of kilometers rather than meters, and at a city scale can be considered to be spatially invariant. In between these two extremes are pollutants such as black carbon, POA, CO, NO_x, NO_2, O_3, and nonexhaust traffic PM.

In addition, it is important to consider that policies to tackle air quality are determined using cost-benefit analysis and so health damage costs are also an important consideration in air pollution exposure. Here too, the health damage costs are predominantly based on annual average concentrations ($PM_{2.5}$ and NO_2) and for O_3 the maximum daily 8 hour averages. When it comes to population exposure and health damage cost calculations, exposure estimates are often done over relatively large scales, for example for a local authority area, city, state, and country, since they rely on population datasets which may only be available at these scales.

In response, in recent outdoor air pollution and health research, there has been a direction of travel towards more finely resolved spatial and temporal estimates of exposure, as well as air pollution predictions from the city and country scales to continental and global scales. Current examples include; using dispersion models at city scales, with spatial predictions of approximately tens of meters and hourly time steps (Beevers et al., 2013); at continental scales, 1 × 1 km annual average concentrations of $PM_{2.5}$ mass and PM components, using combinations of satellite observations, ground-based observations and global models (Van Donkelaar, Martin, & Burnett, 2019), and global $PM_{2.5}$ concentrations (Van Donkelaar et al., 2016). Global data of this kind has been used in major studies such as the global burden of disease (GBD) (Cohen, Brauer, et al., 2017), placing air pollution as fifth ranked mortality risk factor in 2015. However, there is also recognition that since people spend the vast majority of their time indoors, that indoor environments and pollution sources also play an important role in people's exposure, with travel microenvironments, such as in the car, bus, train, and underground making important contributions to people's daily exposure (Smith et al., 2016).

Exposure pathways

As we will see in the coming sections, a dominant exposure pathway for traffic-related air pollution is living close to roads. In these locations, and despite modern homes becoming more airtight, there is the exchange of outdoor air pollutants into the indoor environment. This is important since, in cities like London, people spend greater than 90% of their time indoors, although note that this includes time at work, so may be in a very different environment. And since people spend considerable time indoors this means that the impact of indoor sources, like cooking, smoking, wood/coal/gas fires and cleaning products, becomes important for their overall exposure.

However, air pollution outdoors and indoors at home does not entirely describe people's day to day exposures either, and to improve our understanding of this, use has been made of more detailed exposure models which account for people's time activity, i.e., using better knowledge of where and when people spend their time, and the different outdoor and indoor microenvironments they are exposed to, as discussed in our basic definition of exposure above. Some models include estimates of what has been termed "passive" travel, in-vehicles (cars, buses, underground systems, and trains) and "active" travel (walking and cycling). Our own research (Smith et al., 2016) has found that for the population as a whole, passive travel is more important in determining people's overall exposure than active travel, since whilst both tend to be in high exposure environments, people generally spend more time in passive transport. This is true even when the dose is calculated, i.e., when the air pollution concentrations are multiplied by the breathing rate, which for cyclist exposures is important. But even cycling is not a major determinant of our total daily exposure because as a population in London we do not cycle for that long. This would not be the case everywhere however, an example being in the Netherlands.

As a consequence of personal activity being important in determining your exposure, population-based exposure estimates hide a large range of individual personal exposures, which often cross the boundary between ambient air pollution and workplace exposure, and associated legislation. The latter legislation generally permits much greater concentrations or has no limits at all and includes professional drivers, workers at airports and on metro systems, allowing exposure of large numbers of employees to high concentrations of air pollution and for significant periods of their lives. These higher exposure limits in occupational legislation are usually rationalized on the basis that working in these environments is voluntary and is paid, and that the workforce generally represents a healthier and more robust cohort than the general population.

Vehicle emissions

It would be impossible to write about exposure to traffic sources without mentioning vehicle emissions, especially given their importance as a source and the recent "VW scandal." The most important emissions from traffic exhausts are NO_2 (and NO_x), particle mass $PM_{2.5}$, PM_{10}, and individual PM components, CO and volatile organic compounds (VOCs). Emissions of these pollutants have declined in recent years due to increasingly stringent legislation requiring exhaust after treatment in particular three-way catalysts on gasoline vehicles, and on diesels, oxidation catalysts and diesel particulate filters and in more recent years NO_x control technologies such as exhaust gas recirculation (EGR), selective catalytic reduction (SCR), and low-NO_x traps (LNT).

These technologies, along with increasingly stringent emission regulations in Europe and the United States (European Parliament, 2016), are capable of delivering large reductions in emissions and while this has been the case for gasoline cars, where three-way catalysts have been shown to be very effective in real-world use, the same is not true for diesel cars (Carslaw, Beevers, Tate, Westmoreland, & Williams, 2011).

As is now well known, the drive cycle used in the EU type-approval laboratory dynamometer tests has been shown not to reflect real-world driving and associated emissions. Moreover, sophisticated engine management software in diesel vehicles can recognize when the diesel car is being tested in the laboratory and emission control techniques can be switched on, and then switched off when not being tested and in real-world use, usually to optimize fuel economy at the expense of NO_x emissions. The consequence has been that there has been virtually no improvement in NO_x emissions from diesel cars in Europe over the past 20 years or so, and probably not in the United States although there, the percentage of diesel cars in the fleet is much smaller than in Europe. The use of so-called "defeat devices" to manage NO_x emissions in real-world and in regulatory testing mode was exposed due to the efforts of the International Council on Clean Transport (ICCT) (Franco, Sánchez, German, & Mock, 2014). This has led to the situation where emissions in real-world driving can be very much larger than the regulatory emission limits so that any exposure assessment using the emission rates in the legislated limits will grossly underestimate the actual emissions of NO_x from diesel cars.

This situation is being rectified with the introduction both of the more aggressive Worldwide Harmonized Light Vehicle Test Procedure (WLTP) test cycle and a real-world emission test (real driving emissions—RDE) in EU legislation and a good summary is available at https://www.dieselnet.com/standards/eu/ld.php. The effectiveness of these new regulations in Europe is likely to improve the NO_x emission performance of Euro 6 diesel cars. The database of emission measurements from remote sensing (Carslaw et al., 2011) and Portable Emission Measurement Systems (PEMS) (European Commission, 2016) is increasing rapidly. At the time of writing, it appears that a simple classification of passenger cars as "Euro 6" or "Euro 6d" is too crude to provide accurate real-world emission estimates. In many cases, this crude categorization may be all that is available, but it should be recognized that there is now a wide range of NO_x emissions from both diesel and gasoline cars depending on the NO_x control technology employed and on the particular manufacturer and model. A recent study (O'Driscoll, Stettler, Molden, Oxley, & ApSimon, 2018) of Euro 5 and 6 passenger cars has shown a wide variation in NO_x emissions from diesel and gasoline passenger cars homologated to Euro 5 and Euro 6, where the best-performing diesel cars show lower NO_x emissions than the worst gasoline vehicles, although on average, NO_x emissions from gasoline cars were lower than those from diesel cars, and NO_x emissions from gasoline hybrid cars—albeit only a small number were tested—were much lower than either the gasoline or diesel vehicles.

As exhaust emissions have declined, nonexhaust emissions of particles have increased in importance to the extent that it is estimated that they now dominate over exhaust PM emissions in the national United Kingdom (UK) emission inventory (UK National Atmospheric Emissions Inventory (NAEI), n.d.) and comprise almost two-thirds of primary PM emission from road transport in the United Kingdom. These emissions arise through tire, brake and clutch wear, road surface wear, and resuspension of road dust. There is concern over these emissions because their particles contain various metals such as copper, iron, zinc which are potentially toxicologically active (Amato et al., 2014). These emissions could become more important in future as in many countries vehicle fleets are being electrified (Timmers & Achten, 2016) and exhaust emissions will reduce to very low or even zero levels, although new technology, such as regenerative braking, could reduce the impact on some of these emissions. There is still debate around whether to attribute road surface wear and resuspension to vehicles but given their role in both sources, it would seem sensible to do so. That said, emissions inventories do not count resuspension as a source for the purposes of compliance with emissions limits since it is considered to be double counting of the emissions, i.e., particulate material which has been emitted (and counted) in the inventories exists as silt on the road surface and is "emitted" many times into the atmosphere through the passage of vehicles.

Traffic-related air pollution

Air pollution in cities

Within cities there are important spatial variations of primary air pollutants arising from traffic sources such as NO_x, NO_2, ultrafine particles (UFP), BC, and POA and these are superimposed on top of continental scale secondary air pollutants described at the bottom of this section. In cities like London, we can demonstrate this in a number of ways. First, a plot of annual average NO_2 concentrations is given in Fig. 6.1. This plot is created using an emissions-dispersion model (Beevers et al., 2013), predicting NO_2 every 20 m across the city. What is clear from this map is the considerable contribution that roads make to the spatial variation in NO_2 concentrations, making the road network clearly visible across the city and with the only Heathrow airport, the large yellow area in the west of the city being a comparable source.

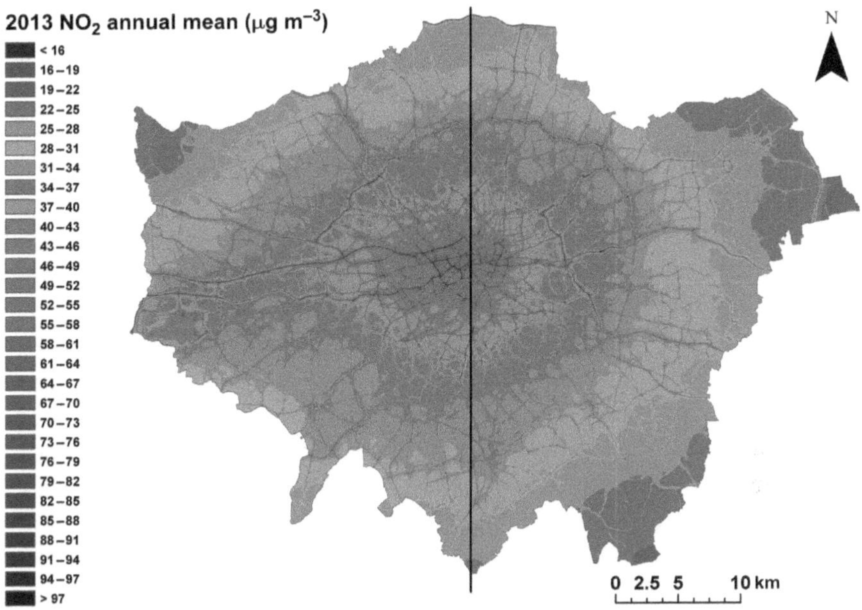

Fig. 6.1 A London annual average NO_2 map ($\mu g\,m^{-3}$).

By extracting the NO_2 concentrations every 20 m along the vertical black line running north to south across the middle of London, the bottom transect plot of Fig. 6.2 is created. The Fig. 6.2 transect plot can be interpreted in the following way; the horizontal line (NO_2_rural) in the plot is the small contribution from outside London and from outside the United Kingdom, the lighter blue line (light gray in print version) represents NO_2 concentration in London, including but not exclusively from road transport, and the peaks in concentration are close to individual roads. So, whilst at background locations (areas not affected by a single local source), concentrations vary

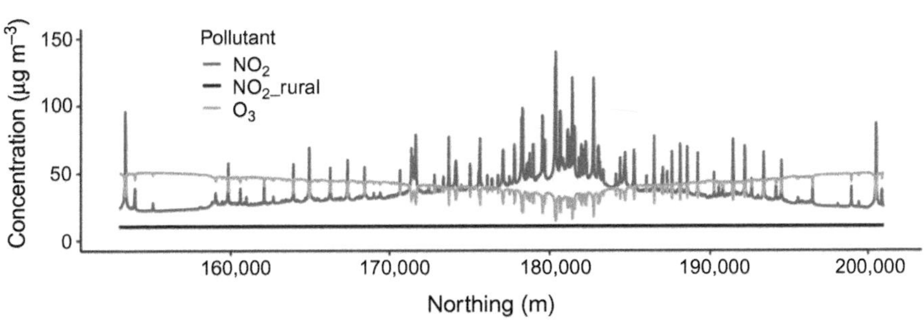

Fig. 6.2 NO_2 and O_3 concentrations ($\mu g\,m^{-3}$) every 20 m along a north-south transect (vertical line in *black*), across London, predicted using an emissions dispersion model.

smoothly across the city and gradually increase the closer you get to central London, the spatial gradients close to road sources are very large with concentrations up to 4 times larger than the local background contribution. But since these very high concentrations reduce within tens of meters from each road source, quickly falling to background concentrations, proximity to these sources is important in determining your exposure.

Ozone concentrations within cities exhibit more complex behavior. During hot summer periods when wind speeds are low, and temperatures are high (typically above about 22°C), hourly ozone concentrations can be elevated in cities and in surrounding rural areas due to the well-known "smog" reactions (Seinfeld & Pandis, 2016). However, these episodes represent only a relatively small fraction of a year and for most of the time ozone concentrations are determined by the photostationary state (PSS) relationship (Seinfeld & Pandis, 2016) between NO, NO_2, and ozone. In cities where NO and NO_2 levels are high, the PSS reduces ozone concentrations to low levels so that cities have lower longer-term (e.g., annual or seasonal) average ozone concentrations compared with a country or region. This is especially true near busy roads where ozone levels can be very low or even approach zero for much of the time. An example of the titrating effect of NO on annual average ozone concentrations is given in the bottom transect plot in Fig. 6.2, where concentrations of Ozone reduce towards the center of the city and even more so close to roads, where there is a sharp decline in concentrations.

Within city, $PM_{2.5}$ concentrations have some similarities to that of NO_2 but also important differences. The contribution that sources beyond London make to total $PM_{2.5}$ concentrations, is given in Fig. 6.3, and are shown to be far bigger than for NO_2 (and NO_x). This is important for

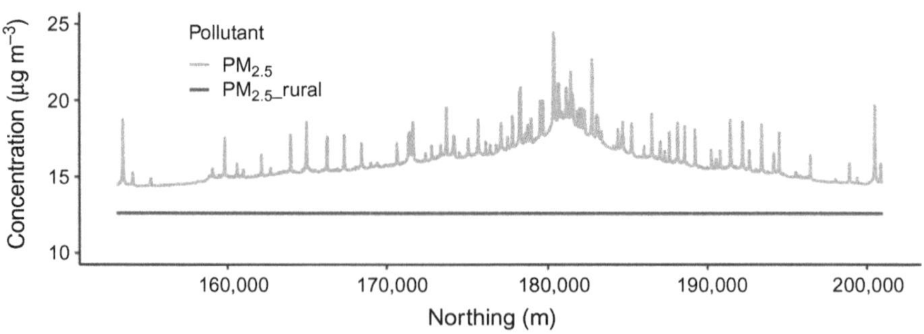

Fig. 6.3 $PM_{2.5}$ concentrations ($\mu g\,m^{-3}$) every 20 m along a north-south transect in London, predicted using an emissions dispersion model.

exposure estimates and also for city policy makers, who have less scope to control $PM_{2.5}$ than NO_2 using city policies alone. However, there is still a gradient of $PM_{2.5}$ towards the center of the city and although smaller in magnitude similar peaks in concentrations appear close to roads. However, since $PM_{2.5}$ is a combination of different PM species, it is important to understand the contribution of different emissions sources (source apportionment), and the contribution that each species (e.g., nitrate, organic aerosol (OA), SOA, sulfate, ammonia) makes. A published example of source apportionment is given at a range of London air pollution measurement sites (Beevers et al., 2013), including sites from suburban background to central London roadsides and for the sources: agriculture, biomass burning, air, rail and ships, domestic and commercial gas combustion, coal and oil burning, and industry, as well as the different components of road transport emissions (exhaust, tire wear, brake wear, surface wear and resuspension of PM from the road surface). These data show that $PM_{2.5}$ from outside London is 4–5 times larger than the contribution from local sources of $PM_{2.5}$ in suburban parts of the city. In addition, $PM_{2.5}$ from outside London is equivalent to the contribution from roads at central London kerbside sites and that the contribution of exhaust and nonexhaust sources is approximately the same.

Temporal changes in traffic-related air pollution

To date exposure estimates for research purposes are either annual or at best daily averages. However, at hourly temporal scales and close to traffic sources, exposure to air pollution is influenced by the variation in emissions by an hour of the day, day of the week and by season, with an additional and important effect of meteorology, driving dispersion of the local pollutants. This is demonstrated through analysis of roadside NO_x concentrations, a good tracer for traffic sources, as well as a surrogate for other vehicle-related primary pollutants such as UFP, BC, POA, and nonexhaust PM emissions. Looking at roadside and background sites in London (Fig. 6.4A), you get a clear hour of day effect, showing for roadsides, a relatively constant high concentration throughout the day from an overnight low, with some, but not all, having peaks during rush hour periods. At a central London urban background site (Fig. 6.4A), there is also an overnight low, but more clearly defined rush hour peaks in concentration between which there is a significant reduction in concentrations, due to enhanced dispersion in the afternoon, and finally, the variation in rural NO_x concentrations (Fig. 6.4B), which is similar to the background site.

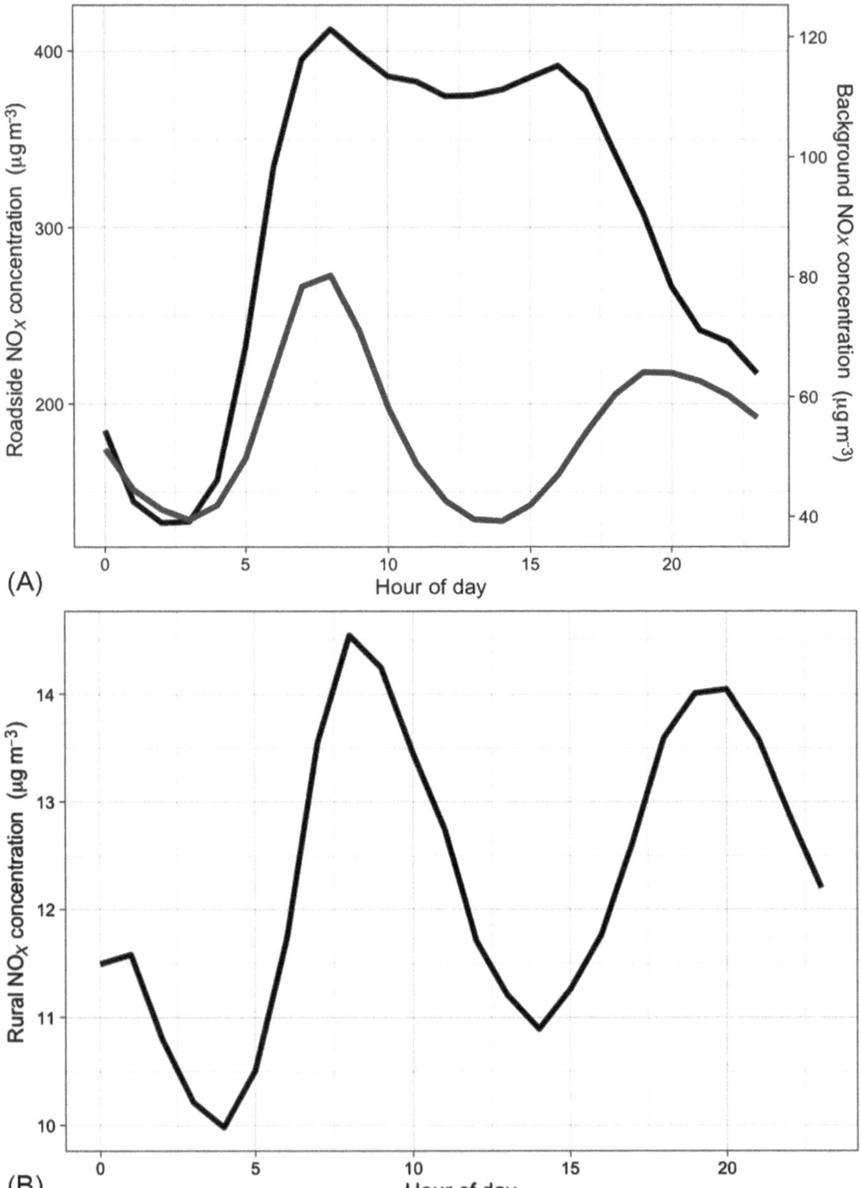

Fig. 6.4 (A) NO_x concentrations ($\mu g\,m^{-3}$) (2009–17) at Marylebone roadside site (*black* trace) and Kensington and Chelsea central London urban background site (*blue* trace; *gray* in print version) and (B) NO_x concentrations ($\mu g\,m^{-3}$) (2009–17) at Harwell rural site.

What is clear from all of these microenvironments is that time of day is very important in determining exposure, suggesting the need for exposure models that are run at hourly time resolutions.

Continental scale air pollution from transport emissions

Often considered to be a local pollutant source, exposure to traffic-related air pollution also occurs across continents in the form of secondary pollutants such as O_3 and secondary particles (nitrate, sulfate, and SOA). For example, O_3 during summer months is partly determined by emissions of NO_x and VOCs from anthropogenic sources, of which road traffic is one, as well as from other anthropogenic and biogenic sources. NO_x, an important traffic-related pollutant, is subject to atmospheric chemical reactions, leading to the generation of ammonium nitrate particles with atmospheric lifetimes of the order of a week or more thereby exposing people across entire continents, with agriculture, representing a dominant source of ammonia for these reactions. Likewise, precursor emissions of SO_2 from sulfur in fuels such as diesel are later transformed to sulfate particles. Although the sulfur content in fuels has reduced significantly in Europe and the United States over recent decades, as well as moves away from coal use, this will not necessarily be the case around the globe. Finally, a proportion of POA from vehicle exhausts evaporates upon dilution to ambient conditions and then photo-oxidizes to give SOA, so whilst the POA and associated black/elemental carbon vary on small spatial scales, the spatial pattern of SOA is "smoother," given the hours to days taken to create it chemically. Note that this does not mean that traffic is the only source of precursor emissions for particle nitrate and sulfate, and other major NO_x and SO_2 sources such as power generation, industry and shipping, and for organic aerosols, biogenic emissions are also important.

The exposure impact of secondary $PM_{2.5}$ concentrations across the United Kingdom and Ireland, in Fig. 6.5, generally show the highest levels to be in the southeast and central England, reducing towards the northwest. However, the spatial variation is more complex than this and is determined to a significant degree by ammonia emissions from agriculture, which are apparent right across the central area of England. Looking more closely there are also areas of local minimum concentrations of $PM_{2.5}$ in towns and cities and reflecting local minima in ammonia, nitrate, and sulfate aerosol. Note also that sulfate maxima occur across the United Kingdom, associated with industry, large power generation sources, and with the contribution from shipping also apparent. Finally, the contribution of $PM_{2.5}$ organic aerosol (POA and SOA)

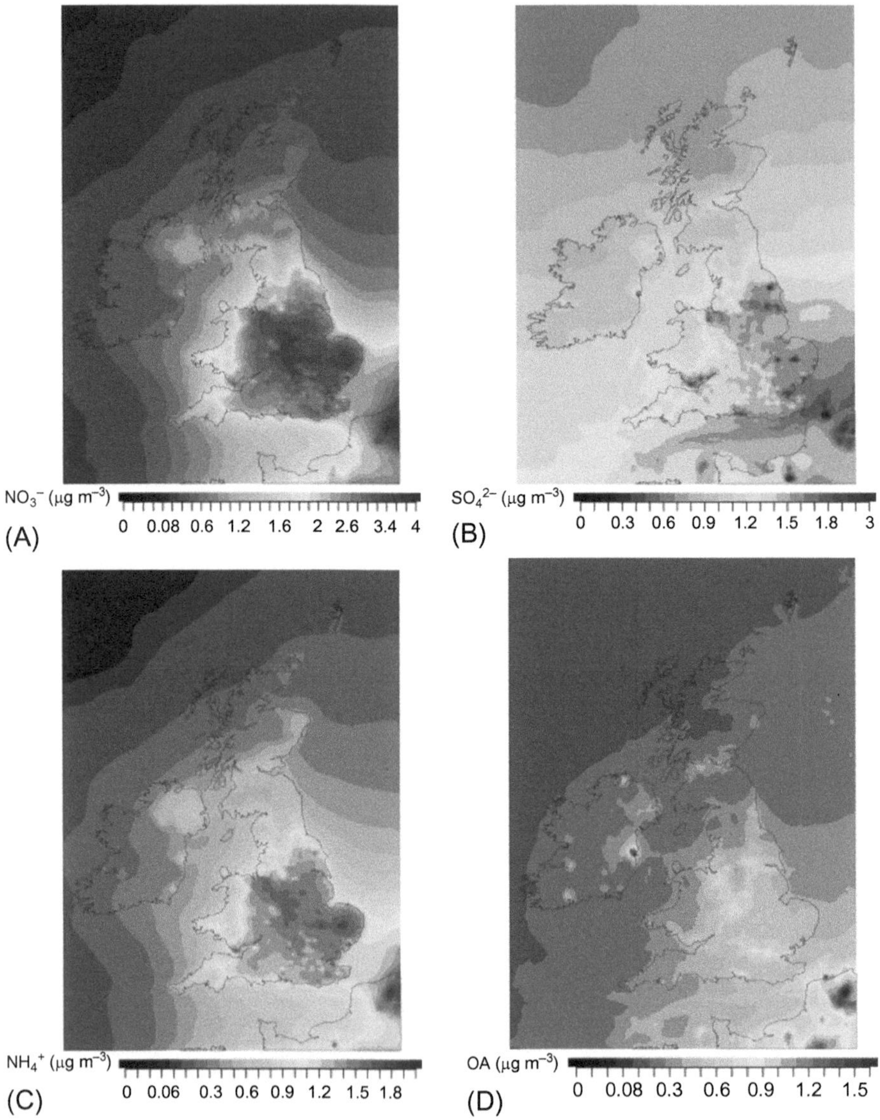

Fig. 6.5 2011 annual average $PM_{2.5}$ ($\mu g\, m^{-3}$) predicted using an emissions dispersion model: (A) nitrate, (B) sulfate, (C) ammonium, and (D) organic aerosols.

is apparent over cities and along transport corridors. In all cases, however, the important point in terms of exposure is that the spatial scales over which these pollutant concentrations vary is the order of tens of kilometers and that investigation of the associations between these pollutants and ill health should be made between cities, countries, and at continental scales.

Exposure assessment methods

Fixed-site monitoring data

The seminal work looking at air pollution effects on mortality comes from the Harvard Six Cities study (Dockery et al., 1993) and the American Cancer Society Study (Pope III et al., 2002). Both studies represented exposure to outdoor air pollution using "centrally located air-monitoring stations" measuring total suspended particle matter, $PM_{2.5}$ and PM_{10}, SO_2, ozone, suspended sulfates and aerosol acidity at points fixed in space. No attempt was made to analyze the data further, to ascertain the degree to which traffic contributed to the total concentrations measured or to understand the within city variability of pollution exposure, relying instead on the contrast between cities.

An early response to the lack of within city exposure variation was *Geo-statistical interpolation,* often using kriging or inverse distance weighting as methods to produce within city surfaces of air pollution concentrations. These methods often resulted in a smoothly varying pollution surface that is not representative of the steep concentration gradients close to traffic sources and were limited by the number of measurement sites. Recent developments in small sensor technology mean that there is potential to place many more sensors in cities. In turn, this means that analyzing these data using geo-statistics to describe the spatiotemporal trends in air pollution remains a possibility, albeit not without problems, given the challenge of measuring air pollutants, with sufficient quality and number of locations using these sensors.

Land use regression (LUR)

LUR builds regression models to describe the air pollution concentrations at a number of fixed measurement sites, using land use characteristics, traffic information, population density and altitude, and using these empirical models to predict between site locations. Examples include predicting PM_{10} and $PM_{2.5}$ in 20 European study areas (Eeftens et al., 2012), and in 36 European study areas for NO_2 and NO_x (Beelen et al., 2013). Traffic variables are often included in the LUR models by describing the road type or traffic density within a fixed distance of each measurement site, sometimes splitting the vehicle density by vehicle type. Once the relationship between land use and the traffic variables is established, the statistical LUR model is used to predict air pollution exposures throughout the study area. LUR models have also been useful in describing the spatial variation of pollutants, where using other methods like dispersion modeling remains difficult (e.g., UFP) or for entirely new exposure metrics such as oxidative potential (Yanosky, Tonne, Beevers, Wilkinson, & Kelly, 2012).

In the case of LUR models of UFP, the predictions are not without their problems, and whilst they have been created in a number of European, United States, and Canadian cities, all have modest performance (R^2~0.3–0.67). Also, due to a lack of routine measurements of UFP, these models are limited to the use of measurement data over very short time periods, examples of which include, using 30-minute samples, 3 times each at 160 sites in Basel, Heraklion, Amsterdam, Maastricht, and Utrecht, Norwich, Sabadell, and Turin (van Nunen et al., 2017); using 85 hours of mobile measurements in Minneapolis (Hankey & Marshall, 2015); using a single hourly measurement at 80 locations in Vancouver (Abernethy, Allen, McKendry, & Brauer, 2013); 1 week of monitoring outside 50 homes in Amsterdam (Hoek et al., 2011); 30 minutes monitoring at 81 sites in Amsterdam and 80 in Rotterdam (Montagne et al., 2015); and based on mobile measurements, using cars (6 hours per day for 5 days) and bicycles (6 hours per day for 23 days) in Montreal (Weichenthal et al., 2016). As such, the application of these data in population health studies is still an area in need of further research and in particular the need for more long-term UFP measurements.

More generally, LUR is limited in its ability to provide detailed source apportionment, since model variables, often used to create the models, such as "traffic density within 100 m," is very difficult to interpret, and furthermore, being based on regression methods, LUR models are unable to predict future air pollution or test policy "what-if" scenarios aimed at reducing air pollution.

Satellite remote sensing + ground-based measurements + atmospheric models + LUR

Whilst the majority of the satellite + atmospheric model predictions have focused on $PM_{2.5}$, using aerosol optical depth (AOD) measurements (Di et al., 2016), there have also been predictions of O_3 (Di, Rowland, Koutrakis, & Schwartz, 2017) and NO_2 (Novotny, Bechle, Millet, & Marshall, 2011). Satellite AOD is an indirect measure of the extinction of light within the atmospheric column. Therefore, predictions of surface-level $PM_{2.5}$ is reliant on resolving the AOD measure vertically in the atmosphere, often using global atmospheric models such as GeosChem, which predict PM in multiple layers throughout the atmosphere, in combination with ground-based measurements and land use data for use in LUR models (De Hoogh et al., 2016).

Using these methods, impressive continental and global datasets of $PM_{2.5}$ have been created (Van Donkelaar et al., 2016, 2019).

However, currently, satellite data has a number of important limitations: First, with a spatial resolution of $\sim 1 \times 1$ km, satellite AOD alone is limited in its ability to describe the spatial detail of traffic-related air pollution concentrations within cities and is also limited by the existence of clouds, requiring imputation of missing values. At present, satellites measuring AOD orbit the earth, so only pass over a given location once per day for seconds at a time. Given the wide variation in diurnal traffic pollutants, this means that detailed exposures must be estimated by supplementing satellite data with other information. Finally, the AOD PM, NO_2, and O_3 measures are not traffic specific, i.e., they cannot provide the traffic contribution to the total pollution modeled, or estimate the contribution of other sources, and cannot forecast forward in time or test 'what-if' scenarios. In an attempt to provide a richer dataset of PM components and a better understanding of sources, predictions of $PM_{2.5}$ mass are combined into their component parts using atmospheric models, often themselves run at large spatial scales approximately tens of kilometers (Van Donkelaar et al., 2019).

By combining satellite remote sensing, atmospheric models and LUR methods, exposure predictions could become more spatially detailed. Specifically, by including traffic variables in LUR models (Novotny et al., 2011), more appropriate exposure to traffic sources could be estimated than using satellite-based measurements alone. However, again, by virtue of the statistical nature of these methods, future scenario testing and source apportionment of air pollution into its source-related components remain limited for these exposure methods.

Atmospheric dispersion models

Atmospheric dispersion models combine meteorology, pollutant emissions, atmospheric chemistry, and transport of pollutants, as well as removal processes such as wet and dry deposition, to predict air pollution concentrations at any location and at the high temporal resolution, typically hourly. There is a range of dispersion models suited to different modeling tasks and with a range of sophistication and expertise required from the user. Since air pollution is transboundary in nature and that the emissions from traffic (and other sources) can have important influences hundreds of kilometers away, a Chemical Transport Model (CTM) is an important tool. Kukkonen et al. (2012), describes in detail examples of CTMs including but not limited to: CMAQ, CAMx, EMEP, Chimere, Lotos-Euros, and whilst differences exist between the models they all aim to describe the physics and chemistry in the atmosphere and importantly, unlike all of the other methods of predicting exposure to air pollution, do not require measurements as part of the modeling process. That is not to say that measurements are not needed, however, and a great deal of effort has been made to evaluate these models against both routine monitoring data, and specific campaign measurements, for example, AQMEII (http://aqmeii.jrc.ec.europa.eu/), Fairmode (https://fairmode.jrc.ec.europa.eu/) and in the United Kingdom, DEFRA's Model intercomparison exercise (https://uk-air.defra.gov.uk/research/air-quality-modelling?view=intercomparison).

Another important skill that CTMs have is that these models are able to predict the components of PM air pollution (nitrate, sulfate, EC (BC), UFP, OC, ammonium), often within different size ranges, combining them to create the PM mass estimates used in air pollution health research, for policy development and for multipollutant exposure estimates, to combine these, within the same model, with predictions of NO_x, NO_2, and O_3. Despite major uncertainties in particle number emissions and transformation processes, Kukkonen et al. (2016) also predicted particle number concentrations in a number of European cities. In addition, CTM's are also able to test "what if" scenarios, provide source apportionment for different emissions sources and predict future concentrations under a range of climate scenarios (Williams et al., 2018).

However, CTMs require a high degree of expertise and computer time to run and are data hungry, in that they require large emissions datasets, are associated with meteorological models, which are themselves time consuming to run, and boundary conditions from global models. Given their complexity, it is not surprising then that their results are subject to a range of uncertainties, and have a number of areas that require improvements, including, the adequacy of emissions inventories, access to use of suitable boundary conditions, improved understanding of the physical and chemical processes and better ground-based data with which to evaluate the model results.

Whilst CTM's predict at hourly time steps, they are limited spatially, down to $\sim 1 \times 1$ km, and so at a more local scale, within cities, there has been interest in augmenting CTMs with local scale dispersion models and LUR-type analysis. For local scale dispersion modeling, Gaussian models are a common tool, which assumes that the concentration of pollutants downwind of a source can be described by a Gaussian curve, whose characteristics can be described by horizontal and vertical standard deviations.

This is a reasonable assumption for ground-level sources in open flat terrain, but for tall stacks with buildings and roads surrounded by buildings, alternatives are required, including skewed and combined Gaussian curves for the former and specific street canyon models for the latter. There are a number of alternative models one might consider using, and for predicting particle concentrations, a number are described in Holmes and Morawska (2006) and for street canyons, by Vardoulakis, Fisher, Pericleous, and Gonzalez-Flesca (2003).

So, to correctly describe the effect of road transport emissions on air pollution regionally and locally, a combination of local and CTM modeling methods have been/are being developed. Foremost amongst these are urban or μEMEP and CMAQ-urban (Beevers, Kitwiroon, Williams, & Carslaw, 2012), which in the case of the latter enables model predictions at the hourly temporal resolution, 20-m spatial resolution for all pollutants of concern over the country scale and as 20 m annual averages, at continental scales.

Microenvironmental personal exposure models

Estimating people's exposure outside at their home address is used widely in both health research and in calculating health impacts. However, this seems unrealistic, not least because we do not spend long periods of time standing outside of our house and also the estimate does not include the large range of emissions sources to which we are exposed. To address this paradox, there are a number of models aimed at assessing the 'real' exposure of people in a range of microenvironments, both indoors and outdoors, and that reflect people's movements during their daily lives. These models attempt to characterize people's movements geographically, the mode of transport they use, the timing of any journeys taken, the time spent indoors, and indoor characteristics, such as the method of heating/cooking, number of occupants and whether there is a smoker in the house. These models are necessarily highly temporally (~hourly) and spatially detailed (tens of meters) and focus mainly on cities, requiring some of the most detailed exposure modeling currently undertaken.

The exposure models used tend to be either LUR or dispersion models, although some make use of personal sensors. To establish the movement of people, a range of data sources and spatial analysis is required, such as the use of census data, describing people's movements, travel surveys, designed to assess the need and method of travel in a city, and global positioning system (GPS) devices, including mobile phones, worn by the exposed population. The use of census data requires detailed routing analysis and for this, a number of relatively easily accessible web routing programs can be used, including for example Open Route Service API, the Project OSRM API, and Google Routes. Whilst this can be a time-consuming task, using census/survey data has the advantage of providing estimates for entire populations, which contrast with the use of GPS/mobile phone data which is often limited to relatively small numbers of people. Mobile phones are capable of providing an ideal solution to this problem since they can reflect both individual movements at population-level, however, there are a number of important barriers to the use of phones, chief amongst them is the difficulty in obtaining individual data, and this has so far limited their use in health and exposure studies. Recently, alternatives to describing people's daily activities, such as Agent-Based Models, are being used (Price, Brandon, Dionisio, Tornero-Velez, & Isaacs, 2016), with the advantage that they can also address changes in behavior, although evaluation of these methods still requires the use of census/survey data.

A key question for personal exposure models is the degree to which traffic, and transport in general, influences our average exposure. Early US models addressed exposure probabilistically, with the Stochastic Human Exposure and Dose Simulation (SHEDS) model (Burke, Zufall, & Ozkaynak, 2001) combining indoor, outdoor, and travel microenvironments, concluding that in Philadelphia, daily exposure to $PM_{2.5}$ was influenced by indoor sources, ambient PM, and was less influenced by activity patterns. In contrast, Setton et al. (2011), using individual-level exposure estimates, concluded that bias in exposure estimates in Vancouver and Southern California showed that ignoring daily mobility patterns can contribute to bias towards the null hypothesis in epidemiological studies. A number of studies have assessed small numbers of people: De Nazelle et al. (2013) used mobile phone data on 36 individuals, concluding that although travel represented 6% of people's time it represented 24% of their inhaled NO_2, Dons et al. (2011) assessed 8 couples exposure using measurements of black carbon (a homemaker and a full-time worker), and found that the exposure of partners differed by up to 30% and was influenced by transport. Nieuwenhuijsen et al. (2015) recruited 54 school children finding that their personal BC exposure when commuting was over twice as much as when they were at home. Dons, Int Panis, Van Poppel, Theunis, and Wets (2012) also found that the 62 individuals studied spent 6% of their time in transport but that this represented 21% of their exposure and 30% of their inhaled dose of black carbon. Dons also concluded that exposure to black carbon was highest for car and bus passengers and halved when traveling by bike or on foot, although when accounting for breathing rates, the dose for the latter, was twice as high.

Finally, the indoor microenvironment and using indoor chemical models (Terry, Carslaw, Ashmore, Dimitroulopoulou, & Carslaw, 2014) to describe it, has had relatively little attention. This is an important exposure pathway and is associated with smoking, cooking, heating, and the use of indoor domestic products, although traffic and other outdoor sources also contribute to human exposure through the exchange between outdoors and indoors.

Personal monitoring data using small/mobile sensors

Recently, there has been a proliferation of small sensor devices for the measurement of air pollution. This has been led by companies, and in particular tech start-ups, and due to the often low cost of these devices, has been linked with members of the general population. University research departments and metrology standards organizations have to some degree been caught out by the number and speed of rollout of these devices and are currently designing test procedures to ensure that the quality of the measurements taken is well understood. For example, Pang, Shawb, Gillot, and Lewis (2018) tested the cross interference of five different electrochemical gas sensors (O_3, SO_2, CO, NO, NO_2) against water vapor and co-pollutants (O_3, SO_2, CO, NO, NO_2, and CO_2). O_3, SO_2, and CO were positively dependent upon relative humidity, NO negatively dependent and NO_2 had no trend. However, co-pollutant and relative humidity interferences could be corrected and afterwards, 5-minute pollutant averages compared well with reference techniques. Similarly, Popoola, Stewart, Mead, and Jones (2016) showed the need for temperature correction of electrochemical sensors.

There is a great attraction of these devices beyond their low cost, and that is their ease of deployment, apparent ease of obtaining data and the specific nature of the measurements, be they fixed to lamp posts in your area, carried as a personal monitor or in your mobile phone. High-density deployment of these sensors has the possibility of providing the high spatially and temporally resolved data needed to characterize air pollution, especially in cities, and to characterize personal exposure and associated health effects. Interpreting the minute-by-minute personal exposure measurements, which are highly variable close to sources such as road traffic, is difficult, however, especially against annual average health standards, although these comparisons are made. Jerrett et al. (2017) used small sensors in Barcelona, to test whether exposure measurement error for epidemiological studies could be reduced, comparing hourly and 30-minute concentrations with government monitors and research instruments. The deployment showed only moderate correlations for NO and CO and performed less well for NO_2, although the sensors were able to detect the high microenvironments close to roads.

With the aim of providing very detailed measurements close to roads, vehicles have been instrumented and driven around city centers (Apte et al., 2017; Messier et al., 2018). The vehicle-based measurements use higher quality instrumentation than is used for personal monitoring and the results have been shown to provide datasets at the spatial resolution of individual roads. However, the data is limited by the number of times the vehicle passes the same location to obtain a stable average concentration and there has been a lack of comparison of these results against fixed monitoring, as well as the limitation of not providing the relative contribution of traffic and other sources to the total concentrations measured. Furthermore, producing the data requires considerable resources and it is not clear that it has significant advantages compared with fixed roadside monitoring and dispersion/LUR modeling of the kind described above. Overall, there is ongoing and rapid change in small fixed and mobile sensors, and the associated data processing algorithms, and these will become an important part of describing people's exposure in the coming years.

Conclusions

In conclusion, ambient air pollution contributes substantially to ill health globally, and if we are to improve our understanding of the pollutants and sources responsible, it is important to continue developing human exposure methods. In this chapter, we have commented on the challenges of developing exposure models for the next generation of health research and highlighted the importance of road traffic as a source in determining the spatial and temporal scale of air pollution as a local pollutant in cities, but also at continental scales, through emissions of precursor species such as NO_x. The direction of travel is towards more finely resolved spatial and temporal estimates of exposure, exemplified by developments in microenvironmental exposure models, including indoor environments, personal exposure using small sensors, and on road mobile sensors. There is also ever more detailed city and country scale model predictions, using LUR and CTM models. In addition, there has been an increase in exposure estimates from the city and country scales to continental and global scales, using satellite observations in combination with global dispersion models and more recently with LUR.

However, whilst these methods are providing impressive exposure datasets, we should continue to prioritize high-quality ground-based measurements, which are essential, and lacking in large parts of the globe, as well as to develop tools that can test air pollution policies, now and in the future.

References

Abernethy, R. C., Allen, R. W., McKendry, I. G., & Brauer, M. (2013). A land use regression model for ultrafine particles in Vancouver, Canada. *Environmental Science & Technology*, *47*, 5217–5225. https://doi.org/10.1021/es304495s.

Amato, F., Cassee, F. R., Denier van der Gon, H. A., Gehrig, R., Gustafsson, M., Hafner, W., et al. (2014). Urban air quality: The challenge of traffic non-exhaust emissions. *Journal of Hazardous Materials*, *275*, 31–36.

Apte, J. S., Messier, K. P., Gani, S., Brauer, M., Kirchstetter, T. W., Lunden, M. M., et al. (2017). High-resolution air pollution mapping with Google street view cars: Exploiting big data. *Environmental Science & Technology*, *51*, 6999–7008. https://doi.org/10.1021/acs.est.7b00891.

Beelen, R., Hoek, G., Vienneau, D., Eeftens, M., Dimakopoulou, K., Pedeli, X., et al. (2013). Development of NO_2 and NO_x land use regression models for estimating air pollution exposure in 36 study areas in Europe e the ESCAPE project. *Atmospheric Environment*, *72*, 10–23. https://doi.org/10.1016/j.atmosenv.2013.02.037.

Beevers, S. D., Kitwiroon, N., Williams, M. L., & Carslaw, D. C. (2012). One way coupling of CMAQ and a road source dispersion model for fine scale air pollution predictions. *Atmospheric Environment*, *59*, 47–58.

Beevers, S. D., Kitwiroon, N., Williams, M. L., Kelly, F. J., Ross Anderson, H., & Carslaw, D. C. (2013). Air pollution dispersion models for human exposure predictions in London. *Journal of Exposure Science & Environmental Epidemiology*, *23*(6), 647–653.

Burke, J. M., Zufall, M. J., & Ozkaynak, H. (2001). A population exposure model for particulate matter: Case study results for $PM_{2.5}$ in Philadelphia, PA. *Journal of Exposure Analysis and Environmental Epidemiology*, *11*, 470–489.

Carslaw, D. C., Beevers, S. D., Tate, J. E., Westmoreland, E. J., & Williams, M. L. (2011). Recent evidence concerning higher NO_x emissions from passenger cars and light duty vehicles. *Atmospheric Environment*, *45*, 7053–7063.

Cohen, A. J., Brauer, M., et al. (2017). Estimates and 25-year trends of the global burden of disease attributable to ambient air pollution: An analysis of data from the global burden of diseases study 2015. *Lancet*, *389*, 1907–1918. https://doi.org/10.1016/S0140-6736(17)30505-6.

De Hoogh, K., Gulliver, J., van Donkelaar, A., Martin, R. V., Marshall, J. D., Bechle, M. J., et al. (2016). Development of West-European $PM_{2.5}$ and NO_2 land use regression models incorporating satellite-derived and chemical transport modelling data. *Environmental Research*, *151*, 1–10. https://doi.org/10.1016/j.envres.2016.07.005.

De Nazelle, A., Seto, E., Donaire-Gonzalez, D., Mendez, M., Matamala, J., Nieuwenhuijsen, M. J., et al. (2013). Improving estimates of air pollution exposure through ubiquitous sensing technologies. *Environmental Pollution*, *176*, 92–99.

Di, Q., Kloog, I., Koutrakis, P., Lyapustin, A., Wang, Y., & Schwartz, J. (2016). Assessing $PM_{2.5}$ exposures with high spatio temporal resolution across the continental United States. *Environmental Science & Technology*, *50*, 4712–4721.

Di, Q., Rowland, S., Koutrakis, P., & Schwartz, J. (2017). A hybrid model for spatially and temporally resolved ozone exposures in the continental United States. *Journal of the Air & Waste Management Association (1995)*, *67*, 39–52.

Dockery, D. W., Pope, A., III, Xiping, X., Spengler, J. D., Ware, J. H., Fay, M. E., et al. (1993). An association between air pollution and mortality in six U.S. cities. *The New England Journal of Medicine*, *329*(24).
Dons, E., Int Panis, L., Van Poppel, M., Theunis, J., & Wets, G. (2012). Personal exposure to black carbon in transport microenvironments. *Atmospheric Environment*, *55*, 392–398.
Dons, E., Int Panis, L., Van Poppel, M., Theunis, J., Willems, H., Torfs, R., et al. (2011). Impact of time activity patterns on personal exposure to black carbon. *Atmospheric Environment*, *45*, 3594–3602. https://doi.org/10.1016/j.atmosenv.2011.03.064.
Eeftens, M., Beelen, R., de Hoogh, K., Bellander, T., Cesaroni, G., Cirach, M., et al. (2012). Development of land use regression models for $PM_{2.5}$, $PM_{2.5}$ absorbance, PM_{10} and PM_{coarse} in 20 European study areas; results of the ESCAPE project. *Environmental Science & Technology*, *46*, 11195–11205. https://doi.org/10.1021/es301948k.
European Commission. (2016). *Portable emission measurement system*. Available at: https://ec.europa.eu/jrc/en/vela/portable-emissions-measurement-systems. (Accessed11 March 2019).
European Parliament. (2016). *Comparative study on the differences between the EU and US legislation on emissions in the automotive sector*. Available at: http://www.europarl.europa.eu/supporting-analyses. (Accessed10 March 2019).
European Union Air Quality Standards (AQS). https://eur-lex.europa.eu/legal-content/EN/TXT/?qid=1486474738782&uri=CELEX:02008L0050-20150918.
Franco, V., Sánchez, F. P., German, J., & Mock, P. (2014). *Real-world exhaust emissions from modern diesel cars a meta-analysis of PEMS emissions data from EU (Euro 6) and US (Tier 2 bin 5/ULEV ii) diesel passenger cars. Part 1: Aggregated results, ICCT*. Available at: https://www.theicct.org/sites/default/files/publications/ICCT_PEMS-study_diesel-cars_20141013_0.pdf. (Accessed10 March 2019).
Hankey, S., & Marshall, J. D. (2015). Land use regression models of on-road particulate air pollution (particle number, black carbon, $PM_{2.5}$, particle size) using mobile monitoring. *Environmental Science & Technology*, *49*, 9194–9202. https://doi.org/10.1021/acs.est.5b01209.
Henschel, S., & Chan, G. (2013). *Health risks of air pollution in Europe (HRAPIE)*. http://www.euro.who.int/en/health-topics/environment-and-health/air-quality/publications/2013/health-risks-of-air-pollution-in-europe-hrapie-project.-new-emerging-risks-to-health-from-air-pollution-results-from-the-survey-of-experts.
Hoek, G., Beelen, R., Kos, G., Dijkema, M., van der Zee, S. C., Fischer, P. H., et al. (2011). Land use regression model for ultrafine particles in Amsterdam. *Environmental Science & Technology*, *45*, 622–628.
Holmes, N. S., & Morawska, L. (2006). A review of dispersion modelling and its application to the dispersion of particles: An overview of different dispersion models available. *Atmospheric Environment*, *40*, 5902–5928. https://doi.org/10.1016/j.atmosenv.2006.06.003.
International Agency for Research on Cancer. (2012). *Diesel engine exhaust carcinogenic*. https://www.iarc.fr/wp-content/uploads/2018/07/pr213_E.pdf.
Jerrett, M., Donaire-Gonzalez, D., Popoola, O., Jones, R., Cohen, R. C., Almanza, E., et al. (2017). Validating novel air pollution sensors to improve exposure estimates for epidemiological analyses and citizen science. *Environmental Research*, *158*, 286–294. https://doi.org/10.1016/j.envres.2017.04.023.
Kukkonen, J., Karl, M., Keuken, M. P., Denier van der Gon, H. A. C., Denby, B. R., Singh Douros, V. J., et al. (2016). Modelling the dispersion of particle numbers in five European cities. *Geoscientific Model Development*, *9*, 451–478. https://doi.org/10.5194/gmd-9-451-2016.
Kukkonen, J., Olsson, T., Schultz, D. M., Baklanov, A., Klein, T., Miranda, A. I., et al. (2012). A review of operational, regional-scale, chemical weather forecasting models in Europe. *Atmospheric Chemistry and Physics*, *12*, 1–87. https://doi.org/10.5194/acp-12-1-2012.

Messier, K. P., Chambliss, S. E., Gani, S., Alvarez, R., Brauer, M., Choi, J. J., et al. (2018). Mapping air pollution with Google street view cars: Efficient approaches with mobile monitoring and land use regression. *Environmental Science & Technology, 52*(21), 12563–12572. https://doi.org/10.1021/acs.est.8b03395.

Montagne, D. R., Hoek, G., Klompmaker, J. O., Wang, M., Meliefste, K., & Brunekreef, B. (2015). Land use regression models for ultrafine particles and black carbon based on short-term monitoring predict past spatial variation. *Environmental Science & Technology, 49*, 8712–8720. https://doi.org/10.1021/es505791g.

Nieuwenhuijsen, M. J., Donaire-Gonzalez, D., Rivas, I., de Castro, M., Cirach, M., Hoek, G., et al. (2015). Variability in and agreement between modelled and personal continuously measured black carbon levels using novel smartphone and sensor technologies. *Environmental Science & Technology, 49*, 2977–2982. https://doi.org/10.1021/es505362x.

Novotny, E. V., Bechle, M. J., Millet, D. B., & Marshall, J. D. (2011). National satellite-based land-use regression: NO_2 in the United States. *Environmental Science & Technology, 45*, 4407–4414. https://doi.org/10.1021/es103578x.

O'Driscoll, R., Stettler, M. E. J., Molden, N., Oxley, T., & ApSimon, H. M. (2018). Real world CO_2 and NO_x emissions from 149 Euro 5 and 6 diesel, gasoline and hybrid passenger cars. *The Science of the Total Environment*, 621, 282–290. https://doi.org/10.1016/j.scitotenv.2017.11.271.

Pang, X., Shawb, M. D., Gillot, S., & Lewis, A. C. (2018). The impacts of water vapour and co-pollutants on the performance of electrochemical gas sensors used for air quality monitoring. *Sensors and Actuators B, 266*, 674–684. https://doi.org/10.1016/j.snb.2018.03.144.

Pope, C. A., III, Burnett, R. T., Thun, M. J., Calle, E. E., Krewski, D., Ito, K., et al. (2002). Lung cancer, cardiopulmonary mortality, and long-term exposure to fine particulate air pollution. *Journal of the American Medical Association, 287*(9), 1132–1141.

Popoola, O. A. M., Stewart, G. B., Mead, M. I., & Jones, R. L. (2016). Development of a baseline-temperature correction methodology for electrochemical sensors and its implications for long-term stability. *Atmospheric Environment, 147*, 330–343. https://doi.org/10.1016/j.atmosenv.2016.10.024.

Price, P., Brandon, N., Dionisio, K., Tornero-Velez, R., & Isaacs, K. (2016). A framework for the use of agent based modeling to simulate inter- and intraindividual variation in human behaviors. In *2016 Annual International Society of Exposure Science Meeting, Utrecht, Netherlands, October 09–13*.

Seinfeld, J. H., & Pandis, S. N. (2016). *Atmospheric chemistry and physics: From air pollution to climate change* (3rd ed.)., ISBN:978-1-118-94740-1.

Setton, E., Marshall, J. D., Brauer, M., Lundquist, K. R., Hystad, P., Keller, P., et al. (2011). The impact of daily mobility on exposure to traffic-related air pollution and health effect estimates. *Journal of Exposure Science and Environmental Epidemiology, 21*, 42–48.

Smith, J. D., Mitsakou, C., Kitwiroon, N., Barratt, B. M., Walton, H. A., Taylor, J. G., et al. (2016). London hybrid exposure model: Improving human exposure estimates to NO_2 and $PM_{2.5}$ in an urban setting. *Environmental Science & Technology, 50*, 11760–11768. https://doi.org/10.1021/acs.est.6b01817.

Terry, A. C., Carslaw, N., Ashmore, M., Dimitroulopoulou, S., & Carslaw, D. C. (2014). Occupant exposure to indoor air pollutants in modern European offices: An integrated modelling approach. *Atmospheric Environment, 82*, 9–16. https://doi.org/10.1016/j.atmosenv.2013.09.042.

Timmers, V. R. J. H., & Achten, P. A. J. (2016). Non-exhaust PM emissions from electric vehicles. *Atmospheric Environment, 135*, 10–17. https://doi.org/10.1016/j.atmosenv.2016.03.017.

UK National Atmospheric Emissions Inventory (NAEI). http://naei.beis.gov.uk/. (Accessed18 March 2019).

United States Environmental Protection Agency, National Ambient Air Quality Standards (NAAQS). https://www.epa.gov/criteria-air-pollutants/naaqs-table.

Van Donkelaar, A., Martin, R. V., Brauer, M., Hsu, N. C., Kahn, R. A., Levy, R. C., et al. (2016). Global estimates of fine particulate matter using a combined geophysical-statistical method with information from satellites, models, and monitors. *Environmental Science & Technology, 50*(7), 3762–3772.

Van Donkelaar, A., Martin, R. V., & Burnett, R. T. (2019). Regional estimates of chemical composition of fine particulate matter using a combined geoscience-statistical method with information from satellites, models, and monitors. *Environmental Science & Technology*. https://doi.org/10.1021/acs.est.8b06392. Article ASAP.

van Nunen, E., Vermeulen, R., Tsai, M.-Y., Probst-Hensch, N., Ineichen, A., Davey, M., et al. (2017). Land use regression models for ultrafine particles in six European areas. *Environmental Science & Technology, 51*, 3336–3345. https://doi.org/10.1021/acs.est.6b05920.

Vardoulakis, S., Fisher, B. E. A., Pericleous, K., & Gonzalez-Flesca, N. (2003). Modelling air quality in street canyons: A review. *Atmospheric Environment, 37*, 155–182.

Weichenthal, S., Van Ryswyk, K., Goldstein, A., Bagg, S., Shekkarizfard, M., & Hatzopoulou, M. (2016). A land use regression model for ambient ultrafine particles in Montreal, Canada: A comparison of linear regression and a machine learning approach. *Environmental Research, 146*, 65–72. https://doi.org/10.1016/j.envres.2015.12.016.

Williams, M. L., Lott, M. C., Kitwiroon, N., Dajnak, D., Walton, H., Holland, M., et al. (2018). The Lancet countdown on health benefits from the UK Climate Change Act—A modelling study for Great Britain. *Lancet Planetary Health, 2*, 202–213.

World Health Organisation. (2005). *Air quality guidelines—Global update 2005*. https://www.who.int/phe/health_topics/outdoorair/outdoorair_aqg/en/.

World Health Organisation. (2013). *Review of evidence on health aspects of air pollution (REVIHAAP)*. http://www.euro.who.int/en/health-topics/environment-and-health/air-quality/publications/2013/review-of-evidence-on-health-aspects-of-air-pollution-revihaap-project-final-technical-report.

Yanosky, J. D., Tonne, C. C., Beevers, S. D., Wilkinson, P., & Kelly, F. J. (2012). Modelling exposures to the oxidative potential of PM_{10}. *Environmental Science & Technology, 46*(14), 7612–7620. https://doi.org/10.1021/es3010305.

CHAPTER 7

Air pollution epidemiology

Zorana Jovanovic Andersen
Section of Environmental Health, Department of Public Health, Faculty of Health and Medical Sciences, University of Copenhagen, Copenhagen, Denmark

Abbreviations

ACS	American Cancer Society
APHEA	air pollution and health: a European approach
AQG	air quality guideline
BC	black carbon
CO	carbon monoxide
COPD	chronic obstructive respiratory disease
EPA	Environmental Protection Agency
HIA	Health impact assessment
ESCAPE	European study of cohorts for air pollution effects
GAM	generalized additive model
GBD	global burden of diseases
NMMAPS	National mortality, morbidity and air pollution studies
NAAQS	National Air Quality Standards
NLCS	Netherlands Cohort Study on Diet and Cancer
NO_2	nitrogen dioxide
NO_x	nitrogen oxide
MCC	multicountry multicity
O_3	ozone
Pb	lead
PD	Parkinson's disease
PM	particulate matter
$PM_{2.5}$	particulate matter with diameter 2.5 $\mu g/m^3$
PM_{10}	particulate matter with diameter 10 $\mu g/m^3$
SAPALDIA	study on air pollution and lung disease in adults
SDG	sustainable development goals
SO_2	sulfur dioxide
TRAP	traffic-related air pollution
TSP	total suspended particles
UFP	ultrafine particles
WHO	World Health Organization

Traffic-Related Air Pollution
https://doi.org/10.1016/B978-0-12-818122-5.00007-7

Introduction

Air pollution epidemiology is a part of a larger field of environmental epidemiology, which is defined as the study of the health consequences of exposures that (which) occur in the general environment (air, water, soil, and diet) (Lim, Vos, Flaxman, et al., 2012). Outdoor air pollution consists a complex mixture of pollutants from various different sources (traffic, heating, industrial activity, etc.), including particulate matter with diameter < 10 μm (PM_{10}), particulate matter with diameter < 2.5 μm ($PM_{2.5}$), particulate matter with diameter < 0.1 μm (ultrafine particles (UFPs)) and gases: nitrogen dioxide (NO_2), nitrogen oxide (NO_x), sulfur dioxide (SO_2), carbon monoxide (CO), ozone (O_3), black carbon (BC), among others. Traffic is a major source of air pollution to urban populations, and thus traffic-related air pollution (TRAP) is the most studied exposure in air pollution epidemiology. Ambient air pollution is a ubiquitous exposure, affecting everyone to a certain extent. Over 90% of the world's population is exposed to $PM_{2.5}$ levels above 10 μg/m^3, which is the World Health Organization (WHO) Air Quality Guideline (AQG) limit value (World Health Organization, 2006). We breathe over 10,000 L of air every day, passively inhaling considerable amount of pollutants during our lifetimes. The main mechanisms by which air pollution affects human health are via inflammation and oxidative stress, which adversely affects a number of organs in the human body, leading to an increasing risk of a number of diseases and premature mortality. Thus, due to its ubiquitous nature, and a number of health outcomes that are affected, air pollution is one of the biggest environmental exposures and the most studied environmental pollutant. The latest Global Burden of Disease (GBD) Study with a focus on air pollution estimated that 4.2 million deaths and 101.3 million lost years of healthy life worldwide could be attributed to outdoor $PM_{2.5}$ in 2015 (Cohen, Brauer, Burnett, et al., 2017). This GBD approach includes mortality from five causes of death, which have been causally linked to air pollution: ischemic heart disease, cerebrovascular disease, chronic obstructive pulmonary disease (COPD), lower respiratory infections, and lung cancer (Cohen et al., 2017). The most recent analyses of the GBD Risk Factors Study from 2017 have estimated the burden of disease attributable to 84 risk factors in 195 countries (Stanaway, Afshin, Gakidou, et al., 2018). Exposure to $PM_{2.5}$ (indoor and outdoor) ranked as the fifth highest ranking risk factor for death after dietary risks, high blood pressure, tobacco and high fasting-plasma glucose. Exposure to $PM_{2.5}$ ranked higher than body mass index (ranked 6th), alcohol use (ranked 9th), or low physical activity (ranked 12th) (Stanaway et al., 2018). Consequently, concerns about air pollution are reflected in the Sustainable Development Goals (SDGs) (United Nations, 2017). Air pollution in cities is cited as an indicator for urban sustainable development (SDG 11), access to clean energy, particularly clean household fuels and technologies, is highlighted as an indicator for sustainable energy in SDG 7, and mortality due to air pollution (ambient and household) is used as an indicator for the health SDG goal in SDG 3.

Air pollution epidemiology is concerned with the health effects related to exposure to air pollution, and is typically divided into two large areas: (1) short-term studies—studies of health effects related to acute or short-term exposures to air pollution, lasting over several hours, days, or weeks and (2) long-term studies—studies of health effects of chronic or long-term exposures to air pollution, lasting several months, years, decades, or an entire lifetime. Short-term studies give answer to the question of whether short-term exposure to high levels of air pollution (day with high level of air pollution, several hour exposure in heavy traffic, etc.) can trigger adverse effects, such as an exacerbation of an existing disease, etc. Typical health effects from short-term or acute exposure to air pollution include cough, headache, irritation of the eyes, nose, and throat, allergic reactions, asthma symptom aggravation, wheezing, and complication of symptoms in elderly subjects with chronic respiratory and cardiovascular disease, possibly leading to hospitalization or even death. Long-term air pollution studies give an answer to the question of whether every day air pollution concentrations also result in adverse effects, when accumulated over long periods of time. Typical health effects resulting from long-term or chronic exposure to air pollution are chronic diseases, such as asthma in children, and in adults, incidence of COPD, cardiovascular disease, stroke, diabetes, lung cancer, breast cancer, pneumonia, or death. This chapter will give a brief overview of the history of air pollution epidemiology, introduce different types of studies used in air pollution epidemiology, with major examples.

Short history of air pollution epidemiology

A common feature for all environmental exposures is that data are observed, and usually involve low-level exposures of the general public, which are difficult to measure and difficult to link to disease. Thus, early realizations of the adverse effects of environmental pollutants came often from "natural experiments," or disasters, where extremely high levels of a pollutant led to a dramatic increase in mortality or morbidity. Similarly, early realizations that air pollution may have adverse effects on human health came from short and intense air pollution episodes, where stagnant and cold weather conditions mixed with emissions from the industry and combustion from traditional fossil fuel (wood, coal), created sharp increases in the concentration of air pollutants, leading to detrimental health consequences, in terms of unusually high number of deaths, hospital admissions, etc. The first documented air pollution episode occurred in the Meuse Valley in Belgium, in 1930, where over 4 days in early December, thick fog due to heavy industrial emissions from steelworks, zinc smelters, glass manufacturers, fertilizer, and explosive plants, developed around Liege in the valley of the river Meuse. During these 4 days in the area with about 35,000 inhabitants, more than 60 people died, which was 10 times higher than the expected mortality rate for the area, raising considerable media attention (Nemery, Hoet, & Nemmar, 2001). Although this event provided the first evidence of causality between air pollution and mortality, another similar episode 20 years later, became the landmark in air pollution epidemiology, because of the scale of disaster, and because it provided data to researchers for the first time to perform detailed analyses of the association between air pollution and mortality. This was "The Great London Smog" of 1952, the most famous air pollution episode, also known as "The Killer Fog," where unusually cold and calm weather trapped the heavy, motionless, thick layer of smoky, dusty fumes from the region's coal stoves and local factories in the London Thames basin (Fig. 7.1) (Logan, 1953). This severe air pollution event affected London for

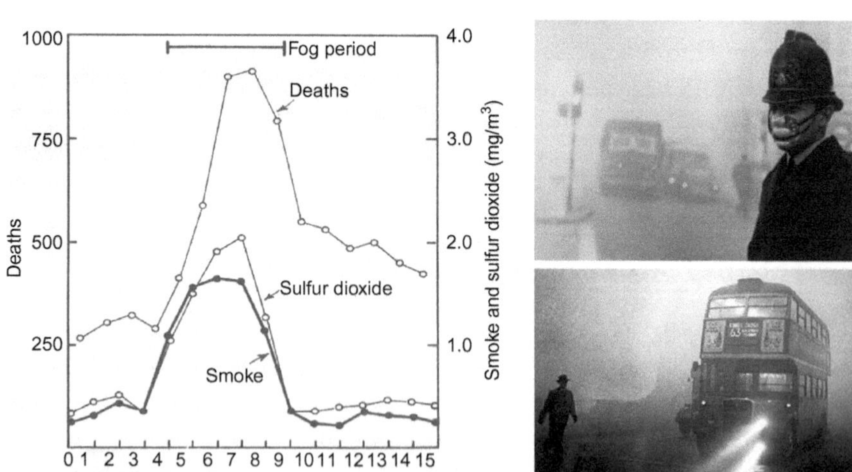

Fig. 7.1 The Great London Smog episode of 1952 resulted in 12,000 deaths.

4 days from 5th to 9th of December, during which time the heavy smog caused major disruption in traffic by reducing visibility, penetrating indoor areas, and causing adverse health effects. The episode led to a fivefold increase in SO_2 levels, to that of several thousands of $\mu g/m^3$, and subsequently, threefold increase in the number of deaths above the expected, with an estimated excess death toll of over 4000 (Wilkins, 1954). Recent reanalysis of the original London Fog data acknowledged that the prolonged effects of the London air pollution episode contributed to higher death rates for several months later, resulting in a total of 12,000 deaths that can be attributed to the smog episode (Bell & Davis, 2001). A similar air pollution episode occurred on October 27–31, 1948 in Donora, Pennsylvania, United States, a steel mill town on the Monongahela River, where 20 people died and over 6000 people suffered from respiratory problems, in the town with the population of 1400 (Hood & Davis, 2004). These air pollution episodes of the 1940s–50s have made an important impact on science, public perception of air pollution, and government regulation, and led to some important legislation on air pollution.

The pollution levels from the 1940s and 1950s episodes decreased steadily in the 1960s and 1970s in Western Europe and the United States, mainly due to effective legislation that led to a dramatic reduction in the use of coal and introduction of central heating in cities. Along with economic development, traffic-related air pollution has gained increasing importance in the 1970s and 1908s, but still, adverse effects were considered unlikely at air pollution concentrations seen at this time. In the 1980s, a few air pollution episodes in Western Europe were caused by long-range transported air pollution from Eastern Europe. Some of these episodes in 1985 led to an increases in mortality and in hospital admissions in Germany and lung function changes in the Netherlands (Wichmann, 2004; Wichmann, Mueller, Allhoff, et al., 1989). The health effects of this short episode were, however, difficult to detect, since the concentrations of SO_2 and PM were in hundreds, instead of thousands of $\mu g/m^3$, which were seen in the early episodes of the 1930s–50s. Studies that evaluated air pollution disasters or episodes present examples of the earliest and methodologically simplest epidemiological studies, which typically compared mortality before, during, and after air pollution episodes. These studies still have major contemporary relevance in the areas of the world with high and rising air pollution levels such as in China and India, which experience some of the highest air pollution levels in the world, and where major cities frequently experience air pollution episodes as seen in Europe and the United States in the 1930s and 1940s.

Epidemiological designs for studying short-term health effects of air pollution

The study design used to study health effects of acute exposures to air pollution is called time-series study. Short-term or time-series air pollution study design examines time series of day-to-day variations in air pollution over long periods of time as determinants of day-to-day variations in mortality, hospital admissions, and other public health indicators. The tool to analyze time-series data was introduced by Schwartz and Marcus (1990) leading to an exponential increase in studies of short-term effects of air pollution ever since. The Poisson generalized additive model (GAM) time-series model became the standard tool as it could easily utilize routinely collected data from air pollution monitors, hospital, and death registries. In these types of studies, typically based in a single large city, time-series (typically several years) data of daily variations in air pollution levels (24-h mean, 8-h mean, 8-h maximum, or maximum pollutant concentrations, measured at a central monitor) are associated to daily variations in health outcomes (daily count of deaths or hospitalizations in that city), and adjusted for population-level confounders that affect the health outcome of interest, such as weather (temperature, humidity), season, day of the week, influenza epidemics, public holidays, pollen counts, etc. In these models, it is important to account for other unmeasured confounders such as trends in survival due to improvements in medical care, changes in population size, trends in occurrence of major diseases, population habit changes with respect to lifestyle, drinking and smoking rates, etc. These unmeasured time trends are modeled by calendar time, which is used as a surrogate for an unmeasured variable that may have causal effects on health outcomes. The Poisson time-series model quickly became the standard tool for its many advantages, as they are relatively cheap and feasible and as they can efficiently use existing data from air pollution monitoring programs and mortality and hospitalization records.

Another design used in analyzing time-series data is a case-crossover design, where air pollution concentration on the day a person has experienced a health outcome (hospitalization) is compared to other weekdays within the same month, when the person did not experience the health outcome. The data are then analyzed by a conditional logistic regression model. This design is attractive and has gained popularity over time-series GAM approach, as it is a case-only design, where persons are used as their own control, and because it provides personal risk estimates. In an example from a study on short-term exposure to air pollution and childhood asthma in Copenhagen (Fig. 7.2), a time series of daily mean concentrations of NO_2 ($\mu g/m^3$) from a central

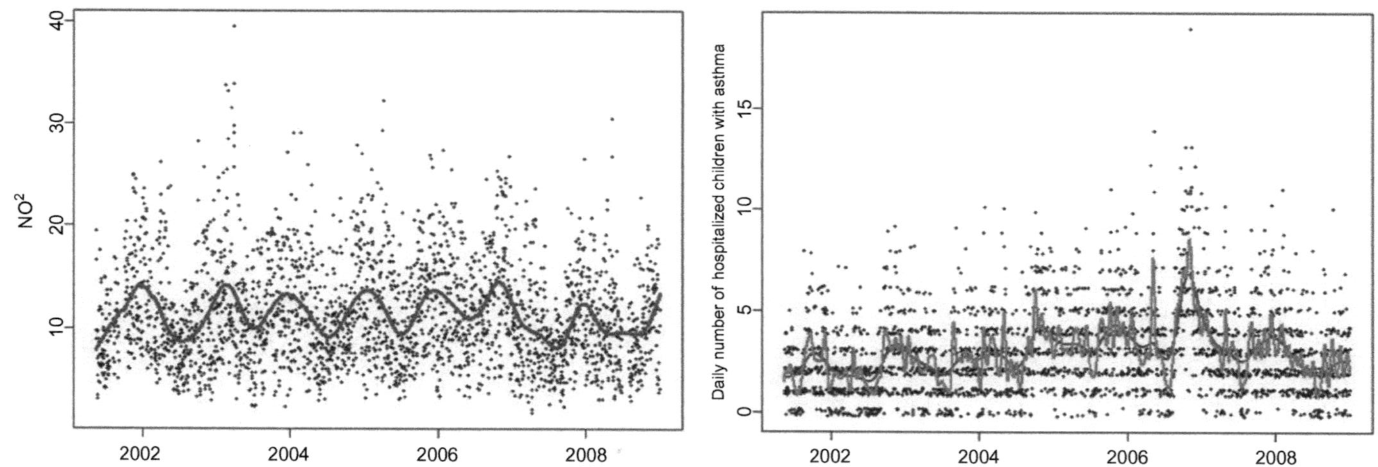

Fig. 7.2 Time series of daily mean levels of NO_2 from background monitor (left) and number of hospital admissions for asthma in children (0–18 years old) in Copenhagen between 2001 and 2009.

background monitor in Copenhagen from 2001 until 2009, was linked to daily number of asthma hospitalizations in children aged 18 years or younger, in 9 Copenhagen hospitals, during the same period. In this study, the authors found that for each 6 ppb increase in mean NO_2 levels over 5 days, there was a 10% increase in risk of childhood asthma hospitalization in the following day (Iskandar et al., 2012).

By the late 1990s, abundant single site time-series studies established that air pollution levels, even at much lower levels than that seen during air pollution episodes, were associated with increased rates of mortality and morbidity in cities in the United States and Europe and other developed regions. More recently, time-series studies in China and other sites in South-East Asia with some of the highest air pollution levels globally have been an invaluable tool for showing health effects related to air pollution exposure in this part of the world, where air pollution epidemiology is still in infancy, but important for pushing forward public awareness and political action for stricter regulation and cleaner air. Furthermore, application of a time-series model has been extended by a hierarchical model (Dominici, Samet, & Zeger, 2000) which allowed for analyses of data from many sites under a common framework, giving pooled estimates, and possibility for effect modification analyses. Two such large collaborative multicity studies, Air Pollution and Health: a European Approach (APHEA) (Katsouyanni, Zmirou, Spix, et al., 1995) in Europe and National Mortality, Morbidity, and Air Pollution Studies (NMMAPS) (Samet, Dominici, Zeger, Schwartz, & Dockery, 2000; Samet, Zeger, Dominici, et al., 2000) in the United States have provided many new insights that were especially useful to policy makers. The APHEA was extended with more recent data in the APHEA-2 study, which covered populations of 43 million people from 29 European cities, followed over 5 years in the mid-1990s, in its mortality study (Katsouyanni, Touloumi, Samoli, et al., 2001), and 38 million people from eight cities, followed over 3–9 years, in the hospital admissions part of the study (Atkinson, Anderson, Sunyer, et al., 2001; Le Tertre, Medina, Samoli, et al., 2002; Sunyer, Atkinson, Ballester, et al., 2003; Sunyer, Ballester, Le Tertre, et al., 2003). Similarly, the NMMAPS mortality study covered the 20 largest cities in the United States, with 50 million inhabitants, and follow-up from 1984 to 1994 (Samet, Zeger, Dominici, et al., 2000), while the hospital admissions study covered about 2 million people from 10 American cities (Zanobetti, Schwartz, & Dockery, 2000). The two studies have provided remarkably similar results with respect to all-cause mortality, where the combined effect for Europe and the United States was respectively 0.6% (95% CI:0.4–0.8)

(Katsouyanni et al., 2001) and 0.5% (0.1–0.9) (Samet, Dominici, Curriero, Coursac, & Zeger, 2000) increase in daily mortality for each 10 $\mu g/m^3$ increase in PM_{10}. Although these relative rates are small, the burden of disease attributable to air pollution is substantial considering the large populations exposed to air pollution and the large number of persons to whom the relative rates of mortality and morbidity apply. The results from NMMAPS had a profound impact on policy makers and played a central role in the US Environmental Protection Agency's (EPA) development of National Air Quality Standards (NAAQS) for the six 'criteria' pollutants defined by EPA (O_3, PM, CO, lead (Pb), SO_2, NO_2). The most recent multicity study, the multicountry multicity (MCC) Collaborative Research Network is an international global collaboration of research teams working on a program aiming to produce epidemiological evidence on the associations between air pollution, climate, and health. The MCC project collected data on air pollution and mortality from cities from all around the world, and has in a recent publication based on the data from 652 cities found 0.44% (95% CI: 0.39–0.43) and 0.68% (95% CI: 0.59–0.77) increase in daily all-cause mortality for each 10 $\mu g/m^3$ increase in 2-day moving average of PM_{10} and $PM_{2.5}$, respectively (Liu, Chen, Sera, et al., 2019).

Epidemiological designs for studying long-term health effects of air pollution

While the studies above indicate that short-term exposure to elevated air pollution levels are associated with short-term changes in cardiorespiratory health, they do not tell us how much life is shortened due to air pollution, how does air pollution affect long-term mortality rates in a given population, or whether air pollution can lead to the development of a chronic disease. Studies of chronic- or long-term exposure to air pollution evaluate effects of typically low or moderate exposure to air pollution (in Western Europe and the United States, although emerging studies from China examine effects of high levels of air pollution) that persist over long periods of time, as well as cumulative effects of repeated exposure to substantially elevated levels of air pollution. This can be done in an ecological design, cross-sectional studies comparing mortality rates between different populations or in cohort, or case-control studies.

Early evidence on the health effects of long-term exposure to air pollution came from ecologic, cross-sectional studies. In an example from a study from 1987, the US vital statistics and available ambient air pollution data for criteria pollutants for the year 1980 were used to estimate the association between various air pollutants and total mortality (Özkaynak & Thurston, 1987). In this cross-sectional analyses, data on air pollution from 1980 were linked to data on mortality in the same year, without temporality between exposure and outcome, which raised debate about the usefulness of this approach for risk assesment (Evans, Tosteson, & Kinney, 1984). A similar cross-sectional ecological design was used in another ground-breaking study on air pollution and infant mortality in the Czech Republic in the period 1986–88 for 46 districts. Here, the mean levels of total suspended particles (TSP) and nitrogen oxide (NO_x) for each district were linked to infant mortality rates for that district, adjusting for district socioeconomic characteristics, and abortion rates (Bobak & Leon, 1992). In both of these types of studies, cross-sectional design implies that exposure to air pollution and health outcome are assessed at a single point of time, without temporality between exposure and outcome, precluding conclusions about causality. Furthermore, these studies are examples of ecological studies, which explore the association between disease and estimated exposures in population groups (districts, cities, etc.) and produce risk estimates at the population level, and not the individual level, as no individual level data are available on air pollution exposure, health outcome, or other personal characteristics (lifestyle, etc.). Cross-sectional studies with air pollution data at the individual level are also used for outcomes such as prevalence of a disease, with example of a study on air pollution and prevalence of diabetes (Dijkema, Mallant, Gehring, et al., 2011), and for outcomes such as biomarkers of disease progression that are collected at a single point of time, such as blood pressure and hypertension (Liu, Chen, Zhao, et al., 2017).

One of the earliest and the most famous long-term air pollution studies is the Harvard Six Cities study, a prospective cohort study of 8111 adults from six American cities, which compared mortality rates for each city with mean air pollution levels in each city. The study reported the highest mortality in the cities with the highest particle levels, with a linear dose-response relationship, detecting 26% higher mortality in the most (Steubenville, Ohio) as compared to the least (Portage, Wisconsin) polluted city (Fig. 7.3) (Dockery, Pope, Xu, et al., 1993). Recent updated reanalysis of the Six Cities study until 2009 has shown that despite decreases in air pollution levels in the United States to levels well below the US annual standard of 12 $\mu g/m^3$, the estimated effects of $PM_{2.5}$ did not change overtime, suggesting stable toxicity of $PM_{2.5}$ (Lepeule, Laden, Dockery, & Schwartz, 2012), and no safe threshold. In the Six Cities studies, mean air pollution for an entire city was

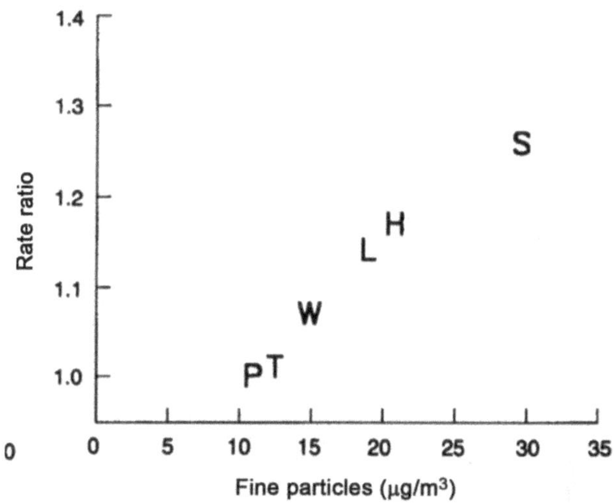

Fig. 7.3 Estimated adjusted rate ratios for total mortality and mean $PM_{2.5}$ levels (1980–85) in the Six Cities study (Özkaynak & Thurston, 1987). *H*, Harriman, TN; *L*, St. Louise, MO; *P*, Portage, WI (reference); *S*, Steubenville, OH; *T*, Topeka, KS; *W*, Watertown, MA.

assigned for all inhabitants, and mortality rate ratios compared between cities, implying that there were no individual levels of air pollution exposure available for each individual in each of the six cities.

A typical long-term exposure study is a cohort study, where air pollution models are used to predict air pollution levels for cohort participants at the individual level, typically at the residential address, at the year of recruitment into the study, or with historical data on residential address history and air pollution levels. As an example, in the Danish Nurse Cohort, individual concentrations of NO_2, $PM_{2.5}$, PM_{10} were estimated at each nurse's residence (28,731 nurses), as an annual mean from 1990 until 2014. A Cox regression model was then applied to estimate whether between person differences in air pollution levels explain between person risk of developing type 2 diabetes (Fig. 7.4) (Hansen, Ravnskjær, Loft, et al., 2016). In other words, nurses were followed up from the cohort baseline in 1993 or 1999 for up to 20 years, to examine whether nurses who live in areas with high pollution have higher risk of developing type 2 diabetes than those living in areas with low air pollution. In this study, the authors found that each $3\,\mu g/m^3$ increase in $PM_{2.5}$ was associated with 11% increase in the risk of developing type 2 diabetes (Hansen et al., 2016). In this type of studies, it is very important to consider other factors related to mortality and morbidity,

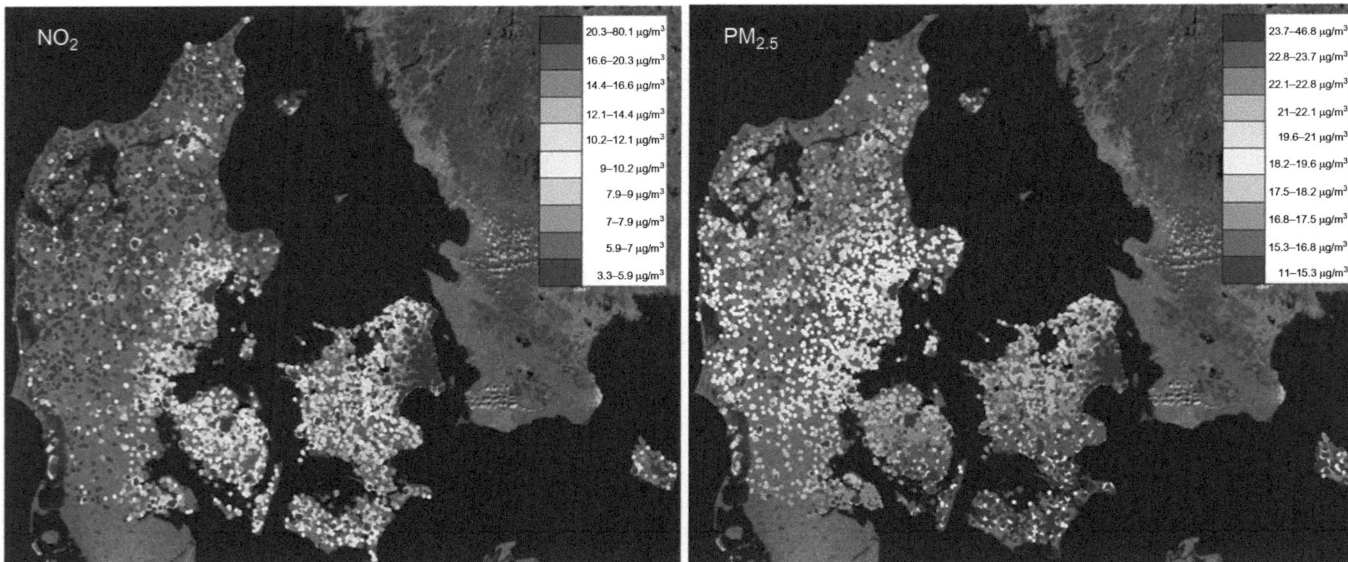

Fig. 7.4 Annual mean levels of NO_2 (left) and $PM_{2.5}$ (right) in μg/m^3, estimated by the Danish DEHM/UBM/AirGIS modeling system at the residences of the 28,731 nurses from the Danish Nurse Cohort, at the cohort baseline in 1993 or 1999 (mean levels of NO_2 were 12.6 and $PM_{2.5}$ were 19.6 μg/m^3).

such as smoking, physical activity, body mass index, alcohol use, diet, etc., which may confound the association between air pollution and a health outcome. Because they typically have these data available, cohort studies are considered the gold standard in air pollution epidemiology and provide the best evidence of health effects of air pollution, taking into account other factors besides air pollution which may explain higher morbidity and mortality in areas with high levels of air pollution, allowing for estimation of an independent effect of air pollution. Another important feature of cohort studies is the temporality between exposure and effect that exposure to air pollution comes before the health effect of interest is observed.

Mortality is one of the most studied long-term exposure outcomes, and the outcome that carries the most weight in the evaluation of the evidence on health effects related to air pollution, which is used in the GBD estimation. If lifelong exposure to air pollution contributes to an increased risk of a number of chronic and infectious diseases, from childhood throughout lifetime, then the ultimate consequence is that air pollution exposure shortens one's life, leading to premature death. A number of cohort studies have evaluated the association between long-term exposure to air pollution and mortality, typically all-cause natural mortality (where death due to external causes, such as accidents, suicides, etc. is excluded, as these are not believed to be plausibly related to air pollution). Cause-specific mortality has also been evaluated, including cardiovascular mortality, respiratory (excluding lung cancer) mortality, lung cancer mortality, and more recently diabetes mortality. This substantial amount of evidence has shown that long-term exposure to air pollution increases risk of premature mortality. Most studies on long-term exposure to air pollution and mortality are based on well-known cohort studies from the United States, Canada, and Europe including the American Nurses' Study (Puett, Hart, Yanosky, et al., 2009), American Cancer Society (ACS) Study (Pope, Burnett, Thun, et al., 2002), California Teachers' Study (Ostro, Lipsett, Reynolds, et al., 2010), Health professionals follow-up study (Puett, Hart, Suh, Mittleman, & Laden, 2011), the Women's Health Initiative Observational Study (Miller, Siscovick, Sheppard, et al., 2007), the Netherlands Cohort Study on Diet and Cancer (NLCS) (Beelen, Raaschou-Nielsen, Stafoggia, et al., 2014), the Canadian National Breast Screening Study (Villeneuve, Weichenthal, Crouse, et al., 2015), among others. Metaanalyses of studies on long-term exposure to air pollution and mortality by Hoek et al. found a 6% (95% confidence interval (CI): 4%–8%) and 11% (5%–16%) excess risk in all-cause and cardiovascular disease mortality per 10 $\mu g/m^3$ increase in $PM_{2.5}$, and weaker associations with nonmalignant respiratory disease mortality (3%; 95% CI: −6%–13%) (Hoek, Krishnan, Beelen, et al., 2013). The large multicenter European Study of Cohorts for Air Pollution Effects (ESCAPE) from 2014 included 22 cohorts from 13 European countries and detected a 7% increased risk of all-cause natural mortality (95% CI: 2%–13%) for each 5 μg/m^3 increase in $PM_{2.5}$ (Beelen et al., 2014). More recently, the latest developments in modeling of air pollution with combination of measurements with satellite data have facilitated modeling of air pollution exposure estimates for entire countries. This led to a new approach, the use of large nationwide administrative data in the study of air pollution health effects, allowing for large statistical power, but lacking data on lifestyle risk factors (smoking, body mass index, physical activity), which may be important confounders of the association between air pollution and a health outcome. An example is a large US Medicare mortality study with over 60 million US citizens (Di, Wang, Zanobetti, et al., 2017) which found a 7.3% (95% CI 7.1–7.5) increase in all-cause mortality for each 10 μg/m^3 increase in $PM_{2.5}$, and even stronger effects at air pollution levels below 12 μg/m^3 of $PM_{2.5}$ (13.6%; 95% CI, 13.1–14.1), which is the US EPA's limit value for $PM_{2.5}$.

Another type of design used to study the association between long-term exposure to air pollution and health is a case-control study, which is typically used in studies of rare diseases, such as brain tumors (Poulsen, Sørensen, Andersen, Ketzel, & Raaschou-Nielsen, 2016), Parkinson's disease (PD) (Ritz, Lee, Hansen, et al., 2016), etc., where cohort studies would not have sufficient power to detect associations. In these type of studies, cases, for example, all patients with PD, would be matched with multiple controls who did not develop PD, and air pollution exposure would be estimated at residences of cases and controls to estimate risk of development of PD due to air pollution (Ritz et al., 2016). Panel or longitudinal studies with data on repeated measurements of a health outcome are a valuable design in showing how changes in air pollution can contribute in changes in a health outcome overtime. Two landmark studies from (the) Swiss Study on Air Pollution and Lung Disease in Adults (SAPALDIA), cohort with repeated measurements of lung function, associated improvements in air pollution levels with attenuated lung function decline (Downs, Schindler, Liu, et al., 2007) and reduced rates of respiratory symptoms (Schindler, Keidel, Gerbase, et al., 2009), providing important clues about the dynamics behind long-term air pollution exposures and development of chronic respiratory disease, and giving strong support to inference on causality.

Experimental designs in air pollution epidemiology

Although environmental epidemiology is mainly relying on observed data that have been typically collected for purposes other than studies of air pollution health effects, experimental designs provide invaluable insights into the health effects related to air pollution and strengthen the evidence for causality between air pollution and adverse health effects. Experimental designs include controlled experimental exposure studies used in examining short-term effects of air pollution, often in real-world exposure settings, where health outcome collected immediately after exposure is compared in subjects exposed to high and low air pollution levels. An example includes the Oxford street study, where 60 adult volunteers with mild or moderate asthma spent 2h walking along one of London's busiest roads, Oxford street, and through a nearby park with low air pollution levels, Hyde Park, while undertaking air pollution measurements with a portable sensors, and providing lung function measurements before, during, and after each walk (McCreanor, Cullinan, Nieuwenhuijsen, et al., 2007). The study found that in adults with asthma, walking for 2h at a leisurely pace along Oxford street, with diesel-powered vehicles as the primary air pollution source, resulted in a significant reduction in lung function. This study provides important support to the results of cohort and cross-sectional studies that traffic-related air pollution worsens lung function. A recent study has repeated the same design including heart patients and cardiovascular outcomes (Sinharay, Gong, Barratt, et al., 2018). A similar experimental design was used in a study with 28 volunteers with cycling in 2 exposures settings (quiet square without traffic and on a pedestrian bridge above busy highway) in Barcelona for evaluating whether benefits of exercise in a polluted environment outweigh the adverse effects of added exposure to air pollution during exercise, due to increased inhalation rates (Kubesch, De Nazelle, Guerra, et al., 2015). The authors found that cycling decreased systolic blood pressure in both low and high air pollution exposure levels, documenting that short-term benefits of exercise seem to outweigh the harms due to air pollution exposure.

Natural experimental studies take advantage of an unusual event resulting in extremely high or low air pollution levels that provide unique opportunities of studying how sudden changes in air pollution levels, whether it be increase or decrease, impact health. The Great London Smog Example is an example of a natural experiment study. Another example includes the study of a steel mill closure in Utah Valley over a year period in 1987–88, which documented that 3 times lower levels of PM_{10} during the steel mill closure period resulted into 2–3 times lower hospital admission rates for respiratory disease in children, as compared to the period before and after the steel mill's closure (Pope, 1989). A similar approach was used in the studies evaluating health effects of traffic control measures during the 1996 Olympic Games in Atlanta in 1996 (Peel et al., 2010), the effect of coal ban in Dublin in 1990 on mortality rates (Clancy, Goodman, Sinclair, & Dockery, 2002), among others.

Summary

Air pollution epidemiology as a tool for mapping the health burden related to air pollution is extremely valuable in the evaluation of air pollution control policies and disease prevention. The burden of air pollution is high and likely to increase, as a number of new diseases and outcomes that are under investigation currently will be added to the total burden, and air pollution epidemiology plays an essential role here. Epidemiological studies from countries with low levels of air pollution have demonstrated adverse health effects even below current European and US limit values and WHO guidelines for $PM_{2.5}$ and NO_2. These results will play a major role in future evaluations of air quality standards. All standard tools used in the air pollution epidemiology will remain highly valuable and relevant. However, the field must adapt to seize the opportunities presented by new rapidly evolving technological developments and data availability. Although fossil fuels and combustion-related air pollutants will be less relevant in the future, along with the increasing popularity of electric vehicles, there will be more focus on nonexhaust vehicular particulate emissions from tire and road wear and their health effects. The development of new air pollution sensors for personal monitoring, use of satellite data in air pollution predictions for entire countries, in combination with OMICs for analyses of new health biomarkers, will present new complex big data that demand new statistical and epidemiological approaches. Future environmental epidemiologists will face a world shaped by longer lifespans but also larger burdens of chronic health conditions and multi-comorbidity, continued urbanization, and global environmental changes that will all shape the field and change priorities. The field will address a changing world by focusing on healthy aging, evidence gaps, especially in susceptible populations and middle and low-income countries, and by developing approaches to better handle the complexity of new data and more formalized analysis.

References

Atkinson, R. W., Anderson, H. R., Sunyer, J., et al. (2001). Acute effects of particulate air pollution on respiratory admissions: Results from APHEA 2 project. *American Journal of Respiratory and Critical Care Medicine*, https://doi.org/10.1164/ajrccm.164.10.2010138.

Beelen, R., Raaschou-Nielsen, O., Stafoggia, M., et al. (2014). Effects of long-term exposure to air pollution on natural-cause mortality: An analysis of 22 European cohorts within the multicentre ESCAPE project. *Lancet*, *383*(9919), https://doi.org/10.1016/S0140-6736(13)62158-3.

Bell, M. L., & Davis, D. L. (2001). Reassessment of the lethal London fog of 1952: Novel indicators of acute and chronic consequences of acute exposure to air pollution. *Environmental Health Perspectives*, https://doi.org/10.2307/3434786.

Bobak, M., & Leon, D. A. (1992). Air pollution and infant mortality in the Czech Republic, 1986-88. *Lancet*, https://doi.org/10.1016/0140-6736(92)93017-H.

Clancy, L., Goodman, P., Sinclair, H., & Dockery, D.W. (2002). Effect of air-pollution control on death rates in Dublin, Ireland: An intervention study. *Lancet*, https://doi.org/10.1016/S0140-6736(02)11281-5.

Cohen, A. J., Brauer, M., Burnett, R., et al. (2017). Estimates and 25-year trends of the global burden of disease attributable to ambient air pollution: an analysis of data from the Global Burden of Diseases Study 2015. *Lancet (London, England)*, *389*(10082), 1907–1918. https://doi.org/10.1016/S0140-6736(17)30505-6.

Di, Q., Wang, Y., Zanobetti, A., et al. (2017). Air pollution and mortality in the medicare population. *The New England Journal of Medicine*, https://doi.org/10.1056/NEJMoa1702747.

Dijkema, M. B. A., Mallant, S. F., Gehring, U., et al. (2011). Long-term exposure to traffic-related air pollution and type 2 diabetes prevalence in a cross-sectional screening-study in the Netherlands. *Environmental Health*, *10*, 76. https://doi.org/10.1186/1476-069X-10-76.

Dockery, D. W., Pope, C. A., Xu, X., et al. (1993). An association between air pollution and mortality in six U.S. cities. *The New England Journal of Medicine*, https://doi.org/10.1056/NEJM199312093292401.

Dominici, F., Samet, J. M., & Zeger, S. L. (2000). Combining evidence on air pollution and daily mortality from the 20 largest US cities: A hierarchical modelling strategy. *Journal of the Royal Statistical Society. Series A, Statistics in Society*, https://doi.org/10.1111/1467-985x.00170.

Downs, S. H., Schindler, C., Liu, L. J. S., et al. (2007). Reduced exposure to PM10 and attenuated age-related decline in lung function. *The New England Journal of Medicine*, https://doi.org/10.1056/NEJMoa073625.

Evans, J. S., Tosteson, T., & Kinney, P. L. (1984). Cross-sectional mortality studies and air pollution risk assessment. *Environment International*, *10*(1), 55–83. https://doi.org/10.1016/0160-4120(84)90233-2.

Hansen, A. B., Ravnskjær, L., Loft, S., et al. (2016). Long-term exposure to fine particulate matter and incidence of diabetes in the Danish Nurse Cohort. *Environment International*, https://doi.org/10.1016/j.envint.2016.02.036.

Hoek, G., Krishnan, R. M., Beelen, R., et al. (2013). Long-term air pollution exposure and cardio-respiratory mortality: A review. *Environmental Health A Global Access Science Source*, https://doi.org/10.1186/1476-069X-12-43.

Hood, C., & Davis, D. L. (2004). When smoke ran like water: Tales of environmental deception and the battle against pollution. *Environmental History*, https://doi.org/10.2307/3985785.

Iskandar, A., Andersen, Z. J., Bonnelykke, K., Ellermann, T., Andersen, K. K., & Bisgaard, H. (2012). Coarse and fine particles but not ultrafine particles in urban air trigger hospital admission for asthma in children. *Thorax*, *67*(3), 252–257. https://doi.org/10.1136/thoraxjnl-2011-200324.

Katsouyanni, K., Touloumi, G., Samoli, E., et al. (2001). Confounding and effect modification in the short-term effects of ambient particles on total mortality: Results from 29 European cities within the APHEA2 project. *Epidemiology*, https://doi.org/10.1097/00001648-200109000-00011.

Katsouyanni, K., Zmirou, D., Spix, C., et al. (1995). Short-term effects of air pollution on health: A European approach using epidemiological time-series data. The APHEA project: Background, objectives, design. *The European Respiratory Journal*, *8*, 1030–1038.

Kubesch, N., De Nazelle, A., Guerra, S., et al. (2015). Arterial blood pressure responses to short-term exposure to low and high traffic-related air pollution with and without moderate physical activity. *European Journal of Preventive Cardiology*, https://doi.org/10.1177/2047487314555602.

Le Tertre, A., Medina, S., Samoli, E., et al. (2002). Short-term effects of particulate air pollution on cardiovascular diseases in eight European cities. *Journal of Epidemiology and Community Health*, https://doi.org/10.1136/jech.56.10.773.

Lepeule, J., Laden, F., Dockery, D., & Schwartz, J. (2012). Chronic exposure to fine particles and mortality: An extended follow-up of the Harvard six cities study from 1974 to 2009. *Environmental Health Perspectives*, https://doi.org/10.1289/ehp.1104660.

Lim, S. S., Vos, T., Flaxman, A. D., et al. (2012). A comparative risk assessment of burden of disease and injury attributable to 67 risk factors and risk factor clusters in 21 regions, 1990–2010: A systematic analysis for the Global Burden of Disease Study 2010. *Lancet (London, England)*, *380*(9859), 2224–2260. https://doi.org/10.1016/S0140-6736(12)61766-8.

Liu, C., Chen, R., Sera, F., et al. (2019). Ambient particulate air pollution and daily mortality in 652 cities. *The New England Journal of Medicine*, https://doi.org/10.1056/NEJMoa1817364.

Liu, C., Chen, R., Zhao, Y., et al. (2017). Associations between ambient fine particulate air pollution and hypertension: A nationwide cross-sectional study in China. *The Science of the Total Environment*, https://doi.org/10.1016/j.scitotenv.2017.01.133.

Logan, W. P. D. (1953). Mortality in the London fog incident, 1952. *Lancet*, https://doi.org/10.1016/S0140-6736(53)91012-5.

McCreanor, J., Cullinan, P., Nieuwenhuijsen, M. J., et al. (2007). Respiratory effects of exposure to diesel traffic in persons with asthma. *The New England Journal of Medicine*, https://doi.org/10.1056/NEJMoa071535.

Miller, K. A., Siscovick, D. S., Sheppard, L., et al. (2007). Long-term exposure to air pollution and incidence of cardiovascular events in women. *The New England Journal of Medicine*, https://doi.org/10.1056/NEJMoa054409.

Nemery, B., Hoet, P. H. M., & Nemmar, A. (2001). The Meuse Valley fog of 1930: An air pollution disaster. *Lancet*, https://doi.org/10.1016/S0140-6736(00)04135-0.

Ostro, B., Lipsett, M., Reynolds, P., et al. (2010). Long-term exposure to constituents of fine particulate air pollution and mortality: Results from the California teachers study. *Environmental Health Perspectives*, https://doi.org/10.1289/ehp.0901181.

Özkaynak, H., & Thurston, G. D. (1987). Associations between 1980 U.S. mortality rates and alternative measures of airborne particle concentration. *Risk Analysis*, https://doi.org/10.1111/j.1539-6924.1987.tb00482.x.

Peel, J. L., Klein, M., Flanders, W. D., Mulholland, J. A., Tolbert, P. E., HEI Health Review Committee. (2010). Impact of improved air quality during the 1996 summer Olympic Games in Atlanta on multiple cardiovascular and respiratory outcomes. *Research Report. Health Effects Institute*, 3–33 [148].

Pope, C. A. (1989). Respiratory disease associated with community air pollution and a steel mill, Utah Valley. *American Journal of Public Health*, https://doi.org/10.2105/ajph.79.5.623.

Pope, C. A., Burnett, R. T., Thun, M. J., et al. (2002). Lung cancer, cardiopulmonary mortality, and long-term exposure to fine particulate air pollution. *Journal of the American Medical Association*, *287*(9), 1132–1141. https://doi.org/10.1016/j.scitotenv.2011.03.017.

Poulsen, A. H., Sørensen, M., Andersen, Z. J., Ketzel, M., & Raaschou-Nielsen, O. (2016). Air pollution from traffic and risk for brain tumors: A nationwide study in Denmark. *Cancer Causes & Control*, *27*(4), 473–480. https://doi.org/10.1007/s10552-016-0721-x.

Puett, R. C., Hart, J. E., Suh, H., Mittleman, M., & Laden, F. (2011). Particulate matter exposures, mortality, and cardiovascular disease in the health professionals follow-up study. *Environmental Health Perspectives*, https://doi.org/10.1289/ehp.1002921.

Puett, R. C., Hart, J. E., Yanosky, J. D., et al. (2009). Chronic fine and coarse particulate exposure, mortality, and coronary heart disease in the nurses' health study. *Environmental Health Perspectives*, https://doi.org/10.1289/ehp.0900572.

Ritz, B., Lee, P.-C., Hansen, J., et al. (2016). Traffic-related air pollution and Parkinson's disease in Denmark: A case-control study. *Environmental Health Perspectives*, *124*(3), 351–356. https://doi.org/10.1289/ehp.1409313.

Samet, J. M., Dominici, F., Curriero, F. C., Coursac, I., & Zeger, S. L. (2000). Fine particulate air pollution and mortality in 20 U.S. cities, 1987–1994. *The New England Journal of Medicine*, https://doi.org/10.1056/NEJM200012143432401.

Samet, J. M., Dominici, F., Zeger, S. L., Schwartz, J., & Dockery, D. W. (2000). The National morbidity, mortality, and air pollution study. Part I: Methods and methodologic issues. *Research Report. Health Effects Institute*, *94*(Pt 1), 5–84.

Samet, J. M., Zeger, S. L., Dominici, F., et al. (2000). The National morbidity, mortality, and air pollution study. Part II: Morbidity and mortality from air pollution in the United States. *Research Report. Health Effects Institute*, *94*(Pt 2), 5–79.

Schindler, C., Keidel, D., Gerbase, M.W., et al. (2009). Improvements in PM10 exposure and reduced rates of respiratory symptoms in a cohort of swiss adults (SAPALDIA). *American Journal of Respiratory and Critical Care Medicine*, https://doi.org/10.1164/rccm.200803-388OC.

Schwartz, J., & Marcus, A. (1990). Mortality and air pollution in London: A time series analysis. *American Journal of Epidemiology*, https://doi.org/10.1093/oxfordjournals.aje.a115473.

Sinharay, R., Gong, J., Barratt, B., et al. (2018). Respiratory and cardiovascular responses to walking down a traffic-polluted road compared with walking in a traffic-free area in participants aged 60 years and older with chronic lung or heart disease and age-matched healthy controls: A randomised, crossover study. *Lancet*, https://doi.org/10.1016/S0140-6736(17)32643-0.

Stanaway, J. D., Afshin, A., Gakidou, E., et al. (2018). Global, regional, and national comparative risk assessment of 84 behavioural, environmental and occupational, and metabolic risks or clusters of risks for 195 countries and territories, 1990-2017: A systematic analysis for the global burden of disease Study. *Lancet*, https://doi.org/10.1016/S0140-6736(18)32225-6.

Sunyer, J., Atkinson, R., Ballester, F., et al. (2003). Respiratory effects of sulphur dioxide: A hierarchical multicity analysis in the APHEA 2 study. *Occupational and Environmental Medicine*, https://doi.org/10.1136/oem.60.8.e2.

Sunyer, J., Ballester, F., Le Tertre, A., et al. (2003). The association of daily sulfur dioxide air pollution levels with hospital admissions for cardiovascular diseases in Europe (the Aphea-II study). *European Heart Journal*, https://doi.org/10.1016/S0195-668X(02)00808-4.

United Nations. (2017). *The sustainable development goals report*. United Nations Publisher. https://doi.org/10.18356/3405d09f-en.

Villeneuve, P. J., Weichenthal, S. A., Crouse, D., et al. (2015). Long-term exposure to fine particulate matter air pollution and mortality among Canadian women. *Epidemiology*, https://doi.org/10.1097/EDE.0000000000000294.

Wichmann, H. E. (2004). What can we learn today from the central European smog episode of 1985 (and earlier episodes)? *International Journal of Hygiene and Environmental Health*, https://doi.org/10.1078/1438-4639-00322.

Wichmann, H. E., Mueller, W., Allhoff, P., et al. (1989). Health effects during a smog episode in West Germany in 1985. *Environmental Health Perspectives*, https://doi.org/10.1289/ehp.897989.

Wilkins, E. T. (1954). Air pollution aspects of the London fog of December 1952. *Quarterly Journal of the Royal Meteorological Society*, https://doi.org/10.1002/qj.49708034420.

World Health Organization. (2006). *Air quality guidelines. Global update 2005.* https://doi.org/10.1007/BF02986808.

Zanobetti, A., Schwartz, J., & Dockery, D. W. (2000). Airborne particles are a risk factor for hospital admissions for heart and lung disease. *Environmental Health Perspectives*, https://doi.org/10.1289/ehp.001081071.

CHAPTER 8

Systematic reviews and metaanalyses of air pollution epidemiological studies

Juleen Lam[a], Hanna M. Vesterinen[b], and Tracey J. Woodruff[b]
[a]Department of Health Sciences, California State University East Bay, Hayward, CA, United States
[b]Program on Reproductive Health and the Environment, University of California, San Francisco, CA, United States

Abbreviations

AMSTAR	assessment of multiple systematic reviews
ARRIVE	animal research: reporting of in vivo experiments
ASD	autism spectrum disorder
CASP	critical appraisal skills program
EPA	US Environmental Protection Agency
GRADE	grading of recommendations assessment, development, and evaluation
HEI	Health Effects Institute
LRAT	literature review appraisal toolkit
NTP	National Toxicology Program
NRC	National Research Council
PECO	participants/population, exposure, comparator, and outcomes
PRISMA	preferred reporting items for systematic reviews and metaanalyses
STROBE	strengthening the reporting of observational studies in epidemiology
TRAP	traffic-related air pollution
WHO	World Health Organization

Introduction

Decision-making to prevent harmful exposures to public health rely on high-quality scientific evidence. In particular for ubiquitous environmental exposures such as air pollution, robust methods to synthesize what is known about links between exposures and harmful health effects are necessary to inform public policy and prevent harm. However, environmental health decisions have historically relied on expert-based narrative reviews to summarize the collection of scientific information (Woodruff & Sutton, 2014). Narrative reviews are comprehensive overviews of the scientific literature that covers a wide range of issues within a given topic. This approach has drawn criticisms for a variety of different

Traffic-Related Air Pollution
https://doi.org/10.1016/B978-0-12-818122-5.00008-9

reasons, including lacking the ability to rigorously evaluate the evidence in a systematic, transparent, and reproducible manner that can produce a bottom-line summary of the evidence. With evidence that harmful chemical exposures are increasing and contributing to adverse health outcomes, the ability for decision makers to base policies and actions on the best available scientific information is crucial to ensure adequate public health protection.

This issue is not unique to the field of environmental health. Notably, clinical medicine faced similar critiques >40 years ago for its reliance on narrative reviews to guide recommendations for treatment or intervention decisions for patients (Rennie & Chalmers, 2009). In response, an explicit approach to harness expertise in a rigorous, transparent, and systematic framework was developed, validated, and empirically tested for several decades. The result was the process of systematic review, defined as a review of literature focused on a specific question that uses explicit, prespecified methods to identify, select, assess, and synthesize scientific evidence (Institute of Medicine, 2011). Prominent systematic review methods such as the Cochrane Collaboration (Higgins & Green, 2019) and Grading of Recommendations Assessment, Development, and Evaluation (GRADE) (Guyatt et al., 2008) are now regularly relied upon in health care to inform decisions that lead to cost savings and improved health outcomes (Fox, 2010). Systematic review methods are currently viewed in clinical sciences as well-established approaches for assessing data to reach recommendations about health care and interventions (Guyatt et al., 2011).

The integration of systematic review methods within the field of environmental health has been expanding over the last 10 years, starting with the Navigation Guide Systematic Review method in the early 2000s (Woodruff & Sutton, 2014). Systematic reviews in environmental health have been evolved to account for differences that exist between clinical medicine and environmental health sciences in the types of evidence available to decision makers. Clinical medicine typically relies on randomized human clinical trials; these data are often available because of regulatory requirements, such as the approval process for new pharmaceuticals. In contrast, the vast majority of chemicals in US commerce have entered the marketplace with limited testing for adverse health outcomes—thus, environmental health practitioners must rely on post hoc experimental animal, observational human, or in vitro mechanistic studies. Randomized controlled trials are virtually precluded because of feasibility and ethical considerations. Thus, the nature of the evidence review in clinical medicine (e.g., GRADE) does not appropriately account for observational human studies.

An additional complication is the context of decision-making which is vastly different in the two fields—in clinical medicine, new pharmaceuticals or interventions are evaluated by weighing the clinical risk-benefit, abiding by regulatory and medical ethical requirements that human exposure to treatments does not occur in the absence of potential benefits that outweigh the risks, which is primarily done premarket (Guyatt et al., 2008). In contrast, environmental chemicals typically do not undergo the same comparison of risks and benefits. Environmental chemical and pollution exposures are typically already ongoing in the public and delays to remove harmful exposures result in continued exposures and result in adverse health effects. Because of these collective differences, systematic review procedures developed in clinical sciences cannot be directly applied to decision-making in environmental health.

Systematic reviews in environmental health have been developed and tested over the past decade (National Research Council, 2014; National Toxicology Program (NTP), 2019; Rooney, Boyles, Wolfe, Bucher, & Thayer, 2014; Woodruff & Sutton, 2014) Some of these systematic review frameworks have been applied to multiple case studies in the field of environmental health (Johnson et al., 2016; Lam et al., 2014, 2016, 2017; NTP, 2019; National Toxicology Program, 2013, 2016, 2018, 2019; Vesterinen et al., 2015). The methods share many commonalities in the fundamental steps of a systematic review, though there are slight differences in the details of the applications (Table 8.1).

Systematic reviews and air pollution

Air pollution in general is a particularly complex exposure to assess systematically for health outcomes, in part due to varying pathophysiological mechanisms affecting apical end points as well as variances due to specific air pollution chemical components, such as complex mixtures (PM_{10} or $PM_{2.5}$) vs specific components (e.g., metals). A recent example of an application of a systematic review framework developed specifically for the field of environmental health (Woodruff & Sutton, 2014) to a research question related to air pollution generally is summarized in Box 8.1. This systematic review (Lam et al., 2016) was the first systematic review and metaanalysis of the literature on the association between autism spectrum disorder (ASD)

Table 8.1 Fundamental steps of a systematic review.

Steps	Details
Specify a focused research question	A specific research question relevant to decision makers about whether human exposure to a chemical or class of chemicals is a health risk
Prepare a protocol	A protocol is a plan or set of steps to be followed in a study. A protocol for a systematic review should describe the rationale for the review, the objectives, and the methods that will be used to locate, select, and critically appraise studies, and to collect and analyze data from the included studies
Search for evidence	Conduct a comprehensive systematic search of the scientific literature covering multiple databases, with details of the search strategy documented in the protocol for reproducibility
Select evidence	All references identified from the search are screened for relevance to the research question, based on the PECO statement, an acronym for: Population or Participants (P), Exposure (E), Comparator (C), and Outcome (O), which represent the key elements in the question of the effects of an intervention or exposure
Extract data	Relevant data (as outlined in the protocol) from included studies are extracted into a database to facilitate evaluation of individual and overall study quality
Synthesize the data	Data from studies deemed to be combinable (as outlined in the protocol) are synthesized, quantitatively if possible in a metaanalysis—a quantitative statistical analysis that integrates results of chosen studies. Studies not combinable in a metaanalysis can be visually displayed or qualitatively evaluated systematically
Assess risk of bias of individual studies	Each included study is assessed for risk of bias (or internal validity), a measure of the credibility of study findings that reflects the ability of a study's design and conduct to protect against systematic errors that may bias (over or underestimate) the results or estimate of effect from the true value
Rate the overall body of evidence	The body of overall evidence is rated by considering the strengths and weaknesses of studies with similar study design features. Different frameworks utilize different approaches for this step, including rating the confidence or quality of studies, and level or strength of evidence. Conclusions in the form of a clear, bottom-line statement regarding the body of evidence in relation to the research question are presented

BOX 8.1 Air pollution systematic review case study.

Research question: is developmental air pollution exposure associated with Autism Spectrum Disorder (ASD)?

PECO STATEMENT

P(articipants): Humans

E(xposure): Any developmental exposure to air pollution that occurred prior to the ASD assessment.

C(omparator): Humans exposure to lower levels of air pollution than the more highly exposed humans.

O(utcome): Any clinical diagnosis or other continuous or dichotomous scale assessment of ASD.

Protocol: Available on PROSPERO, a University of York database for registering prespecified systematic review protocols (http://www.crd.york.ac.uk/PROSPERO/; CRD # 42015017890).

Search for evidence: Searched PubMed, ISI Web of Science, Biosis Previews, Embase, Google Scholar, Toxline, additional toxicological websites, and gray literature databases using search terms outlined in the protocol. Conducted "snowball searching" to hand search the reference lists of all included studies and review articles identified during screening, and using Web of Science to search articles that cited the included studies. A group of recognized experts in the field were solicited to review included studies to identify any potentially missed studies. A total of 1155 references were retrieved.

Select evidence: Studies were screened in duplicate in DRAGON software (now referred to as litstream: ICF International, http://www.icfi.com/insights/products-and-tools/dragon-online-tool-systematic-review) and DistillerSR (Evidence Partners; available at: http://www.systematic-review.net), through title and abstract screening followed by full text screening. 96 studies were included after title and abstract review, 20 studies were included after full text review, and 3 studies were identified after consulting experts. A total of 23 studies were included in the systematic review.

Extract data: Data fields prespecified in the protocol were extracted independently by two reviewers into a DRAGON database. Corresponding authors were contacted to obtain information missing from the published paper.

Synthesize data: Predetermined features in the protocol were used to evaluate study comparability. Different air pollutant chemicals or classes or compounds were analyzed separately. Studies measuring PM_{10} (six studies) and $PM_{2.5}$ (three studies) prior to the assessment of ASD were deemed combinable in a metaanalysis. A metaanalysis of PM_{10} studies identified an overall effect estimate of OR = 1.07 (95% CI: [1.06, 1.08]) per 10-$\mu g/m^3$ increase in PM_{10}, indicating a statistically significant relationship between increasing PM_{10} exposure and resulting odds of ASD diagnosis. A metaanalysis of $PM_{2.5}$

(Continued)

BOX 8.1 Air pollution systematic review case study—cont'd

studies identified an overall effect estimate of OR = 2.32 (95% CI: [2.15, 2.51]) per 10-μg/m^3 increase in $PM_{2.5}$. All other studies were visually displayed on scatterplots by chemical air pollutant to visually inspect results and evaluate associations. A general trend towards positive effects (increasing exposure associated with increasing autism risks) was observed, although there were limited data.

Assess risk of bias of individual studies: Overall, most studies were rated as "low" or "probably low" risk of bias in most domains other than confounding and exposure assessment. Most studies receiving "probably high" or "high" ratings for potential confounding were due to failure to adjust for many of the important confounders outlined in the protocol. Exposure assessment risk of bias was rated separately for 103 different air pollution chemicals and chemical classes.

Rate the overall body of evidence: The quality of human evidence was rated as "moderate," based primarily on concerns regarding "high" or "probably high" risk of bias for exposure assessment methods. The final overall strength of evidence was rated as "limited," due to concerns of limited number of studies and presence of unexplained heterogeneity. The final conclusion was "limited evidence of toxicity" for the association between early life exposure to air pollution and diagnosis of ASD.

(From Lam, J., Sutton, P., Kalkbrenner, A., Windham, G., Halladay, A., Koustas, E., et al. (2016). A systematic review and meta-analysis of multiple airborne pollutants and autism spectrum disorder. PLoS One. 11(9), e0161851.)

and air pollution and ultimately concluded that there was "limited evidence of toxicity." While there was a relatively rich database of human evidence, we found several critical gaps in this literature. These included challenges in assessing exposure to the complex mixtures that make up air pollution with different types of pollutants being studied (e.g., criteria air pollutants vs hazardous air pollutants (as defined in the US Clean Air Act)), heterogeneity in study design, reporting effect estimates on different scales that could not readily be standardized for comparison, and presence of heterogeneity that could not be explained by study characteristics, thus leading the diverse group of scientists to this conclusion (Lam et al., 2016). An important strength of conducting this systematic review (and systematic reviews in general) is that it allowed us to identify more precisely the research gaps which can be used to prioritize future studies to improve the ability to increase the strength of conclusions.

Traffic-related air pollution systematic reviews

Systematic reviews are becoming the norm in environmental health and have been recommended or conducted by the National Academy of Sciences in four separate reports (National Academies of Sciences Engineering and Medicine, 2017, 2018, 2019; National Research Council, 2014). Additionally, systematic reviews have been increasingly adopted by federal and international agencies (National Toxicology Program (NTP), 2019; U.S. Environmental Protection Agency, 2018; World Health Organization, 2017). However, the quality of systematic reviews vary and do not always adhere to best practices (National Academies of Sciences Engineering and Medicine, 2019). Systematic reviews of traffic-related air pollution have been evolving. In 2010, the Health Effects Institute (HEI) published a critical review of the literature on traffic-related air pollution (TRAP) from fossil fuel engine combustion emissions and health effects, identifying potential associations with asthma (incident asthma and asthma exacerbation), reduced lung function, myocardial infarction, atherosclerosis, and cardiovascular mortality (Health Effects Institute (HEI), 2010), though this was not a systematic review. Since the publication of this critical review, several systematic reviews have been published regarding the human health effects of TRAP. Accordingly, to assess the state of the field, identify best practices and opportunities for improvement, in this chapter, we provide an overview of the state of systematic reviews on traffic-related air pollution (TRAP) and human health effects (as of March 2019).

We use systematic methods to identify existing systematic reviews and evaluate their quality to assess the current status of reviews in the field—specifically to address whether systematic reviews are adequately being conducted and if not, what gaps must be addressed to improve evidence evaluation.

Summary of included studies

Systematic reviews were identified from a broad search in PubMed, Embase, and Web of Science and screened to identify studies that (1) were a systematic review; (2) investigated chemical components of traffic-related air pollution exposures; (3) reported on any health-related outcome. Our electronic search of the scientific literature (from the inception of the databases searched until March 2019) resulted in 58 systematic reviews relevant to TRAP exposures and human health outcomes. Systematic reviews were published between 2010 and 2019 but not all were identified as such by the study authors; the majority of studies were identified as being a systematic review ($n=52$) by the study authors, of which 27 also included a metaanalysis for quantitative synthesis of data. One study identified as only a metaanalysis without reporting that it was also a systematic review; another study identified as a "catalog and synthesis," and four studies did not report what type of review they were but were considered eligible based on their methodology.

Study population

Reviews were categorized as using adults ($n=21$), children ($n=11$), or all ages ($n=14$). A further 12 reviews did not specify their population directly, although for the majority of these ($n=9$) it could be reasonably inferred that the population would likely be adults based on the outcomes of interest (e.g., hypertension).

Of the reviews including adult subjects, 14 were specific as to the type of population: 2 used only males, 3 included only pregnant women, 6 focused solely on occupational exposure or included occupational exposed adults (e.g., includes commercial drivers, outdoor workers exposed to air pollution, e.g., drivers/petrol pump attendants), 1 examined commuters, 1 included only subjects with preexisting disease, and 1 included in utero exposure.

This variability in study population generally illustrates another common problem in systematic reviews related to challenges in conducting quantitative data synthesis of evidence. Included studies may have heterogeneous study populations resulting from the study design, for instance, due to the specific outcome being investigated and the population being affected. Although this increases the evidence base for the variety of potential health effects within the population that is attributable to an environmental exposure, this ultimately limits the ability to combine study results when few studies investigating similar populations are available. This is further compounded by the fact that often data are reported collectively for the entire study population, potentially with wide ranges in age, socioeconomic status, or other individual characteristics, which results in the need to contact study authors to obtain additional data, such as the stratification of study results by key characteristics such as age.

Exposures

Measures of TRAP exposure were diverse among the 58 included reviews (Fig. 8.1). These include: TRAP-related ambient air pollutants (SO_2 ($n=8$), CO ($n=7$), NO_2 ($n=12$), O_3 ($n=7$), NO_x ($n=6$), black or elemental carbon

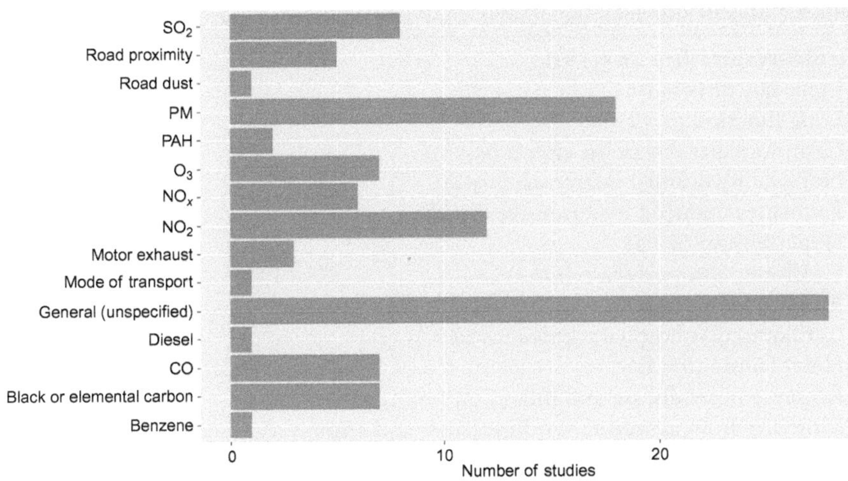

Fig. 8.1 Number of systematic reviews reporting on each TRAP-related exposure (some studies report more than one exposure and may be represented multiple times).

(n=7), poly aromatic hydrocarbon (PAH; n=7)), road proximity (n=5), road dust (n=1), diesel (n=1), benzene (n=1), motor exhaust (n=4), mode of transport (n=1), or unspecified (i.e., the study said it was evaluating exposure to TRAP, but the specific exposures were not reported; n=28). This demonstrates the challenges in assessing exposure to the complex mixtures that defines TRAP exposures as well as air pollution in general. In particular, this limits the ability to combine information from different studies via metaanalysis. The heterogeneity can be further exacerbated by variability in exposure assessment method used (i.e., monitoring, modeling, or use of surrogate measures such as distance to freeway). To increase the usefulness of study results, investigators must directly address these challenges, for instance, by incorporating newer technologies (such as novel exposomic tools) to maximize the accuracy and applicability of exposure measurements to the target population, developing tools that can be used to compare across exposure measures, and/or creating standards for measuring exposures in more consistent ways.

Health outcomes

There was substantial diversity across the health outcomes reported in the 58 included reviews (Table 8.2). The most common were related to cardiorespiratory diseases and mortality (n=19), asthma (n=5), fertility (n=4), diabetes risk (n=4), childhood cancers (n=4), and cognition (n=3).

Table 8.2 Health outcomes included in TRAP systematic reviews.

Cardio-respiratory ($n = 19$):
- Lung cancer ($n=3$)
- Lung function ($n=1$)
- Cardiovascular disease or effects ($n=5$)
- Nonfatal myocardial infarction ($n=1$)
- Pregnancy-induced hypertensive disorders ($n=1$)
- Hypertension ($n=1$)
- Cardiorespiratory disease ($n=1$)
- Atherosclerosis ($n=1$)
- Cardiovascular hospital admissions ($n=1$)
- Heart failure ($n=1$)
- Ischemic heart disease and infarction ($n=1$)
- Mortality from all-causes, cardiovascular and respiratory disease ($n=1$)
- Morbidities of respiratory tract diseases ($n=1$)

Asthma ($n=5$)

General health effects ($n=5$)[a]

Fertility ($n=4$):
- Male fertility ($n=2$)
- Fertility reproductive health ($n=1$)
- Gestational fetal growth ($n=1$)

Diabetes risk ($n=4$):
- Type 2 diabetes ($n=3$)
- Risk or either type 1, 2 or gestational ($n=1$)

Childhood cancer ($n=4$):
- Leukemia ($n=3$)
- Any cancer ($n=1$)

Cognition ($n=3$)

Congenital heart disease or abnormalities ($n=2$)

Morbidity, mortality and hospital admissions ($n=3$)[b]

Genetic ($n=3$):
- DNA or oxidative damage ($n=2$)
- Immune response genes ($n=1$)

Other ($n=6$):
- Parkinson's disease risk ($n=1$)
- Neuroimaging ($n=1$)
- Rheumatoid arthritis ($n=1$)
- Tuberculosis outcomes ($n=1$)
- Years of life expectancy ($n=1$)
- Autism spectrum disorder ($n=1$)

[a]Much like our systematic review was not specific on the health outcome, these were too. For example, one systematic review included in our systematic review identified studies pertaining to white blood cell count, lung cancer risk and mortality ratio.

[b]One study described morbidity outcomes as relating to: "cardiovascular, respiratory, cancer, diabetes, hospitalization, neurological, and pregnancy-birth."

This wide range of adverse health outcomes illustrates the severity of potential health effects related to TRAP.

However, although several health outcomes were generally common across the reviews, the specific outcome measured varied quite a bit. For instance, although 19 studies measured cardiorespiratory health outcomes, breaking this down to more specific outcome resulted in five reviews measuring cardiovascular disease or effects, three reviews measuring lung cancer, and one review measuring the range of other health outcomes (such as lung function, nonfatal myocardial infarction, hypertension, heart failure, etc.). Thus, even considering a relatively large number of systematic reviews, comparability across the varied set of health outcomes is rather limited. This illustrates a common challenge with environmental health exposures where suspected links to a wide range of health outcomes requires a large body of evidence for sufficient investigation of each potential effect.

However, it is not necessarily a limitation that few systematic reviews are available for specific health outcomes. A well-conducted systematic review that offers a robust and comprehensive evaluation of the existing body of evidence may be sufficient to support decision-making about environmental chemicals. To achieve this, the systematic review must be of high quality, with valid and consistent methodology applied to evaluate the body of evidence.

Review methodology

We identified how many of the reviews reported that they used a quality assessment tool, though some of the tools reported are more checklists or have other problems for assessing quality. Of the 58 reviews 25 reported on using a study quality, with the most commonly used tool ($n=7$) being the Newcastle-Ottawa Scale (Wells et al., 2019). The Newcastle-Ottawa Scale is a quality assessment tool jointly developed by the Universities of Newcastle (Australia) and Ottawa (Canada) to evaluate nonrandomized studies included in systematic reviews, such as cohort and case-control epidemiology studies (Wells et al., 2019). Although widely used and endorsed by the Cochrane Collaboration (Higgins et al., 2019), this tool has faced criticisms for low interrater reliability, difficulty in using, and lack of ability to accurately identify studies with biased results (Hartling et al., 2013). Other reported quality assessment tools were also used to evaluate study quality, including the National Toxicology Program (NTP) Office of Health Assessment and Translation systematic review framework (National Toxicology Program (NTP), 2019), the Strengthening the Reporting of

Observational Studies in Epidemiology (STROBE) checklist (Von Elm et al., 2008), and the Critical Appraisal Skills Program (CASP) checklist (Critical Appraisal Skills Programme U.K, 2013). However, the majority of systematic reviews ($n=33$) either did not assess study quality or only informally assessed it without the use of a structured assessment tool. Consistency in evaluating study quality is important for ensuring that studies are evaluated in the same way, increasing confidence in the overall conclusions of the analysis. However, as we have identified there is variation in how systematic reviews evaluate included studies, and this is a problem that is observed in systematic reviews in other fields.

Of the 58 included reviews only 30 (51%) included a quantitative data analysis component: the majority performed a random-effects metaanalysis alone ($n=17$); 9 performed both a random and fixed effects metaanalysis; 1 performed metaregression; 1 performed a count of significant and nonsignificant results; and 2 performed both random-effects metaanalysis and Bayesian metaanalysis. Less than half of the reviews (21) statistically evaluated the presence of publication bias, the majority of which used funnel plotting, with or without Egger's regression or trim and fill method. Thus, half of the included systematic reviews lacked any quantitative data synthesis, even though a large number of studies were included in many of the reviews (average of 39 included studies within each systematic review). This illustrates a common issue with systematic reviews—many lack quantitative synthesis of data via metaanalysis, which can strengthen the capacity to detect an association beyond individual study findings (given often limited statistical power in individual studies) and strengthen confidence in the findings.

A key challenge to conducting a metaanalysis is that many individual studies are either too heterogeneous to combine (i.e., differences in study design, timing of exposure and outcome measurement, etc.) or lack sufficient data reporting to combine studies (e.g., may report a *p*-value when an estimate of sample variability such as a confidence interval is needed). Often, it is necessary for systematic review authors to contact study authors to obtain additional data or information, which is not always available, particularly with older studies. Retroactively obtaining additional data can also be time consuming with potentially low success rates—for instance, many corresponding authors may no longer have access to the data in question, particularly in research labs where graduate students may have generated and analyzed much of the data who are no longer present. This speaks to the need to consider approaches for archiving or aggregating data after studies are concluded, for instance, through free and publicly accessible centralized database platforms such as Open Science Database (http://www.opensciencedb.com/) or Zenodo (https://zenodo.org/).

A broader issue with ensuring adequate reporting within the publication of a study is tied to the overall need to follow more standard reporting requirements. This is related to issues regarding lack of sufficient data to combine studies in a metaanalysis highlighted above, but is also tied to the ability to sufficiently evaluate potential biases within a study by including key elements of a study's design or implementation within the published manuscripts. Again, this often necessitates contacting study authors to obtain additional information, which can be time consuming or unsuccessful. Several high-impact journals have recently adopted checklists for the reporting of elements necessary to describe studies comprehensively and transparently, such as the Animal Research: Reporting of In Vivo Experiments (ARRIVE) guidelines for experimental animal studies (Kilkenny et al., 2010) or STROBE guidelines for observational human studies (Von Elm et al., 2008). Additionally, in 2016 Environment International hired an editor explicitly for overseeing systematic reviews and has a policy of peer reviewing and publishing protocols prior to peer reviewing systematic reviews (Whaley, Letcher, Covaci, & Alcock, 2016). This may help to ensure the prospective reporting of necessary details for incorporating study results into future literature reviews. However, to effectively address this issue, greater standardization of study methods including approaches to measure exposures and outcomes and incorporate important confounders would be appropriate (although challenging).

Lastly, only 3 of the 58 reviews reported conclusions on the overall strength of the body of evidence (De Marchis, Verso, Tramuto, Amodio, & Picciotto, 2018; Demetriou et al., 2012; Morales-Suarez-Varela, Peraita-Costa, & Llopis-Gonzalez, 2017). Lacking a clear and concise bottom-line conclusion regarding the strength of the evidence (i.e., concluding statements used by the US Environmental Protection Agency (EPA) such as "known to be toxic," "possibly toxic," etc. (U.S. Environmental Protection Agency, 1991, 1996)) can hinder the utility of systematic reviews for decisions-making on environmental chemicals as there is a lack of a bottom-line summary that helps make the science become actionable in further decision-making.

Evaluating systematic review quality

The Policy from Science Project Literature Review Appraisal Toolkit (LRAT) was used to assess the credibility of each systematic review

(Policy from Science, 2014). LRAT is one of the first published appraisal toolkits specifically tailored for environmental health systematic reviews. Its creation was based on modification of several well-known toolkits for appraising the methodological quality of medical science literature reviews, including Cochrane (Higgins et al., 2019), the Preferred Reporting Items for Systematic Reviews and Metaanalyses (PRISMA) (Moher et al., 2009), and Assessment of Multiple Systematic Reviews (AMSTAR) (Shea et al., 2007). The goal of LRAT is to provide a structured appraisal process for evaluating the strengths and weaknesses of literature reviews and guide and, ultimately, a decision regarding the level of confidence in the overall conclusions of each review. While LRAT is not specific to systematic reviews, it provides a framework that can be applied to systematic reviews.

LRAT comprises nine questions (Table 8.3), which receive three potential ratings ("Satisfactory," "Unclear," and "Unsatisfactory") that require documentation of the rationale/justification for each rating. Each domain contains detailed descriptions to guide the rater, including specific criteria for each rating. In general, the ratings are as follows, with slight deviations in interpretation tailored to each domain:

- *Satisfactory:* Conducted according to a clear, valid, and consistent procedure.
- *Unclear:* Insufficient documentation to allow evaluation.
- *Unsatisfactory:* Positive evidence of invalid or inconsistent procedure.

The toolkit outlines the following recommended approach to evaluate each domain:

1. *Search for evidence of an invalid or inconsistent procedure. If found, the domain may be rated as "Unsatisfactory."*
2. *If no evidence of unsatisfactory methodology was found, either the documentation is insufficient (in which case the domain may be rated as "Unclear") or there is no obvious problem with the methodology (in which case the domain may be rated as "Satisfactory").*
3. *Brief justification for each rating.*

Judging the 58 included systematic reviews using LRAT, the quality of the included reviews was generally unsatisfactory (Fig. 8.2). Several domains performed somewhat well individually, with the majority of reviews being rated as satisfactory. For the question on whether they reported a clear question and objectives (Q2), the majority ($n=34$) of included reviews performed satisfactorally. However, this was not the case for all reviews—in particular, 11 studies were rated as unsatisfactory for this domain. For instance, Boothe, Boehmer, Wendel, and Yip (2014) stated a clear

Table 8.3 Nine domains of the literature review assessment toolkit (LRAT) (Policy from Science, 2014).

LRAT domain	Details
Q1. Objective of the review	Does the review answer a clear question of sufficient relevance?
Q2. Use of a protocol	Does the review follow a preconceived plan for finding and analyzing evidence relevant to its objective?
Q3. Interests and contributions	Are the interests of the authors of the review and records of how each author contributed to the review process clear?
Q4. Search strategy	Did the review locate all the evidence relevant to the review's objective?
Q5. Selection process	Did the review employ a screening process that identified all the studies of relevance to the review's objective?
Q6. Appraisal of directness (external validity) of evidence	Did the review present and consistently apply a valid scheme for giving greater weight to findings of studies of more direct relevance to the review objective?
Q7. Appraisal of methodological quality (internal validity) of evidence	Did the review present and consistently apply a valid scheme for giving greater weight to findings of studies of more robust methodological quality?
Q8. Synthesis of evidence	Did the authors combine, according to a valid methodology, the results, directness and methodological quality of evidence into a statement of what is and is not known regarding the answer to the review objective?
Q9. Summation of findings	Do the concluding and summary sections of the review present a succinct and accurate summary of the findings of the review?

objective "This specific study examines the association between residential traffic exposure and childhood cancer" but did not present a focused research question of clear relevance. A well-constructed research question should be focused with narrow parameters and ideally fits into the PECO (Participants/Population, Exposure, Comparator, and Outcomes) question framework that ultimately guides decisions regarding knowledge synthesis. In comparison, Mills et al. states a research question of "whether the evidence from epidemiological time-series studies suggests adverse associations between short-term exposure to nitrogen dioxide and increased numbers of daily deaths and emergency admissions to the hospital"

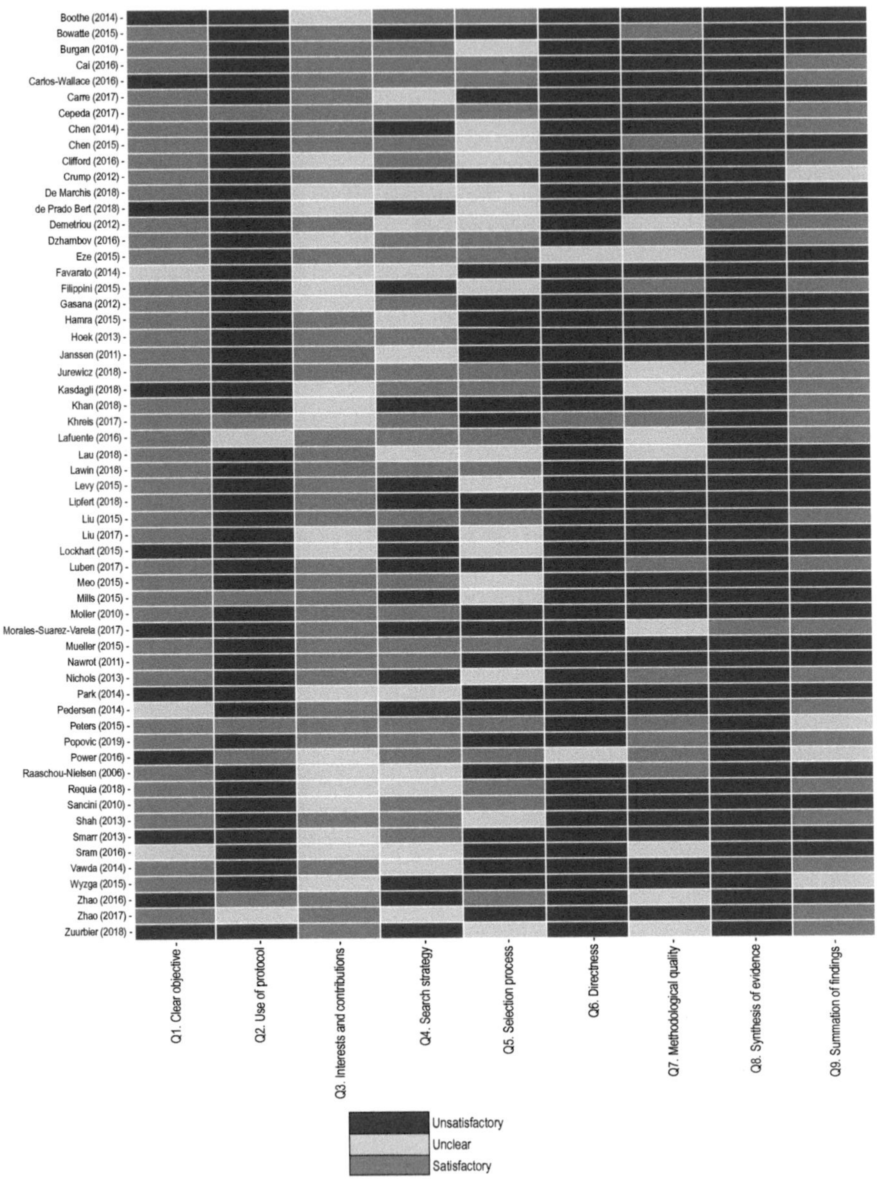

Fig. 8.2 Heat map of assigned ratings using the literature review assessment toolkit (LRAT) to evaluate quality of systematic reviews.

(Mills, Atkinson, Kang, Walton, & Anderson, 2015). Presentation of a clear systematic review research question is critical because it guides the structure of all components of the systematic review, and also ensures that there is no duplication of existing or ongoing reviews (Whitlock et al., 2010).

For the question on search strategy (Q4), the majority of studies had satisfactory methods ($n=27$); however those that were unsatisfactory ($n=24$) typically fell short because of failure to: specify databases search or only searching one database; provide exact search terms used to identify relevant studies; or justify rationale for implementing limtations on search date (e.g., literature published within the past 5 years). The search strategy is an integral component of the systematic review, as it dictates the ability to find evidence directly relevant to the study question. Lacking critical details regarding exactly how the search was conducted reduces confidence in the systematic review results because it is not apparent whether all studies relevant to the study question were included. Missing relevant articles may be significant, especially if the search fails to identify data that might alter conclusions about the health effects of an environmental exposure.

For the question on interests and contributions (Q3), the majority of reviews rated satisfactory ($n=37$). However, the rest were rated unclear ($n=21$) because the reviews did not report sufficient information to judge the interests and participation of each author. The question of conflicting interests and contributions is critically important to put the findings of a review in its full context. In the evaluation of studies, consideration of financial conflicts of interest has been proposed as an individual risk of bias domain (Bero, 2013) or as a consideration to inform risk of bias judgments in existing domains (for instance, if investigators with financial conflicts of interest resulted in selective reporting of a favorable result, due to either selecting reporting of particular outcome measurements or of particular analysis (Boutron et al., 2019)). This is based on empirical data from various research fields illustrating that on average, a study's source of funding influences its outcome (Barnes & Bero, 1998; Bero, Oostvogel, Bacchetti, & Lee, 2007; Lundh, Lexchin, Mintzes, Schroll, & Bero, 2018; Popelut, Valet, Fromentin, Thomas, & Bouchard, 2010). Similarly, systematic reviews reporting financial conflicts of interest more often have favorable conclusions and tend to be of lower methodological quality than reviews without such conflicts of interest (Hansen, Lundh, Rasmussen, & Hrobjartsson, 2019). Although it may not be conclusive that financial conflicts of interest influence the results and conclusions of systematic reviews (Hansen et al., 2019), it is critically important that systematic reviews clearly report interests and contributions for full transparency.

Overall, almost all of the reviews performed unsatisfactorily on at least 5 of the 9 domains, in particular those relating to the use of protocol (Q2, $n=50$), directness (Q6, $n=55$), methodological quality (Q7, $n=37$), and synthesis of evidence (Q9, $n=56$). Systematic reviews by Khreis et al. (2017) and Peters, Peters, Booth, and Mudway (2015) are considered of highest quality, with only two domains rated as unsatisfactory and one domain rated as unclear. Systematic review by Khreis et al. (2017) had domains 5 and 8 rated as unsatisfactory and domain 3 rated as unclear. That by Peters et al. (2015) had domains 6 and 8 rated as unsatisfactory and domain 9 rated as unclear. Several other reviews were rated unsatisfactory or unclear in only 3–4 domains (Cai et al., 2016; Cepeda et al., 2017; Dzhambov & Dimitrova, 2017; Jurewicz, Dziewirska, Radwan, & Hanke, 2018; Lafuente, Garcia-Blaquez, Jacquemin, & Checa, 2016; Liu et al., 2015; Popovic et al., 2019).

We did not formally attempt to conduct a quantitative synthesis on the health outcomes from the included systematic reviews; however, the self-reported conclusions from the included studies were fairly unanimous in suggesting that air pollution is negatively associated with various health outcomes. Interestingly, many of the studies discussed issues similar to those discussed here; for example, Jurewicz et al. (2018) concluded from 22 studies on various air pollutants that these had a negative effect on at least one of the semen parameters, but that the heterogeneity in exposures and outcomes made drawing conclusions more challenging.

Respiratory effects were among the most commonly evaluated, unsurprising considering the well-established connection between respiratory health and air pollution, with consistent conclusions that TRAP is adversely associated with outcomes such as childhood asthma. What was less clear was the magnitude of the overall effect, and the effects for specific pollutants. Studies often reported a negative but not statistically significant effect—again, most commonly due to the variation in exposures and outcomes in the underlying studies, with some pollutants more often studied than others.

Implications for policy- and decision-making

Traffic-related air pollution poses a challenge to both health and climate due to its widespread exposure and identified health effects. Both decision makers and the public need a rigorous and reliable bottom-line summary of the evidence in order to understand, and act on, the relevant human health harms from exposures. Summarizing the state of the evidence on environmental exposures and health effects has been a fundamental aspect of the decision- and policy-making. However, this process has been hampered and delayed, by both difficulty in interpreting individual studies and lack of a consistent, coherent method for evaluating and summarizing the evidence. Often, conclusions from scientific papers take the form of "there are some things we know about this question and some things we do not and we need more research." While there are always uncertainties in the science, the lack of clarity on the state of the evidence makes it both difficult to incorporate into the policy framework and lengthens the decision time, as inconsistent summary of the science leads to a conclusion that more studies are needed to confidently identify a relationship before action can be taken.

The introduction of systematic reviews is a key step in addressing this problem, as it allows the scientific evidence to be both more clearly and consistently presented and evaluated but also provides a consistent bottom-line summary of the science. Additionally, aggregating the data in a consistent quantitative manner, such as via metaanalysis, is a key step for using in the policy process via estimating the burden of disease and supporting benefit-cost analysis.

Summary and conclusions

Systematic reviews that are of high quality have provided a succinct bottom-line summary of the strength of the evidence. Having a bottom-line summary of the strength of evidence is critical for decision makers to be able to factor in the science (e.g., state of the science) with other factors that are important for decision-making such as magnitude of the problem, extent of exposure, options for mitigation, and benefits and costs. Indeed, high-quality reviews also provide results from metaanalysis that are in many cases critical for decision makers, as they can be used for evaluating burden of disease, benefits and costs, and other quantitative relationships between TRAP and health effects that are often a key input into the policy process. While there has been much important progress, there is also much work to be done to improve our ability to use the extensive body of science that has been published to date. As we have identified in our review of systematic reviews, not all systematic reviews are equal, and one that is named a systematic review does not automatically mean that the review is sufficient and of high quality. But when we have high-quality systematic reviews, these can be used to support evidence-based decision-making. Systematic reviews are also critical for more precisely identifying research gaps to prioritize for future studies. As we see more widespread adoption of systematic reviews, we need to continue to monitor and improve their quality to ensure that we are using the available science to improve the public's health.

References

Barnes, D. E., & Bero, L. A. (1998). Why review articles on the health effects of passive smoking reach different conclusions. *JAMA, 279*(19), 1566–1570.

Bero, L. A. (2013). Why the cochrane risk of bias tool should include funding source as a standard item. *Cochrane Database of Systematic Reviews, 12*, ED000075.

Bero, L., Oostvogel, F., Bacchetti, P., & Lee, K. (2007). Factors associated with findings of published trials of drug-drug comparisons: Why some statins appear more efficacious than others. *PLoS Medicine, 4*(6), e184.

Boothe, V. L., Boehmer, T. K., Wendel, A. M., & Yip, F. Y. (2014). Residential traffic exposure and childhood leukemia: A systematic review and meta-analysis. *American Journal of Preventive Medicine, 46*(4), 413–422.

Boutron, I., Page, M. J., Higgins, J. P. T., Altman, D. G., Lundh, A., & Hróbjartsson, A. (2019). Chapter 7: Considering bias and conflicts of interest among the included studies. In H. JPT, J. Thomas, J. Chandler, M. Cumpston, T. Li, M. J. Page, & V. A. Welch (Eds.), *Cochrane handbook for systematic reviews of interventions version 6.0 (updated July 2019).*

Cai, Y., Zhang, B., Ke, W., Feng, B., Lin, H., Xiao, J., et al. (2016). Associations of short-term and Long-term exposure to ambient air pollutants with hypertension: A systematic review and meta-analysis. *Hypertension, 68*(1), 62–70.

Cepeda, M., Schoufour, J., Freak-Poli, R., Koolhaas, C. M., Dhana, K., Bramer, W. M., et al. (2017). Levels of ambient air pollution according to mode of transport: A systematic review. *The Lancet Public Health, 2*(1), e23–e34.

Critical Appraisal Skills Programme U.K. (2013). *CASP tools checklist* [cited 2019 11/01]; Available from: https://casp-uk.net/casp-tools-checklists/.

De Marchis, P., Verso, M. G., Tramuto, F., Amodio, E., & Picciotto, D. (2018). Ischemic cardiovascular disease in workers occupationally exposed to urban air pollution—A systematic review. *Annals of Agricultural and Environmental Medicine, 25*(1), 162–166.

Demetriou, C. A., Raaschou-Nielsen, O., Loft, S., Moller, P., Vermeulen, R., Palli, D., et al. (2012). Biomarkers of ambient air pollution and lung cancer: A systematic review. *Occupational and Environmental Medicine, 69*(9), 619–627.

Dzhambov, A. M., & Dimitrova, D. D. (2017). Children's blood pressure and its association with road traffic noise exposure—A systematic review with meta-analysis. *Environmental Research, 152*, 244–255.

Fox, D. M. (2010). *The convergence of science and governance: Research, health policy, and American states*. Univ of California Press.

Guyatt, G. H., Oxman, A. D., Akl, E. A., Kunz, R., Vist, G., & Brozek, J. (2011). GRADE guidelines: 1. Introduction—GRADE evidence profiles and summary of findings tables. *Journal of Clinical Epidemiology, 64*, 383–394.

Guyatt, G. H., Oxman, A. D., Vist, G. E., Kunz, R., Falck-Ytter, Y., Alonso-Coello, P., et al. (2008). GRADE: An emerging consensus on rating quality of evidence and strength of recommendations. *BMJ, 336*(7650), 924–926.

Hansen, C., Lundh, A., Rasmussen, K., & Hrobjartsson, A. (2019). Financial conflicts of interest in systematic reviews: Associations with results, conclusions, and methodological quality. *Cochrane Database of Systematic Reviews, 8*, MR000047.

Hartling, L., Milne, A., Hamm, M. P., Vandermeer, B., Ansari, M., Tsertsvadze, A., et al. (2013). Testing the Newcastle Ottawa scale showed low reliability between individual reviewers. *Journal of Clinical Epidemiology, 66*(9), 982–993.

Health Effects Institute (HEI). (2010). *Traffic-related air pollution: A critical review of the literature on emissions, exposure, and health effects.* Washington, DC: Health Effects Institute. Panel on the Health Effects of Traffic-Related Air Pollution.

Higgins, J. P. T., Chandler, T. J., Cumpston, M., Li, T., Page, M. J., & Welch, V. A. E. (2019). *Cochrane handbook for systematic reviews of interventions* (2nd ed.). Chichester (UK): John Wiley & Sons.

Higgins, J., & Green, S. (2019). *Cochrane handbook for systematic reviews of interventions.* Chichester (UK).

Institute of Medicine. (2011). *Finding what works in health care: Standards for systematic reviews.* Washington, DC.

Johnson, P., Koustas, E., Vesterinen, H. M., Sutton, P., Atchley, D. S., Kim, A., et al. (2016). Application of the navigation guide systematic review methodology to the evidence for developmental and reproductive toxicity of triclosan. *Environment International, 92–93,* 716–728.

Jurewicz, J., Dziewirska, E., Radwan, M., & Hanke, W. (2018). Air pollution from natural and anthropic sources and male fertility. *Reproductive Biology and Endocrinology, 16*(1), 109.

Khreis, H., Kelly, C., Tate, J., Parslow, R., Lucas, K., & Nieuwenhuijsen, M. (2017). Exposure to traffic-related air pollution and risk of development of childhood asthma: A systematic review and meta-analysis. *Environment International, 100,* 1–31.

Kilkenny, C., Browne, W., Cuthill, I. C., Emerson, M., Altman, D. G., NC3Rs Reporting Guidelines Working Group. (2010). Animal research: Reporting in vivo experiments: The ARRIVE guidelines. *British Journal of Pharmacology, 160*(7), 1577–1579.

Lafuente, R., Garcia-Blaquez, N., Jacquemin, B., & Checa, M. A. (2016). Outdoor air pollution and sperm quality. *Fertility and Sterility, 106*(4), 880–896.

Lam, J., Koustas, E., Sutton, P., Johnson, P. I., Atchley, D. S., Sen, S., et al. (2014). The navigation guide—Evidence-based medicine meets environmental health: Integration of animal and human evidence for PFOA effects on fetal growth. *Environmental Health Perspectives, 122*(10), 1040–1051.

Lam, J., Lanphear, B. P., Bellinger, D., Axelrad, D. A., Mcpartland, J., Sutton, P., et al. (2017). Developmental PBDE exposure and IQ/ADHD in childhood: A systematic review and meta-analysis. *Environmental Health Perspectives, 125*(8).

Lam, J., Sutton, P., Kalkbrenner, A., Windham, G., Halladay, A., Koustas, E., et al. (2016). A systematic review and meta-analysis of multiple airborne pollutants and autism spectrum disorder. *PLoS One, 11*(9), 1–27.

Liu, X., Lian, H., Ruan, Y., Liang, R., Zhao, X., Routledge, M., et al. (2015). Association of Exposure to particular matter and carotid intima-media thickness: A systematic review and meta-analysis. *International Journal of Environmental Research and Public Health, 12*(10), 12924–12940.

Lundh, A., Lexchin, J., Mintzes, B., Schroll, J. B., & Bero, L. (2018). Industry sponsorship and research outcome: Systematic review with meta-analysis. *Intensive Care Medicine, 44*(10), 1603–1612.

Mills, I. C., Atkinson, R. W., Kang, S., Walton, H., & Anderson, H. R. (2015). Quantitative systematic review of the associations between short-term exposure to nitrogen dioxide and mortality and hospital admissions. *BMJ Open, 5*(5), e006946.

Moher, D., Liberati, A., Tetzlaff, J., Altman, D. G., PRISMA Group. (2009). Preferred reporting items for systematic reviews and meta-analyses: The PRISMA statement. *PLoS Medicine, 6*(7), e1000097.

Morales-Suarez-Varela, M., Peraita-Costa, I., & Llopis-Gonzalez, A. (2017). Systematic review of the association between particulate matter exposure and autism spectrum disorders. *Environmental Research, 153,* 150–160.

National Academies of Sciences Engineering and Medicine. (2017). *Application of systematic review methods in an overall strategy for evaluating low-dose toxicity from endocrine active chemicals.* Washington, DC.

National Academies of Sciences Engineering and Medicine. (2018). *Progress toward transforming the integrated risk information system (IRIS) program: A 2018 evaluation*. Washington, DC.
National Academies of Sciences Engineering and Medicine. (2019). *Review of DOD's approach to deriving an occupational exposure level for trichloroethylene*. Washington, DC.
National Research Council. (2014). *Review of EPA's integrated risk information system (IRIS) process*. Washington, DC: National Academies of Science, Engineering and Medicine.
National Toxicology Program. (2013). *Monograph on developmental effects and pregnancy outcomes associated with cancer chemotherapy use during pregnancy*. Washington, DC: National Institute of Environmental Health Sciences.
National Toxicology Program. (2016). *Monograph on immunotoxicity associated with exposure to perfluorooctanoic acid (PFOA) or perfluorooctane sulfonate (PFOS)*. Washington, DC: National Institute of Environmental Health Sciences.
National Toxicology Program. (2018). *Systematic review of long-term neurological effects following acute exposure to the organophosphorus nerve agent sarin*. Washington, DC: National Institute of Environmental Health Sciences.
National Toxicology Program. (2019). *Monograph on the systematic review of occupational exposure to cancer chemotherapy agents and adverse health outcomes*. Washington, DC: National Institute of Environmental Health Sciences.
National Toxicology Program (NTP). (2019). *Handbook for Conducting Systematic Reviews for Health Effects Evaluations* [cited 2019 11–01]; Available from: https://ntp.niehs.nih.gov/ntp/ohat/pubs/handbookmarch2019_508.pdf.
NTP. (2019). *NTP Monograph on the systematic review of traffic-related air pollution and hypertensive disorders of pregnancy*. Washington, DC: National Institute of Environmental Health Sciences.
Peters, R., Peters, J., Booth, A., & Mudway, I. (2015). Is air pollution associated with increased risk of cognitive decline? A systematic review. *Age and Ageing*, *44*(5), 755–760.
Policy from Science. (2014). *Literature review appraisal toolkit* [cited 2019 11–01]; Available from: http://policyfromscience.com/lrat/about-the-lra-toolkit/.
Popelut, A., Valet, F., Fromentin, O., Thomas, A., & Bouchard, P. (2010). Relationship between sponsorship and failure rate of dental implants: A systematic approach. *PLoS One*, *5*(4), e10274.
Popovic, I., Soares Magalhaes, R. J., Ge, E., Marks, G. B., Dong, G. H., Wei, X., et al. (2019). A systematic literature review and critical appraisal of epidemiological studies on outdoor air pollution and tuberculosis outcomes. *Environmental Research*, *170*, 33–45.
Rennie, D., & Chalmers, I. (2009). Assessing authority: A comparison of results of meta-analyses of randomized control trials and recommendations of clinical experts: Treatments for myocardial infarction. *JAMA*, *301*(17), 1819–1821.
Rooney, A. A., Boyles, A. L., Wolfe, M. S., Bucher, J. R., & Thayer, K. A. (2014). Systematic review and evidence integration for literature-based environmental health science assessments. *Environmental Health Perspectives*, *122*(7), 711–718.
Shea, B. J., Grimshaw, J. M., Wells, G. A., Boers, M., Andersson, N., Hamel, C., et al. (2007). Development of AMSTAR: A measurement tool to assess the methodological quality of systematic reviews. *BMC Medical Research Methodology*, 7, 10.
U.S. Environmental Protection Agency. (1991). *Guidelines for developmental toxicity risk assessment*. Washington, DC.
U.S. Environmental Protection Agency. (1996). *Guidelines for reproductive toxicity risk assessment*. Washington, DC.
U.S. Environmental Protection Agency. (2018). *Application of systematic review in tsca risk evaluations*. Washington, DC: Office of Chemical Safety and Pollution Prevention (OCSPP), Office of Pollution Prevention and Toxics (OPPT).
Vesterinen, H. M., Johnson, P. I., Atchley, D. S., Sutton, P., Lam, J., Zlatnik, M. G., et al. (2015). Fetal growth and maternal glomerular filtration rate: A systematic review. *Journal of Maternal Fetal Neonatal Medicine*, *28*(18), 2176–2181.

Von Elm, E., Altman, D. G., Egger, M., Pocock, S. J., Gotzsche, P. C., Vandenbroucke, J. P., et al. (2008). The strengthening the reporting of observational studies in epidemiology (STROBE) statement: Guidelines for reporting observational studies. *Journal of Clinical Epidemiology, 61*(4), 344–349.

Wells, G. A., Shea, B., O'connell, D., Peterson, J., Welch, V., & Losos, M. (2019). *The Newcastle-Ottawa Scale (NOS) for assessing the quality if nonrandomized studies in meta-analyses.* [cited 2019 11–01]; Available from: http://www.ohri.ca/programs/clinical_epidemiology/oxford.htm.

Whaley, P., Letcher, R. J., Covaci, A., & Alcock, R. (2016). Raising the standard of systematic reviews published in environment international. *Environment International, 97*, 274–276.

Whitlock, E. P., Lopez, S. A., Chang, S., Helfand, M., Eder, M., & Floyd, N. (2010). AHRQ series paper 3: Identifying, selecting, and refining topics for comparative effectiveness systematic reviews: AHRQ and the effective health-care program. *Journal of Clinical Epidemiology, 63*(5), 491–501.

Woodruff, T. J., & Sutton, P. (2014). The navigation guide systematic review methodology: A rigorous and transparent method for translating environmental health science into better health outcomes. *Environmental Health Perspectives, 122*(10), 1007–1014.

World Health Organization. (2017). *WHO methods and data sources for global burden of disease estimates 2000–2015.* Geneva: World Health Organization.

CHAPTER 9

Established and emerging effects of traffic-related air pollution

Mary Fox[a], Kirsten Koehler[b], and Natalie Johnson[c]
[a]Department of Health Policy and Management, Johns Hopkins Bloomberg School of Public Health, Baltimore, MD, United States
[b]Department of Environmental Health and Engineering, Johns Hopkins Bloomberg School of Public Health, Baltimore, MD, United States
[c]Department of Environmental and Occupational Health, Texas A&M University School of Public Health, College Station, TX, United States

Abbreviations

ALL	acute lymphoblastic leukemia
AML	acute myeloid leukemia
BC	black carbon
CNS	central nervous system
CO	carbon monoxide
COPD	chronic obstructive pulmonary disease
CVD	cardiovascular disease
EC	elemental carbon
ECAT	elemental carbon attributable to traffic
EJ	environmental justice
EPA	Environmental Protection Agency
GIS	geographic information system
HEI	Health Effects Institute
IQ	intelligence quotient
IQR	inter-quartile range
IVF	in vitro fertilization
m	meter
$\mu g/m^3$	microgram per cubic meter
NAAQS	National Ambient Air Quality Standards
NO_x	nitrogen oxides including NO and NO_2
NTD	neural tube defect
O_3	ozone
PAH	polycyclic aromatic hydrocarbon
PM	particulate matter including PM_{10} and $PM_{2.5}$
PTB	preterm birth
SES	socioeconomic status
TRAP	traffic-related air pollution
US	United States
USA	United States of America
VOC	volatile organic compound
yr	year

Traffic-Related Air Pollution
https://doi.org/10.1016/B978-0-12-818122-5.00009-0

Introduction

Transportation sector emissions are one of the major sources of air pollutants. The World Health Organization estimates that annually about 4 million deaths worldwide are related to air pollution exposures including deaths from ischemic heart disease and stroke, respiratory diseases, and lung cancer (World Health Organization, 2018a). Road traffic is also a significant contributor to community noise; noise exposure is independently associated with health effects such as ischemic heart disease, sleep disturbance, and stress, among others (World Health Organization, 2018b). The range of health effects that have been associated with traffic-related air pollution are described below. Well-established effects will be briefly summarized with attention to newer findings and research directions. Emerging effects will be examined more closely with a focus on sensitive subpopulations and health effects over the life course.

Health effects of TRAP: Well-established effects

Exposure considerations

Traffic-related air pollution (TRAP) is a mixture of pollutants derived during the combustion of gasoline or diesel fuel in the vehicle engine and can also include particles from wear and tear of other mechanical components. TRAP is comprised of carbon monoxide (CO), nitrogen oxides (NO, NO_2, sometimes abbreviated as NO_x), hydrocarbons, mobile-source air toxics, and particulate matter (PM) (Health Effects Institute, 2010). However, vehicles are not the only sources of these pollutants, many of which are common from all fossil fuel combustion. Furthermore, different species have different lifetimes in the atmosphere, which can lead to different patterns around sources. Understanding the potential impacts of TRAP on health are complicated by the complex exposure assessment process. This is because: (1) there are many pollutants that can be considered (e.g., particulate matter, black carbon content of particulate matter, nitrogen oxides, organic species, metals, etc.), (2) exposures are highly variable in time and space, and (3) exposures likely depend on person-specific activities and mobility patterns. In the United States, as in other places, the existing air pollution monitoring networks operating to show compliance with the National Ambient Air Quality Standards (NAAQS) established by the Environmental Protection Agency (EPA) are generally not well suited to meet the needs of estimating human exposures to TRAP. Such monitors are sited to represent typical exposures for a community or city and may not be sited near major roads where TRAP exposures are higher. These challenges can result in exposure misclassification that can complicate epidemiologic analysis of the association between TRAP and various health outcomes.

The earliest studies considering potential health effects of TRAP relied on proximity metrics (e.g., distance to nearest major roadway) to estimate exposures for epidemiologic analysis. More advanced models, including dispersion models and land-use regression, consider data on land use, traffic data, meteorology, geography, and atmospheric chemical transformations providing more realistic estimates of TRAP over time and space. Measurements of exposures to different TRAP constituents are the gold standard and likely reduce exposure misclassification in epidemiologic studies, but such measurements are expensive and can be infeasible for studies with large sample sizes that are often needed for epidemiologic analysis, particularly for outcomes with lower prevalence. It is also typically not feasible to measure all pollutants associated with vehicle emissions and these pollutants may have other sources than just vehicles, complicating exposure assessment. This limitation is especially true for studies using $PM_{2.5}$ measurements for exposure assessment. $PM_{2.5}$ concentrations are impacted by many sources and may not be a good surrogate for TRAP. Other pollutants including black carbon (BC), elemental carbon (EC), ultrafine particles (particles with diameter less than 100 nm), and NO_x may serve as better markers for TRAP. Ground-level ozone (O_3) is formed in the atmosphere through reactions of NO_x and volatile organic compounds (VOCs); given the importance of vehicle emissions to NO_x concentrations, O_3 can be related to traffic exposures, but is also influenced by VOCs and sunlight. A more detailed discussion of TRAP exposure assessment can be found in Chapter 6.

Established health effects

In 2010, The Health Effects Institute (HEI) published a comprehensive report on the health effects associated with TRAP (Health Effects Institute, 2010). The report focused on the primary emissions from vehicles (those emitted directly from vehicles without chemical transformation in the atmosphere) in primarily urban areas and the associated health effects. The report concluded that the contribution of motor vehicles to ambient air pollution was highly variable (and differs by pollutant); for example, only 5% of ambient fine particulate matter ($PM_{2.5}$) was from motor vehicles in Pittsburg, USA; similarly, it was only 6% in Beijing, China, but 55% in Los Angeles, USA, and 53% in Barcelona, Spain. The report furthermore concluded that those living within 300–500 m of a highway or major roadway are those most likely to be impacted and that an estimated 30%–45% of people in North America live in those impacted areas. The panel of experts concluded that only the association between TRAP and exacerbations of asthma in asthmatic children had sufficient evidence for a causal association. However, they also concluded that there was suggestive, but not sufficient, evidence for causal associations between TRAP and all-cause mortality, cardiovascular mortality, cardiovascular morbidity, asthma prevalence and incidence in children, and decrements in lung function in children and adults (Health Effects Institute, 2010). Four major categories of health effects have been consistently associated with TRAP including mortality, cardiovascular and related cardiometabolic effects, some cancers, and respiratory diseases. Table 9.1 lists health effects associated with TRAP exposure for these established health effect categories.

Several studies have shown a relationship between TRAP and all-cause mortality. The HEI report included a 2007 study that found a 2.3% increase in all-cause mortality per IQR increase in traffic particle exposure from a GIS-based exposure model in the Boston metropolitan area (Maynard, Coull, Gryparis, & Schwartz, 2007). Beelen et al. found that traffic intensity on the nearest roadway was associated with mortality in a Dutch cohort (Beelen, Hoek, van den Brandt, et al., 2008).

Research on cardiovascular disease has continued including large studies showing association with increased blood pressure (Cosselman, Krishnan, Oron, et al., 2012; Delfino, Tjoa, Gillen, et al., 2010; Fuks, Weinmayr, Foraster, et al., 2014), cardiovascular events (Alexeeff, Roy, Shan, et al., 2018), coronary heart disease (Gan et al., 2011; Hoffmann, Moebus, Dragano, et al., 2009;

Table 9.1 Established health effects associated with TRAP exposure.

Mortality	Cardiovascular and related cardiometabolic diseases	Cancers	Respiratory
All cause Cardiovascular	Blood pressure Coronary heart disease Diabetes Heart rate variability Metabolic syndrome Myocardial infarction Obesity	Lung Childhood leukemia Child CNS tumors Embryonal tumors	Asthma and allergy Lung function decrements

CNS, central nervous system.

Kan et al., 2008), heart rate variability (Baja, Schwartz, Wellenius, et al., 2010; Weichenthal et al., 2011), and cardiac autonomic function (Wu et al., 2011). There has also been a consideration of cardiometabolic disease, including metabolic syndrome, obesity, and diabetes. Recent reviews found generally consistent findings between traffic-related particulate matter and these cardiometabolic end points (Ahmed, Jiang, Chen, & Lin, 2018; Eze, Hemkens, Bucher, et al., 2015). Ahmed et al. highlighted the growing evidence supporting systemic inflammation and oxidative stress as important pathways for cardiovascular and cardiometabolic disease (Ahmed et al., 2018).

Research has also continued to investigate the role of TRAP on cancer development. Among adults, exposure to TRAP (NO_2, in particular) was associated with a 4% increase in lung cancer per $10\,\mu g/m^3$ increase in NO_2 in a systematic review and meta-analysis (Hamra et al., 2015). A meta-analysis of childhood cancer studies found a positive association between postnatal (but not prenatal) TRAP exposure and childhood leukemia (Boothe, Boehmer, Wendel, & Yip, 2014). A later review supported higher estimates in general for exposures in the postnatal period compared to the prenatal period (Filippini, Heck, Malagoli, Del Giovane, & Vinceti, 2015). Specifically, NO_2 and benzene exposures showed an increased leukemia risk. Carlos-Wallace et al. also found associations between childhood leukemia and different metrics of benzene exposure in their meta-analysis (Carlos-Wallace, Zhang, Smith, Rader, & Steinmaus, 2016). In regard to TRAP, the relative risk was higher for acute myeloid leukemia (AML) in comparison to acute lymphoblastic leukemia (ALL).

Reviews for other cancers cite numerous limitations preventing a firm conclusion on the association with TRAP (Raaschou-Nielsen & Reynolds, 2006). Danysh et al. recently found associations suggestive of a link between traffic-related hazardous air pollutants (i.e., 1,3-butadiene and diesel exhaust particulate matter) and the incidence of childhood central nervous system tumors (Danysh, Mitchell, Zhang, Scheurer, & Lupo, 2015). Furthermore, Kumar et al. observed an increased odds of embryonal tumors in children born to mothers living within 500 m of a major roadway (Kumar, Lupo, Pompeii, & Danysh, 2018). These recent findings underscore the need for further investigation into understanding the role of TRAP exposure early in life and the development of pediatric cancers.

Since the HEI (2010) report was completed, hundreds of papers each year have evaluated associations between TRAP and various health outcomes. Many have focused on expanding the evidence base for the health outcomes identified by the HEI Report. Recent meta-analyses of birth cohort studies found an association between TRAP and asthma development up to age 18 (Bowatte, Lodge, Lowe, et al., 2015; Khreis et al., 2017). One of these papers also found an association between TRAP and both food and aeroallergen sensitization (Bowatte et al., 2015). Exposure to TRAP in early life at school ages was also associated with decrements in childhood lung function that persists at least until adolescence and possibly into adulthood (Schultz, Litonjua, & Melén, 2017). Gene-environment interactions and epigenetics may play an important role in the relationship between TRAP and allergy (Carlsten & Rider, 2017).

Emerging health effects of TRAP

A snapshot of the literature on emerging effects

A search of reviews in PubMed with terms for TRAP and health covering 2009–2018 was done to explore research topics following the HEI report mentioned above. In the past few years, reviews found in PubMed represent research on health effects beyond mortality, respiratory, cancer, and cardiovascular and cardiometabolic diseases to include a growing literature on pregnancy, birth outcomes and development, neurotoxicity, and various neurological diseases and disorders, summarized in Table 9.2 and in order by life stage as presented below. Other topics observed in this literature are also described in the Special Topics section.

Emerging effects of TRAP in early life

Pregnancy complications and preterm birth

The relationship between TRAP and development of gestational diabetes has been hypothesized based on the evidence supporting air pollution exposure and adult onset of type 2 diabetes mellitus (Janghorbani, Momeni, &

Table 9.2 Emerging health effects associated with TRAP exposure.

Pregnancy complications	Birth outcomes	Fertility	Neurological effects	Eye effects
Gestational diabetes Preeclampsia	Birth defects Fetal growth Preterm birth	Infertility Reduced rate of live births	Alzheimer's disease Autism Cognitive dysfunction and cognitive decline Neurodevelopment Parkinson's disease Dementia	Blurred vision Irritation Redness Tearing

Mansourian, 2014; Wang et al., 2014). An initial study in pregnant women without known diabetes living in the Boston area found an association between traffic density and impaired glucose tolerance, but not gestational diabetes (Fleisch, Gold, Rifas-Shiman, et al., 2014). In another study, maternal preconception and early pregnancy exposure to NO_x was associated with gestational diabetes risk (Robledo, Mendola, Yeung, et al., 2015). Other studies, however, have cited inconsistent associations or associations in unexpected ("protective") directions. Padula et al. unexpectedly found inverse associations between maternal TRAP exposure during the first two trimesters and gestational diabetes (Padula et al., 2019). Similarly, Wu and colleagues identified decreases in many TRAP metrics that were associated with an increased risk of gestational diabetes, as well as preeclampsia (Wu et al., 2016). Investigators suggested that underreporting, particularly in low socioeconomic groups, may contribute to these unexpected associations. Cumulatively, these studies reinforce the need for additional evidence regarding the association between gestational diabetes and maternal TRAP exposure, including more detailed source and speciation information about pollutants.

Preeclampsia, a risky pregnancy complication characterized by high blood pressure, has also been linked with maternal TRAP exposure. Preeclampsia risk has been associated with PM_{10} brake dust and combined traffic-related sources (Dadvand, Ostro, Amato, et al., 2014). In a retrospective study, TRAP exposure was associated with risk of late onset preeclampsia (Wu et al., 2016). Higher TRAP exposure has been associated with preeclampsia risk, particularly during the third trimester and among younger or older women (<20 or ≥ 40) (Pereira et al., 2013). Comparably, elemental carbon attributable to traffic (ECAT) exposure at 20 weeks of gestation has been associated with higher systolic blood pressure late in pregnancy (Sears, Braun, Ryan, et al., 2018). However, a few studies have found no association (van den Hooven, Jaddoe, de Kluizenaar, et al., 2009) or an inverse association, as mentioned previously (Wu et al., 2016). In a systematic review and meta-analysis of 17 articles evaluating the impact of NO_x, PM (PM_{10}, $PM_{2.5}$), CO, O_3, proximity to major roads, and traffic density, most studies reported that air pollution increased risk for pregnancy-induced hypertensive disorders (Pedersen, Stayner, Slama, et al., 2014). Notably, estimated effect sizes have been shown to depend on how both temporal and spatial variations were incorporated into exposure assessment, signifying the difficulty in comparison across studies (Wu et al., 2011).

Importantly, hypertensive disorders during pregnancy may underlie essential clinical outcomes, mainly preterm birth (PTB). Yorifuji and colleagues proposed this as a mechanism responsible for the association between TRAP and PTB based on data showing that TRAP increased the risk of preeclampsia and preterm premature rupture of fetal membranes before the onset of labor (Yorifuji, Naruse, Kashima, Murakoshi, & Doi, 2015). Premature (i.e., preterm) birth is defined as delivery at fewer than 37 weeks of gestation. Studies evaluating the association between preterm birth and air pollution, often from traffic emissions, are abundant due to the ease of collecting outcome information from birth records. A recent, large retrospective cohort study showed that NO_2 may increase PTB risk, especially for exposures during the third trimester, the month and the week before delivery (Ji, Meng, Liu, et al., 2019). Consistently underscored in individual studies and reviews is the importance of exposure assessment, particularly improving spatial and temporal resolution to prevent exposure misclassification and reduce heterogeneity. Moreover, accounting for important confounders, such as race/ethnicity, socioeconomic factors, exposure to tobacco smoke or additional maternal exposures/comorbidities, could impact effect sizes. Indeed, maternal comorbidities influence PTB and may interact with TRAP exposure to increase risk. Mendola et al. showed that mothers with asthma may experience a higher risk for PTB after TRAP exposure, highlighting the need to control for maternal characteristics (Mendola, Wallace, Hwang, et al., 2016).

Fetal growth

TRAP exposures during pregnancy have been inversely associated with some fetal growth measures. In the Netherlands, van den Hooven et al. found that the higher NO_2 exposures were associated with shorter fetal femur length in the second and third trimesters and that PM_{10} and NO_2 exposures were associated with smaller head circumference in the third trimester (van den Hooven, Pierik, de Kluizenaar, et al., 2012). Ritz et al., following a Los Angeles, California birth cohort, found that higher NO_x exposures in the third trimester were associated with biparietal diameter reductions (smaller fetal head size) (Ritz, Qiu, Lee, et al., 2014). In a large European cohort study, increased NO_2 and traffic density on the nearest street were associated with increased risk of low birthweight at term (Pedersen, Giorgis-Allemand, Bernard, et al., 2013).

Birth defects

Birth defects are prevalent in an estimated 6% of live births worldwide (Christianson, Howson, & Modell, 2006). Heart, orofacial, and neural tube defects (NTDs) represent some of the most common congenital defects.

Associations between TRAP and birth defects have been suggested; however, the existing evidence is limited. Vrijheid et al. found increased risk of certain heart defects with increased exposure to NO_2 (Vrijheid, Martinez, Manzanares, et al., 2011). Padula et al. reported differing results for selected birth defects and TRAP exposure (Padula, Tager, Carmichael, et al., 2013). When investigators took into consideration social factors, CO, NO, and NO_2 were more strongly associated with NTDs among US-born Hispanic mothers (Padula, Yang, Carmichael, et al., 2017). Other potential modifiers have been investigated, including dietary factors. Interactions between methionine intake and higher maternal NO_2 exposure have been shown to impact risk of heart defects (Stingone, Luben, Carmichael, et al., 2017). These types of studies underscore the importance of accounting for potential modifiers in birth defects studies.

Neurotoxicity

Kilian and Kitazawa reviewed several studies of in utero exposures to NO_x or proximity to roadway (Kilian & Kitazawa, 2018). The in utero exposures were associated with decrements in various measures of cognitive function and neurodevelopment including psychomotor development, mental scores, verbal and nonverbal IQ in children between ages 1 and 14 years (Kilian & Kitazawa, 2018).

Exposures to BC or NO_2 at school and/or home in school-age children (5–11 years) were associated with decreased intelligence test performance and motor function (Kilian & Kitazawa, 2018). Higher EC exposures in first year of life were associated with hyperactivity in 7-year-olds (Sram, Veleminsky Jr, Veleminsky, & Stejskalova, 2017); residential proximity to freeways during gestation and higher NO_2 (and PM) exposures were associated with increased autism risk (Volk, Hertz-Picciotto, Delwiche, Lurmann, & McConnell, 2011; Volk, Lurmann, Penfold, Hertz-Picciotto, & McConnell, 2013). Higher EC and NO_2 were also associated with slower brain maturation in primary school children (de Prado, Mercader, Pujol, Sunyer, & Mortamais, 2018).

Effects of TRAP in adulthood

Human reproduction

TRAP has emerged as an important exposure impacting numerous reproductive health end points. Accumulating evidence on TRAP and fertility is suggestive of causality. Carré et al. recently reviewed the role of air pollution in infertility (Carre, Gatimel, Moreau, Parinaud, & Leandri, 2017).

Overall, exposures focused on PM, O_3, polycyclic aromatic hydrocarbons (PAHs), and NO_2, representing pollutants emitted from both transportation and industry. Most pertinent to TRAP, this review highlighted three studies showing a positive association between traffic-related air pollution metrics (i.e., $PM_{2.5}$, coarse PM, NO_2, and proximity to major roads) and reduced fertility rates (Mahalingaiah, Hart, Laden, et al., 2016; Nieuwenhuijsen, Basagana, Dadvand, et al., 2014; Slama et al., 2013). Notably, Mahalingaiah et al., using data from 116,430 female nurses as part of the Nurses' Health Study II cohort, observed the strongest association between cumulative average exposures and infertility risk, suggesting that chronic exposure may be of greater significance versus acute exposure (Mahalingaiah et al., 2016). In addition, Kioumourtzoglou et al. observed lower live birth rates associated with increasing NO_2 in two separate cohorts, in Boston and Tel Aviv (Kioumourtzoglou, Raz, Wilson, et al., 2019).

Infertility diagnoses in epidemiological studies can be complex due to the difficulty in determining the exact dates of outcome, thereby limiting the linkage with exposure data. Well-characterized cohorts of women undergoing in vitro fertilization (IVF) have recently afforded the opportunity to investigate tighter windows of exposure and well-defined reproductive outcomes. Legro et al. found increased NO_2 concentrations consistently associated with lower live birth rates (Legro, Sauer, Mottla, et al., 2010). Using residential proximity to major roadways and traffic as proxies for TRAP, Gaskins et al. found that residential proximity to major roadways was related to reduced likelihood of live births following IVF treatment (Gaskins, Hart, Minguez-Alarcon, et al., 2018). Additionally, there were suggestive associations between roadway proximity and higher estradiol trigger concentrations and lower endometrial thickness (factors that can influence IVF success). Female hormone levels, clearly important in ovulation and fertility, have been studied in populations exposed to TRAP. Female traffic control officers in Rome had significantly lower estradiol levels during follicular and luteal phases of the cycle, but not in the ovulatory phase, in comparison to officers in indoor positions (Tomei, Ciarrocca, Fortunato, et al., 2006). This study addresses one hypothesis of hormone disruption regarding mechanisms underlying TRAP exposure and infertility. Carré et al. reviewed additional mechanisms, including induction of reactive oxygen species, cell DNA alterations, and epigenetic modifications (Carre et al., 2017).

In addition, an impact on female and male gametogenesis has been suggested. Few studies focus on oocytes and folliculogenesis, which are mainly restricted to IVF studies, whereas more researchers have investigated the effect of pollutants on spermatogenesis. These results are often contradictory, which is likely due to study design, exposure misclassification, and confounding. The majority of studies have been cross-sectional or retrospective, and to date, few prospective studies have interrogated TRAP and male fertility. More recently, in a cohort of 797 men attending a fertility clinic in Massachusetts, residential distance to major roadways was not related to sperm characteristics or serum reproductive hormone levels (Gaskins et al., 2018). Since the number of longitudinal studies are limited and results from cross-sectional and retrospective studies are inconsistent, additional evidence on TRAP and men's reproductive health is needed.

Eye effects

Jung et al. reviewed TRAP exposures and eye effects including eye irritation, redness, tearing, and blurred vision (Jung, Mehta, & Tong, 2018). These authors searched Entrez PubMed over a 10-year period from 2007 to 2017, finding seven cross-sectional studies; several of the studies found associations between NO_2, PM_{10}, or $PM_{2.5}$ exposures and eye irritation symptoms. Concerns such as no reference group were noted for a number of the studies reviewed (Jung et al., 2018).

Effects of TRAP in later life

Neurotoxicity

In later life (> 50 years), BC, NO_2, and proximity to roadway were associated with decreased cognitive function (Clifford, Lang, Chen, Anstey, & Seaton, 2016). A study in Sweden found that higher exposures to NO_x were associated with increased risk of Alzheimer's disease and vascular dementia (Oudin, Forsberg, Adolfsson, et al., 2016); in Denmark, Ritz et al. found that higher NO_2 exposures were associated with increased risk of Parkinson's disease in urban populations (but not in rural residents) (Ritz, Lee, Hansen, et al., 2016).

Health effects of TRAP: Special topics

Susceptible populations

Ongoing research has also focused on identifying populations that are susceptible to the effects of TRAP. For respiratory outcomes, those with asthma and COPD are more susceptible to TRAP exposures (Andersen, Bønnelykke, Hvidberg, et al., 2012; Patel, Chillrud, Correa, et al., 2010).

For cardiovascular outcomes, obesity, preexisting heart disease, diabetes, and those with lower socioeconomic status may be at enhanced risk (Baja et al., 2010; Dragano, Hoffmann, Moebus, et al., 2009). There may also be differences by sex. For example, TRAP exposures were associated with wheeze, night cough, and rhinitis symptoms in boys, but not in girls (Rancière, Bougas, Viola, & Momas, 2017). Other studies have also found stronger effects among boys for pneumonia (Chang, Liu, & Huang, 2018), attention measures (Chiu, Bellinger, Coull, et al., 2013), and asthma symptoms (Gonzalez-Barcala et al., 2013). Certain worker populations may also be vulnerable given their higher exposures to TRAP. These can include traffic police and conductors, taxi drivers, professional motorcyclists, gas station attendants, and street vendors (Brucker, Moro, Charão, et al., 2013; Carvalho, Carneiro, Barbosa Jr, et al., 2018; Choudhary & Tarlo, 2014; Garshick, Laden, Hart, et al., 2008; Huang, Lai, Chen, et al., 2012).

Combined exposures and cumulative risk

A theme that appeared in the recent literature was that of combined exposures and multi-stressor interactions. This theme is consistent with the concepts of cumulative risk, a developing area of environmental and occupational health (Fox et al., 2018; Fox, Brewer, & Martin, 2017). The specific traffic-related pollutants addressed were BC, diesel exhaust, and NO_2 (along with some other air pollutants) combined with exposure to noise, psychosocial stress, socioeconomic status (SES), and aeroallergens. Cardiovascular and cardiometabolic outcomes were most often addressed in these papers (Fuller, Feeser, Sarnat, & O'Neill, 2017; Munzel, Sorensen, Gori, et al., 2017; Stansfeld, 2015; Tetreault, Perron, & Smargiassi, 2013); others focused on respiratory (Carlsten, 2018) and neuroendocrine development (Cowell & Wright, 2017).

These papers document evidence of increases in risk with combination exposures often but not uniformly across different studies and exposures. For example:

- noise and air pollutant exposures seem to be independent contributors to CVD with possible small interactive effects;
- interaction of air pollutant exposures and stress or SES on CVD risk found often but not always;
- negative life events and high BC exposure of mothers affected attention-concentration in boys at age 6;
- combination exposure to aeroallergens and NO_2 increases risk of new asthma and asthma exacerbation.

Given the ubiquity of complex combination exposures, further research in this area should be a priority. Specific topics of interest include identifying the most robust and reliable methods to measure or represent SES and research that includes both individual and community or area-level measures of psychosocial stress and SES.

Environmental justice

The US Environmental Protection Agency defines environmental justice (EJ) as: "… the fair treatment and meaningful involvement of all people regardless of race, color, national origin, or income, with respect to the development, implementation, and enforcement of environmental laws, regulations, and policies" (U.S. Environmental Protection Agency, 2019). Evidence of environmental injustice arose in the 1980s, often related to the existence or siting of hazardous waste sites, landfills, industrial facilities, etc. (Bullard, 1990). Environmental injustice and related exposures are considered to contribute to health disparities along with racism, psychosocial stress, low SES, and lack of health insurance and limited access to health care (Bullard, Mohai, Saha, & Wright, 2008; Committee on Environmental Justice, Institute of Medicine, National Academy of Sciences, 2000; Fox, Groopman, & Burke, 2002; Gee & Payne-Sturges, 2004).

EJ and TRAP

Examining and addressing environmental injustices with respect to TRAP continues to be an active area of exposure research and prevention practice for development of interventions. Research continues on differential exposures, health risks, and TRAP impacts including identifying disparities across race, ethnicity, and socioeconomic status groups. For example, locations of existing schools and siting of new schools in relation to heavily trafficked roads is a concern. In the Sacramento, California area, Gaffron and Niemeier found $PM_{2.5}$ emissions from road traffic correlated with schools with a higher percent share of Black, Hispanic, and multiethnic students and students eligible for reduced-price meals (Gaffron & Niemeier, 2015). Studies in multiple cities in the US and Canada found mobile-source and traffic-related exposures higher in minority or low socioeconomic status populations (Martenies, Milando, Williams, & Batterman, 2017; Pratt, Vadali, Kvale, & Ellickson, 2015). In Sao Paulo, Brazil, Ribeiro et al. found that traffic density and NO_2 exposures were associated with higher respiratory cancer incidence and mortality rates citywide and that the highest rates were in areas with the lowest SES (Ribeiro, Downward, de Freitas, et al., 2019).

Tonne et al. found in their study estimating both residential and personal exposures that differences were not uniform across ethnic or socioeconomic groups; they urge caution about potential for exposure misclassification in epidemiological research (Tonne, Milà, Fecht, et al., 2018).

Practical approaches to reducing traffic-related exposure disparities include a variety of community-level actions such as addition of green space and vegetative barriers, creating active transport opportunities; urban planning focused on transport and health including traffic system design and regulation to prevent building schools or residences near traffic sources (Brugge, Patton, Bob, et al., 2015; Mladenovic & McPherson, 2016).

Summary and conclusions

- TRAP continues to be an important contributor to environmental exposures and has been associated with a wide range of human health effects.
- A large body of research has been developed over decades documenting associations between TRAP and respiratory diseases including asthma and allergy, cardiovascular and cardiometabolic conditions, lung and childhood cancers.
- Recent studies have found associations between TRAP exposures and many health effects across the life span including Alzheimer's disease, autism, birth outcomes, dementia, eye effects, pregnancy and reproductive complications, neurodevelopment, and Parkinson's disease.
- Populations more highly exposed include those living, studying, or working near major roadways.
- Populations that may be more sensitive to TRAP exposures include pregnant women and the developing fetus; those with preexisting health conditions such as asthma, heart disease, and diabetes; and those of lower SES.
- Research needs include: understanding TRAP-related exposure and health disparities in sensitive subpopulations; attention to exposure misclassification and complex multi-pollutant or combination exposures; further health-related research on outcomes including childhood cancers, eye effects, and pregnancy complications. Our improved understanding of the health impacts of TRAP offers opportunities for collaboration between the transportation and health sectors to build healthier communities, reducing TRAP exposures and improving public health (Koehler, Latshaw, Matte, et al., 2018).

Suggested readings

The following resources provide further details for interested readers:

Boothe, V.L., et al. (2014). Residential traffic exposure and childhood leukemia: A systematic review and meta-analysis. American Journal of Preventive Medicine. 46(4), 413–22. https://doi.org/10.1016/j.amepre.2013.11.004.

Carré, J., et al. (2017). Does air pollution play a role in infertility? A systematic review. Environmental Health. 16(1), 82. https://doi.org/10.1186/s12940-017-0291-8.

Clifford, A., et al. (2016). Exposure to air pollution and cognitive functioning across the life course—A systematic literature review. Environmental Research. 147, 383–98. https://doi.org/10.1016/j.envres.2016.01.018.

Health Effects Institute. (2010). Traffic-related air pollution: A critical review of the literature on emissions, exposure, and health effects. https://www.healtheffects.org/publication/traffic-related-air-pollution-critical-review-literature-emissions-exposure-and-health.

References

Ahmed, C., Jiang, H., Chen, J., & Lin, Y. (2018). Traffic-related particulate matter and cardiometabolic syndrome: A review. *Atmosphere*, *9*(9), 336. https://search.proquest.com/docview/2125266205. https://doi.org/10.3390/atmos9090336.

Alexeeff, S. E., Roy, A., Shan, J., et al. (2018). High-resolution mapping of traffic related air pollution with Google street view cars and incidence of cardiovascular events within neighborhoods in Oakland, CA. *Environmental Health: A Global Access Science Source*, *17*(1), 38. https://www.ncbi.nlm.nih.gov/pubmed/29759065. https://doi.org/10.1186/s12940-018-0382-1.

Andersen, Z. J., Bønnelykke, K., Hvidberg, M., et al. (2012). Long-term exposure to air pollution and asthma hospitalisations in older adults: A cohort study. *Thorax*, *67*(1), 6–11. https://www.ncbi.nlm.nih.gov/pubmed/21890573. https://doi.org/10.1136/thoraxjnl-2011-200711.

Baja, E. S., Schwartz, J. D., Wellenius, G. A., et al. (2010). Traffic-related air pollution and QT interval: Modification by diabetes, obesity, and oxidative stress gene polymorphisms in the Normative Aging Study. *Environmental Health Perspectives*, *118*(6), 840–846. https://www.jstor.org/stable/40661602. https://doi.org/10.1289/ehp.0901396.

Beelen, R., Hoek, G., van den Brandt, P. A., et al. (2008). Long-term effects of traffic-related air pollution on mortality in a Dutch cohort (NLCS-AIR study). *Environmental Health Perspectives*, *116*(2), 196–202. https://doi.org/10.1289/ehp.10767.

Boothe, V. L., Boehmer, T. K., Wendel, A. M., & Yip, F. Y. (2014). Residential traffic exposure and childhood leukemia. *American Journal of Preventive Medicine*, *46*(4), 413–422. https://www.clinicalkey.es/playcontent/1-s2.0-S0749379713006193. https://doi.org/10.1016/j.amepre.2013.11.004.

Bowatte, G., Lodge, C., Lowe, A. J., et al. (2015). The influence of childhood traffic-related air pollution exposure on asthma, allergy and sensitization: A systematic review and a meta-analysis of birth cohort studies. *Allergy*, *70*(3), 245–256. https://doi.org/10.1111/all.12561.

Brucker, N., Moro, A. M., Charão, M. F., et al. (2013). Biomarkers of occupational exposure to air pollution, inflammation and oxidative damage in taxi drivers. *Science of the Total Environment, 463–464*, 884–893. https://www.sciencedirect.com/science/article/pii/S004896971300750X. https://doi.org/10.1016/j.scitotenv.2013.06.098.

Brugge, D., Patton, A. P., Bob, A., et al. (2015). Developing community-level policy and practice to reduce traffic-related air pollution exposure. *Environmental Justice, 8*(3), 95–104. https://www.liebertpub.com/doi/abs/10.1089/env.2015.0007. https://doi.org/10.1089/env.2015.0007.

Bullard, R. D. (1990). *Dumping in dixie*. Boulder: Westview Press.

Bullard, R. D., Mohai, P., Saha, R., & Wright, B. (2008). Toxic wastes and race at twenty: Why race still matters after all of these years. *Environmental Law, 38*(2), 371–411. https://www.jstor.org/stable/43267204.

Carlos-Wallace, F. M., Zhang, L., Smith, M. T., Rader, G., & Steinmaus, C. (2016). Parental, in utero, and early-life exposure to benzene and the risk of childhood leukemia: A meta-analysis. *American Journal of Epidemiology, 183*(1), 1–14. https://doi.org/10.1093/aje/kwv120.

Carlsten, C. (2018). Synergistic environmental exposures and the airways capturing complexity in humans: An underappreciated world of complex exposures. *Chest, 154*(4), 918–924. https://doi.org/S0012-3692(18)30915-2.

Carlsten, C., & Rider, C. (2017). Traffic-related air pollution and allergic disease: An update in the context of global urbanization. *Current Opinion in Allergy and Clinical Immunology, 17*(2), 85–89. https://www.ncbi.nlm.nih.gov/pubmed/28141628. https://doi.org/10.1097/ACI.0000000000000351.

Carre, J., Gatimel, N., Moreau, J., Parinaud, J., & Leandri, R. (2017). Does air pollution play a role in infertility?: A systematic review. *Environmental Health, 16*(1), 8. https://doi.org/10.1186/s12940-017-0291-8.

Carvalho, R., Carneiro, M., Barbosa, F., Jr., et al. (2018). The impact of occupational exposure to traffic-related air pollution among professional motorcyclists from Porto Alegre, Brazil, and its association with genetic and oxidative damage. *Environmental Science and Pollution Research, 25*(19), 18620–18631. https://www.ncbi.nlm.nih.gov/pubmed/29704180. https://doi.org/10.1007/s11356-018-2007-1.

Chang, J., Liu, W., & Huang, C. (2018). Residential ambient traffic in relation to childhood pneumonia among urban children in Shandong, China: A cross-sectional study. *International Journal of Environmental Research and Public Health, 15*(6), 1076. https://www.ncbi.nlm.nih.gov/pubmed/29799501. https://doi.org/10.3390/ijerph15061076.

Chiu, Y. M., Bellinger, D. C., Coull, B. A., et al. (2013). Associations between traffic-related black carbon exposure and attention in a prospective birth cohort of urban children. *Environmental Health Perspectives, 121*(7), 859–864. https://www.ncbi.nlm.nih.gov/pubmed/23665743. https://doi.org/10.1289/ehp.1205940.

Choudhary, H., & Tarlo, S. (2014). Airway effects of traffic-related air pollution on outdoor workers. *Current Opinion in Allergy and Clinical Immunology, 14*(2), 106–112. https://www.ncbi.nlm.nih.gov/pubmed/24451913. https://doi.org/10.1097/ACI.0000000000000038.

Christianson, A., Howson, C. P., & Modell, B. (2006). *Global report on birth defects: The hidden toll of dying and disabled children*. https://www.marchofdimes.org/global-report-on-birth-defects-the-hidden-toll-of-dying-and-disabled-children-full-report.pdf.

Clifford, A., Lang, L., Chen, R., Anstey, K. J., & Seaton, A. (2016). Exposure to air pollution and cognitive functioning across the life course—A systematic literature review. *Environmental Research, 147*, 383–398. https://doi.org/10.1016/j.envres.2016.01.018.

Committee on Environmental Justice, Institute of Medicine, National Academy of Sciences. (2000). *Toward environmental justice*. Washington, DC: National Academies Press. http://www.nap.edu/6034.

Cosselman, K., Krishnan, R., Oron, A., et al. (2012). Blood pressure response to controlled diesel exhaust exposure in human subjects. *Hypertension, 59*(5), 943–948. https://www.ncbi.nlm.nih.gov/pubmed/22431582. https://doi.org/10.1161/HYPERTENSIONAHA.111.186593.

Cowell, W. J., & Wright, R. J. (2017). Sex-specific effects of combined exposure to chemical and non-chemical stressors on neuroendocrine development: A review of recent findings and putative mechanisms. *Current Environmental Health Reports, 4*(4), 415–425. https://doi.org/10.1007/s40572-017-0165-9.

Dadvand, P., Ostro, B., Amato, F., et al. (2014). Particulate air pollution and preeclampsia: A source-based analysis. *Occupational and Environmental Medicine, 71*(8), 570–577. https://doi.org/10.1136/oemed-2013-101693.

Danysh, H. E., Mitchell, L. E., Zhang, K., Scheurer, M. E., & Lupo, P. J. (2015). Traffic-related air pollution and the incidence of childhood central nervous system tumors: Texas, 2001–2009. *Pediatric Blood & Cancer, 62*(9), 1572–1578. https://doi.org/10.1002/pbc.25549.

de Prado, B. P., Mercader, E. M. H., Pujol, J., Sunyer, J., & Mortamais, M. (2018). The effects of air pollution on the brain: A review of studies interfacing environmental epidemiology and neuroimaging. *Current Environmental Health Reports*. https://doi.org/10.1007/s40572-018-0209-9.

Delfino, R. J., Tjoa, T., Gillen, D. L., et al. (2010). Traffic-related air pollution and blood pressure in elderly subjects with coronary artery disease. *Epidemiology, 21*(3), 396–404. https://www.jstor.org/stable/25662878. https://doi.org/10.1097/EDE.0b013e3181d5e19b.

Dragano, N., Hoffmann, B., Moebus, S., et al. (2009). Traffic exposure and subclinical cardiovascular disease: Is the association modified by socioeconomic characteristics of individuals and neighbourhoods? Results from a multilevel study in an urban region. *Occupational and Environmental Medicine, 66*(9), 628–635. https://www.jstor.org/stable/27797631. https://doi.org/10.1136/oem.2008.044032.

Eze, I. C., Hemkens, L. G., Bucher, H. C., et al. (2015). Association between ambient air pollution and diabetes mellitus in Europe and North America: Systematic review and meta-analysis. *Environmental Health Perspectives, 123*(5), 381–389. https://doi.org/10.1289/ehp.1307823.

Filippini, T., Heck, J. E., Malagoli, C., Del Giovane, C., & Vinceti, M. (2015). A review and meta-analysis of outdoor air pollution and risk of childhood leukemia. *Journal of Environmental Science and Health. Part C, Environmental Carcinogenesis & Ecotoxicology Reviews, 33*(1), 36–66. https://doi.org/10.1080/10590501.2015.1002999.

Fleisch, A. F., Gold, D. R., Rifas-Shiman, S. L., et al. (2014). Air pollution exposure and abnormal glucose tolerance during pregnancy: The project Viva cohort. *Environmental Health Perspectives, 122*(4), 378–383. https://doi.org/10.1289/ehp.1307065.

Fox, M. A., Brewer, L. E., & Martin, L. (2017). An overview of literature topics related to current concepts, methods, tools, and applications for cumulative risk assessment (2007–2016). *International Journal of Environmental Research and Public Health, 14*(4), 389. https://www.ncbi.nlm.nih.gov/pubmed/28387705. https://doi.org/10.3390/ijerph14040389.

Fox, M. A., Groopman, J. D., & Burke, T. (2002). Evaluating cumulative risk assessment for environmental justice: A community case study. *Environmental Health Perspectives, 110*(2), 203–209. https://doi.org/sc271_5_1835.

Fox, M. A., Spicer, K., Chosewood, L. C., Susi, P., Johns, D. O., & Dotson, G. S. (2018). Implications of applying cumulative risk assessment to the workplace. *Environment International, 115*, 230–238. https://doi.org/S0160-4120(17)31986-4.

Fuks, K. B., Weinmayr, G., Foraster, M., et al. (2014). Arterial blood pressure and long-term exposure to traffic-related air pollution: An analysis in the European Study of Cohorts for Air Pollution Effects (ESCAPE). *Environmental Health Perspectives, 122*(9), 896–905. https://www.narcis.nl/publication/RecordID/oai:dspace.library.uu.nl:1874%2F305129. https://doi.org/10.1289/ehp.1307725.

Fuller, C. H., Feeser, K. R., Sarnat, J. A., & O'Neill, M. S. (2017). Air pollution, cardiovascular endpoints and susceptibility by stress and material resources: A systematic review of the evidence. *Environmental Health*, *16*(1). https://doi.org/10.1186/s12940-017-0270-0.

Gaffron, P., & Niemeier, D. (2015). School locations and traffic emissions—Environmental (in)justice findings using a new screening method. *International Journal of Environmental Research and Public Health*, *12*(2), 2009–2025. https://www.ncbi.nlm.nih.gov/pubmed/25679341. https://doi.org/10.3390/ijerph120202009.

Gan, W. Q., Koehoorn, M., Davies, H. W., Demers, P. A., Tamburic, L., & Brauer, M. (2011). Long-term exposure to traffic-related air pollution and the risk of coronary heart disease hospitalization and mortality. *Environmental Health Perspectives*, *119*(4), 501–507. https://doi.org/10.1289/ehp.1002511.

Garshick, E., Laden, F., Hart, J. E., et al. (2008). Lung cancer and vehicle exhaust in trucking industry workers. *Environmental Health Perspectives*, *116*(10), 1327–1332. https://www.jstor.org/stable/25071183. https://doi.org/10.1289/ehp.11293.

Gaskins, A. J., Hart, J. E., Minguez-Alarcon, L., et al. (2018). Residential proximity to major roadways and traffic in relation to outcomes of in vitro fertilization. *Environment International*, *115*, 239–246. https://doi.org/S0160-4120(18)30352-0.

Gee, G. C., & Payne-Sturges, D. C. (2004). Environmental health disparities: A framework integrating psychosocial and environmental concepts. *Environmental Health Perspectives*, *112*(17), 1645–1653. https://doi.org/10.1289/ehp.7074.

Gonzalez-Barcala, F. J., Pertega, S., Garnelo, L., Castro, T. P., Sampedro, M., Lastres, J. S., et al. (2013). Truck traffic related air pollution associated with asthma symptoms in young boys: A cross-sectional study. *Public Health*, *127*(3), 275–281. https://www.clinicalkey.es/playcontent/1-s2.0-S0033350612004775. https://doi.org/10.1016/j.puhe.2012.12.028.

Hamra, G. B., Laden, F., Cohen, A. J., Raaschou-Nielsen, O., Brauer, M., & Loomis, D. (2015). Lung cancer and exposure to nitrogen dioxide and traffic: A systematic review and meta-analysis. *Environmental Health Perspectives*, *123*(11), 1107–1112. https://doi.org/10.1289/ehp.1408882.

Health Effects Institute. (2010). Panel on the health effects of traffic-related air pollution. In *Traffic-related air pollution: A critical review of the literature on emissions, exposure, and health effects*. United States https://www.healtheffects.org/publication/traffic-related-air-pollution-critical-review-literature-emissions-exposure-and-health.

Hoffmann, B., Moebus, S., Dragano, N., et al. (2009). Residential traffic exposure and coronary heart disease: Results from the Heinz Nixdorf recall study. *Biomarkers*, *14*(Suppl 1), 74–78. https://doi.org/10.1080/13547500902965096.

Huang, H. B., Lai, C. H., Chen, G. W., et al. (2012). Traffic-related air pollution and DNA damage: A longitudinal study in Taiwanese traffic conductors. *PLoS One*, 7(5), e37412. https://www.ncbi.nlm.nih.gov/pubmed/22629390. https://doi.org/10.1371/journal.pone.0037412.

Janghorbani, M., Momeni, F., & Mansourian, M. (2014). Systematic review and metaanalysis of air pollution exposure and risk of diabetes. *European Journal of Epidemiology*, *29*(4), 231–242. https://doi.org/10.1007/s10654-014-9907-2.

Ji, X., Meng, X., Liu, C., et al. (2019). Nitrogen dioxide air pollution and preterm birth in Shanghai, China. *Environmental Research*, *169*, 79–85. https://doi.org/S0013-9351(18)30578-4.

Jung, S. J., Mehta, J. S., & Tong, L. (2018). Effects of environment pollution on the ocular surface. *The Ocular Surface*, *16*(2), 198–205. https://doi.org/S1542-0124(17)30224-0.

Kan, H., Heiss, G., Rose, K. M., Whitsel, E. A., Lurmann, F., & London, S. J. (2008). Prospective analysis of traffic exposure as a risk factor for incident coronary heart disease: The atherosclerosis risk in communities (ARIC) study. *Environmental Health Perspectives*, *116*(11), 1463–1468. https://doi.org/10.1289/ehp.11290.

Khreis, H., Kelly, C., Tate, J., Parslow, R., Lucas, K., & Nieuwenhuijsen, M. (2017). Exposure to traffic-related air pollution and risk of development of childhood asthma: A systematic review and meta-analysis. *Environment International, 100*, 1–31. https://doi.org/S0160-4120(16)30783-8.

Kilian, J., & Kitazawa, M. (2018). The emerging risk of exposure to air pollution on cognitive decline and Alzheimer's disease—Evidence from epidemiological and animal studies. *Biomedical Journal, 41*(3), 141–162. https://doi.org/S2319-4170(18)30080-5.

Kioumourtzoglou, M.A., Raz, R., Wilson, A., et al. (2019). Traffic-related air pollution and pregnancy loss. *Epidemiology, 30*(1), 4–10. https://doi.org/10.1097/EDE.0000000000000918.

Koehler, K., Latshaw, M., Matte, T., et al. (2018). Building healthy community environments: A public health approach. *Public Health Reports, 133*(1_suppl), 43S. https://doi.org/10.1177/0033354918798809.

Kumar, S.V., Lupo, P. J., Pompeii, L. A., & Danysh, H. E. (2018). Maternal residential proximity to major roadways and pediatric embryonal tumors in offspring. *International Journal of Environmental Research and Public Health, 15*(3), E505. https://doi.org/10.3390/ijerph15030505.

Legro, R. S., Sauer, M.V., Mottla, G. L., et al. (2010). Effect of air quality on assisted human reproduction. *Human Reproduction, 25*(5), 1317–1324. https://doi.org/10.1093/humrep/deq021.

Mahalingaiah, S., Hart, J. E., Laden, F., et al. (2016). Adult air pollution exposure and risk of infertility in the Nurses' Health Study II. *Human Reproduction, 31*(3), 638–647. https://doi.org/10.1093/humrep/dev330.

Martenies, S. E., Milando, C.W., Williams, G. O., & Batterman, S.A. (2017). Disease and health inequalities attributable to air pollutant exposure in Detroit, Michigan. *International Journal of Environmental Research and Public Health, 14*(10), 1243. https://www.ncbi.nlm.nih.gov/pubmed/29048385. https://doi.org/10.3390/ijerph14101243.

Maynard, D., Coull, B. A., Gryparis, A., & Schwartz, J. (2007). Mortality risk associated with short-term exposure to traffic particles and sulfates. *Environmental Health Perspectives, 115*(5), 751–755. https://doi.org/10.1289/ehp.9537.

Mendola, P., Wallace, M., Hwang, B. S., et al. (2016). Preterm birth and air pollution: Critical windows of exposure for women with asthma. *The Journal of Allergy and Clinical Immunology, 138*(2), 440.e5. https://doi.org/10.1016/j.jaci.2015.12.1309.

Mladenovic, M., & McPherson, T. (2016). Engineering social justice into traffic control for self-driving vehicles? *Science and Engineering Ethics, 22*(4), 1131–1149. https://www.ncbi.nlm.nih.gov/pubmed/26231407. https://doi.org/10.1007/s11948-015-9690-9.

Munzel, T., Sorensen, M., Gori, T., et al. (2017). Environmental stressors and cardio-metabolic disease: Part I-epidemiologic evidence supporting a role for noise and air pollution and effects of mitigation strategies. *European Heart Journal, 38*(8), 550–556. https://doi.org/10.1093/eurheartj/ehw269.

Nieuwenhuijsen, M. J., Basagana, X., Dadvand, P., et al. (2014). Air pollution and human fertility rates. *Environment International, 70*, 9–14. https://doi.org/10.1016/j.envint.2014.05.005.

Oudin, A., Forsberg, B., Adolfsson, A. N., et al. (2016). Traffic-related air pollution and dementia incidence in northern Sweden: A longitudinal study. *Environmental Health Perspectives, 124*(3), 306–312. https://www.ncbi.nlm.nih.gov/pubmed/26305859. https://doi.org/10.1289/ehp.1408322.

Padula, A. M., Tager, I. B., Carmichael, S. L., et al. (2013). Traffic-related air pollution and selected birth defects in the San Joaquin Valley of California. *Birth Defects Research. Part A, Clinical and Molecular Teratology, 97*(11), 730–735. https://doi.org/10.1002/bdra.23175.

Padula, A. M., Yang, W., Carmichael, S. L., et al. (2017). Air pollution, neighborhood acculturation factors, and neural tube defects among hispanic women in California. *Birth Defects Research, 109*(6), 403–422. https://doi.org/10.1002/bdra.23602.

Padula, A. M., Yang, W., Lurmann, F. W., Balmes, J., Hammond, S. K., & Shaw, G. M. (2019). Prenatal exposure to air pollution, maternal diabetes and preterm birth. *Environmental Research, 170*, 160–167. https://doi.org/S0013-9351(18)30659-5.

Patel, M. M., Chillrud, S. N., Correa, J. C., et al. (2010). Traffic-related particulate matter and acute respiratory symptoms among New York City area adolescents. *Environmental Health Perspectives, 118*(9), 1338–1343. https://doi.org/10.1289/ehp.0901499.

Pedersen, M., Giorgis-Allemand, L., Bernard, C., et al. (2013). Ambient air pollution and low birthweight: A European cohort study (ESCAPE). *The Lancet Respiratory Medicine, 1*(9), 695–704. https://doi.org/10.1016/S2213-2600(13)70192-9.

Pedersen, M., Stayner, L., Slama, R., et al. (2014). Ambient air pollution and pregnancy-induced hypertensive disorders: A systematic review and meta-analysis. *Hypertension, 64*(3), 494–500. https://doi.org/10.1161/HYPERTENSIONAHA.114.03545.

Pereira, G., Haggar, F., Shand, A. W., Bower, C., Cook, A., & Nassar, N. (2013). Association between pre-eclampsia and locally derived traffic-related air pollution: A retrospective cohort study. *Journal of Epidemiology and Community Health, 67*(2), 147–152. https://doi.org/10.1136/jech-2011-200805.

Pratt, G. C., Vadali, M. L., Kvale, D. L., & Ellickson, K. M. (2015). Traffic, air pollution, minority and socio-economic status: Addressing inequities in exposure and risk. *International Journal of Environmental Research and Public Health, 12*(5), 5355–5372. https://www.ncbi.nlm.nih.gov/pubmed/25996888. https://doi.org/10.3390/ijerph120505355.

Raaschou-Nielsen, O., & Reynolds, P. (2006). Air pollution and childhood cancer: A review of the epidemiological literature. *International Journal of Cancer, 118*(12), 2920–2929. https://doi.org/10.1002/ijc.21787.

Rancière, F., Bougas, N., Viola, M., & Momas, I. (2017). Early exposure to traffic-related air pollution, respiratory symptoms at 4 years of age, and potential effect modification by parental allergy, stressful family events, and sex: A prospective follow-up study of the PARIS birth cohort. *Environmental Health Perspectives, 125*(4), 737–745. https://www.ncbi.nlm.nih.gov/pubmed/27219743. https://doi.org/10.1289/EHP239.

Ribeiro, A. G., Downward, G. S., de Freitas, C. U., et al. (2019). Incidence and mortality for respiratory cancer and traffic-related air pollution in São Paulo, Brazil. *Environmental Research, 170*, 243–251. https://www.sciencedirect.com/science/article/pii/S0013935118306637. https://doi.org/10.1016/j.envres.2018.12.034.

Ritz, B., Lee, P., Hansen, J., et al. (2016). Traffic-related air pollution and Parkinson's disease in Denmark: A case-control study. *Environmental Health Perspectives, 124*(3), 351–356. https://www.ncbi.nlm.nih.gov/pubmed/26151951. https://doi.org/10.1289/ehp.1409313.

Ritz, B., Qiu, J., Lee, P. C., et al. (2014). Prenatal air pollution exposure and ultrasound measures of fetal growth in Los Angeles, California. *Environmental Research, 130*, 7–13. https://doi.org/10.1016/j.envres.2014.01.006.

Robledo, C. A., Mendola, P., Yeung, E., et al. (2015). Preconception and early pregnancy air pollution exposures and risk of gestational diabetes mellitus. *Environmental Research, 137*, 316–322. https://doi.org/10.1016/j.envres.2014.12.020.

Schultz, E. S., Litonjua, A. A., & Melén, E. (2017). Effects of long-term exposure to traffic-related air pollution on lung function in children. *Current Allergy and Asthma Reports, 17*(6), 41. https://www.ncbi.nlm.nih.gov/pubmed/28551888. https://doi.org/10.1007/s11882-017-0709-y.

Sears, C. G., Braun, J. M., Ryan, P. H., et al. (2018). The association of traffic-related air and noise pollution with maternal blood pressure and hypertensive disorders of pregnancy in the HOME study cohort. *Environment International, 121*(Pt 1), 574–581. https://doi.org/S0160-4120(18)31119-X.

Slama, R., Bottagisi, S., Solansky, I., Lepeule, J., Giorgis-Allemand, L., & Sram, R. (2013). Short-term impact of atmospheric pollution on fecundability. *Epidemiology, 24*(6), 871–879. https://doi.org/10.1097/EDE.0b013e3182a702c5.

Sram, R. J., Veleminsky, M., Jr., Veleminsky, M. S., & Stejskalova, J. (2017). The impact of air pollution to central nervous system in children and adults. *Neuro Endocrinology Letters, 38*(6), 389–396. https://doi.org/NEL380617R01.

Stansfeld, S. A. (2015). Noise effects on health in the context of air pollution exposure. *International Journal of Environmental Research and Public Health, 12*(10), 12735–12760. https://doi.org/10.3390/ijerph121012735.

Stingone, J. A., Luben, T. J., Carmichael, S. L., et al. (2017). Maternal exposure to nitrogen dioxide, intake of methyl nutrients, and congenital heart defects in offspring. *American Journal of Epidemiology, 186*(6), 719–729. https://doi.org/10.1093/aje/kwx139.

Tetreault, L. F., Perron, S., & Smargiassi, A. (2013). Cardiovascular health, traffic-related air pollution and noise: Are associations mutually confounded? A systematic review. *International Journal of Public Health, 58*(5), 649–666. https://doi.org/10.1007/s00038-013-0489-7.

Tomei, G., Ciarrocca, M., Fortunato, B. R., et al. (2006). Exposure to traffic pollutants and effects on 17-beta-estradiol (E2) in female workers. *International Archives of Occupational and Environmental Health, 80*(1), 70–77. https://doi.org/10.1007/s00420-006-0105-8.

Tonne, C., Milà, C., Fecht, D., et al. (2018). Socioeconomic and ethnic inequalities in exposure to air and noise pollution in London. *Environment International, 115*, 170–179. https://www.sciencedirect.com/science/article/pii/S0160412017321256. https://doi.org/10.1016/j.envint.2018.03.023.

U.S. Environmental Protection Agency. (2019). *Environmental justice*. https://www.epa.gov/environmentaljustice. (Accessed18 March 2019).

van den Hooven, E. H., Jaddoe, V. W., de Kluizenaar, Y., et al. (2009). Residential traffic exposure and pregnancy-related outcomes: A prospective birth cohort study. *Environmental Health, 8*, 59. https://doi.org/10.1186/1476-069X-8-59.

van den Hooven, E. H., Pierik, F. H., de Kluizenaar, Y., et al. (2012). Air pollution exposure during pregnancy, ultrasound measures of fetal growth, and adverse birth outcomes: A prospective cohort study. *Environmental Health Perspectives, 120*(1), 150–156. https://doi.org/10.1289/ehp.1003316.

Volk, H. E., Hertz-Picciotto, I., Delwiche, L., Lurmann, F., & McConnell, R. (2011). Residential proximity to freeways and autism in the CHARGE study. *Environmental Health Perspectives, 119*(6), 873–877. https://doi.org/10.1289/ehp.1002835.

Volk, H. E., Lurmann, F., Penfold, B., Hertz-Picciotto, I., & McConnell, R. (2013). Traffic-related air pollution, particulate matter, and autism. *JAMA Psychiatry, 70*(1), 71–77. https://doi.org/10.1001/jamapsychiatry.2013.266.

Vrijheid, M., Martinez, D., Manzanares, S., et al. (2011). Ambient air pollution and risk of congenital anomalies: A systematic review and meta-analysis. *Environmental Health Perspectives, 119*(5), 598–606. https://doi.org/10.1289/ehp.1002946.

Wang, B., Xu, D., Jing, Z., Liu, D., Yan, S., & Wang, Y. (2014). Effect of long-term exposure to air pollution on type 2 diabetes mellitus risk: A systemic review and meta-analysis of cohort studies. *European Journal of Endocrinology, 171*(5), 173. https://doi.org/10.1530/EJE-14-0365.

Weichenthal, S., Kulka, R., Dubeau, A., Martin, C., Wang, D., & Dales, R. (2011). Traffic-related air pollution and acute changes in heart rate variability and respiratory function in urban cyclists. *Environmental Health Perspectives, 119*(10), 1373–1378. https://www.jstor.org/stable/41263046. https://doi.org/10.1289/ehp.1003321.

World Health Organization. (2018a). *Ambient (outdoor) air pollution*. https://www.who.int/news-room/fact-sheets/detail/ambient-(outdoor)-air-quality-and-health. (Accessed 15 May 2020).

World Health Organization. (2018b). *Environmental noise guidelines for the European region*. http://www.euro.who.int/en/health-topics/environment-and-health/noise/publications/2018/environmental-noise-guidelines-for-the-european-region-2018. (Accessed15 May 2020).

Wu, S., Deng, F., Niu, J., Huang, Q., Liu, Y., & Guo, X. (2011). The relationship between traffic-related air pollutants and cardiac autonomic function in a panel of healthy adults: A further analysis with existing data. *Inhalation Toxicology*, *23*(5), 289–303. https://www.ncbi.nlm.nih.gov/pubmed/21506879. https://doi.org/10.3109/08958378.2011.568976.
Wu, M., Ries, J. J., Proietti, E., Vogt, D., Hahn, S., & Hoesli, I. (2016). Development of late-onset preeclampsia in association with road densities as a proxy for traffic-related air pollution. *Fetal Diagnosis and Therapy*, *39*(1), 21–27. https://doi.org/10.1159/000381802.
Yorifuji, T., Naruse, H., Kashima, S., Murakoshi, T., & Doi, H. (2015). Residential proximity to major roads and obstetrical complications. *Science of the Total Environment*, *508*, 188–192. https://doi.org/10.1016/j.scitotenv.2014.11.077.

CHAPTER 10

Evidence from toxicological and mechanistic studies

Mark R. Miller and Jennifer B. Raftis
University/BHF Centre for Cardiovascular Science, University of Edinburgh, Edinburgh, United Kingdom

Abbreviations

AhR	aryl hydrocarbon receptors
BBB	blood-brain barrier
CAPs	concentrated ambient particles
CO	carbon monoxide
CO_2	carbon dioxide
COPD	chronic obstructive pulmonary disease
DE	diesel exhaust
DEP	diesel exhaust particulate
ECG	electrocardiogram
eNOS	endothelial nitric oxide synthase
ET-1	endothelin-1
FEV_1	forced expiratory volume in one second
FVC	forced vital capacity
GSTM-1	glutathione-S-transferase-mu-1
HRV	heart rate variability
ICAM-1	intercellular adhesion molecule-1
LDL	low-density lipoprotein
LEZ	low-emission zone
LOX-1	lectin-like oxidized low-density lipoprotein receptor
NE-PM	nonexhaust traffic-derived PM
NO	nitric oxide
NO_2	nitrogen dioxide
NO_x	nitrogen oxides
O_3	ozone
oxLDL	oxidised low-density lipoprotein
PAH	polyaromatic hydrocarbon
PM	particulate matter
PM_{10}	particulate matter with a diameter $<10\,\mu m$
$PM_{2.5}$	particulate matter with a diameter $<2.5\,\mu m$
$PM_{0.1}$	particulate matter with a diameter $<100\,nm$
SO_2	sulfur dioxide
t-PA	tissue plasminogen activator
TRAP	traffic-related air pollution

Traffic-Related Air Pollution
https://doi.org/10.1016/B978-0-12-818122-5.00010-7

VCAM-1	vascular cell adhesion molecule-1
VE-PM	vehicle exhaust-derived PM
vWF	von Willebrand factor
WHO	World Health Organisation

Background

Air pollution imposes a phenomenal burden on health. It is estimated that air pollution is responsible for several million premature deaths worldwide every single year (Burnett et al., 2018; Lim et al., 2012; World Health Organization, 2011). Ambient (outdoor) air pollution is the number one environmental risk factor, and the fifth biggest risk factor overall, for all-cause mortality (Cohen et al., 2017). The morbidity accompanying air pollution is similarly great (3.1% of global disability-adjusted life years), with a loss in life expectancy ranging from a few months in modestly polluted countries (e.g., Europe or the United States) to several years in heavily polluted developing and heavily populated countries (e.g., India and China) (Apte, Brauer, Cohen, Ezzati, & Pope III., 2018). One of the reasons for the scale of these figures is that exposure to air pollution is almost ubiquitous; it has been estimated that > 90% of the world's population lives in areas of above World Health Organization (WHO) guidelines (World Health Organization, 2018).

Air pollution is a systemic problem, not only on a geographical level but also physiologically in that it has effects throughout the body. It is intuitive that inhalation of air pollution has effects on the lung, however, in recent years research has revealed a plethora of other targets, including the liver, kidney, brain, gut, circulating cells, as well as exacerbation of metabolic syndrome/diabetes and reproductive effects on sterility and from maternal exposure (Schraufnagel et al., 2019a, 2019b). In particular, it is now recognized that exposure to air pollution has striking cardiovascular effects. Indeed, 40%–60% of the premature mortality from air pollution is linked to cardiovascular causes (e.g., ischemic heart disease and cerebrovascular disease) (Burnett et al., 2018; Lelieveld et al., 2019).

While air pollution arises from many different sources, it is urban ambient air pollution that has received the most research attention, due to the high density of urban populations, increasing urbanization of societies worldwide and the recognized toxicity of certain urban emissions such as those from industry and traffic. Urban air pollution is a complex mixture of different gases, liquids, and particles (see Chapters 1-6). Gases include sulfur dioxide (SO_2), carbon dioxide (CO_2) and monoxide (CO), ozone (O_3), and nitrogen dioxide (NO_2) (Dickey & Part, 2000). Semivolatile liquids within air pollution include methane, benzene, naphthalene, formaldehyde, polyaromatic hydrocarbons, and alkanes (Kansal, 2009; Weitekamp, Stevens, Stewart, Bhave, & Gilmour, 2020). Urban air is rich in particulate matter (PM), which has a varied profile depending on its source (see below). Environmental PM tends to be categorized, measured and regulated in relation to particle size: coarse particles (PM_{10}; particles with a diameter of 10 μm or less) and fine particles ($PM_{2.5}$; diameter of 2.5 μm or less) (Fig. 10.1A). A third category, ultrafine particles (diameter of < 100 nm, also termed "nanoparticles"), is believed to be especially important for health (see below) although it is not currently practical to widely measure ultrafine PM using monitoring networks in the environment. Diesel exhaust (DE), in particular, has been singled out as a pollutant of interest due to comparatively higher emissions of both NO_2 and ultrafine PM compared to other vehicles (Hooftman, Oliveira, Messagie, Coosemans, & van Mierlo, 2016) (Fig. 10.1B). Exceedances in NO_2 levels at roadside monitoring networks in many cities, and high profile scandals in improper reporting of vehicle emissions by car manufacturers, have further focused attention on diesel emissions (Burki, 2015).

Epidemiological evidence supports the emphasis on traffic-derived air pollution (TRAP). A huge body of epidemiological data has linked urban air pollution to morbidity and mortality and the significant contribution traffic-derived pollutants make to urban air (see Chapters 7–9). Accordingly, similar findings have been seen for health outcomes associated with exposure to traffic (e.g., the distance of residential address from a major road; in individuals with occupations that have high levels of traffic exposure). While it is difficult to pinpoint these effects to TRAP alone, studies that have attempted to tease apart the effects of pollution from that of possible confounding variables (e.g., noise, stress, levels of activity, socioeconomic status, etc.) have found associations with pollution to remain consistent.

This chapter discusses the evidence for health effects of TRAP, focusing on the biological mechanisms underlying potential health effects by describing key data from exposures in vitro assays, cellular and animal models and, where available, controlled exposure to pollutants in human subjects. The chapter centers on PM from vehicle exhaust (VE-PM), which is the most commonly studied pollutant from vehicles although the role

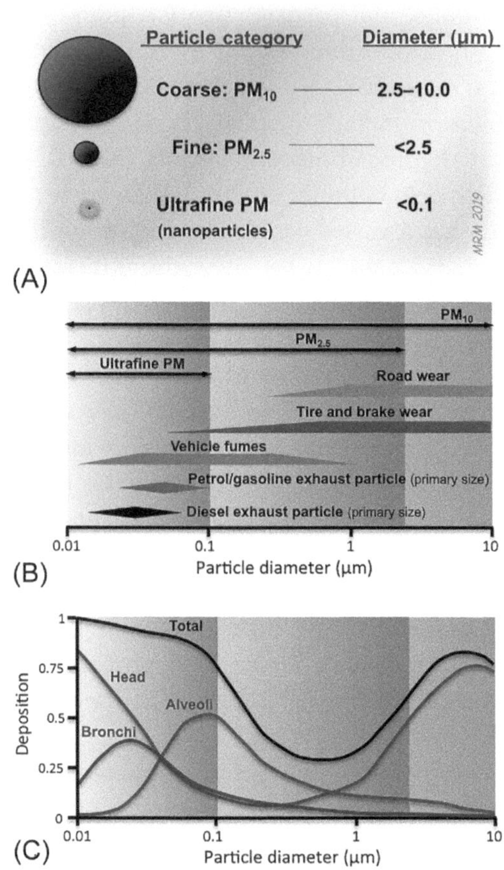

Fig. 10.1 Size categories of particulate matter (PM) in air pollution. (A) Relative sizes of PM_{10}, $PM_{2.5}$, and ultrafine PM. (B) Relative size range of vehicle emissions. Note, sizes are approximate and will vary greatly between different engines, fuels, driving conditions, temperature, etc. Primary size = size of a single particle (rather than an agglomerate as found in air) (C) Differential deposition of particle size fractions in the respiratory tract. Note that the greatest deposition of particles in the alveoli is for ultrafine particles. *((A) Modified from Miller, M. R., & Newby, D. E. (2019). Air pollution and cardiovascular disease: Car sick. Cardiovascular Research, 116, 279–294.)*

of gaseous co-pollutants and nonexhaust PM from traffic will be briefly discussed. While the chapter will discuss the effects of TRAP throughout the body, the emphasis is given to the actions of VE-PM such as diesel exhaust particles (DEP) on the cardiovascular system as an exemplar of the many ways TRAP can exert multiple detrimental effects on organs through wide-ranging mechanistic pathways.

Direct actions of vehicle exhaust emissions in biological systems

A variety of gases are emitted from vehicle exhaust, including CO_2, CO, O_3, nitrogen, and nitrogen oxides (NO_x) (Weitekamp et al., 2020). Nitrogen and carbon dioxide are not considered to be directly harmful to human health. CO is a recognized poison with the potential to induce death by reducing the ability of the blood to carry oxygen, however, the dispersion of CO in the atmosphere on leaving vehicle tailpipes largely limits its immediate health effects. While O_3 is not directly produced by fuel combustion, it can be present in exhaust emissions after secondary reactions (such as with NO_x and hydrocarbons) with sunlight. Both O_3 and NO_x, particularly NO_2, participate in the oxidation of biological molecules and induces inflammation (Blomberg, 2000; Gamon & Wille, 2016). This is readily observed in the lung (e.g., by depletion of antioxidants and activation of inflammatory cells). Both gases are soluble to a degree, but they are unlikely to penetrate tissue to an extent that would cause significant direct effects outside the lung although they may have indirect actions on extrapulmonary organs (e.g., through inflammatory processes; see below).

Vehicle exhaust PM has a high capacity to cause oxidation and inflammation. The surface of carbon-based particles is capable of generating oxygen free radicals that can oxidize biological molecules and promote cellular oxidative stress (an imbalance in the amount of oxidative-mediators beyond what the cell is capable of removing by cellular antioxidant systems, leading to harm to the cell and promotion of disease). The generation of free radicals from particle surfaces can be assessed by acellular biochemical assays, including oxidation of other molecules, binding to fluorescent compounds and electron spin resonance (Ikeda, Suzuki, Watarai, Sagai, & Tomita, 1995; Miller, Shaw, & Langrish, 2012; Ovrevik, 2019; Risom, Moller, & Loft, 2005). Additionally, combustion-derived PM also contains a wide range of metallic elements including iron, copper, nickel, vanadium, and many others (Steiner, Bisig, Petri-Fink, & Rothen-Rutishauser, 2016; Vouk & Piver, 1983) (Fig. 10.2). Several of these metals interfere with cellular processes, especially through the metal-mediated catalytic generation of oxygen free radicals (e.g., hydroxyl free radical production via iron through the Fenton reaction), impaired mitochondrial function and activation of cellular sources of other free radicals (Samet, Chen, Pennington, & Bromberg, 2019). Additionally, the surface of VE-PM is coated with a cocktail of organic carbon species, including alkanes, quinones, and polyaromatic hydrocarbons (PAHs). Many of these species have carcinogenic activity and can

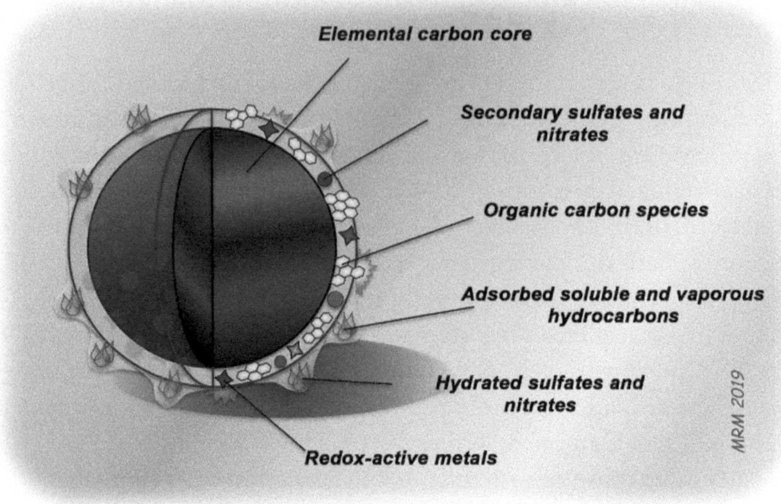

Fig. 10.2 Complex composition of a vehicle exhaust particle. A solid carbon core is surrounded by a mixture of organic and inorganic species. Importantly, the surface of the particle contains a cocktail of reactive organic carbon species and transition metals which are important to the oxidative properties of the particle and toxicity of the particle to cells. *(Modified from Miller, M. R., Shaw, C. A., & Langrish, J.P. (2012). From particles to patients: Oxidative stress and the cardiovascular effects of air pollution. Future Cardiology, 8, 577–602.)*

also undergo redox cycling to generate free radicals (Valavanidis, Fiotakis, & Vlachogianni, 2008). VE-PM can also induce oxidative stress through the stimulation of cellular enzymes and disruption of mitochondrial function (Miller et al., 2012). Additionally, organic species within VE-PM activate a series of different inflammatory pathways, including classical inflammatory signaling cascades, cytokine release, interaction with cytochrome P450 pathways, and activation of aryl hydrocarbon receptors (AhR) (Holme, Brinchmann, Refsnes, Lag, & Ovrevik, 2019; Ovrevik et al., 2017; Schwarze et al., 2013). While oxidative potential and in vitro assays of inflammation using PM can help assess its potential toxicity, the complex interaction of PM with cells and between organs means a wide range of approaches is needed to determine true toxicity, especially outside the lung (Miller et al., 2012; Ovrevik, 2019).

Particles of different size deposit in different regions of the airways (Fig. 10.1C), with smaller particles ($PM_{2.5}$ and ultrafine PM) reaching the alveoli of the lungs where gases exchange occurs (Donaldson,

Stone, Clouter, Renwick, & MacNee, 2001). On deposition in the alveoli of the lungs, macrophages will attempt to phagocytose particles to aid their removal. However, unlike many biological xenobiotics, VE-PM cannot be readily digested by macrophages and particles persist in these cells, leading to release of inflammatory mediators, with the dose and physicochemical properties of the particles determining the subsequent effect on inflammation (Donaldson et al., 2001). These foci of inflammation may induce pathological effects in tissue and initiate/exacerbate the disease. Nano-sized particles may pass through the lung tissue and access the blood circulation (discussed further below), and thus have the potential to exert direct effects on other organs of the body (Ohlwein, Kappeler, Kutlar Joss, Kunzli, & Hoffmann, 2019; Raftis & Miller, 2019). The properties of these particles are likely to be altered during passage through tissues, particularly in respect to which constituents elute from the particle surface and which stay bound. Particles will also adsorb biological molecules onto the particle surface (e.g., "protein corona") that can influence subsequent cellular uptake and responses (Riediker et al., 2019).

Finally, most research investigating the toxicological actions of TRAP focus on exhaust emissions. An understudied but growing, area of interest is the potential harm caused by nonexhaust traffic-derived PM (NE-PM), for example, from brake and tire wear, as well as the resuspension of road dust. Nonexhaust PM can form a significant proportion of overall roadside PM, for example, 60%–75% of PM from transport, albeit 5%–10% of overall PM emissions (both metrics by mass) (AQEG_Air_Quality_Expert_Group, 2019). The relative proportion of NE-PM in overall PM may increase with the uptake of comparatively heavier electric vehicles. Efforts have been made to estimate the potential health risks of these sources of PM, however, it should be noted that these studies frequently assume that particle size and composition is equivalent to other sources of PM (COMEAP, 2015). This is unlikely to be the case: in general nonexhaust PM is of a greater size than exhaust PM (AQEG_Air_Quality_Expert_Group, 2019) (Fig. 10.1B), which could lower its toxicity due a lower particle surface area for a given mass and lead to lesser penetration into the alveolar space and systemic tissues. In contrast, the high metal content of brake-wear PM in particular may mean these particles have a greater capacity to induce oxidative stress within cells. The effects of NE-PM are briefly discussed below, and are likely to represent a topic of heightened research interest in the years ahead.

TRAP and the lung

VE-PM and the lung

The respiratory effects of air pollution are widely discussed elsewhere (e.g., Gotschi, Heinrich, Sunyer, & Kunzli, 2008; Health_Effects_Institute, 2010 and see Chapters 7–9). In brief, short- and long-term exposure is associated with hospital admission for respiratory exacerbations and chronic decline in pulmonary function with the development of respiratory diseases. These include exacerbation of asthma, bronchitis, increased risk of infection, chronic obstructive pulmonary disease (COPD), and respiratory mortality overall (Carlsen et al., 2015; Dominici et al., 2006; Gotschi et al., 2008; Schikowski et al., 2014; Zanobetti & Schwartz, 2009). Indeed, recent estimates from the Global Burden of Disease group have indicated that, globally, air pollution is linked to over 1 million premature deaths from COPD and lung cancer combined every year (Cohen et al., 2017) although the evidence for causality is "suggestive rather than conclusive" (Schikowski et al., 2014). There is substantial evidence that traffic-derived sources contribute to the links between air pollution and respiratory disease (Guan, Zheng, Chung, & Zhong, 2016; Health_Effects_Institute, 2010; Holguin, 2008). Early life exposure has been a major topic of interest. For example, TRAP has been associated with the exacerbation and development of asthma in children (Gauderman et al., 2007; Patel et al., 2011). Modeled traffic exposure during the first year of life was associated with lower forced expiratory volume (FEV_1) at age 8 (Schultz et al., 2012). Walking alongside roads with heavy traffic in London was associated with respiratory symptoms, reduced FEV1 and forced vital capacity (FVC) compared to a similar walk in a city center park, in both volunteers and patients with COPD (McCreanor et al., 2007; Sinharay et al., 2018).

The pathways underlying the effect of TRAP on the lung have revealed a number of mechanisms at play in concert. "Chamber studies," where a volunteer inhales a specified dose of pollutant in a controlled manner, are an effective means to assess the action of individual pollutants in the absence of many of the confounding factors of real-world scenarios. The evidence from controlled exposure studies in human subjects has been more mixed than might be suggested from epidemiological evidence (Behndig et al., 2011; Ghio, Smith, & Madden, 2012). Nonetheless, acute inhalation of DE (typically 300 $\mu g/m^3$ for 1–2 h; levels that would be high for short exposure—representative of cycling in extremely heavy traffic for over an hour—but representative of urban exposure when averaged over 24 h) can impair pulmonary function (e.g., lung expiratory parameters and increased sensitivity to bronchoconstrictors), increase airway inflammation, and promote allergic sensitivity (Carlsten et al., 2016; Nightingale et al., 2000; Nordenhall et al., 2001; Salvi et al., 1999). DE also sensitized the lung to O_3-induced inflammation (Bosson et al., 2008). Interleukin mediators appear to play a prominent role (Holgate et al., 2003; Pourazar et al., 2004).

Preclinical studies are a useful means to address biological mechanisms underpinning the health effects identified in human studies. A range of different models are available from cell cultures (an inexpensive and semi high throughput means to assess the effects of pollutants on cells, albeit these models lack the interplay between cell types and organs, and particle administration is usually nonphysiological), animal instillation experiments (where a droplet of particle suspension is administered to the lungs: a route that avoids the complex nasal passages of rodents that can filter out particles, but inevitably has a high dose rate of exposure, i.e., a rapid administration of a high concentration), or animal inhalation experiments (the most physiological relevant method of administration, but one that requires specialist facilities with high cost and time demands). Preclinical studies have revealed a plethora of biological mechanisms (de Kok, Driece, Hogervorst, & Briede, 2006; Fariss, Gilmour, Reilly, Liedtke, & Ghio, 2013; Guan et al., 2016; Schwarze et al., 2013; Steiner et al., 2016). These include direct stimulation of bronchial smooth muscle, inflammation and disruption of the alveolar epithelium. Oxidative stress, following loss of protective antioxidant compounds in the lung lining fluid, is likely to be a key early event in many of the respiratory effects of inhaled TRAP (Gilmour, Jaakkola, London, Nel, & Rogers, 2006). Many types of inflammatory cells are key to the respiratory effects of TRAP, with sequential and interregulatory processes leading from the initial response to the insult through to fibrosis, and also the potential for resolution. PM, such as DEP, can also have adjuvant activity sensitizing and amplifying the allergic response to other agents (Diaz-Sanchez, Jyrala, Ng, Nel, & Saxon, 2000). From a cellular perspective, dysfunction can be caused by many pathways including oxidative stress, deregulation of antioxidant pathways, changes to cellular metabolism, mitochondrial dysfunction, lipid peroxidation, DNA modification, apoptosis, genotoxicity, epigenetic modifications, impaired phagocyte clearance, microRNA release, and a multitude of inflammatory pathways (Cheng et al., 2020; de Kok et al., 2006; Kampfrath et al., 2011; Ma & Ma, 2002; Maier et al., 2008; Steiner et al., 2016). The autonomic nervous system also plays a role in amplifying inflammatory responses to DEP (McQueen et al., 2007). Organic carbon species on the surface of DEP play a role in their acute effects on airway inflammation, in particular via transient receptor potential channels (Akopian, Fanick, & Brooks, 2016). However, both the particulate core (or tightly bound constituents) and organic surface chemicals play contributing roles in mediating oxidative stress and inflammatory responses in the lung (Ma & Ma, 2002).

Nonexhaust traffic-derived PM and the lung

A limited number of studies have addressed the toxicological effects of NE-PM, from a pulmonary perspective at least. In vitro studies with cultured pulmonary cells (usually epithelial cells models although blood monocytes have also been used to model lung and tissue macrophages) have shown that tire, brake, and fine road debris can have biological activity in these cells, with a capacity to induce inflammation and cytotoxicity (e.g., Barosova et al., 2018; Gualtieri, Mantecca, Cetta, & Camatini, 2008; Gustafsson et al., 2008). Similarly, in vivo exposures have shown that nonexhaust PM can induce lung inflammation in rodent models, however, often there are equal number of other biological parameters measured in the same study that the PM does not significantly alter (e.g., Gottipolu et al., 2008; Mantecca et al., 2009). It should be noted that most studies reporting pulmonary toxicity tend to use high doses of PM (e.g., inhalation of several mg/m^3, instillation of several mg/kg body weight or $> 75\,\mu g/mL$ in cell cultures), likely to be beyond that found at a roadside for a single period of exposure. At the time of writing, there were two inhalation studies with nonexhaust PM; the most physiologically relevant route of administration. The first (Gerlofs-Nijland et al., 2019) found some modest effects at high doses of PM ($9\,mg/m^3$ for 1.5–6 h), with other sources of PM (farming PM, DEP) having greater effects on other parameters. The second (Kreider, Doyle-Eisele, Russell, McDonald, & Panko, 2012) used a longer exposure of a lower dose of tire and road wear PM (28 days at $10–100\,\mu g/m^3$) and, notably, did not observe any toxicological actions. Studies directly comparing nonexhaust to other PM will be important for assessing the potential health risks of this PM.

TRAP and the cardiovascular system

Particulate matter and cardiovascular disease

The landmark "Harvard Six Cities Study" in 1993 found a relationship between exposure to $PM_{2.5}$ and hospital admissions and deaths from cardiovascular disease (Dockery et al., 1993). Subsequent epidemiological studies have shown clear associations between air pollution and many cardiovascular diseases. These include coronary artery disease (Cesaroni et al., 2014; Chen et al., 2013; Hoffmann et al., 2007; Miller et al., 2007), cardiac arrhythmia and arrest (Pope III et al., 2006; Raza et al., 2014; Watkins et al., 2013), heart failure (Atkinson et al., 2013; Bai et al., 2019; Shah et al., 2013), cerebrovascular disease (Lee, Miller, & Shah, 2018; Low et al., 2006; Miller et al., 2007; Miller & Shah, 2016; Shah et al., 2015; Stafoggia et al., 2014), peripheral arterial disease (Hoffmann et al., 2009; Peng et al., 2008; Ward-Caviness et al., 2018), and venous thromboembolism (Baccarelli et al., 2009) (see the seminal American Heart Association statement (Brook et al., 2010) on $PM_{2.5}$ for further details). Many of these associations hold true for exposure metrics associated with traffic, such as the distance of residential address from a major road or black carbon in PM as an indicator of combustion-derived PM (Adar & Kaufman, 2007). Investigations using source apportionment have found that vehicle exhaust, in particular, may drive several cardiovascular actions of $PM_{2.5}$ (Rich et al., 2019). Additionally, acute exposure is linked to cardiovascular events, with individuals presenting with myocardial infarction having been more likely to have been in traffic in the preceding hours (Bhaskaran et al., 2011; Peters et al., 2013; Peters, Dockery, Muller, & Mittleman, 2001).

Atherothombotic disease underpins, or at least contributes to, the vast majority of cardiovascular disease. In this disease, regions of an artery become inflamed leading the formation of fatty plaques on the inner surface of the vessel. These plaques grow in size over years ultimately disturbing the smooth flow of blood through the vessel lumen and, if severe, reducing perfusion of downstream organs. Plaque growth is often asymptomatic until the occurrence of a "cardiovascular event" caused by erosion or rupture of plaques leading to thrombosis (a blood clot) at the site of an exposed plaque or in small arteries blocked by dislodged blood clots (an embolism). The occurrence of an such an event in the coronary arteries of the heart can cause ischemia of the downstream cardiac tissue leading to angina or potentially a heart attack. Occlusion of the arteries within, or leading to, the brain may cause a stroke, or in the peripheral arteries can lead to ischemic damage to the limbs.

Exposure to urban air pollution is associated with atherosclerosis in a range of arterial beds (Hoffmann et al., 2007; Kunzli et al., 2005; Provost, Madhloum, Int Panis, De Boever, & Nawrot, 2015). Atherothrombosis is a progressive disease culminating from many sequential and parallel pathological processes. It is of note that TRAP can exacerbate many of these different steps (see Fig. 10.3), and these will be discussed in the following paragraphs.

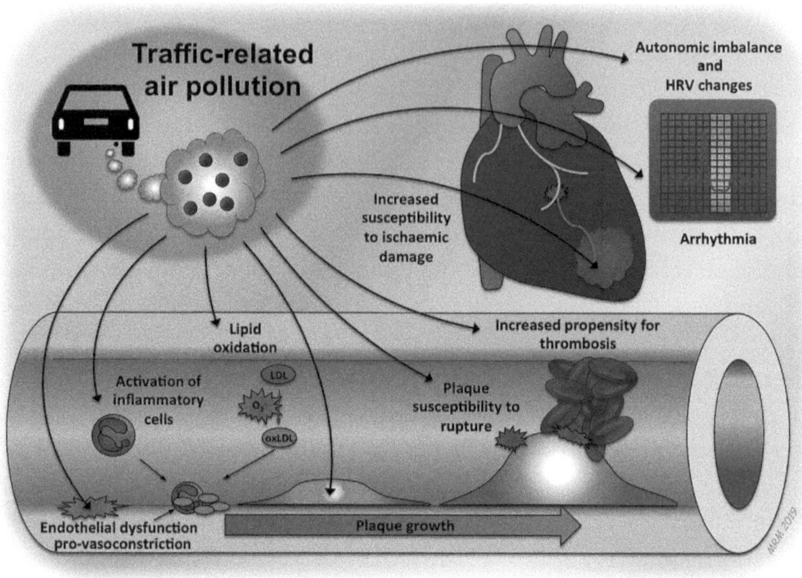

Fig. 10.3 Mechanisms by which TRAP can promote atherothrombotic disease. Abbreviations: $O_2\cdot^-$ = superoxide free radical (shown to represent oxidative stress in general); LDL = low-density lipoprotein; oxLDL = oxidized low-density lipoprotein.

Endothelial dysfunction and vascular contractility

The lumen of blood vessels is lined with a thin layer of cells called the vascular endothelium. These cells have a barrier function between the blood and walls of the artery but also play a far more complex role in regulating cardiovascular homeostasis. Endothelial cells respond to stimuli (such as blood flow and blood-borne mediators) by generating a host of cellular intermediaries that have a paracrine role in regulating other aspects of cardiovascular function. Of note, endothelial cells generate the mediator nitric oxide (NO) which diffuses out of the cell to locally regulate multiple targets, such as the vascular contractility of the underlying smooth muscle cells (that govern the diameter of vessels to control blood flow, pressure, and distribution), inhibition of vascular smooth muscle cell proliferation and migration, inhibition of platelet activation (platelets are anucleated cells in the blood that play a key role in blood clotting) and regulation of the activity of circulating white blood cells (that play a role in inflammation). Loss of endothelial cell function leads to an inclination toward constriction over dilatation, stiffening of blood vessels, and greater susceptibility to other processes leading to vascular disease.

Early studies demonstrated that controlled exposure to ambient $PM_{2.5}$ with O_3 in human subjects was associated with constriction of the brachial arteries, indicating that air pollutants could affect the peripheral circulation (Brook et al., 2002). Subsequently, controlled exposure studies found that acute exposure to DE has prominent effects on the vasculature. Using forearm plethysmography, a 1-h exposure of DE impairs the ability of brachial blood vessels to relax in response to infusions of vasodilator agents (Mills et al., 2005). The pattern of inhibition between vasodilators suggests an impairment of endothelial function and the NO pathway (Langrish et al., 2013; Mills et al., 2005), characteristic of the inhibition associated with oxidative stress. DE also impairs endothelial responses in the skin microvasculature (Wauters et al., 2013). DE attenuated vascular responses rapidly (within 2 h) and, concerningly, this impairment persisted for at least 24 h after the exposure (Tornqvist et al., 2007). DE exhaust can also decrease brachial artery diameter in patients with metabolic syndrome, possibly through increases in circulating endothelin-1 (ET-1; a potent vasoconstrictor molecule) (Peretz et al., 2008). Others have found that, while DE increase vascular sensitivity to the endothelin pathway, these changes were more likely to be caused by decreased NO bioavailability rather than changes in ET-1 levels per se (Langrish et al., 2009).

Removing PM from DE prevents the detrimental effect of the exhaust on endothelial dysfunction in both healthy volunteers (Lucking et al., 2011; Mills et al., 2011) and patients with heart failure (Vieira et al., 2016). Particle composition is important to cardiovascular effects of TRAP. Pure carbon particles (i.e., without any surface chemicals associated with DEP) were not associated with cardiovascular effects (Mills, Miller, et al., 2011). Finally, both exhaust from idling engines and engines running in a "city cycle" can induce detrimental cardiovascular effects (Barath et al., 2010).

Experiments with isolated tissues have shown that DEP can directly induce endothelial dysfunction in the absence of inflammatory cells (Ikeda, Suzuki, et al., 1995; Miller et al., 2009; Moller et al., 2011). Pulmonary exposure to DEP in rodent models in vivo can induce vascular dysfunction with a similar profile of impairment to that seen with controlled exposure to DE in man, albeit depending on the model used and vascular bed studied (Moller et al., 2011). As well as effects on the pulmonary and peripheral vasculature, animal models have shown that PM has detrimental effects on the coronary circulation (Bartoli et al., 2009; Campen et al., 2005; Cherng, Campen, Knuckles, Gonzalez Bosc, & Kanagy, 2009; Lemos et al., 2006).

Arterial stiffness is a hallmark of aging that has detrimental effects on cardiovascular health to promote high blood pressure and the early stages of atherosclerosis (Safar, 2018). Short and long-term exposure to $PM_{2.5}$ is associated with increased arterial stiffness (Mehta et al., 2014; Scheers, Nawrot, Nemery, & Casas, 2018; Zhang et al., 2018) although not all studies have found significant associations (Ljungman et al., 2018; Zanoli et al., 2017). Carotid arterial stiffness was found to be greater in children living next to the main road than those living >750 m away (Iannuzzi et al., 2010). A 2-h walk along streets with heavy traffic in London was associated with increases in indices of arterial stiffness regardless of the health status of the individual (healthy, COPD, or ischemic heart disease) (Sinharay et al., 2018). Effects were associated with black carbon exposure and were not observed during a similar walk in an urban park. A similar study was used for streets in Beijing, also seeing an increase in some (but not all) arterial stiffness parameters (Guan et al., 2018). These effects were not observed if the participants wore a facemask during the walk. Finally, controlled exposure to DE increases arterial stiffness in healthy volunteers and patients with heart failure (Lundback et al., 2009; Vieira, Guimaraes, de Andre, Cruz, et al., 2016).

The stiffening of arteries and the tendency for proconstriction/antidilation will be a contributing factor to increased blood pressure. Hypertension (raised blood pressure) is the leading risk factor not only for loss of life globally (predominantly from mortality from cardiovascular causes), but also as a risk factor for many other noncardiovascular diseases (Cohen et al., 2017). There is a large body of literature demonstrating that particulate air pollution, in general, is accompanied by an increase in blood pressure (Brook & Rajagopalan, 2009; Fuks et al., 2017; Giorgini et al., 2016; Liang et al., 2014; Munzel et al., 2017a; Yang et al., 2018). While the increase appears relatively small (there are large variations, but typically epidemiological studies shown an increase of <5 mmHg per interquartile range of PM), a prolonged increase of this magnitude across a population would be associated with considerable levels of cardiovascular mortality. Exposure to traffic (e.g., residing in close to proximity to a major road, source apportionment for traffic-related constituents of PM, ultrafine PM + NO_2, black carbon, or modeling for traffic source) has been linked to hypertension (Bai et al., 2018; Brook et al., 2016; Fuks et al., 2016; Giorgini et al., 2016; Jhun et al., 2019; Zhong et al., 2016). Controlled exposure to DE is associated with increased systolic blood pressure (Cosselman et al., 2012). Animal studies have shown that elevated blood pressure is evident after concentrated ambient particles (CAPs) or DEP through diverse mechanisms, including endothelial dysfunction, oxidative stress, mitochondrial dysfunction, central mechanisms, DNA methylation, heightened sensitivity of vascular smooth muscle cells to calcium, and circulating mediators such as endothelin and those of the renin-angiotensin system (Brook et al., 2010; Giorgini et al., 2016; Munzel et al., 2017b; Rao et al., 2019; Sanidas et al., 2017; Zhong et al., 2016).

Recruitment of inflammatory cells and oxidized lipids

Loss of endothelial function will contribute to the initial stages of vascular inflammation associated with atherosclerosis. Together with the loss of protective endothelial-derived mediators such as NO, insults to the vascular endothelium can induce a pro-inflammatory phenotype in endothelial cells. This includes the expression of cell surface molecules such as intercellular adhesion molecule-1 (ICAM-1)/vascular cell adhesion molecule-1 (VCAM-1) that attract and tether circulating monocytes to the vessel wall and into the endothelial space, with conversion to macrophages. Furthermore, oxidative stress promotes the oxidation of low-density lipoprotein (LDL to oxLDL) in the blood. ox-LDL is preferentially taken up by macrophages within the subendothelial space forming the beginnings of the fatty streak; the first stages of atherosclerosis.

Inhalation of PM promotes early events in atherogenesis, for example, oxidation of LDL and the adherence of leukocytes to the vascular wall (Yatera et al., 2008). Cell culture studies demonstrate that while DEP has only modest direct effects on endothelial cells (unless at very high concentrations), prior interaction with macrophages leads to a release of inflammatory mediators that can then induce marked inflammatory changes in endothelial cells (Shaw et al., 2011). A recent gene expression study demonstrated that extracts of DEP had extensive effects on inflammatory pathways in macrophages, both in vitro and in vivo (Bhetraratana et al., 2019). Cultured macrophages demonstrate that urban PM and DEP promote inflammation via the AhR and cytochrome P450 pathways although the pattern of effects was complex when addressing which constituents mediated these effects (Vogel et al., 2005). Toll-like receptors may play a role in mediating the inflammatory effects of PM on early stages of atherosclerosis (Campen, Lund, & Rosenfeld, 2012). Similarly, the oxLDL receptor mediates a number of cardiovascular effects of vehicle exhaust emissions in atherosclerotic mice, including infiltration of monocytes and macrophages in the vessel wall (Lund et al., 2011).

The oxidative stress induced by DEP engenders this PM with an ability to promote many stages of atherosclerosis (Miller et al., 2012). DEP is capable of directly oxidizing LDL in vitro (Ikeda, Shitashige, Yamasaki,

Sagai, & Tomita, 1995), as well as promoting the lipid content of macrophages and their conversion to foam cells (Cao et al., 2016). Furthermore, extracts of DEP have been shown to work synergistically with lipids to regulate the expression of genes involved in vascular inflammation (Gong et al., 2007). Interestingly, studies in healthy adults (Li et al., 2019) have also expanded on data from animal studies (Araujo et al., 2008;Yin et al., 2013) showing that $PM_{2.5}$ exposure is linked to impaired high-density lipoprotein function, leading to greater levels of circulating oxLDL.

Plaque development

Mouse models of atherosclerosis have been valuable for addressing the vascular effects of chronic exposure to PM (Moller et al., 2011). Atherosclerosis can take many decades to develop in humans, whereas knockout mouse models are available that, due to removal of specific genes, exhibit high blood cholesterol (especially on high-fat diets) leading to the formation of atherosclerotic plaques within a few weeks. Particles such as DEP increase the burden of atherosclerosis by a range of mechanisms including oxidative stress (Araujo et al., 2008; Bai et al., 2011; Gong et al., 2007; Kampfrath et al., 2011; Lund et al., 2007; Miller et al., 2013; Sun et al., 2005; Wan, Rajagopalan, Sun, & Zhang, 2010;Yin et al., 2013), changes in arachidonic acid metabolites (Lund et al., 2011; Soares et al., 2009;Yin et al., 2013), endothelin-1 pathways (Lund et al., 2009), dysfunctional high-density lipoprotein (Araujo et al., 2008; Li et al., 2013), endothelial nitric oxide synthase (eNOS) uncoupling (Cherng et al., 2011), and signaling through lectin-like oxidized LDL receptors (LOX-1) (Kodavanti et al., 2011; Lund et al., 2011).

Most studies focus on the size of plaques to quantify the effect of PM on atherosclerosis. While plaque growth has important effects of hemodynamic flow of blood, it is the erosion or rupture of plaques that induces a cardiovascular event by causing thrombotic occlusion of the vessel. Current mouse models of atherosclerosis do not exhibit the signs of cardiovascular events; however, plaque characteristics can be explored to assess features of the plaque that are associated with plaque vulnerability in humans. Inhalation of urban PM, vehicle exhaust or pulmonary instillation of DEP have been shown to increase markers of plaque vulnerability such as lipid content, inflammatory cell content, matrix metalloproteinase, oxidative stress (Bai et al., 2011; Campen et al., 2010; Floyd, Chen, Vallanat, & Dreher, 2009; Quan, Sun, Lippmann, & Chen, 2010; Sun et al., 2005, 2008; Suwa et al., 2002;Yatera et al., 2008;Ying et al., 2009), and buried fibrous layers that may be indicative of a "healed" plaque rupture (Cassee et al., 2012; Miller et al., 2013). Such observations would support those of epidemiological studies linking exposure to traffic with hospital admissions for acute myocardial infarction.

Thrombosis

Increased blood coagulability will increase the risk of atherothrombotic events, as well as the likelihood of embolism (the lodging of small thrombosis in a small distant artery). PM exposure is linked to pro-thrombotic pathways (Emmerechts & Hoylaerts, 2012; Franchini & Mannucci, 2011; Robertson & Miller, 2018; Strak et al., 2013). There is a tendency for PM to increase blood fibrinogen, thrombin, von Willebrand factor (vWF) and platelet activity, and decrease ex vivo coagulation times (Robertson & Miller, 2018). Fibrinolysis (the breakdown of clots) is also blunted by PM, with decreases in tissue plasminogen activator (t-PA) release and upregulation of plasminogen activator inhibitor-1 pathways (Langrish et al., 2012; Robertson & Miller, 2018). In relation to TRAP, the distance of residence from a major road was also associated with blood factor VIII and thrombin levels (Emmerechts et al., 2012). Commuting in traffic is associated with increases is coagulation pathways (Zuurbier et al., 2011) and black carbon exposure is associated with increased blood levels of ICAM-1/VCAM-1 in elderly individuals (Bind et al., 2012).

Using a Badimon system (an ex vivo model of thrombosis using human blood flowing over a damaged blood vessel), our group demonstrated that acute exposure to DE promoted blood clotting, the mechanisms of which included activation of platelets (Lucking et al., 2008) and reduced release of the fibrinolytic factor t-PA from the vascular endothelium (Mills et al., 2005). In contrast, another controlled exposure study with DE found little effect of thrombotic markers (Carlsten et al., 2008). While controlled exposure to DE does not appear to be associated with a consistent inflammatory response (Ghio, Sobus, Pleil, & Madden, 2012), controlled exposures to CAPs increase blood plasminogen and markers of acute phase response in individuals with genetic deficiencies in various antioxidant systems (Devlin et al., 2014).

In vivo models have been used to look at the actions of TRAP on coagulation pathways. Exposure of mice to air pollution in a roadside tunnel was linked to a series of prothrombotic markers, for example, endothelial vWF expression, selectin-molecules, and platelet activation (Emmerechts et al., 2012). Tissue factor is also involved in the action of ultrafine particles from a roadside tunnel (Kilinc et al., 2011). DEP, with or without co-exposure to O_3, increased expression of tissue factor (Kodavanti et al., 2011). Mechanical or chemical injury of blood vessel walls can be used to induce a thrombosis at the site of injury, with the time to the formation of the blood clot indicating the thrombogenicity of the blood. Pulmonary exposure of DEP to rats potentiated the thrombotic occlusion of the carotid artery following arterial injury in rats (Tabor et al., 2016). The response was more notable in response to DEP exposure compared to pure carbon nanoparticles, emphasizing the importance of surface chemicals on carbon-based PM. Platelet activation and impaired fibrinolysis were important mechanisms (Tabor et al., 2016), complementing the findings of the clinical exposures to DE (Lucking et al., 2008; Mills et al., 2005). Procoagulant effects of DEP were greater in diabetic mice (Nemmar, Subramaniyan, Yasin, & Ali, 2013). Animal studies also support a role for multiple mechanistic pathways in the prothrombotic effects of PM, including inflammation and oxidative stress (Mutlu et al., 2007; Nemmar, Subramaniyan, & Ali, 2012), tissue factor (Kilinc et al., 2011), fibrinogen binding (Wilson et al., 2010), impaired fibrinolysis (Budinger et al., 2011; Kodavanti et al., 2011), and platelets (Tabor et al., 2016) (reviewed in Robertson & Miller, 2018).

Cardiac effects

A plethora of studies have made use of the noninvasive techniques to measure heart rate variability (HRV) (Buteau & Goldberg, 2016; COMEAP, 2018). HRV is a set of parameters looking at the regularity of the heartbeat using an electrocardiogram, with reductions in HRV parameters indicating a less desirable cardiac rhythm, for example, from altered regulation of the heart by the autonomic nervous system. Overall, there is a trend toward a reduction in HRV parameters with exposure to air pollutants. Although there is a large degree of inconsistency between parameters and studies (Buteau & Goldberg, 2016; Mills et al., 2011), the HRV effects associated with PM would be predictive of a worse prognosis at a population level. The relative ease of measuring HRV has been exploited in smaller panel studies where it is possible to measure personal exposure to pollutants (measured by portable devices, instead of estimates of pollution based on the nearest stationary monitor). These studies have shown that submicron particles (20–1000 nm, so including within the ultrafine range) were accompanied by detrimental changes in HRV (Chan, Chuang, Shiao, & Lin, 2004; Folino et al., 2009). Similar findings were not found in volunteers taking beta-blockers, indicating a role for the autonomic nervous system

(Folino et al., 2009). Black carbon is linked to changes in HRV at sites of high, but not low, traffic (Cole-Hunter et al., 2016). In a natural intervention experiment, Wu and colleagues made use of the Beijing Olympic games when strict air pollution measures were enforced across the city center, including interventions to limit traffic congestion. In these studies, taxi drivers exhibited HRV changes were associated with PM exposure, and varied across the periods studied (before, during, and after the games) (Wu et al., 2010; Wu et al., 2011). Lastly, the use of a facemask had beneficial effects on HRV in healthy volunteers and patients with ischemic heart disease during a 2-h walk alongside roadways in Beijing (Langrish et al., 2009, 2012).

Surprisingly, controlled exposure studies to DE in human subjects have had less consistent effects on HRV (Mills, Finlayson, et al., 2011). The authors suggest that other constituents in urban air pollution are linked to effects on HRV observed in epidemiological studies. Nonetheless, animal studies have found that DEP can induce prominent HRV effects, suggesting that other factors may account for this variation for example, dose-rate, gaseous vehicle emissions, or composition of DEP. Farraj and colleagues have performed detailed preclinical characterization of the cardiac effects of combustion-derived particles (see Perez, Hazari, & Farraj, 2015 for a review).

In addition to HRV, PM also has the capacity to induce arrhythmia (distinct irregularities in heart rhythm electrocardiogram (ECG) beyond that of HRV) (Feng et al., 2019). Isoproterenol-induced models of cardiomyopathy show that inhalation of various types of PM promote arrhythmias, alterations in HRV, and delays in cardiac conductance (Carll et al., 2010, 2013, 2015). The renin-angiotensin system and oxidative stress have also been implicated in the cardiac effects of PM (Ghelfi, Wellenius, Lawrence, Millet, & Gonzalez-Flecha, 2010). Pulmonary exposure to DEP also induced arrhythmia in a rat model of myocardial infarction after coronary ligation (Robertson et al., 2014). The nature of arrhythmia depends on the source of pollution although anthropogenic sources were primarily responsible. The organic extracts of DEP contributed to the arrhythmogenic effects of DEP (Yokota, Ohara, & Kobayashi, 2008).

ECG changes have also been used to demonstrate the effect of PM on cardiac ischemia (Brook et al., 2010; Pekkanen et al., 2002). Controlled exposure to DE induced a depression in the S-T segment of the ECG in patients with ischemic heart disease on exercise, indicating that DE worsened the cardiac ischemic stress across key regions of the heart (Mills et al., 2007). Ultrafine CAPs had similar effects in patients with metabolic syndrome who were null for glutathione-S-transferase-mu-1 (GSTM1) allele (a prominent antioxidant gene), but not a comparative group from the general population (Devlin et al., 2014). Pulmonary exposure to DEP also induced arrhythmia and a greater degree of myocardial infarction in a rat model of myocardial infarction after coronary ligation (Robertson et al., 2014). Pharmacological inhibition of pulmonary sensory receptors or neural pathways diminished these effects, demonstrating a role for the neural systems in the cardiac effects of this air pollutant (Hazari et al., 2011; Robertson et al., 2014).

PM also influences the progression of cardiac conditions that occur in response to insults to the heart. Prolonged exposure (5 years) is associated with dilatation to the ventricles of the heart, an indicator and predictor of the development of heart failure (Aung et al., 2018). An impressive paper by Wold and colleagues demonstrates that 9 months exposure to ambient PM promotes myocardial hypertrophy and loss of cardiac function in mice (Wold et al., 2012). Parameters affected by PM included a switch of cardiomyocytes to a fibrotic phenotype, decreased fractional shortening (a measure of ventricular contraction), mitral valve inflow, and changes cardiomyocyte calcium handling. Oxidative stress was implicated in these effects. Direct exposure of cardiac myocytes to DEP can alter myocardial contractility and calcium handling (Gorr et al., 2015), a finding that may have relevance if particles can enter the blood (see below) and reach the heart (Calderon-Garciduenas et al., 2019).

Metaanalyses have demonstrated that PM is associated with heart failure globally (Shah et al., 2013). Residing near a major road also alters left ventricle contractility in systole although other parameters related to remodeling were not affected (Weaver et al., 2017). Additionally, source apportionment of PM linked to diesel emissions is associated with congestive heart failure (Rich et al., 2019). Lastly, controlled exposure to DE had a detrimental effect on cardiopulmonary parameters and exercise capacity in heart failure patients (Vieira, Guimaraes, de Andre, Saldiva, & Bocchi, 2016). These effects could be prevented by filtering the PM from the exhaust. That TRAP can promote both acute myocardial infarction and heart failure in experimental studies fits with a recent epidemiological study associating these conditions with ultrafine particles and NO_2 (Bai et al., 2019).

Risk factors for cardiovascular disease: Metabolic disease

We have outlined above (see section on vascular function) that TRAP exposure is linked to increased blood pressure; a major risk factor for cardiovascular disease. In addition, there is a growing body of research showing that exposure to air pollution has effects on metabolism that are linked to diabetes, another major risk factor for cardiovascular disease (Eze et al., 2015; Lim & Thurston, 2019; Liu et al., 2019). Blood glucose levels are another key risk factor for cardiovascular mortality (Cohen et al., 2017) and cardiovascular complications such as microvascular dysfunction are a hallmark of diabetes. Exposure to traffic has been shown to be associated with higher fasting blood glucose levels, insulin insensitivity, lipid abnormalities, islet beta-cell function, and early development of diabetes (Alderete et al., 2018; Beyerlein et al., 2015; Li et al., 2018). Black carbon is associated with both raised blood pressure and insulin resistance (Brook et al., 2016). Oxidative stress and inflammation are likely to be prominent mechanisms in the metabolic effects of air pollution, just as they are for cardiovascular disease (Lim & Thurston, 2019). Furthermore, associations between air pollution and diabetes have been found for exposure at many different stages of life, from young to elderly, and before and during pregnancy (Lim & Thurston, 2019). Animal studies strengthen the evidence between air pollution and diabetes (reviewed in Rao, Patel, Puett, & Rajagopalan, 2015). Mechanistic pathways addressed those in humans described above, but also include inflammation of adipose tissue (Mendez et al., 2013), changes to thermogenesis (Liu et al., 2014), alterations in leptin pathways (Xu et al., 2011), central effects via the hypothalamus, the nuclear factor kappa B pathway, and peripheral inflammation (Liu et al., 2014). At present there are limited studies investigating the metabolic effects of TRAP specifically although DEP has been shown to induce pathological effects in the pancreas via oxidative stress (Nemmar et al., 2014) and have greater coagulant effects in diabetic mice than healthy mice (Nemmar et al., 2013).

Gaseous pollutants

In the sections above, we have mostly concentrated on PM in TRAP. While beyond the scope of this chapter to discuss in any detail, it is important to state that gaseous pollutants can also affect the cardiovascular system. Epidemiological studies have shown that NO_2 (a common constituent of vehicle exhaust, especially from diesel vehicles) has been shown to be associated with blood pressure (Yang et al., 2018), carotid atherosclerosis (Rivera et al., 2013), ischemic heart disease (Eum et al., 2019), myocardial infarction (Bai et al., 2019; Mustafic et al., 2012), arterial stiffness (Zhang et al., 2018), thrombotic markers (Chen, Chan, & Su, 2017), metabolic syndrome/diabetes (Bai et al., 2018; Beyerlein et al., 2015; Toledo-Corral et al., 2018), right ventricular hypertrophy (Leary et al., 2014), the development of heart failure (Aung et al., 2018; Bai et al., 2019; Shah et al., 2013; Sorensen et al., 2017), cerebrovascular disease (Eum et al., 2019; Shah et al., 2015), and cardiovascular mortality in general (Faustini, Rapp, & Forastiere, 2014; Mills, Atkinson, Kang, Walton, & Anderson, 2015). We should note the studies listed here are a small selection from those that explore NO_2, and overall there is a substantial mix of positive and negative studies looking at associations with cardiovascular disease.

Controlled DE exposures in human subjects have also addressed the question of gases vs particles. A retrofit "particle trap" on the engine exhaust efficiently reduces particle mass in the DE and completely prevented the thrombotic actions of DE (Lucking et al., 2011). Filtering of particles from DE also prevented the vascular impairment observed with whole exhaust (Mills, Miller, et al., 2011). This observation was supported by a study (Langrish et al., 2010) whereby volunteers were exposed to pure NO_2 at concentrations representative of whole exhaust; no acute cardiovascular effects were observed. The differences between the results of epidemiological and controlled exposure studies may be due to the length of exposure (i.e., acute vs chronic) or susceptibility of the population studied. A review of the preclinical data on the role of NO_2 in the cardiovascular actions of air pollution is beyond the scope of this review. However, we would like to highlight the studies performed by Campen and colleagues showing that the atherosclerotic effects of vehicle exhaust remain after filtering of particulates from the exhaust (e.g., Campen et al., 2010; Lund et al., 2007; Seilkop, Campen, Lund, McDonald, & Mauderly, 2012). The findings suggest that gaseous constituents can drive the pro-atherosclerotic effects of vehicle exhaust, as can the particulate components in the absence of gases. Others have suggested that there may be chemical interactions between gases and particles within emissions to promote their toxicity beyond that of each in isolation. However, current attempts to dissect these interactions yielded complex findings (while several different constituents could induce biological effects, co-exposure of gases and particles had varying effects depending on the type of PM used and the biological parameter being studied) (Quan et al., 2010), thus further work is needed to address this hypothesis.

Nonexhaust traffic-derived PM

At the time of writing, there are no studies that have investigated the cardiovascular effects of traffic-derived nonexhaust PM. Isolated studies have looked for changes in blood markers following pulmonary administration of NE-PM to rodents, without showing striking changes in blood biochemistry, inflammation, or thrombotic markers (Gerlofs-Nijland et al., 2019; Kreider et al., 2012). A single study looked at instillation of tire PM and found no effect on cardiac enzymes (Gottipolu et al., 2008). Given the paucity of research and increasing ratio of NE-PM:VE-PM from traffic, it will be important to establish if relevant doses of NE-PM can induce systemic inflammation and whether this has subsequent effects on the cardiovascular system and other organs.

The lung to the peripheral organs

Both the initial pulmonary response to inhaled PM and the multifaceted nature of cardiovascular impairment have now been well characterized. However, the biological processes that link these two organ systems, or that between the lung and other systemic organs (see below), still remains to be fully established (Miller, 2014). Four broad mechanisms have arisen to account for the linking mechanism (Fig. 10.4) (Miller & Newby, 2019).

Firstly, inhalation of highly active or persistent pollutants may lead to a release of inflammatory mediators from the lung that pass into the circulation whereby they can directly influence cardiovascular function (Seaton, MacNee, Donaldson, & Godden, 1995). There is a wealth of literature showing that pollutant exposure leads to increases in blood biomarkers of systemic inflammation and oxidative stress (Brook et al., 2010; Forbes et al., 2009; Goodman et al., 2015; Robertson et al., 2012), however, there is a considerable inconsistency between different biomarkers and studies. Furthermore, the time-course inflammatory response often does not match other systemic effects. Nonetheless, there is a clear role for both inflammation and oxidative stress in multiple stages of the mode of action of inhaled PM (Miller, 2014; Miller et al., 2012) and these pathways represent a key means to amplify the signal from PM (Shaw et al., 2011) even if they are not the critical underlying cause.

Secondly, while conventional inflammatory mediators may not fully fit the bill, there is convincing evidence that other unrecognized blood-borne mediators could play a key role in linking these organ systems. Certainly, treatment of cells or animals with diluted serum from exposed animals/volunteers induces cardiovascular effects that cannot be explained by conventional cytokines (Channell, Paffett, Devlin, Madden, & Campen, 2012; Schisler et al., 2015). The identities of these unidentified blood-borne mediators are currently under exploration, with early evidence suggesting that oxidized lipids (Brower et al., 2016; Kampfrath et al., 2011) or

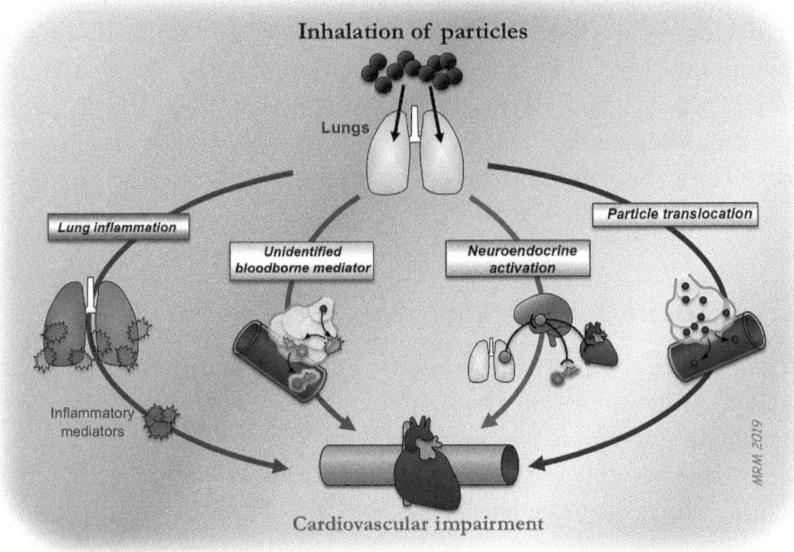

Fig. 10.4 Pathways through which inhaled particles can affect the cardiovascular system and other peripheral organs. Four main pathways have been proposed; (1) induction of pulmonary inflammation, the mediators of which pass into the circulation; (2) the passage of unidentified mediators, that are not traditional cytokines, from the lung into the circulation; (3) activation of neural afferents which can alter the activity of the autonomic nervous system or release of endocrines; (4) direct passage of particles (or chemicals eluting from particles) into the circulation to directly impair cardiovascular function. *(Modified from Miller, M. R., & Newby, D. E. (2019). Air pollution and cardiovascular disease: Car sick. Cardiovascular Research, 116, 279–294.)*

peptide fragment could be promising candidates (Mostovenko et al., 2019) (although the latter has only been currently shown for inhaled carbon nanotubes rather than environmental pollutants).

The third pathway is that inhaled pollutants have systemic effects through regulation of the autonomic nervous system. Inhaled pollutants activate alveolar receptors, such as vanilloid receptors, triggering sensory afferents that feed into the autonomic nervous system. Peripheral organs are then affected through parasympathetic:sympathetic imbalance, changes to baroreceptor sensitivity or centrally controlled neuroendocrine regulation (Kodavanti, 2016; Perez et al., 2015; Pope III et al., 1999). Certainly, the cardiac effects of PM (e.g., HRV) reveal an increase in sympathetic dominance of the heart; effects of which can be diminished by beta-blockers (see above). Additionally, in animal models, pharmacological inhibition of sensory receptors can attenuate the arrhythmias induced by DE (Hazari et al., 2011) and the potentiating effects of DEP on ischemic damage to the heart (Robertson et al., 2014). Polyaromatic hydrocarbons on the surface of DEP have been shown to play a role in the activation of sensory afferents (Robinson et al., 2018). PM and gaseous pollutants can also induce systemic effects through the hypothalamus-pituitary-adrenal axis and/or through alterations in endocrine mediators such as catecholamines, corticotrophin-releasing hormone, cortisol/corticosterone, and cortisone (Kodavanti, 2016; Li et al., 2017; Liu, Fonken, et al., 2014; Niu et al., 2018).

The fourth pathway is "particle translocation." A proportion of nano-sized PM is small enough to cross the lung epithelial barrier and enter the circulation, whereby it can be carried to other organs of the body (Oberdorster et al., 2002). Studies in animals have provided convincing evidence for this possibility (Choi et al., 2010; Furuyama, Kanno, Kobayashi, & Hirano, 2009; Husain et al., 2015; Kreyling et al., 2002) (and see also Miller et al., 2017a). Recently this evidence has been bolstered by findings from human subjects inhaling gold nanoparticles (Miller et al., 2017b). In this study, gold could be detected in the blood and urine of volunteers within 24-h after inhalation, and was still present 3 months later, indicating that these particles persist in the body for a very long time. Parallel animal studies used a range of sizes of gold nanoparticles to demonstrate that the size cutoff for translocation was around 30 nm, so within the range of that of particles in the exhaust of modern diesel vehicles. Importantly, translocated gold in mice preferentially accumulated in atherosclerotic arteries compared to arteries without the disease and reached areas of carotid vascular disease in patients with a history of stroke (Miller et al., 2017b). Recent studies by the laboratories of Calderon-Garciduenas and Maher have used advanced electron microscopy techniques to identify iron-based particles in the brain (Calderon-Garciduenas et al., 2018; Maher et al., 2016) and heart (Calderon-Garciduenas et al., 2019) of cadavers from heavily polluted Mexico City. The chemical structure and smooth rounded surface suggest that these particles were derived from high-temperature processes, and friction-induced generation of particles from brake-wear could be a likely source. Interestingly, cellular damage was observed in tissues where particles were found. The ability of combustion-derived particles to generate free radicals and activate inflammatory cells suggests that should PM such as DEP translocate in a similar manner, then these particles would be likely to exacerbate disease.

The systemic effects of TRAP

The widespread detrimental effects of TRAP on cardiovascular function exemplify the capacity for air pollution to harm the body. Indeed, one of the growing trends in air pollution research is the discovery of further extrapulmonary effects of pollution in different organs (comprehensively reviewed in Schraufnagel et al., 2019a, 2019b) (Fig. 10.5). Besides a host of pulmonary and cardiovascular diseases, air pollution (in general, rather than TRAP specifically) has now been associated with diabetes (Munzel et al., 2017a; Rajagopalan & Brook, 2012; Sun et al., 2009; Xu et al., 2010), liver disease (Kim, Park, Lim, Lee, & Kim, 2014), chronic kidney disease (Al Suleimani et al., 2017; Xu, Nie, Ding, & Hou, 2018), inflammatory bowel disease (Abegunde, Muhammad, Bhatti, & Ali, 2016), osteoporosis (Prada et al., 2017), skin diseases (Kim, Cho, & Park, 2016), autoimmune diseases

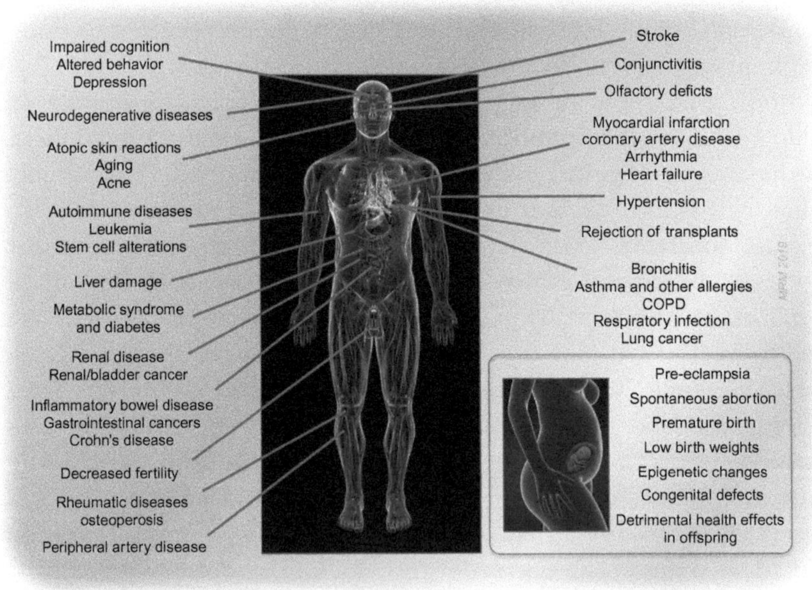

Fig. 10.5 Emerging evidence showing that air pollution has effects throughout the body. Over the last few decades, it has become apparent that air pollution can induce effects beyond the pulmonary and cardiovascular system. This schematic highlights a number of examples of extrapulmonary effects of air pollution in general, although in most cases there is strong evidence for the role of particulates specifically. Abbreviations: COPD = chronic obstructive pulmonary disease. *(Modified from Miller, M. R., & Newby, D. E. (2019). Air pollution and cardiovascular disease: Car sick. Cardiovascular Research, 116, 279–294. See Schraufnagel et al. (2019a, 2019b) for an excellent overview of these effects.)*

(Zhao et al., 2019), and a wide range of cancers (Andersen et al., 2017; Janitz et al., 2016; Nagel et al., 2018; Pope III et al., 2002). Air pollution exposure will likely affect repair and regeneration through inflammation and circulating stem cells (Cui, Sun, & Liu, 2016; DeJarnett et al., 2015) and has also been associated with the rejection of organ transplants (Ruttens et al., 2017).

The effects of air pollution on the central nervous system are a major topic for current research. Air pollution has been shown to be associated with cognitive decline through a range of different tests for reasoning, memory, and semantic and phonemic fluency, as well as psychomotor and intellectual development in children (Allen et al., 2017; Kilian & Kitazawa, 2018; Peters et al., 2019; Power, Adar, Yanosky, & Weuve, 2016; Schraufnagel et al., 2019b). The effects on cognition also extend to depression, antisocial behavior, aggression, and teenage psychosis (Haynes et al., 2011; Newbury et al., 2019). The possibility that air pollution is linked to chronic cerebrovascular conditions was raised many years ago, however, recent research has bolstered links with dementia and Parkinson's disease (Peters et al., 2019; Sram, Veleminsky Jr., Veleminsky Sr., & Stejskalova, 2017; The Lancet Neurology, 2018). Animal models have been used to show similar cognitive impairments (Allen et al., 2017), however, given the complexities of these behaviors and conditions, it is unsurprising that dissecting underlying mechanisms is challenging. Nonetheless, the buildup of amyloid-beta plaques, inflammation, and oxidative stress are broad mechanisms (Kilian & Kitazawa, 2018). Of note, DE inhalation increased amyloid plaque formation in the brains of mice, an effect that was accompanied by impairment in motor function (Hullmann et al., 2017). Nano-sized PM may be able to directly access the brain through the blood-brain barrier (BBB) or olfactory bulb, and then exert detrimental effects on the CNS through inflammation and oxidative stress. These actions of PM are likely to contribute to the permeability of the BBB themselves. Indeed inhalation of vehicle exhaust in mice induced changes in the cerebral vasculature that suggest increased BBB permeability (Suwannasual, Lucero, McDonald, & Lund, 2018). There is an increasing awareness of the role lipid imbalance plays in Alzheimer's disease (Poirier et al., 2014), thus metabolic and oxidative effects of air pollution could exacerbate cerebrovascular disorders. Additionally, other blood-borne mediators such as ET-1 (Palmer, Tayler, & Love, 2013) and angiotensin-II (Miners, van Helmond, Raiker, Love, & Kehoe, 2010) may play a contributing role to the disease process.

A large literature has also amassed showing that air pollution (including traffic-derived emissions) is associated with stroke (Babadjouni et al., 2017; Lee et al., 2018; Shah et al., 2015). Many of the cardiovascular effects of air pollution could promote the risk of stroke (Miller & Newby, 2019). Atherothrombosis of arteries in the brain (and those leading the brain e.g., the carotid) will play a role in the development of ischemic stroke, and loss of vessel distensibility (the capacity of vessels to relax or be stretched) of cerebral blood vessels together with increased systolic blood pressure may predispose toward hemorrhagic stroke. In line with this, inhalation of DE in mice increased matrix metalloproteinase activity in the cerebral vasculature (Suwannasual et al., 2018). Finally, atrial fibrillation increases the risk of embolization, thus the arrhythmogenic effects of air pollution could also play a contributing role risk of ischemic stroke (Miller & Shah, 2016).

As well as links to infertility (Carre, Gatimel, Moreau, Parinaud, & Leandri, 2017), early life exposure to air pollution (e.g., prenatal, gestation, and in childhood) has been linked to poor health. Maternal exposure is associated with preterm birth, low birth weight, stillbirth spontaneous abortion, congenital defects, and with a greater incidence of various health conditions later in life (Baldacci et al., 2018; Grippo et al., 2018; Klepac, Locatelli, Korosec, Kunzli, & Kukec, 2018; Li et al., 2017; Wang & Pinkerton, 2007). Childhood exposure to air pollution is associated with exacerbation of asthma (Jung, Chen, Tang, & Hwang, 2019), childhood leukemia (Filippini, Heck, Malagoli, Del Giovane, & Vinceti, 2015), obesity (Kim et al., 2018), attention disorders (Myhre et al., 2018), and autism (Flores-Pajot, Ofner, Do, Lavigne, & Villeneuve, 2016). Once again, it is likely that several underlying mechanisms will account for these effects. Prenatal and gestation exposure to air pollution has been shown to be linked to epigenetic mechanisms (e.g., DNA methylation, histone modification, and alterations in noncoding RNA) providing a means through which maternal exposure could have effects in subsequent generations (Saenen et al., 2019). In relation to TRAP, epigenetic changes have been associated with maternal exposure to black carbon (Neven et al., 2018) and distance of mother's address from a major road (Kingsley et al., 2016). Two mouse studies suggest that in utero exposure of mice to DE may lead to a greater susceptibility of offspring to heart failure (Goodson et al., 2017; Weldy, Liu, Liggitt, & Chin, 2014). A rabbit study has also shown that in utero exposure to DE leads to alterations in tissue lipid levels in subsequent generations (Rousseau-Ralliard et al., 2019). Besides cardiovascular dysfunction, exposure to DE is likely to have a detrimental effect on the maternal environment, through alterations in oxidative stress, inflammation, circulating mediators (e.g., lipid imbalance and raised stress hormones), preeclampsia, and impaired placenta-foetal blood flow (Saenen et al., 2019; Weldy et al., 2014; Xue, Zhu, Lin, & Talbott, 2018). Indeed, studies in rodents have shown that exposure to roadside PM during pregnancy changes the structural integrity of the umbilical cord (Veras et al., 2012) and gestational exposure to DE decreases placental blood flow and foetal capillary numbers (Valentino et al., 2016) as well as inducing hemorrhage of the placental vascular spaces (Weldy et al., 2014). Lastly, two very recent reports have found suspect PM in the placenta of nonsmoking mothers living in polluted areas (Liu, 2019) and on the foetal side of the placenta (Bove et al., 2019). These findings align with rodent studies showing that blood-borne nanoparticles can reach the foetus (Hougaard et al., 2015; Semmler-Behnke et al., 2014).

Conclusions, considerations, and implications

It is inarguable that air pollution is associated with considerable health effects. Traffic-derived emissions are a prominent source of fine and ultrafine PM and NO_2 in our environment, and these sources will contribute greatly to the observed health effects of air pollution. There has been a significant expansion of the evidence for the cardiorespiratory effects of TRAP, and recent years have seen a concerning list of other detrimental effects of air pollution in different organs of the body and for a range of diseases. The mechanisms underlying these effects are multifactorial with a complex degree of interaction between pathways. Indeed, a comprehensive review of potential mechanisms would need to encompass a whole tome rather than a single chapter of a book. But there are central mechanisms driving these processes; in particular oxidative stress and inflammation, as well as the increasing plausibility that different air pollution constituents can penetrate out of the lung. The pathways by which inhaled pollutants can have effects around the body (through pro-oxidative/pro-inflammatory mediators, by alterations in neuroendocrine systems and by translocation of particles into the circulation) are beginning to coalesce although the relative importance of each pathway in different health effects needs clarification. The particle translocation mechanism is of particular relevance to TRAP given the high proportion of nano-sized particles in these emissions. Nonetheless, while the reactivity of gaseous pollutants means they are unlikely to have direct effects on systemic organs, there is sufficient evidence to suggest that chronic exposure gases like NO_2 (a common pollutant in vehicle exhaust) will impart detrimental health effects.

Establishing the biological mechanisms by which TRAP has detrimental effects on the body centers around preclinical models in animals and cells. As with all scientific models, there are inherent limitations in using these approaches to study air pollution. Outside of human studies, in vivo models are currently the most physiological representative way to study exposure to TRAP. Inhalation of pollutants represents the gold standard method of administration in terms of replicating real-world scenarios, however, inhalation exposure facilities are complex and expensive to establish and run. Ideally, exposures would use ambient levels of air pollution over prolonged periods, but it is clearly not possible to do so over a human lifetime and often not the time span of disease development. Consequently, higher doses within shorter periods have to be used and there will always be debate as to whether the biological effects observed in these studies are relevant. Species differences will always remain a concern, both in terms of physiology and pathophysiology. Furthermore, the issue of dose relevance is also complicated by the complex nasal turbinates of rodents whereby there are concerns that a greater proportion of PM is deposited higher in the respiratory tract (rather than the alveoli) in comparison to human breathing. Instillation experiments can be used to bypass the nasal cavities, although these models are more susceptible to criticism over high dose and dose rate, agglomeration of particles in solution, and possibly poorer dispersion of particles throughout the lung. As well as ethical reasons, there is a need to develop in vitro models to allow screening of TRAP from different sources, especially when trying to determine which constituents drive biological effects. Currently there is no single in vitro assay (or combination of assays) that can accurately predict in vivo actions in the lung, and even less so for other organ systems. Nonetheless, in vitro models can be useful in addressing specific mechanistic questions and for broadly assessing relative toxicity from different sources of PM, for example, exhaust from different engines or different running characteristics. Once again, the dose relevance of assays is a concern. Extrapolative estimates can be used for cell cultures of pulmonary cells, but the relevance of dose for nonpulmonary cell types is difficult to assess. Lastly, controlled exposure studies in human volunteers provide an excellent means to explore the health effects of air pollutants without many of the confounding variables of epidemiological studies. They are not without their limitations, for example, restricted techniques available with the sensitivity to detect biological changes; usually limited to noninvasive methods; restricted to relatively acute exposures. Nonetheless, they have been used to great effect in determining the mechanisms of vehicle exhaust, particularly on the cardiovascular system (Mills et al., 2009).

Tackling the issue of air pollution inevitably centers on removing the sources of air pollution. If the solution is "simply" getting rid of air pollution, do we really need to know the biological mechanisms underlying the health effects of air pollution? We would argue that we do. Firstly, understanding the biological mechanisms underlying the health effects of pollution is a key step in making the case for causality, building on associations arising from epidemiological studies. Additionally, identification of specific biological mechanisms can be used to assess who may be especially susceptible to the effects of air pollution (e.g., age and patient groups) and establish better models to understand which pollutants represent the greatest harm. There is also the option to test whether pharmacological interventions or dietary supplements could prevent the effect of air pollutants (Barthelemy, Sanchez, Miller, & Khreis, 2020). While we would not advocate the use of medicines over strategies to reduce air pollutants, there may be a place for interventions that could ameliorate the effects of high air pollution episodes, or for highly susceptible individuals, during the (inevitable prolonged) time it takes to implement policies and strategies to reduce pollutants in our societies.

Advances in our understanding of the biological effects of air pollution need to be applied to real-world questions. Recent controversies in the reporting of vehicle emissions under true driving conditions have further raised awareness of the pollution caused by vehicles. Ascertaining the health effects of vehicle emissions generated from real driving conditions should hopefully follow. This is also true in relation to testing of emissions of modern engines. A substantial proportion of the evidence for the health effects of vehicle exhaust derives from older engines that will not only emit considerably higher levels of pollutants, but they will also exhibit differences in the composition of pollutants in the exhaust. The potential toxicity of PM from modern engines needs further investigation to establish if differences in size and physicochemical characteristics change their effects on health parameters. Similarly, with the rapid electrification of vehicles, there is a need to better ascertain the toxicity of nonexhaust vehicle emissions, which will represent a greater proportion of traffic-derived PM as exhaust emissions decline. Interactions between air pollution and associated environmental stressors such as traffic noise also require further attention (Munzel et al., 2017b).

There is justifiable attention on reducing emissions from vehicles through a range of policy measures. The consequences and effectiveness of such strategies also need further consideration. For example, the success of low-emission zones (LEZs) in cities will inevitably be judged based on the levels of pollution after implementation. However, there are many sources of PM in urban environments and small areas of restricted traffic control may be followed by only a modest reduction in PM in terms of mass. This may overlook the fact LEZs could reduce numbers of ultrafine particle emissions to a greater extent than that shown by mass alone. Given the prominent health effects of combustion-derived ultrafine PM (and the large numbers of people exposed to pollution in city centers), this could well be accompanied by unseen improvements in health in the long-term—effects that are challenging to measure in real-world studies—even in the face of seemingly mediocre successes in reducing pollution in terms of the conventional mass metrics of PM_{10} and $PM_{2.5}$. Lastly, the electrification of transport offers great promise, especially since exhaust emissions from those vehicles are essentially zero. However, there is concern that the greater weight of battery-operated vehicles may increase nonexhaust emissions from traffic via tire, brake, and road wear. The toxicity of NE-PM will be an important area of research in the upcoming years. From the available evidence, while it appears that NE-PM does have the capacity to induce pulmonary inflammation and oxidative stress, there is uncertainty as to the relative toxicity of these particles compared to exhaust-PM, and if effects from these (usually larger) particles are seen in other organs of the body.

In summary, a large body of mechanistic research now supports epidemiological associations showing that TRAP is associated with wide-ranging detrimental effects throughout the body. Plausible biological mechanisms support the case for causality in many of these associations. This body of work emphasizes the need to place vehicle emissions high on the agenda of policies to reduce air pollution. The realization and effective implementation of such policies will reduce the detrimental effects of TRAP and will likely be accompanied by significant improvements in health in many societies, especially if realized at a global level.

Acknowledgments

MRM is supported by the British Heart Foundation (SP/15/8/31575; CH/09/002). JR is supported by the Chief Scientist Office (Scotland) (CGA/19/16).

Conflicts of interest

The authors declare that they have no conflicts of interest.

References

Abegunde, A. T., Muhammad, B. H., Bhatti, O., & Ali, T. (2016). Environmental risk factors for inflammatory bowel diseases: Evidence based literature review. *World Journal of Gastroenterology*, *22*, 6296–6317.

Adar, S. D., & Kaufman, J. D. (2007). Cardiovascular disease and air pollutants: Evaluating and improving epidemiological data implicating traffic exposure. *Inhalation Toxicology*, *19*(Suppl. 1), 135–149.

Akopian, A. N., Fanick, E. R., & Brooks, E. G. (2016). TRP channels and traffic-related environmental pollution-induced pulmonary disease. *Seminars in Immunopathology*, *38*, 331–338.

Al Suleimani, Y. M., Al Mahruqi, A. S., Al Za'abi, M., Shalaby, A., Ashique, M., Nemmar, A., et al. (2017). Effect of diesel exhaust particles on renal vascular responses in rats with chronic kidney disease. *Environmental Toxicology*, *32*, 541–549.

Alderete, T. L., Chen, Z., Toledo-Corral, C. M., Contreras, Z. A., Kim, J. S., Habre, R., et al. (2018). Ambient and traffic-related air pollution exposures as novel risk factors for metabolic dysfunction and type 2 diabetes. *Current Epidemiology Reports*, *5*, 79–91.

Allen, J. L., Klocke, C., Morris-Schaffer, K., Conrad, K., Sobolewski, M., & Cory-Slechta, D. A. (2017). Cognitive effects of air pollution exposures and potential mechanistic underpinnings. *Current Environmental Health Reports*, *4*, 180–191.

Andersen, Z. J., Stafoggia, M., Weinmayr, G., Pedersen, M., Galassi, C., Jorgensen, J. T., et al. (2017). Long-term exposure to ambient air pollution and incidence of postmenopausal breast cancer in 15 European cohorts within the ESCAPE project. *Environmental Health Perspectives*, *125*, 107005.

Apte, J. S., Brauer, M., Cohen, A., Ezzati, M., Pope, C. A., III. (2018). Ambient $PM_{2.5}$ reduces global and regional life expectancy. *Environmental Science & Technology Letters*, *5*, 546–551.

AQEG_Air_Quality_Expert_Group. (2019). *Non-exhaust emissions from road traffic*. https://uk-air.defra.gov.uk/library/aqeg/publications (Accessed October 16, 2019).

Araujo, J. A., Barajas, B., Kleinman, M., Wang, X., Bennett, B. J., Gong, K. W., et al. (2008). Ambient particulate pollutants in the ultrafine range promote early atherosclerosis and systemic oxidative stress. *Circulation Research*, *102*, 589–596.

Atkinson, R. W., Carey, I. M., Kent, A. J., van Staa, T. P., Anderson, H. R., & Cook, D. G. (2013). Long-term exposure to outdoor air pollution and incidence of cardiovascular diseases. *Epidemiology*, *24*, 44–53.

Aung, N., Sanghvi, M. M., Zemrak, F., Lee, A. M., Cooper, J. A., Paiva, J. M., et al. (2018). Association between ambient air pollution and cardiac morpho-functional phenotypes: Insights from the UK biobank population imaging study. *Circulation*, *138*, 2175–2186.

Babadjouni, R. M., Hodis, D. M., Radwanski, R., Durazo, R., Patel, A., Liu, Q., et al. (2017). Clinical effects of air pollution on the central nervous system: A review. *Journal of Clinical Neuroscience*, *43*, 16–24.

Baccarelli, A., Martinelli, I., Pegoraro, V., Melly, S., Grillo, P., Zanobetti, A., et al. (2009). Living near major traffic roads and risk of deep vein thrombosis. *Circulation*, *119*, 3118–3124.

Bai, L., Chen, H., Hatzopoulou, M., Jerrett, M., Kwong, J. C., Burnett, R. T., et al. (2018). Exposure to ambient ultrafine particles and nitrogen dioxide and incident hypertension and diabetes. *Epidemiology*, *29*, 323–332.

Bai, N., Kido, T., Kavanagh, T. J., Kaufman, J. D., Rosenfeld, M. E., van Breemen, C., et al. (2011). Exposure to diesel exhaust up-regulates iNOS expression in ApoE knockout mice. *Toxicology and Applied Pharmacology*, *255*, 184–192.

Bai, N., Kido, T., Suzuki, H., Yang, G., Kavanagh, T. J., Kaufman, J. D., et al. (2011). Changes in atherosclerotic plaques induced by inhalation of diesel exhaust. *Atherosclerosis, 216*, 299–306.

Bai, L., Weichenthal, S., Kwong, J. C., Burnett, R. T., Hatzopoulou, M., Jerrett, M., et al. (2019). Associations of long-term exposure to ultrafine particles and nitrogen dioxide with increased incidence of congestive heart failure and acute myocardial infarction. *American Journal of Epidemiology, 188*, 151–159.

Baldacci, S., Gorini, F., Santoro, M., Pierini, A., Minichilli, F., & Bianchi, F. (2018). Environmental and individual exposure and the risk of congenital anomalies: a review of recent epidemiological evidence. *Epidemiologia e Prevenzione, 42*, 1–34.

Barath, S., Mills, N. L., Lundback, M., Tornqvist, H., Lucking, A. J., Langrish, J. P., et al. (2010). Impaired vascular function after exposure to diesel exhaust generated at urban transient running conditions. *Particle and Fibre Toxicology*, 7, 19.

Barosova, H., Chortarea, S., Peikertova, P., Clift, M. J. D., Petri-Fink, A., Kukutschova, J., et al. (2018). Biological response of an in vitro human 3D lung cell model exposed to brake wear debris varies based on brake pad formulation. *Archives of Toxicology, 92*, 2339–2351.

Barthelemy, J., Sanchez, K., Miller, M. R., & Khreis, H. (2020). New opportunities to mitigate the burden of disease caused by traffic related air pollution: Antioxidant-rich diets and supplements. *International Journal of Environmental Research and Public Health, 17*.

Bartoli, C. R., Wellenius, G. A., Coull, B. A., Akiyama, I., Diaz, E. A., Lawrence, J., et al. (2009). Concentrated ambient particles alter myocardial blood flow during acute ischemia in conscious canines. *Environmental Health Perspectives, 117*, 333–337.

Behndig, A. F., Larsson, N., Brown, J. L., Stenfors, N., Helleday, R., Duggan, S. T., et al. (2011). Proinflammatory doses of diesel exhaust in healthy subjects fail to elicit equivalent or augmented airway inflammation in subjects with asthma. *Thorax, 66*, 12–19.

Beyerlein, A., Krasmann, M., Thiering, E., Kusian, D., Markevych, I., D'Orlando, O., et al. (2015). Ambient air pollution and early manifestation of type 1 diabetes. *Epidemiology, 26*, e31–e32.

Bhaskaran, K., Hajat, S., Armstrong, B., Haines, A., Herrett, E., Wilkinson, P., et al. (2011). The effects of hourly differences in air pollution on the risk of myocardial infarction: Case crossover analysis of the MINAP database. *BMJ, 343*, d5531.

Bhetraratana, M., Orozco, L. D., Hong, J., Diamante, G., Majid, S., Bennett, B. J., et al. (2019). Diesel exhaust particles dysregulate multiple immunological pathways in murine macrophages: Lessons from microarray and sc RNA-seq technologies. *Archives of Biochemistry and Biophysics, 678*, 108116.

Bind, M. A., Baccarelli, A., Zanobetti, A., Tarantini, L., Suh, H., Vokonas, P., et al. (2012). Air pollution and markers of coagulation, inflammation, and endothelial function: Associations and epigene-environment interactions in an elderly cohort. *Epidemiology, 23*, 332–340.

Blomberg, A. (2000). Airway inflammatory and antioxidant responses to oxidative and particulate air pollutants—Experimental exposure studies in humans. *Clinical and Experimental Allergy, 30*, 310–317.

Bosson, J., Barath, S., Pourazar, J., Behndig, A. F., Sandstrom, T., Blomberg, A., et al. (2008). Diesel exhaust exposure enhances the ozone-induced airway inflammation in healthy humans. *The European Respiratory Journal, 31*, 1234–1240.

Bove, H., Bongaerts, E., Slenders, E., Bijnens, E. M., Saenen, N. D., Gyselaers, W., et al. (2019). Ambient black carbon particles reach the fetal side of human placenta. *Nature Communications, 10*, 3866.

Brook, R. D., Brook, J. R., Urch, B., Vincent, R., Rajagopalan, S., & Silverman, F. (2002). Inhalation of fine particulate air pollution and ozone causes acute arterial vasoconstriction in healthy adults. *Circulation, 105*, 1534–1536.

Brook, R. D., & Rajagopalan, S. (2009). Particulate matter, air pollution, and blood pressure. *Journal of the American Society of Hypertension*, *3*, 332–350.
Brook, R. D., Rajagopalan, S., Pope, C. A., III, Brook, J. R., Bhatnagar, A., Diez-Roux, A. V., et al. (2010). Particulate matter air pollution and cardiovascular disease: An update to the scientific statement from the American Heart Association. *Circulation*, *121*, 2331–2378.
Brook, R. D., Sun, Z., Brook, J. R., Zhao, X., Ruan, Y., Yan, J., et al. (2016). Extreme air pollution conditions adversely affect blood pressure and insulin resistance: The air pollution and cardiometabolic disease study. *Hypertension*, *67*, 77–85.
Brower, J. B., Doyle-Eisele, M., Moeller, B., Stirdivant, S., McDonald, J. D., & Campen, M. J. (2016). Metabolomic changes in murine serum following inhalation exposure to gasoline and diesel engine emissions. *Inhalation Toxicology*, *28*, 241–250.
Budinger, G. R., McKell, J. L., Urich, D., Foiles, N., Weiss, I., Chiarella, S. E., et al. (2011). Particulate matter-induced lung inflammation increases systemic levels of PAI-1 and activates coagulation through distinct mechanisms. *PLoS One*, *6*, e18525.
Burki, T. K. (2015). Diesel cars and health: The Volkswagen emissions scandal. *The Lancet Respiratory Medicine*, *3*, 838–839.
Burnett, R., Chen, H., Szyszkowicz, M., Fann, N., Hubbell, B., Pope, C. A., III, et al. (2018). Global estimates of mortality associated with long-term exposure to outdoor fine particulate matter. *Proceedings of the National Academy of Sciences of the United States of America*, *115*, 9592–9597.
Buteau, S., & Goldberg, M. S. (2016). A structured review of panel studies used to investigate associations between ambient air pollution and heart rate variability. *Environmental Research*, *148*, 207–247.
Calderon-Garciduenas, L., Gonzalez-Maciel, A., Mukherjee, P. S., Reynoso-Robles, R., Perez-Guille, B., Gayosso-Chavez, C., et al. (2019). Combustion- and friction-derived magnetic air pollution nanoparticles in human hearts. *Environmental Research*, *176*, 108567.
Calderon-Garciduenas, L., Gonzalez-Maciel, A., Reynoso-Robles, R., Delgado-Chavez, R., Mukherjee, P. S., Kulesza, R. J., et al. (2018). Hallmarks of Alzheimer disease are evolving relentlessly in Metropolitan Mexico City infants, children and young adults. APOE4 carriers have higher suicide risk and higher odds of reaching NFT stage V at </=40 years of age. *Environmental Research*, *164*, 475–487.
Campen, M. J., Babu, N. S., Helms, G. A., Pett, S., Wernly, J., Mehran, R., et al. (2005). Nonparticulate components of diesel exhaust promote constriction in coronary arteries from ApoE-/- mice. *Toxicological Sciences*, *88*, 95–102.
Campen, M. J., Lund, A. K., Knuckles, T. L., Conklin, D. J., Bishop, B., Young, D., et al. (2010). Inhaled diesel emissions alter atherosclerotic plaque composition in ApoE(-/-) mice. *Toxicology and Applied Pharmacology*, *242*, 310–317.
Campen, M. J., Lund, A., & Rosenfeld, M. (2012). Mechanisms linking traffic-related air pollution and atherosclerosis. *Current Opinion in Pulmonary Medicine*, *18*, 155–160.
Cao, Y., Long, J., Ji, Y., Chen, G., Shen, Y., Gong, Y., et al. (2016). Foam cell formation by particulate matter (PM) exposure: A review. *Inhalation Toxicology*, *28*, 583–590.
Carll, A. P., Haykal-Coates, N., Winsett, D. W., Hazari, M. S., Ledbetter, A. D., Richards, J. H., et al. (2015). Cardiomyopathy confers susceptibility to particulate matter-induced oxidative stress, vagal dominance, arrhythmia and pulmonary inflammation in heart failure-prone rats. *Inhalation Toxicology*, *27*, 100–112.
Carll, A. P., Haykal-Coates, N., Winsett, D. W., Rowan, W. H., III, Hazari, M. S., Ledbetter, A. D., et al. (2010). Particulate matter inhalation exacerbates cardiopulmonary injury in a rat model of isoproterenol-induced cardiomyopathy. *Inhalation Toxicology*, *22*, 355–368.
Carll, A. P., Lust, R. M., Hazari, M. S., Perez, C. M., Krantz, Q. T., King, C. J., et al. (2013). Diesel exhaust inhalation increases cardiac output, bradyarrhythmias, and parasympathetic tone in aged heart failure-prone rats. *Toxicological Sciences*, *131*, 583–595.

Carlsen, H. K., Modig, L., Levinsson, A., Kim, J. L., Toren, K., Nyberg, F., et al. (2015). Exposure to traffic and lung function in adults: A general population cohort study. *BMJ Open*, *5*, e007624.

Carlsten, C., Blomberg, A., Pui, M., Sandstrom, T., Wong, S.W., Alexis, N., et al. (2016). Diesel exhaust augments allergen-induced lower airway inflammation in allergic individuals: A controlled human exposure study. *Thorax*, *71*, 35–44.

Carlsten, C., Kaufman, J. D., Trenga, C. A., Allen, J., Peretz, A., & Sullivan, J. H. (2008). Thrombotic markers in metabolic syndrome subjects exposed to diesel exhaust. *Inhalation Toxicology*, *20*, 917–921.

Carre, J., Gatimel, N., Moreau, J., Parinaud, J., & Leandri, R. (2017). Does air pollution play a role in infertility? A systematic review. *Environmental Health*, *16*, 82.

Cassee, F. R., Campbell, A., Boere, A. J., McLean, S. G., Duffin, R., Krystek, P., et al. (2012). The biological effects of subacute inhalation of diesel exhaust following addition of cerium oxide nanoparticles in atherosclerosis-prone mice. *Environmental Research*, *115*, 1–10.

Cesaroni, G., Forastiere, F., Stafoggia, M., Andersen, Z. J., Badaloni, C., Beelen, R., et al. (2014). Long term exposure to ambient air pollution and incidence of acute coronary events: Prospective cohort study and meta-analysis in 11 European cohorts from the ESCAPE Project. *BMJ*, *348*, f7412.

Chan, C. C., Chuang, K. J., Shiao, G. M., & Lin, L.Y. (2004). Personal exposure to submicrometer particles and heart rate variability in human subjects. *Environmental Health Perspectives*, *112*, 1063–1067.

Channell, M. M., Paffett, M. L., Devlin, R. B., Madden, M. C., & Campen, M. J. (2012). Circulating factors induce coronary endothelial cell activation following exposure to inhaled diesel exhaust and nitrogen dioxide in humans: Evidence from a novel translational in vitro model. *Toxicological Sciences*, *127*, 179–186.

Chen, S.Y., Chan, C. C., & Su, T. C. (2017). Particulate and gaseous pollutants on inflammation, thrombosis, and autonomic imbalance in subjects at risk for cardiovascular disease. *Environmental Pollution*, *223*, 403–408.

Chen, H., Goldberg, M. S., Burnett, R. T., Jerrett, M., Wheeler, A. J., & Villeneuve, P. J. (2013). Long-term exposure to traffic-related air pollution and cardiovascular mortality. *Epidemiology*, *24*, 35–43.

Cheng, M., Wang, B., Yang, M., Ma, J., Ye, Z., Xie, L., et al. (2020). microRNAs expression in relation to particulate matter exposure: A systematic review. *Environmental Pollution*, *260*, 113961.

Cherng, T. W., Campen, M. J., Knuckles, T. L., Gonzalez Bosc, L.V., & Kanagy, N. L. (2009). Impairment of coronary endothelial cell ETB receptor function following short-term inhalation exposure to whole diesel emissions. *American Journal of Physiology. Regulatory, Integrative and Comparative Physiology*, *297*, R640–R647.

Cherng, T. W., Paffett, M. L., Jackson-Weaver, O., Campen, M. J., Walker, B. R., & Kanagy, N. L. (2011). Mechanisms of diesel-induced endothelial nitric oxide synthase dysfunction in coronary arterioles. *Environmental Health Perspectives*, *119*, 98–103.

Choi, H. S., Ashitate, Y., Lee, J. H., Kim, S. H., Matsui, A., Insin, N., et al. (2010). Rapid translocation of nanoparticles from the lung airspaces to the body. *Nature Biotechnology*, *28*, 1300–1303.

Cohen, A. J., Brauer, M., Burnett, R., Anderson, H. R., Frostad, J., Estep, K., et al. (2017). Estimates and 25-year trends of the global burden of disease attributable to ambient air pollution: an analysis of data from the Global Burden of Diseases Study 2015. *Lancet*, *389*, 1907–1918.

Cole-Hunter, T., Weichenthal, S., Kubesch, N., Foraster, M., Carrasco-Turigas, G., Bouso, L., et al. (2016). Impact of traffic-related air pollution on acute changes in cardiac autonomic modulation during rest and physical activity: A cross-over study. *Journal of Exposure Science & Environmental Epidemiology*, *26*, 133–140.

COMEAP. (2015). *Statement on the evidence for differential health effects of particulate matter according to source or components*. https://www.gov.uk/government/publications/particulate-air-pollution-health-effects-of-exposure. (Accessed October 16, 2019).

COMEAP. (2018). *Effects of long-term exposure to ambient air pollution on cardiovascular morbidity: Mechanistic evidence*. https://assets.publishing.service.gov.uk/government/uploads/system/uploads/attachment_data/file/749657/COMEAP_CV_Mechanisms_Report.pdf. (Accessed October 16, 2019).

Cosselman, K. E., Krishnan, R. M., Oron, A. P., Jansen, K., Peretz, A., Sullivan, J. H., et al. (2012). Blood pressure response to controlled diesel exhaust exposure in human subjects. *Hypertension*, *59*, 943–948.

Cui, Y., Sun, Q., & Liu, Z. (2016). Ambient particulate matter exposure and cardiovascular diseases: A focus on progenitor and stem cells. *Journal of Cellular and Molecular Medicine*, *20*, 782–793.

de Kok, T. M., Driece, H. A., Hogervorst, J. G., & Briede, J. J. (2006). Toxicological assessment of ambient and traffic-related particulate matter: A review of recent studies. *Mutation Research*, *613*, 103–122.

DeJarnett, N., Yeager, R., Conklin, D. J., Lee, J., O'Toole, T. E., McCracken, J., et al. (2015). Residential proximity to major roadways is associated with increased levels of AC133+ circulating angiogenic cells. *Arteriosclerosis, Thrombosis, and Vascular Biology*, *35*, 2468–2477.

Devlin, R. B., Smith, C. B., Schmitt, M.T., Rappold, A. G., Hinderliter, A., Graff, D., et al. (2014). Controlled exposure of humans with metabolic syndrome to concentrated ultrafine ambient particulate matter causes cardiovascular effects. *Toxicological Sciences*, *140*, 61–72.

Diaz-Sanchez, D., Jyrala, M., Ng, D., Nel, A., & Saxon, A. (2000). In vivo nasal challenge with diesel exhaust particles enhances expression of the CC chemokines rantes, MIP-1alpha, and MCP-3 in humans. *Clinical Immunology*, *97*, 140–145.

Dickey, J. H., & Part, V. I. I. (2000). Air pollution: Overview of sources and health effects. *Disease-a-Month*, *46*, 566–589.

Dockery, D. W., Pope, C. A., III, Xu, X., Spengler, J. D., Ware, J. H., Fay, M. E., et al. (1993). An association between air pollution and mortality in six U.S. cities. *The New England Journal of Medicine*, *329*, 1753–1759.

Dominici, F., Peng, R. D., Bell, M. L., Pham, L., McDermott, A., Zeger, S. L., et al. (2006). Fine particulate air pollution and hospital admission for cardiovascular and respiratory diseases. *JAMA*, *295*, 1127–1134.

Donaldson, K., Stone, V., Clouter, A., Renwick, L., & MacNee, W. (2001). Ultrafine particles. *Occupational and Environmental Medicine*, *58*, 211–216. 199.

Emmerechts, J., De Vooght, V., Haenen, S., Loyen, S., van Kerckhoven, S., Hemmeryckx, B., et al. (2012). Thrombogenic changes in young and old mice upon subchronic exposure to air pollution in an urban roadside tunnel. *Thrombosis and Haemostasis*, *108*, 756–768.

Emmerechts, J., & Hoylaerts, M. F. (2012). The effect of air pollution on haemostasis. *Hämostaseologie*, *32*, 5–13.

Emmerechts, J., Jacobs, L., Van Kerckhoven, S., Loyen, S., Mathieu, C., Fierens, F., et al. (2012). Air pollution-associated procoagulant changes: The role of circulating microvesicles. *Journal of Thrombosis and Haemostasis*, *10*, 96–106.

Eum, K. D., Kazemiparkouhi, F., Wang, B., Manjourides, J., Pun, V., Pavlu, V., et al. (2019). Long-term NO_2 exposures and cause-specific mortality in American older adults. *Environment International*, *124*, 10–15.

Eze, I. C., Hemkens, L. G., Bucher, H. C., Hoffmann, B., Schindler, C., Kunzli, N., et al. (2015). Association between ambient air pollution and diabetes mellitus in Europe and North America: Systematic review and meta-analysis. *Environmental Health Perspectives*, *123*, 381–389.

Fariss, M. W., Gilmour, M. I., Reilly, C. A., Liedtke, W., & Ghio, A. J. (2013). Emerging mechanistic targets in lung injury induced by combustion-generated particles. *Toxicological Sciences*, *132*, 253–267.

Faustini, A., Rapp, R., & Forastiere, F. (2014). Nitrogen dioxide and mortality: Review and meta-analysis of long-term studies. *The European Respiratory Journal, 44*, 744–753.
Feng, B., Song, X., Dan, M., Yu, J., Wang, Q., Shu, M., et al. (2019). High level of source-specific particulate matter air pollution associated with cardiac arrhythmias. *The Science of the Total Environment, 657*, 1285–1293.
Filippini, T., Heck, J. E., Malagoli, C., Del Giovane, C., & Vinceti, M. (2015). A review and meta-analysis of outdoor air pollution and risk of childhood leukemia. *Journal of Environmental Science and Health Part C, Environmental Carcinogenesis & Ecotoxicology Reviews, 33*, 36–66.
Flores-Pajot, M. C., Ofner, M., Do, M. T., Lavigne, E., & Villeneuve, P. J. (2016). Childhood autism spectrum disorders and exposure to nitrogen dioxide, and particulate matter air pollution: A review and meta-analysis. *Environmental Research, 151*, 763–776.
Floyd, H. S., Chen, L. C., Vallanat, B., & Dreher, K. (2009). Fine ambient air particulate matter exposure induces molecular alterations associated with vascular disease progression within plaques of atherosclerotic susceptible mice. *Inhalation Toxicology, 21*, 394–403.
Folino, A. F., Scapellato, M. L., Canova, C., Maestrelli, P., Bertorelli, G., Simonato, L., et al. (2009). Individual exposure to particulate matter and the short-term arrhythmic and autonomic profiles in patients with myocardial infarction. *European Heart Journal, 30*, 1614–1620.
Forbes, L. J., Patel, M. D., Rudnicka, A. R., Cook, D. G., Bush, T., Stedman, J. R., et al. (2009). Chronic exposure to outdoor air pollution and markers of systemic inflammation. *Epidemiology, 20*, 245–253.
Franchini, M., & Mannucci, P. M. (2011). Thrombogenicity and cardiovascular effects of ambient air pollution. *Blood, 118*, 2405–2412.
Fuks, K. B., Weinmayr, G., Basagana, X., Gruzieva, O., Hampel, R., Oftedal, B., et al. (2017). Long-term exposure to ambient air pollution and traffic noise and incident hypertension in seven cohorts of the European study of cohorts for air pollution effects (ESCAPE). *European Heart Journal, 38*, 983–990.
Fuks, K. B., Weinmayr, G., Hennig, F., Tzivian, L., Moebus, S., Jakobs, H., et al. (2016). Association of long-term exposure to local industry- and traffic-specific particulate matter with arterial blood pressure and incident hypertension. *International Journal of Hygiene and Environmental Health, 219*, 527–535.
Furuyama, A., Kanno, S., Kobayashi, T., & Hirano, S. (2009). Extrapulmonary translocation of intratracheally instilled fine and ultrafine particles via direct and alveolar macrophage-associated routes. *Archives of Toxicology, 83*, 429–437.
Gamon, L. F., & Wille, U. (2016). Oxidative damage of biomolecules by the environmental pollutants NO_2 and NO_3. *Accounts of Chemical Research, 49*, 2136–2145.
Gauderman, W. J., Vora, H., McConnell, R., Berhane, K., Gilliland, F., Thomas, D., et al. (2007). Effect of exposure to traffic on lung development from 10 to 18 years of age: A cohort study. *Lancet, 369*, 571–577.
Gerlofs-Nijland, M. E., Bokkers, B. G. H., Sachse, H., Reijnders, J. J. E., Gustafsson, M., Boere, A. J. F., et al. (2019). Inhalation toxicity profiles of particulate matter: A comparison between brake wear with other sources of emission. *Inhalation Toxicology, 31*, 89–98.
Ghelfi, E., Wellenius, G. A., Lawrence, J., Millet, E., & Gonzalez-Flecha, B. (2010). Cardiac oxidative stress and dysfunction by fine concentrated ambient particles (CAPs) are mediated by angiotensin-II. *Inhalation Toxicology, 22*, 963–972.
Ghio, A. J., Smith, C. B., & Madden, M. C. (2012). Diesel exhaust particles and airway inflammation. *Current Opinion in Pulmonary Medicine, 18*, 144–150.
Ghio, A. J., Sobus, J. R., Pleil, J. D., & Madden, M. C. (2012). Controlled human exposures to diesel exhaust. *Swiss Medical Weekly, 142*, w13597.
Gilmour, M. I., Jaakkola, M. S., London, S. J., Nel, A. E., & Rogers, C. A. (2006). How exposure to environmental tobacco smoke, outdoor air pollutants, and increased pollen burdens influences the incidence of asthma. *Environmental Health Perspectives, 114*, 627–633.

Giorgini, P., Di Giosia, P., Grassi, D., Rubenfire, M., Brook, R. D., & Ferri, C. (2016). Air pollution exposure and blood pressure: An updated review of the literature. *Current Pharmaceutical Design, 22*, 28–51.
Gong, K.W., Zhao, W., Li, N., Barajas, B., Kleinman, M., Sioutas, C., et al. (2007). Air-pollutant chemicals and oxidized lipids exhibit genome-wide synergistic effects on endothelial cells. *Genome Biology, 8*, R149.
Goodman, J. E., Prueitt, R. L., Sax, S. N., Pizzurro, D. M., Lynch, H. N., Zu, K., et al. (2015). Ozone exposure and systemic biomarkers: Evaluation of evidence for adverse cardiovascular health impacts. *Critical Reviews in Toxicology, 45*, 412–452.
Goodson, J. M., Weldy, C. S., MacDonald, J. W., Liu, Y., Bammler, T. K., Chien, W. M., et al. (2017). In utero exposure to diesel exhaust particulates is associated with an altered cardiac transcriptional response to transverse aortic constriction and altered DNA methylation. *The FASEB Journal, 31*, 4935–4945.
Gorr, M. W., Youtz, D. J., Eichenseer, C. M., Smith, K. E., Nelin, T. D., Cormet-Boyaka, E., et al. (2015). In vitro particulate matter exposure causes direct and lung-mediated indirect effects on cardiomyocyte function. *American Journal of Physiology Heart and Circulatory Physiology, 309*, H53–H62.
Gotschi, T., Heinrich, J., Sunyer, J., & Kunzli, N. (2008). Long-term effects of ambient air pollution on lung function: A review. *Epidemiology, 19*, 690–701.
Gottipolu, R. R., Landa, E. R., Schladweiler, M. C., McGee, J. K., Ledbetter, A. D., Richards, J. H., et al. (2008). Cardiopulmonary responses of intratracheally instilled tire particles and constituent metal components. *Inhalation Toxicology, 20*, 473–484.
Grippo, A., Zhang, J., Chu, L., Guo, Y., Qiao, L., Zhang, J., et al. (2018). Air pollution exposure during pregnancy and spontaneous abortion and stillbirth. *Reviews on Environmental Health, 33*, 247–264.
Gualtieri, M., Mantecca, P., Cetta, F., & Camatini, M. (2008). Organic compounds in tire particle induce reactive oxygen species and heat-shock proteins in the human alveolar cell line A549. *Environment International, 34*, 437–442.
Guan, T., Hu, S., Han, Y., Wang, R., Zhu, Q., Hu, Y., et al. (2018). The effects of facemasks on airway inflammation and endothelial dysfunction in healthy young adults: A double-blind, randomized, controlled crossover study. *Particle and Fibre Toxicology, 15*, 30.
Guan, W. J., Zheng, X. Y., Chung, K. F., & Zhong, N. S. (2016). Impact of air pollution on the burden of chronic respiratory diseases in China: Time for urgent action. *Lancet, 388*, 1939–1951.
Gustafsson, M., Blomqvist, G., Gudmundsson, A., Dahl, A., Swietlicki, E., Bohgard, M., et al. (2008). Properties and toxicological effects of particles from the interaction between tyres, road pavement and winter traction material. *The Science of the Total Environment, 393*, 226–240.
Haynes, E. N., Chen, A., Ryan, P., Succop, P., Wright, J., & Dietrich, K. N. (2011). Exposure to airborne metals and particulate matter and risk for youth adjudicated for criminal activity. *Environmental Research, 111*, 1243–1248.
Hazari, M. S., Haykal-Coates, N., Winsett, D. W., Krantz, Q. T., King, C., Costa, D. L., et al. (2011). TRPA1 and sympathetic activation contribute to increased risk of triggered cardiac arrhythmias in hypertensive rats exposed to diesel exhaust. *Environmental Health Perspectives, 119*, 951–957.
Health_Effects_Institute. (2010). *Traffic-related air pollution: A critical review of the literature on emission, exposure, and health effects.*
Hoffmann, B., Moebus, S., Kroger, K., Stang, A., Mohlenkamp, S., Dragano, N., et al. (2009). Residential exposure to urban air pollution, ankle-brachial index, and peripheral arterial disease. *Epidemiology, 20*, 280–288.
Hoffmann, B., Moebus, S., Mohlenkamp, S., Stang, A., Lehmann, N., Dragano, N., et al. (2007). Residential exposure to traffic is associated with coronary atherosclerosis. *Circulation, 116*, 489–496.

Holgate, S. T., Sandstrom, T., Frew, A. J., Stenfors, N., Nordenhall, C., Salvi, S., et al. (2003). Health effects of acute exposure to air pollution. Part I: healthy and asthmatic subjects exposed to diesel exhaust. *Research Report Health Effects Institute*, 1–30. discussion 51–67.

Holguin, F. (2008). Traffic, outdoor air pollution, and asthma. *Immunology and Allergy Clinics of North America*, *28*, 577–588. viii-ix.

Holme, J. A., Brinchmann, B. C., Refsnes, M., Lag, M., & Ovrevik, J. (2019). Potential role of polycyclic aromatic hydrocarbons as mediators of cardiovascular effects from combustion particles. *Environmental Health*, *18*, 74.

Hooftman, N., Oliveira, L., Messagie, M., Coosemans, T., & van Mierlo, J. (2016). Environmental analysis of petrol, diesel and electric passenger cars in a Belgian urban setting. *Energies*, *9*, 84.

Hougaard, K. S., Campagnolo, L., Chavatte-Palmer, P., Tarrade, A., Rousseau-Ralliard, D., Valentino, S., et al. (2015). A perspective on the developmental toxicity of inhaled nanoparticles. *Reproductive Toxicology*, *56*, 118–140.

Hullmann, M., Albrecht, C., van Berlo, D., Gerlofs-Nijland, M. E., Wahle, T., Boots, A. W., et al. (2017). Diesel engine exhaust accelerates plaque formation in a mouse model of Alzheimer's disease. *Particle and Fibre Toxicology*, *14*, 35.

Husain, M., Wu, D., Saber, A. T., Decan, N., Jacobsen, N. R., Williams, A., et al. (2015). Intratracheally instilled titanium dioxide nanoparticles translocate to heart and liver and activate complement cascade in the heart of C57BL/6 mice. *Nanotoxicology*, *9*, 1013–1022.

Iannuzzi, A., Verga, M. C., Renis, M., Schiavo, A., Salvatore, V., Santoriello, C., et al. (2010). Air pollution and carotid arterial stiffness in children. *Cardiology in the Young*, *20*, 186–190.

Ikeda, M., Shitashige, M., Yamasaki, H., Sagai, M., & Tomita, T. (1995). Oxidative modification of low density lipoprotein by diesel exhaust particles. *Biological & Pharmaceutical Bulletin*, *18*, 866–871.

Ikeda, M., Suzuki, M., Watarai, K., Sagai, M., & Tomita, T. (1995). Impairment of endothelium-dependent relaxation by diesel exhaust particles in rat thoracic aorta. *Japanese Journal of Pharmacology*, *68*, 183–189.

Janitz, A. E., Campbell, J. E., Magzamen, S., Pate, A., Stoner, J. A., & Peck, J. D. (2016). Traffic-related air pollution and childhood acute leukemia in Oklahoma. *Environmental Research*, *148*, 102–111.

Jhun, I., Kim, J., Cho, B., Gold, D. R., Schwartz, J., Coull, B. A., et al. (2019). Synthesis of Harvard Environmental Protection Agency (EPA) Center studies on traffic-related particulate pollution and cardiovascular outcomes in the Greater Boston Area. *Journal of the Air & Waste Management Association (1995)*, *69*, 900–917.

Jung, C. R., Chen, W. T., Tang, Y. H., & Hwang, B. F. (2019). Fine particulate matter exposure during pregnancy and infancy periods and incident asthma. *The Journal of Allergy and Clinical Immunology*, *143*, 2254–2262.

Kampfrath, T., Maiseyeu, A., Ying, Z., Shah, Z., Deiuliis, J. A., Xu, X., et al. (2011). Chronic fine particulate matter exposure induces systemic vascular dysfunction via NADPH oxidase and TLR4 pathways. *Circulation Research*, *108*, 716–726.

Kansal, A. (2009). Sources and reactivity of NMHCs and VOCs in the atmosphere: A review. *Journal of Hazardous Materials*, *166*, 17–26.

Kilian, J., & Kitazawa, M. (2018). The emerging risk of exposure to air pollution on cognitive decline and Alzheimer's disease—Evidence from epidemiological and animal studies. *Biomedical Journal*, *41*, 141–162.

Kilinc, E., Van Oerle, R., Borissoff, J. I., Oschatz, C., Gerlofs-Nijland, M. E., Janssen, N. A., et al. (2011). Factor XII activation is essential to sustain the procoagulant effects of particulate matter. *Journal of Thrombosis and Haemostasis*, *9*, 1359–1367.

Kim, J. S., Alderete, T. L., Chen, Z., Lurmann, F., Rappaport, E., Habre, R., et al. (2018). Longitudinal associations of in utero and early life near-roadway air pollution with trajectories of childhood body mass index. *Environmental Health*, *17*, 64.

Kim, K. E., Cho, D., & Park, H. J. (2016). Air pollution and skin diseases: Adverse effects of airborne particulate matter on various skin diseases. *Life Sciences*, *152*, 126–134.
Kim, J. W., Park, S., Lim, C. W., Lee, K., & Kim, B. (2014). The role of air pollutants in initiating liver disease. *Toxicology Research*, *30*, 65–70.
Kingsley, S. L., Eliot, M. N., Whitsel, E. A., Huang, Y. T., Kelsey, K. T., Marsit, C. J., et al. (2016). Maternal residential proximity to major roadways, birth weight, and placental DNA methylation. *Environment International*, *92–93*, 43–49.
Klepac, P., Locatelli, I., Korosec, S., Kunzli, N., & Kukec, A. (2018). Ambient air pollution and pregnancy outcomes: A comprehensive review and identification of environmental public health challenges. *Environmental Research*, *167*, 144–159.
Kodavanti, U. P. (2016). Stretching the stress boundary: Linking air pollution health effects to a neurohormonal stress response. *Biochimica et Biophysica Acta*, *1860*, 2880–2890.
Kodavanti, U. P., Thomas, R., Ledbetter, A. D., Schladweiler, M. C., Shannahan, J. H., Wallenborn, J. G., et al. (2011). Vascular and cardiac impairments in rats inhaling ozone and diesel exhaust particles. *Environmental Health Perspectives*, *119*, 312–318.
Kreider, M. L., Doyle-Eisele, M., Russell, R. G., McDonald, J. D., & Panko, J. M. (2012). Evaluation of potential for toxicity from subacute inhalation of tire and road wear particles in rats. *Inhalation Toxicology*, *24*, 907–917.
Kreyling, W. G., Semmler, M., Erbe, F., Mayer, P., Takenaka, S., Schulz, H., et al. (2002). Translocation of ultrafine insoluble iridium particles from lung epithelium to extrapulmonary organs is size dependent but very low. *Journal of Toxicology and Environmental Health Part A*, *65*, 1513–1530.
Kunzli, N., Jerrett, M., Mack, W. J., Beckerman, B., LaBree, L., Gilliland, F., et al. (2005). Ambient air pollution and atherosclerosis in Los Angeles. *Environmental Health Perspectives*, *113*, 201–206.
Langrish, J. P., Bosson, J., Unosson, J., Muala, A., Newby, D. E., Mills, N. L., et al. (2012). Cardiovascular effects of particulate air pollution exposure: Time course and underlying mechanisms. *Journal of Internal Medicine*, *272*, 224–239.
Langrish, J. P., Li, J., Wang, S., Lee, M. M., Barnes, G. D., Lei, G. G., et al. (2012). Reducing particulate air pollution exposure in patients with coronary heart disease improved cardiovascular health. *Environmental Health Perspectives*, *120*, 367–372.
Langrish, J. P., Lundback, M., Barath, S., Soderberg, S., Mills, N. L., Newby, D. E., et al. (2010). Exposure to nitrogen dioxide is not associated with vascular dysfunction in man. *Inhalation Toxicology*, *22*, 192–198.
Langrish, J. P., Lundback, M., Mills, N. L., Johnston, N. R., Webb, D. J., Sandstrom, T., et al. (2009). Contribution of endothelin 1 to the vascular effects of diesel exhaust inhalation in humans. *Hypertension*, *54*, 910–915.
Langrish, J. P., Mills, N. L., Chan, J. K., Leseman, D. L., Aitken, R. J., Fokkens, P. H., et al. (2009). Beneficial cardiovascular effects of reducing exposure to particulate air pollution with a simple facemask. *Particle and Fibre Toxicology*, *6*, 8.
Langrish, J. P., Unosson, J., Bosson, J., Barath, S., Muala, A., Blackwell, S., et al. (2013). Altered nitric oxide bioavailability contributes to diesel exhaust inhalation-induced cardiovascular dysfunction in man. *Journal of the American Heart Association*, *2*, e004309.
Leary, P. J., Kaufman, J. D., Barr, R. G., Bluemke, D. A., Curl, C. L., Hough, C. L., et al. (2014). Traffic-related air pollution and the right ventricle. The multi-ethnic study of atherosclerosis. *American Journal of Respiratory and Critical Care Medicine*, *189*, 1093–1100.
Lee, K. K., Miller, M. R., & Shah, A. S. V. (2018). Air pollution and stroke. *Journal of Stroke*, *20*, 2–11.
Lelieveld, J., Klingmuller, K., Pozzer, A., Poschl, U., Fnais, M., Daiber, A., et al. (2019). Cardiovascular disease burden from ambient air pollution in Europe reassessed using novel hazard ratio functions. *European Heart Journal*, *40*, 1590–1596.
Lemos, M., Mohallen, S. V., Macchione, M., Dolhnikoff, M., Assuncao, J. V., Godleski, J. J., et al. (2006). Chronic exposure to urban air pollution induces structural alterations in murine pulmonary and coronary arteries. *Inhalation Toxicology*, *18*, 247–253.

Li, H., Cai, J., Chen, R., Zhao, Z., Ying, Z., Wang, L., et al. (2017). Particulate matter exposure and stress hormone levels: A randomized, double-blind, crossover trial of air purification. *Circulation, 136*, 618–627.

Li, W., Dorans, K. S., Wilker, E. H., Rice, M. B., Kloog, I., Schwartz, J. D., et al. (2018). Ambient air pollution, adipokines, and glucose homeostasis: The Framingham heart study. *Environment International, 111*, 14–22.

Li, X., Huang, S., Jiao, A., Yang, X., Yun, J., Wang, Y., et al. (2017). Association between ambient fine particulate matter and preterm birth or term low birth weight: An updated systematic review and meta-analysis. *Environmental Pollution, 227*, 596–605.

Li, R., Navab, M., Pakbin, P., Ning, Z., Navab, K., Hough, G., et al. (2013). Ambient ultrafine particles alter lipid metabolism and HDL anti-oxidant capacity in LDLR-null mice. *Journal of Lipid Research, 54*, 1608–1615.

Li, J., Zhou, C., Xu, H., Brook, R. D., Liu, S., Yi, T., et al. (2019). Ambient air pollution is associated with HDL (high-density lipoprotein) dysfunction in healthy adults. *Arteriosclerosis, Thrombosis, and Vascular Biology, 39*, 513–522.

Liang, R., Zhang, B., Zhao, X., Ruan, Y., Lian, H., & Fan, Z. (2014). Effect of exposure to $PM_{2.5}$ on blood pressure: A systematic review and meta-analysis. *Journal of Hypertension, 32*, 2130–2140. discussion 2141.

Lim, C. C., & Thurston, G. D. (2019). Air pollution, oxidative stress, and diabetes: A life course epidemiologic perspective. *Current Diabetes Reports, 19*, 58.

Lim, S. S., Vos, T., Flaxman, A. D., Danaei, G., Shibuya, K., Adair-Rohani, H., et al. (2012). A comparative risk assessment of burden of disease and injury attributable to 67 risk factors and risk factor clusters in 21 regions, 1990-2010: A systematic analysis for the Global Burden of Disease Study 2010. *Lancet, 380*, 2224–2260.

Liu, N. (2019). *Abstract no: PA360, "Late breaking abstract—Do inhaled carbonaceous particles translocate from the lung to the placenta?"*. In N. Liu, et al. (Eds.), *Occupational and environmental lung diseases: Asthma and the airways, 08:30 hrs CEST, Sunday 16 September, Paris Expo Porte de Versailles. European Respiratory Congress.*

Liu, C., Bai, Y., Xu, X., Sun, L., Wang, A., Wang, T. Y., et al. (2014). Exaggerated effects of particulate matter air pollution in genetic type II diabetes mellitus. *Particle and Fibre Toxicology, 11*, 27.

Liu, F., Chen, G., Huo, W., Wang, C., Liu, S., Li, N., et al. (2019). Associations between long-term exposure to ambient air pollution and risk of type 2 diabetes mellitus: A systematic review and meta-analysis. *Environmental Pollution, 252*, 1235–1245.

Liu, C., Fonken, L. K., Wang, A., Maiseyeu, A., Bai, Y., Wang, T. Y., et al. (2014). Central IKKbeta inhibition prevents air pollution mediated peripheral inflammation and exaggeration of type II diabetes. *Particle and Fibre Toxicology, 11*, 53.

Ljungman, P. L. S., Li, W., Rice, M. B., Wilker, E. H., Schwartz, J., Gold, D. R., et al. (2018). Long- and short-term air pollution exposure and measures of arterial stiffness in the Framingham Heart Study. *Environment International, 121*, 139–147.

Low, R. B., Bielory, L., Qureshi, A. I., Dunn, V., Stuhlmiller, D. F., & Dickey, D. A. (2006). The relation of stroke admissions to recent weather, airborne allergens, air pollution, seasons, upper respiratory infections, and asthma incidence, September 11, 2001, and day of the week. *Stroke, 37*, 951–957.

Lucking, A. J., Lundback, M., Barath, S. L., Mills, N. L., Sidhu, M. K., Langrish, J. P., et al. (2011). Particle traps prevent adverse vascular and prothrombotic effects of diesel engine exhaust inhalation in men. *Circulation, 123*, 1721–1728.

Lucking, A. J., Lundback, M., Mills, N. L., Faratian, D., Barath, S. L., Pourazar, J., et al. (2008). Diesel exhaust inhalation increases thrombus formation in man. *European Heart Journal, 29*, 3043–3051.

Lund, A. K., Knuckles, T. L., Obot Akata, C., Shohet, R., McDonald, J. D., Gigliotti, A., et al. (2007). Gasoline exhaust emissions induce vascular remodeling pathways involved in atherosclerosis. *Toxicological Sciences, 95*, 485–494.

Lund, A. K., Lucero, J., Harman, M., Madden, M. C., McDonald, J. D., Seagrave, J. C., et al. (2011). The oxidized low-density lipoprotein receptor mediates vascular effects of inhaled vehicle emissions. *American Journal of Respiratory and Critical Care Medicine, 184*, 82–91.
Lund, A. K., Lucero, J., Lucas, S., Madden, M. C., McDonald, J. D., Seagrave, J. C., et al. (2009). Vehicular emissions induce vascular MMP-9 expression and activity associated with endothelin-1-mediated pathways. *Arteriosclerosis, Thrombosis, and Vascular Biology, 29*, 511–517.
Lundback, M., Mills, N. L., Lucking, A., Barath, S., Donaldson, K., Newby, D. E., et al. (2009). Experimental exposure to diesel exhaust increases arterial stiffness in man. *Particle and Fibre Toxicology, 6*, 7.
Ma, J. Y., & Ma, J. K. (2002). The dual effect of the particulate and organic components of diesel exhaust particles on the alteration of pulmonary immune/inflammatory responses and metabolic enzymes. *Journal of Environmental Science and Health Part C, Environmental Carcinogenesis & Ecotoxicology Reviews, 20*, 117–147.
Maher, B. A., Ahmed, I. A., Karloukovski, V., MacLaren, D. A., Foulds, P. G., Allsop, D., et al. (2016). Magnetite pollution nanoparticles in the human brain. *Proceedings of the National Academy of Sciences of the United States of America, 113*, 10797–10801.
Maier, K. L., Alessandrini, F., Beck-Speier, I., Hofer, T. P., Diabate, S., Bitterle, E., et al. (2008). Health effects of ambient particulate matter—Biological mechanisms and inflammatory responses to in vitro and in vivo particle exposures. *Inhalation Toxicology, 20*, 319–337.
Mantecca, P., Sancini, G., Moschini, E., Farina, F., Gualtieri, M., Rohr, A., et al. (2009). Lung toxicity induced by intratracheal instillation of size-fractionated tire particles. *Toxicology Letters, 189*, 206–214.
McCreanor, J., Cullinan, P., Nieuwenhuijsen, M. J., Stewart-Evans, J., Malliarou, E., Jarup, L., et al. (2007). Respiratory effects of exposure to diesel traffic in persons with asthma. *The New England Journal of Medicine, 357*, 2348–2358.
McQueen, D. S., Donaldson, K., Bond, S. M., McNeilly, J. D., Newman, S., Barton, N. J., et al. (2007). Bilateral vagotomy or atropine pre-treatment reduces experimental diesel-soot induced lung inflammation. *Toxicology and Applied Pharmacology, 219*, 62–71.
Mehta, A. J., Zanobetti, A., Koutrakis, P., Mittleman, M. A., Sparrow, D., Vokonas, P., et al. (2014). Associations between short-term changes in air pollution and correlates of arterial stiffness: The Veterans Affairs Normative Aging Study, 2007-2011. *American Journal of Epidemiology, 179*, 192–199.
Mendez, R., Zheng, Z., Fan, Z., Rajagopalan, S., Sun, Q., & Zhang, K. (2013). Exposure to fine airborne particulate matter induces macrophage infiltration, unfolded protein response, and lipid deposition in white adipose tissue. *American Journal of Translational Research, 5*, 224–234.
Miller, M. R. (2014). The role of oxidative stress in the cardiovascular actions of particulate air pollution. *Biochemical Society Transactions, 42*, 1006–1011.
Miller, M. R., Borthwick, S. J., Shaw, C. A., McLean, S. G., McClure, D., Mills, N. L., et al. (2009). Direct impairment of vascular function by diesel exhaust particulate through reduced bioavailability of endothelium-derived nitric oxide induced by superoxide free radicals. *Environmental Health Perspectives, 117*, 611–616.
Miller, M. R., McLean, S. G., Duffin, R., Lawal, A. O., Araujo, J. A., Shaw, C. A., et al. (2013). Diesel exhaust particulate increases the size and complexity of lesions in atherosclerotic mice. *Particle and Fibre Toxicology, 10*, 61.
Miller, M. R., & Newby, D. E. (2019). Air pollution and cardiovascular disease: Car sick. *Cardiovascular Research, 116*, 279–294.
Miller, M. R., Raftis, J. B., Langrish, J. P., McLean, S. G., Samutrtai, P., Connell, S. P., et al. (2017a). Correction to "inhaled nanoparticles accumulate at sites of vascular disease". *ACS Nano, 11*, 10623–10624.
Miller, M. R., Raftis, J. B., Langrish, J. P., McLean, S. G., Samutrtai, P., Connell, S. P., et al. (2017b). Inhaled nanoparticles accumulate at sites of vascular disease. *ACS Nano, 11*, 4542–4552.

Miller, M. R., & Shah, A. S.V. (2016). Ambient particles and cerebrovascular disease. In S. C. Bondy & A. Campbell (Eds.), *Inflammation, aging and oxidative stress* (pp. 133–160). Switzerland: Springer International.

Miller, M. R., Shaw, C. A., & Langrish, J. P. (2012). From particles to patients: Oxidative stress and the cardiovascular effects of air pollution. *Future Cardiology*, *8*, 577–602.

Miller, K. A., Siscovick, D. S., Sheppard, L., Shepherd, K., Sullivan, J. H., Anderson, G. L., et al. (2007). Long-term exposure to air pollution and incidence of cardiovascular events in women. *The New England Journal of Medicine*, *356*, 447–458.

Mills, I. C., Atkinson, R. W., Kang, S., Walton, H., & Anderson, H. R. (2015). Quantitative systematic review of the associations between short-term exposure to nitrogen dioxide and mortality and hospital admissions. *BMJ Open*, *5*, e006946.

Mills, N. L., Donaldson, K., Hadoke, P. W., Boon, N. A., MacNee, W., Cassee, F. R., et al. (2009). Adverse cardiovascular effects of air pollution. *Nature Clinical Practice Cardiovascular Medicine*, *6*, 36–44.

Mills, N. L., Finlayson, A. E., Gonzalez, M. C., Tornqvist, H., Barath, S., Vink, E., et al. (2011). Diesel exhaust inhalation does not affect heart rhythm or heart rate variability. *Heart*, *97*, 544–550.

Mills, N. L., Miller, M. R., Lucking, A. J., Beveridge, J., Flint, L., Boere, A. J., et al. (2011). Combustion-derived nanoparticulate induces the adverse vascular effects of diesel exhaust inhalation. *European Heart Journal*, *32*, 2660–2671.

Mills, N. L., Tornqvist, H., Gonzalez, M. C., Vink, E., Robinson, S. D., Soderberg, S., et al. (2007). Ischemic and thrombotic effects of dilute diesel-exhaust inhalation in men with coronary heart disease. *The New England Journal of Medicine*, *357*, 1075–1082.

Mills, N. L., Tornqvist, H., Robinson, S. D., Gonzalez, M., Darnley, K., MacNee, W., et al. (2005). Diesel exhaust inhalation causes vascular dysfunction and impaired endogenous fibrinolysis. *Circulation*, *112*, 3930–3936.

Miners, J. S., van Helmond, Z., Raiker, M., Love, S., & Kehoe, P. G. (2010). ACE variants and association with brain Abeta levels in Alzheimer's disease. *American Journal of Translational Research*, *3*, 73–80.

Moller, P., Mikkelsen, L., Vesterdal, L. K., Folkmann, J. K., Forchhammer, L., Roursgaard, M., et al. (2011). Hazard identification of particulate matter on vasomotor dysfunction and progression of atherosclerosis. *Critical Reviews in Toxicology*, *41*, 339–368.

Mostovenko, E., Young, T., Muldoon, P. P., Bishop, L., Canal, C. G., Vucetic, A., et al. (2019). Nanoparticle exposure driven circulating bioactive peptidome causes systemic inflammation and vascular dysfunction. *Particle and Fibre Toxicology*, *16*, 20.

Munzel, T., Sorensen, M., Gori, T., Schmidt, F. P., Rao, X., Brook, J., et al. (2017a). Environmental stressors and cardio-metabolic disease: Part I-epidemiologic evidence supporting a role for noise and air pollution and effects of mitigation strategies. *European Heart Journal*, *38*, 550–556.

Munzel, T., Sorensen, M., Gori, T., Schmidt, F. P., Rao, X., Brook, F. R., et al. (2017b). Environmental stressors and cardio-metabolic disease: Part II-mechanistic insights. *European Heart Journal*, *38*, 557–564.

Mustafic, H., Jabre, P., Caussin, C., Murad, M. H., Escolano, S., Tafflet, M., et al. (2012). Main air pollutants and myocardial infarction: A systematic review and meta-analysis. *JAMA*, *307*, 713–721.

Mutlu, G. M., Green, D., Bellmeyer, A., Baker, C. M., Burgess, Z., Rajamannan, N., et al. (2007). Ambient particulate matter accelerates coagulation via an IL-6-dependent pathway. *The Journal of Clinical Investigation*, *117*, 2952–2961.

Myhre, O., Lag, M., Villanger, G. D., Oftedal, B., Ovrevik, J., Holme, J. A., et al. (2018). Early life exposure to air pollution particulate matter (PM) as risk factor for attention deficit/hyperactivity disorder (ADHD): Need for novel strategies for mechanisms and causalities. *Toxicology and Applied Pharmacology*, *354*, 196–214.

Nagel, G., Stafoggia, M., Pedersen, M., Andersen, Z. J., Galassi, C., Munkenast, J., et al. (2018). Air pollution and incidence of cancers of the stomach and the upper aerodigestive tract in the European study of cohorts for air pollution effects (ESCAPE). *International Journal of Cancer, 143*, 1632–1643.

Nemmar, A., Al-Salam, S., Beegam, S., Yuvaraju, P., Yasin, J., & Ali, B. H. (2014). Pancreatic effects of diesel exhaust particles in mice with type 1 diabetes mellitus. *Cellular Physiology and Biochemistry, 33*, 413–422.

Nemmar, A., Subramaniyan, D., & Ali, B. H. (2012). Protective effect of curcumin on pulmonary and cardiovascular effects induced by repeated exposure to diesel exhaust particles in mice. *PLoS One, 7*, e39554.

Nemmar, A., Subramaniyan, D., Yasin, J., & Ali, B. H. (2013). Impact of experimental type 1 diabetes mellitus on systemic and coagulation vulnerability in mice acutely exposed to diesel exhaust particles. *Particle and Fibre Toxicology, 10*, 14.

Neven, K. Y., Saenen, N. D., Tarantini, L., Janssen, B. G., Lefebvre, W., Vanpoucke, C., et al. (2018). Placental promoter methylation of DNA repair genes and prenatal exposure to particulate air pollution: An ENVIRONAGE cohort study. *The Lancet Planetary Health, 2*, e174–e183.

Newbury, J. B., Arseneault, L., Beevers, S., Kitwiroon, N., Roberts, S., Pariante, C. M., et al. (2019). Association of air pollution exposure with psychotic experiences during adolescence. *JAMA Psychiatry, 76*, 614–623.

Nightingale, J. A., Maggs, R., Cullinan, P., Donnelly, L. E., Rogers, D. F., Kinnersley, R., et al. (2000). Airway inflammation after controlled exposure to diesel exhaust particulates. *American Journal of Respiratory and Critical Care Medicine, 162*, 161–166.

Niu, Y., Chen, R., Xia, Y., Cai, J., Ying, Z., Lin, Z., et al. (2018). Fine particulate matter constituents and stress hormones in the hypothalamus-pituitary-adrenal axis. *Environment International, 119*, 186–192.

Nordenhall, C., Pourazar, J., Ledin, M. C., Levin, J. O., Sandstrom, T., & Adelroth, E. (2001). Diesel exhaust enhances airway responsiveness in asthmatic subjects. *The European Respiratory Journal, 17*, 909–915.

Oberdorster, G., Sharp, Z., Atudorei, V., Elder, A., Gelein, R., Lunts, A., et al. (2002). Extrapulmonary translocation of ultrafine carbon particles following whole-body inhalation exposure of rats. *Journal of Toxicology and Environmental Health Part A, 65*, 1531–1543.

Ohlwein, S., Kappeler, R., Kutlar Joss, M., Kunzli, N., & Hoffmann, B. (2019). Health effects of ultrafine particles: A systematic literature review update of epidemiological evidence. *International Journal of Public Health, 64*, 547–559.

Ovrevik, J. (2019). Oxidative potential versus biological effects: A review on the relevance of cell-free/abiotic assays as predictors of toxicity from airborne particulate matter. *International Journal of Molecular Sciences, 20*.

Ovrevik, J., Refsnes, M., Lag, M., Brinchmann, B. C., Schwarze, P. E., & Holme, J. A. (2017). Triggering mechanisms and inflammatory effects of combustion exhaust particles with implication for carcinogenesis. *Basic & Clinical Pharmacology & Toxicology, 121*(Suppl. 3), 55–62.

Palmer, J. C., Tayler, H. M., & Love, S. (2013). Endothelin-converting enzyme-1 activity, endothelin-1 production, and free radical-dependent vasoconstriction in Alzheimer's disease. *Journal of Alzheimer's Disease, 36*, 577–587.

Patel, M. M., Quinn, J. W., Jung, K. H., Hoepner, L., Diaz, D., Perzanowski, M., et al. (2011). Traffic density and stationary sources of air pollution associated with wheeze, asthma, and immunoglobulin E from birth to age 5 years among New York City children. *Environmental Research, 111*, 1222–1229.

Pekkanen, J., Peters, A., Hoek, G., Tiittanen, P., Brunekreef, B., de Hartog, J., et al. (2002). Particulate air pollution and risk of ST-segment depression during repeated submaximal exercise tests among subjects with coronary heart disease: The exposure and risk assessment for fine and ultrafine particles in ambient air (ULTRA) study. *Circulation, 106*, 933–938.

Peng, R. D., Chang, H. H., Bell, M. L., McDermott, A., Zeger, S. L., Samet, J. M., et al. (2008). Coarse particulate matter air pollution and hospital admissions for cardiovascular and respiratory diseases among medicare patients. *JAMA*, *299*, 2172–2179.

Peretz, A., Sullivan, J. H., Leotta, D. F., Trenga, C. A., Sands, F. N., Allen, J., et al. (2008). Diesel exhaust inhalation elicits acute vasoconstriction in vivo. *Environmental Health Perspectives*, *116*, 937–942.

Perez, C. M., Hazari, M. S., & Farraj, A. K. (2015). Role of autonomic reflex arcs in cardiovascular responses to air pollution exposure. *Cardiovascular Toxicology*, *15*, 69–78.

Peters, A., Dockery, D. W., Muller, J. E., & Mittleman, M. A. (2001). Increased particulate air pollution and the triggering of myocardial infarction. *Circulation*, *103*, 2810–2815.

Peters, R., Ee, N., Peters, J., Booth, A., Mudway, I., & Anstey, K. J. (2019). Air pollution and dementia: A systematic review. *Journal of Alzheimers Disease*, *70*, S145–S163.

Peters, A., von Klot, S., Mittleman, M. A., Meisinger, C., Hormann, A., Kuch, B., et al. (2013). Triggering of acute myocardial infarction by different means of transportation. *European Journal of Preventive Cardiology*, *20*, 750–758.

Poirier, J., Miron, J., Picard, C., Gormley, P., Theroux, L., Breitner, J., et al. (2014). Apolipoprotein E and lipid homeostasis in the etiology and treatment of sporadic Alzheimer's disease. *Neurobiology of Aging*, *35*(Suppl. 2), S3–10.

Pope, C. A., III, Burnett, R. T., Thun, M. J., Calle, E. E., Krewski, D., Ito, K., et al. (2002). Lung cancer, cardiopulmonary mortality, and long-term exposure to fine particulate air pollution. *JAMA*, *287*, 1132–1141.

Pope, C. A., III, Muhlestein, J. B., May, H. T., Renlund, D. G., Anderson, J. L., & Horne, B. D. (2006). Ischemic heart disease events triggered by short-term exposure to fine particulate air pollution. *Circulation*, *114*, 2443–2448.

Pope, C. A., III, Verrier, R. L., Lovett, E. G., Larson, A. C., Raizenne, M. E., Kanner, R. E., et al. (1999). Heart rate variability associated with particulate air pollution. *American Heart Journal*, *138*, 890–899.

Pourazar, J., Frew, A. J., Blomberg, A., Helleday, R., Kelly, F. J., Wilson, S., et al. (2004). Diesel exhaust exposure enhances the expression of IL-13 in the bronchial epithelium of healthy subjects. *Respiratory Medicine*, *98*, 821–825.

Power, M. C., Adar, S. D., Yanosky, J. D., & Weuve, J. (2016). Exposure to air pollution as a potential contributor to cognitive function, cognitive decline, brain imaging, and dementia: A systematic review of epidemiologic research. *Neurotoxicology*, *56*, 235–253.

Prada, D., Zhong, J., Colicino, E., Zanobetti, A., Schwartz, J., Dagincourt, N., et al. (2017). Association of air particulate pollution with bone loss over time and bone fracture risk: Analysis of data from two independent studies. *The Lancet Planetary Health*, *1*, e337–e347.

Provost, E. B., Madhloum, N., Int Panis, L., De Boever, P., & Nawrot, T. S. (2015). Carotid intima-media thickness, a marker of subclinical atherosclerosis, and particulate air pollution exposure: The meta-analytical evidence. *PLoS One*, *10*, e0127014.

Quan, C., Sun, Q., Lippmann, M., & Chen, L. C. (2010). Comparative effects of inhaled diesel exhaust and ambient fine particles on inflammation, atherosclerosis, and vascular dysfunction. *Inhalation Toxicology*, *22*, 738–753.

Raftis, J. B., & Miller, M. R. (2019). Nanoparticle translocation and multi-organ toxicity: A particularly small problem. *Nano Today*, *26*, 8–12.

Rajagopalan, S., & Brook, R. D. (2012). Air pollution and type 2 diabetes: Mechanistic insights. *Diabetes*, *61*, 3037–3045.

Rao, X., Asico, L. D., Zanos, P., Mahabeleshwar, G. H., Singh Gangwar, R., Xia, C., et al. (2019). Alpha2B-adrenergic receptor overexpression in the brain potentiate air pollution-induced behavior and blood pressure changes. *Toxicological Sciences*, *169*, 95–107.

Rao, X., Patel, P., Puett, R., & Rajagopalan, S. (2015). Air pollution as a risk factor for type 2 diabetes. *Toxicological Sciences*, *143*, 231–241.

Raza, A., Bellander, T., Bero-Bedada, G., Dahlquist, M., Hollenberg, J., Jonsson, M., et al. (2014). Short-term effects of air pollution on out-of-hospital cardiac arrest in Stockholm. *European Heart Journal, 35*, 861–868.

Rich, D. Q., Zhang, W., Lin, S., Squizzato, S., Thurston, S. W., van Wijngaarden, E., et al. (2019). Triggering of cardiovascular hospital admissions by source specific fine particle concentrations in urban centers of New York State. *Environment International, 126*, 387–394.

Riediker, M., Zink, D., Kreyling, W., Oberdorster, G., Elder, A., Graham, U., et al. (2019). Particle toxicology and health—Where are we? *Particle and Fibre Toxicology, 16*, 19.

Risom, L., Moller, P., & Loft, S. (2005). Oxidative stress-induced DNA damage by particulate air pollution. *Mutation Research, 592*, 119–137.

Rivera, M., Basagana, X., Aguilera, I., Foraster, M., Agis, D., de Groot, E., et al. (2013). Association between long-term exposure to traffic-related air pollution and subclinical atherosclerosis: The REGICOR study. *Environmental Health Perspectives, 121*, 223–230.

Robertson, S., Gray, G. A., Duffin, R., McLean, S. G., Shaw, C. A., Hadoke, P. W., et al. (2012). Diesel exhaust particulate induces pulmonary and systemic inflammation in rats without impairing endothelial function ex vivo or in vivo. *Particle and Fibre Toxicology, 9*, 9.

Robertson, S., & Miller, M. R. (2018). Ambient air pollution and thrombosis. *Particle and Fibre Toxicology, 15*, 1.

Robertson, S., Thomson, A. L., Carter, R., Stott, H. R., Shaw, C. A., Hadoke, P. W., et al. (2014). Pulmonary diesel particulate increases susceptibility to myocardial ischemia/reperfusion injury via activation of sensory TRPV1 and beta1 adrenoreceptors. *Particle and Fibre Toxicology, 11*, 12.

Robinson, R. K., Birrell, M. A., Adcock, J. J., Wortley, M. A., Dubuis, E. D., Chen, S., et al. (2018). Mechanistic link between diesel exhaust particles and respiratory reflexes. *The Journal of Allergy and Clinical Immunology, 141*. 1074–1084 e 1079.

Rousseau-Ralliard, D., Valentino, S. A., Aubriere, M. C., Dahirel, M., Lallemand, M. S., Archilla, C., et al. (2019). Effects of first-generation in utero exposure to diesel engine exhaust on second-generation placental function, fatty acid profiles and foetal metabolism in rabbits: Preliminary results. *Scientific Reports, 9*, 9710.

Ruttens, D., Verleden, S. E., Bijnens, E. M., Winckelmans, E., Gottlieb, J., Warnecke, G., et al. (2017). An association of particulate air pollution and traffic exposure with mortality after lung transplantation in Europe. *European Respiratory Journal, 49*.

Saenen, N. D., Martens, D. S., Neven, K. Y., Alfano, R., Bove, H., Janssen, B. G., et al. (2019). Air pollution-induced placental alterations: An interplay of oxidative stress, epigenetics, and the aging phenotype? *Clinical Epigenetics, 11*, 124.

Safar, M. E. (2018). Arterial stiffness as a risk factor for clinical hypertension. *Nature Reviews. Cardiology, 15*, 97–105.

Salvi, S., Blomberg, A., Rudell, B., Kelly, F., Sandstrom, T., Holgate, S. T., et al. (1999). Acute inflammatory responses in the airways and peripheral blood after short-term exposure to diesel exhaust in healthy human volunteers. *American Journal of Respiratory and Critical Care Medicine, 159*, 702–709.

Samet, J. M., Chen, H., Pennington, E. R., & Bromberg, P. A. (2019). Non-redox cycling mechanisms of oxidative stress induced by PM metals. *Free Radical Biology & Medicine*, https://doi.org/10.1016/j.freeradbiomed.2019.12.027. in press.

Sanidas, E., Papadopoulos, D. P., Grassos, H., Velliou, M., Tsioufis, K., Barbetseas, J., et al. (2017). Air pollution and arterial hypertension. A new risk factor is in the air. *Journal of the American Society of Hypertension, 11*, 709–715.

Scheers, H., Nawrot, T. S., Nemery, B., & Casas, L. (2018). Changing places to study short-term effects of air pollution on cardiovascular health: A panel study. *Environmental Health, 17*, 80.

Schikowski, T., Mills, I. C., Anderson, H. R., Cohen, A., Hansell, A., Kauffmann, F., et al. (2014). Ambient air pollution: A cause of COPD? *The European Respiratory Journal, 43*, 250–263.

Schisler, J. C., Ronnebaum, S. M., Madden, M., Channell, M., Campen, M., & Willis, M. S. (2015). Endothelial inflammatory transcriptional responses to an altered plasma exposome following inhalation of diesel emissions. *Inhalation Toxicology, 27*, 272–280.

Schraufnagel, D. E., Balmes, J. R., Cowl, C. T., De Matteis, S., Jung, S. H., Mortimer, K., et al. (2019a). Air pollution and noncommunicable diseases: A review by the Forum of International Respiratory Societies' Environmental Committee, part 1: The damaging effects of air pollution. *Chest, 155*, 409–416.

Schraufnagel, D. E., Balmes, J. R., Cowl, C. T., De Matteis, S., Jung, S. H., Mortimer, K., et al. (2019b). Air pollution and noncommunicable diseases: A review by the Forum of International Respiratory Societies' Environmental Committee, part 2: Air pollution and organ systems. *Chest, 155*, 417–426.

Schultz, E. S., Gruzieva, O., Bellander, T., Bottai, M., Hallberg, J., Kull, I., et al. (2012). Traffic-related air pollution and lung function in children at 8 years of age: A birth cohort study. *American Journal of Respiratory and Critical Care Medicine, 186*, 1286–1291.

Schwarze, P. E., Totlandsdal, A. I., Lag, M., Refsnes, M., Holme, J. A., & Ovrevik, J. (2013). Inflammation-related effects of diesel engine exhaust particles: Studies on lung cells in vitro. *BioMed Research International, 2013*, 685142.

Seaton, A., MacNee, W., Donaldson, K., & Godden, D. (1995). Particulate air pollution and acute health effects. *Lancet, 345*, 176–178.

Seilkop, S. K., Campen, M. J., Lund, A. K., McDonald, J. D., & Mauderly, J. L. (2012). Identification of chemical components of combustion emissions that affect pro-atherosclerotic vascular responses in mice. *Inhalation Toxicology, 24*, 270–287.

Semmler-Behnke, M., Lipka, J., Wenk, A., Hirn, S., Schaffler, M., Tian, F., et al. (2014). Size dependent translocation and fetal accumulation of gold nanoparticles from maternal blood in the rat. *Particle and Fibre Toxicology, 11*, 33.

Shah, A. S., Langrish, J. P., Nair, H., McAllister, D. A., Hunter, A. L., Donaldson, K., et al. (2013). Global association of air pollution and heart failure: A systematic review and meta-analysis. *Lancet, 382*, 1039–1048.

Shah, A. S., Lee, K. K., McAllister, D. A., Hunter, A., Nair, H., Whiteley, W., et al. (2015). Short term exposure to air pollution and stroke: Systematic review and meta-analysis. *BMJ, 350*, h1295.

Shaw, C. A., Robertson, S., Miller, M. R., Duffin, R., Tabor, C. M., Donaldson, K., et al. (2011). Diesel exhaust particulate—Exposed macrophages cause marked endothelial cell activation. *American Journal of Respiratory Cell and Molecular Biology, 44*, 840–851.

Sinharay, R., Gong, J., Barratt, B., Ohman-Strickland, P., Ernst, S., Kelly, F. J., et al. (2018). Respiratory and cardiovascular responses to walking down a traffic-polluted road compared with walking in a traffic-free area in participants aged 60 years and older with chronic lung or heart disease and age-matched healthy controls: A randomised, crossover study. *Lancet, 391*, 339–349.

Soares, S. R., Carvalho-Oliveira, R., Ramos-Sanchez, E., Catanozi, S., da Silva, L. F., Mauad, T., et al. (2009). Air pollution and antibodies against modified lipoproteins are associated with atherosclerosis and vascular remodeling in hyperlipemic mice. *Atherosclerosis, 207*, 368–373.

Sorensen, M., Wendelboe Nielsen, O., Sajadieh, A., Ketzel, M., Tjonneland, A., Overvad, K., et al. (2017). Long-term exposure to road traffic noise and nitrogen dioxide and risk of heart failure: A cohort study. *Environmental Health Perspectives, 125*, 097021.

Sram, R. J., Veleminsky, M., Jr., Veleminsky, M., Sr., & Stejskalova, J. (2017). The impact of air pollution to central nervous system in children and adults. *Neuro Endocrinology Letters, 38*, 389–396.

Stafoggia, M., Cesaroni, G., Peters, A., Andersen, Z. J., Badaloni, C., Beelen, R., et al. (2014). Long-term exposure to ambient air pollution and incidence of cerebrovascular events: Results from 11 European cohorts within the ESCAPE project. *Environmental Health Perspectives, 122*, 919–925.

Steiner, S., Bisig, C., Petri-Fink, A., & Rothen-Rutishauser, B. (2016). Diesel exhaust: Current knowledge of adverse effects and underlying cellular mechanisms. *Archives of Toxicology, 90*, 1541–1553.

Strak, M., Hoek, G., Godri, K. J., Gosens, I., Mudway, I. S., van Oerle, R., et al. (2013). Composition of PM affects acute vascular inflammatory and coagulative markers—The RAPTES project. *PLoS One, 8*, e58944.

Sun, Q., Wang, A., Jin, X., Natanzon, A., Duquaine, D., Brook, R. D., et al. (2005). Long-term air pollution exposure and acceleration of atherosclerosis and vascular inflammation in an animal model. *JAMA, 294*, 3003–3010.

Sun, Q., Yue, P., Deiuliis, J. A., Lumeng, C. N., Kampfrath, T., Mikolaj, M. B., et al. (2009). Ambient air pollution exaggerates adipose inflammation and insulin resistance in a mouse model of diet-induced obesity. *Circulation, 119*, 538–546.

Sun, Q., Yue, P., Kirk, R. I., Wang, A., Moatti, D., Jin, X., et al. (2008). Ambient air particulate matter exposure and tissue factor expression in atherosclerosis. *Inhalation Toxicology, 20*, 127–137.

Suwa, T., Hogg, J. C., Quinlan, K. B., Ohgami, A., Vincent, R., & van Eeden, S. F. (2002). Particulate air pollution induces progression of atherosclerosis. *Journal of the American College of Cardiology, 39*, 935–942.

Suwannasual, U., Lucero, J., McDonald, J. D., & Lund, A. K. (2018). Exposure to traffic-generated air pollutants mediates alterations in brain microvascular integrity in wildtype mice on a high-fat diet. *Environmental Research, 160*, 449–461.

Tabor, C. M., Shaw, C. A., Robertson, S., Miller, M. R., Duffin, R., Donaldson, K., et al. (2016). Platelet activation independent of pulmonary inflammation contributes to diesel exhaust particulate-induced promotion of arterial thrombosis. *Particle and Fibre Toxicology, 13*, 6.

The Lancet Neurology. (2018). Air pollution and brain health: An emerging issue. *Lancet Neurology, 17*, 103.

Toledo-Corral, C. M., Alderete, T. L., Habre, R., Berhane, K., Lurmann, F. W., Weigensberg, M. J., et al. (2018). Effects of air pollution exposure on glucose metabolism in Los Angeles minority children. *Pediatric Obesity, 13*, 54–62.

Tornqvist, H., Mills, N. L., Gonzalez, M., Miller, M. R., Robinson, S. D., Megson, I. L., et al. (2007). Persistent endothelial dysfunction in humans after diesel exhaust inhalation. *American Journal of Respiratory and Critical Care Medicine, 176*, 395–400.

Valavanidis, A., Fiotakis, K., & Vlachogianni, T. (2008). Airborne particulate matter and human health: Toxicological assessment and importance of size and composition of particles for oxidative damage and carcinogenic mechanisms. *Journal of Environmental Science and Health Part C, Environmental Carcinogenesis & Ecotoxicology Reviews, 26*, 339–362.

Valentino, S. A., Tarrade, A., Aioun, J., Mourier, E., Richard, C., Dahirel, M., et al. (2016). Maternal exposure to diluted diesel engine exhaust alters placental function and induces intergenerational effects in rabbits. *Particle and Fibre Toxicology, 13*, 39.

Veras, M. M., Guimaraes-Silva, R. M., Caldini, E. G., Saldiva, P. H., Dolhnikoff, M., & Mayhew, T. M. (2012). The effects of particulate ambient air pollution on the murine umbilical cord and its vessels: A quantitative morphological and immunohistochemical study. *Reproductive Toxicology, 34*, 598–606.

Vieira, J. L., Guimaraes, G.V., de Andre, P. A., Cruz, F. D., Saldiva, P. H., & Bocchi, E. A. (2016). Respiratory filter reduces the cardiovascular effects associated with diesel exhaust exposure: A randomized, prospective, double-blind, controlled study of heart failure: The FILTER-HF Trial. *JACC Heart Failure, 4*, 55–64.

Vieira, J. L., Guimaraes, G.V., de Andre, P. A., Saldiva, P. H., & Bocchi, E. A. (2016). Effects of reducing exposure to air pollution on submaximal cardiopulmonary test in patients with heart failure: Analysis of the randomized, double-blind and controlled FILTER-HF trial. *International Journal of Cardiology*, *215*, 92–97.

Vogel, C. F., Sciullo, E., Wong, P., Kuzmicky, P., Kado, N., & Matsumura, F. (2005). Induction of proinflammatory cytokines and C-reactive protein in human macrophage cell line U937 exposed to air pollution particulates. *Environmental Health Perspectives*, *113*, 1536–1541.

Vouk, V. B., & Piver, W. T. (1983). Metallic elements in fossil fuel combustion products: Amounts and form of emissions and evaluation of carcinogenicity and mutagenicity. *Environmental Health Perspectives*, *47*, 201–225.

Wan, G., Rajagopalan, S., Sun, Q., & Zhang, K. (2010). Real-world exposure of airborne particulate matter triggers oxidative stress in an animal model. *International Journal of Physiology, Pathophysiology and Pharmacology*, *2*, 64–68.

Wang, L., & Pinkerton, K. E. (2007). Air pollutant effects on fetal and early postnatal development. *Birth Defects Research Part C, Embryo Today*, *81*, 144–154.

Ward-Caviness, C. K., Kraus, W. E., Blach, C., Haynes, C. S., Dowdy, E., Miranda, M. L., et al. (2018). Associations between residential proximity to traffic and vascular disease in a cardiac catheterization cohort. *Arteriosclerosis, Thrombosis, and Vascular Biology*, *38*, 275–282.

Watkins, A., Danilewitz, M., Kusha, M., Masse, S., Urch, B., Quadros, K., et al. (2013). Air pollution and arrhythmic risk: The smog is yet to clear. *The Canadian Journal of Cardiology*, *29*, 734–741.

Wauters, A., Dreyfuss, C., Pochet, S., Hendrick, P., Berkenboom, G., van de Borne, P., et al. (2013). Acute exposure to diesel exhaust impairs nitric oxide-mediated endothelial vasomotor function by increasing endothelial oxidative stress. *Hypertension*, *62*, 352–358.

Weaver, A. M., Wellenius, G. A., Wu, W. C., Hickson, D. A., Kamalesh, M., & Wang, Y. (2017). Residential distance to major roadways and cardiac structure in African Americans: Cross-sectional results from the Jackson heart study. *Environmental Health*, *16*, 21.

Weitekamp, C. A., Stevens, T., Stewart, M. J., Bhave, P., & Gilmour, M. I. (2020). Health effects from freshly emitted versus oxidatively or photochemically aged air pollutants. *The Science of the Total Environment*, *704*, 135772.

Weldy, C. S., Liu, Y., Liggitt, H. D., & Chin, M. T. (2014). In utero exposure to diesel exhaust air pollution promotes adverse intrauterine conditions, resulting in weight gain, altered blood pressure, and increased susceptibility to heart failure in adult mice. *PLoS One*, *9*, e88582.

Wilson, D. W., Aung, H. H., Lame, M. W., Plummer, L., Pinkerton, K. E., Ham, W., et al. (2010). Exposure of mice to concentrated ambient particulate matter results in platelet and systemic cytokine activation. *Inhalation Toxicology*, *22*, 267–276.

Wold, L. E., Ying, Z., Hutchinson, K. R., Velten, M., Gorr, M. W., Velten, C., et al. (2012). Cardiovascular remodeling in response to long-term exposure to fine particulate matter air pollution. *Circulation. Heart Failure*, *5*, 452–461.

World Health Organization. (2011). *Air quality and health. WHO factsheets.* http://www.who.int/mediacentre/factsheets/fs313/en/. (Accessed October 16, 2019).

World Health Organization. (2018). *Air pollution.*

Wu, S., Deng, F., Niu, J., Huang, Q., Liu, Y., & Guo, X. (2010). Association of heart rate variability in taxi drivers with marked changes in particulate air pollution in Beijing in 2008. *Environmental Health Perspectives*, *118*, 87–91.

Wu, S., Deng, F., Niu, J., Huang, Q., Liu, Y., & Guo, X. (2011). Exposures to $PM_{2.5}$ components and heart rate variability in taxi drivers around the Beijing 2008 Olympic games. *The Science of the Total Environment*, *409*, 2478–2485.

Xu, X., Liu, C., Xu, Z., Tzan, K., Zhong, M., Wang, A., et al. (2011). Long-term exposure to ambient fine particulate pollution induces insulin resistance and mitochondrial alteration in adipose tissue. *Toxicological Sciences*, *124*, 88–98.

Xu, X., Nie, S., Ding, H., & Hou, F. F. (2018). Environmental pollution and kidney diseases. *Nature Reviews. Nephrology*, *14*, 313–324.

Xu, X., Yavar, Z., Verdin, M., Ying, Z., Mihai, G., Kampfrath, T., et al. (2010). Effect of early particulate air pollution exposure on obesity in mice: Role of p47phox. *Arteriosclerosis, Thrombosis, and Vascular Biology*, *30*, 2518–2527.

Xue, T., Zhu, T., Lin, W., & Talbott, E. O. (2018). Association between hypertensive disorders in pregnancy and particulate matter in the contiguous United States, 1999-2004. *Hypertension*, *72*, 77–84.

Yang, B.Y., Qian, Z., Howard, S.W., Vaughn, M. G., Fan, S. J., Liu, K. K., et al. (2018). Global association between ambient air pollution and blood pressure: A systematic review and meta-analysis. *Environmental Pollution*, *235*, 576–588.

Yatera, K., Hsieh, J., Hogg, J. C., Tranfield, E., Suzuki, H., Shih, C. H., et al. (2008). Particulate matter air pollution exposure promotes recruitment of monocytes into atherosclerotic plaques. *American Journal of Physiology Heart and Circulatory Physiology*, *294*, H944–H953.

Yin, F., Lawal, A., Ricks, J., Fox, J. R., Larson, T., Navab, M., et al. (2013). Diesel exhaust induces systemic lipid peroxidation and development of dysfunctional pro-oxidant and proinflammatory high-density lipoprotein. *Arteriosclerosis, Thrombosis, and Vascular Biology*, *33*, 1153–1161.

Ying, Z., Kampfrath, T., Thurston, G., Farrar, B., Lippmann, M., Wang, A., et al. (2009). Ambient particulates alter vascular function through induction of reactive oxygen and nitrogen species. *Toxicological Sciences*, *111*, 80–88.

Yokota, S., Ohara, N., & Kobayashi, T. (2008). The effects of organic extract of diesel exhaust particles on ischemia/reperfusion-related arrhythmia and on pulmonary inflammation. *The Journal of Toxicological Sciences*, *33*, 1–10.

Zanobetti, A., & Schwartz, J. (2009). The effect of fine and coarse particulate air pollution on mortality: A national analysis. *Environmental Health Perspectives*, *117*, 898–903.

Zanoli, L., Lentini, P., Granata, A., Gaudio, A., Fatuzzo, P., Serafino, L., et al. (2017). A systematic review of arterial stiffness, wave reflection and air pollution. *Molecular Medicine Reports*, *15*, 3425–3429.

Zhang, S., Wolf, K., Breitner, S., Kronenberg, F., Stafoggia, M., Peters, A., et al. (2018). Long-term effects of air pollution on ankle-brachial index. *Environment International*, *118*, 17–25.

Zhao, C. N., Xu, Z., Wu, G. C., Mao, Y. M., Liu, L. N., Qian, W., et al. (2019). Emerging role of air pollution in autoimmune diseases. *Autoimmunity Review*, *18*, 607–614.

Zhong, J., Cayir, A., Trevisi, L., Sanchez-Guerra, M., Lin, X., Peng, C., et al. (2016). Traffic-related air pollution, blood pressure, and adaptive response of mitochondrial abundance. *Circulation*, *133*, 378–387.

Zuurbier, M., Hoek, G., Oldenwening, M., Meliefste, K., Krop, E., van den Hazel, P., et al. (2011). In-traffic air pollution exposure and CC16, blood coagulation, and inflammation markers in healthy adults. *Environmental Health Perspectives*, *119*, 1384–1389.

CHAPTER 11

Biomarkers and omics of health effects associated with traffic-related air pollution

Christiana Demetriou[a] **and Paolo Vineis**[b]
[a]Department of Primary Care and Population Health, University of Nicosia Medical School, Nicosia, Cyprus
[b]MRC-PHE Centre for Environment and Health, Imperial College, London, United Kingdom

List of abbreviations

APP	ambient air pollution
BMI	body mass index
CAFEH	community assessment of freeway exposure and health
CCVD	cerebro-cardiovascular disease
CHD	coronary heart disease
CHEAR	children's health exposure analysis resource
Cu	copper
ECC	Estarreja chemical complex
EDCs	endocrine disrupting chemicals
ENVIRONAGE	environmental influence on early aging study
EPIC	European prospective investigation into cancer and nutrition cohort
ESCAPE	European study of cohorts for air pollution effects
EWAS	exposome-wide association studies
Fe	iron
GLM	generalized linear models
GWAS	genome-wide association studies
HEALS	health and environment-wide associations based on large population surveys project
HELIX	human early life exposome study
HPLC MS	high-performance liquid chromatography
IARC	International Agency for Research on Cancer
IL-8	interleukin-8
ILS	Italian longitudinal study
INMA	infancia y medio ambiente—(environment and childhood) project
K	potassium
LC	liquid chromatography
LIFE	lifestyle interventions and independence for elders study
MITM	"meet-in-the-middle"
MS	mass spectrometry
NHANES	national health and nutrition examination survey
Ni	nickel

Traffic-Related Air Pollution
https://doi.org/10.1016/B978-0-12-818122-5.00011-9

NMR	nuclear magnetic resonance
NO_2	nitric dioxide
NO_x	oxides of nitrogen
PAHs	polycyclic aromatic hydrocarbons
PBL	peripheral blood leukocytes
PCBs	polychlorinated biphenyls
PCOS	polycystic ovary syndrome
PFAS	per- and polyfluoroalkyl substances
PISCINA II	swimming pool in Greek, an Exposomics study
PM_{10}	particulate matter particles, diameter of less than $10\,\mu m$
$PM_{2.5}$	particulate matter particles, diameter of less than $2.5\,\mu m$
PM_x	particulate matter particles, undetermined size
PROBE	PROgramme for Biomonitoring general population Exposure
S	sulfur
Si	silicon
T2D	type 2 diabetes
TCE	trichloroethylene
TRAP	traffic-related air pollution
UFPs	ultrafine particles
USA	United States of America
V	vanadium
Zn	zinc

The exposome and environmental pollutants

We live in an era in which human activity has the greatest impact on the environment and climate. This has prompted scientists to label this era as the "Anthropocene" (Lewis & Maslin, 2015; Monastersky, 2015). In the "Anthropocene" there is a reciprocal relationship between humans and environment, in the sense that humans are profoundly influenced by their own environmental impacts. Therefore, it is not surprising that environmental factors, for example, urban air quality, feature among the top global health risks (Burnett et al., 2018; Cohen, Brauer, et al., 2017; World Health Organization, 2009). In particular, traffic-related air pollution (TRAP) is one of the main contributors to such risks.

However, the identification of hazardous environmental pollutants and their sources—including air pollutants—is complex, particularly in relation to chronic, noncommunicable diseases (Snyder et al., 2013). The key contributors to this complexity are the *diversity of hazards that may exist, the typically low levels of environmental contaminants, long latency periods, and largely unknown modes of action.* The unraveling of environmental causes of disease is also limited by the technical difficulties in defining, and accurately measuring exposures and their sources (e.g., TRAP), and by considerable spatial, temporal, and intraindividual variation. The complex and partially unknown interactions with underlying genetic and other factors that modulate susceptibility and response to environmental exposures further complicate the process of delineating and understanding environmental hazards. The concept of the "exposome" was introduced to address these difficulties and allow an empowerment of environmental research, by improving measurements of external stressors and of internal biological changes, taking advantage of advancements in high-throughput technologies called "omics" (Snyder et al., 2013). However, the application of the exposome approach to TRAP is still in its infancy.

Air is one of the main vehicles of exposure to chemicals and therefore potential toxicants, together with food and water. Most epidemiological research so far has been on air as a whole (e.g., TRAP), or single ("criterion") pollutants such as oxides of nitrogen (NO_x) or particulate matter (PM_x). However, an exposome approach can reveal the implication of unknown air pollutants in disease onset, considering that pollutants are probably many thousands.

Previous evidence on biomarkers and air pollution: The example of carcinogenesis

The association between ambient air pollution exposure (APP)—including TRAP—and lung cancer risk, which has been investigated in several prospective studies, demonstrates generally consistent results indicating that long-term exposure to air pollution is associated with lung cancer incidence and mortality (Demetriou et al., 2012; Demetriou & Vineis, 2015). Pooling together results from 19 prospective cohort studies, published up until 2013, demonstrated a significantly increased risk of lung cancer *mortality* associated with $PM_{2.5}$ and PM_{10}, which was consistent across most subgroups (Cui, Huang, Han, Song, & Chen, 2015). Hazard ratios for lung cancer *incidence* were also increased, but did not reach statistical significance, probably due to the small number of studies investigating PM_x exposure in relation to incidence rather than mortality. This metaanalysis also included a pooled analysis of prospective data from 17 cohort studies based in nine European countries (from the European Study of Cohorts for Air Pollution Effects: ESCAPE), in which PM_{10} was found to be statistically significantly associated with lung cancer risk, while $PM_{2.5}$ was significantly associated with lung adenocarcinomas incidence (Raaschou-Nielsen et al., 2013). The main, but not exclusive, source of exposure in these studies was TRAP.

Looking at more recent investigations conducted in Europe, using data from 14 cohort studies in eight European countries, where air pollution was assessed using land-use regression models for eight elements (Cu, Fe, K, Ni, S, Si, V, and Zn) in size fractions of $PM_{2.5}$ and $PM_{10.}$ Statistically significant associations with lung cancer were found for $PM_{2.5,}$ Cu, PM_{10} Zn, PM_{10} S, PM_{10} Ni, and PM_{10} K (Raaschou-Nielsen et al., 2016). As part of the Italian Longitudinal Study (ILS), individuals exposed to fine particulate matter with an aerodynamic diameter ≤ 2.5 μm ($PM_{2.5}$) and nitrogen dioxide (NO_2) at baseline were followed-up and individual hospitalizations were recorded. Exposure to both chemicals was significantly associated with hospitalizations for lung cancer (Gandini et al., 2018). The Dutch Environmental Longitudinal Study investigated exposure to particulates with an aerodynamic diameter ≤ 10 μm (PM_{10}) and NO_2 in 7 million adults. Results demonstrated that for each 10 μg/m^3 increase, both PM_{10} and NO_2 were significantly associated with lung cancer mortality (Fischer et al., 2015). Again, these studies concerned exposure mainly to TRAP.

Outside Europe, in a study in Israel assessing whether TRAP is associated with cancers previously linked to TRAP exposure (including lung, breast, prostate, kidney, and bladder), significant positive associations were found (Cohen et al., 2018). In the same study, exposure to nitrogen oxide was associated with increased lung, bladder, kidney, or prostate cancer incidence (Cohen, Levy, et al., 2017). In turn, in China, in a prospective investigation of ~ 190,000 individuals, long-term exposure to $PM_{2.5}$ was significantly associated with lung cancer mortality (Yin et al., 2017). In Canada, among participants of the Canadian National Breast Screening Study, long-term exposure to $PM_{2.5}$, even at the lower levels in the distribution, was significantly associated with an elevated risk of lung cancer (Tomczak et al., 2016). For participants in the Nurses' Health Study, in the USA, a 10 μg/m^3 increase in long-term exposure to PM_{10} or $PM_{2.5}$ was significantly associated with lung cancer, and the associations became stronger when restricting the analysis to never-smokers and to former smokers who had quit at least 10 years before (Puett et al., 2014).

There are still several open questions on the long-term effects of TRAP, including the shape of the dose-response relationship, the impact of multiple pollutants in the air mixture, the existence of more susceptible populations, and also a strengthening of causal reasoning by addressing mechanisms.

Biomarkers can enhance research on the health effects of air pollution by improving exposure assessment, increasing the understanding of mechanisms, and enabling the investigation of individual susceptibility. In a review published in 2015 (Demetriou & Vineis, 2015), we assessed, among other markers,

DNA adducts as biomarkers of exposure to air pollution and early biological effect, and DNA methylation as a biomarker of early biological change. In that context, we discussed critical issues arising from their incorporation in AAP health impact evaluations, such as confounding, individual susceptibilities, timing, intensity and duration of exposure, and the investigated tissue. DNA adducts and DNA methylation were treated as paradigms. Confounding arises from the multiple sources of exposure, that can be intercorrelated, such as indoor exposure (e.g., cooking) or tobacco smoke. Whereas the adduct technology is not usually able to disentangle the impact of these different sources, the expectation from methylation studies is of a greater specificity, with a tobacco signature, for example, left on specific genes such as AHRR. Usually DNA adducts showed a dose-response relationship with TRAP exposure, while the evidence from methylation studies is very limited.

In that review, we concluded that the application of DNA adducts and DNA methylation, as biomarkers of exposure and early biological effects, in large prospective studies on air pollution, has the potential to increase the accuracy of exposure assessments and shed light on possible mechanisms of carcinogenesis. In particular, specific polycyclic aromatic hydrocarbon (PAH)-related adducts may allow a better quantitative assessment of exposure and thus reduce misclassification. Also, DNA adducts express longer-term markers of exposure compared to other markers like 1-nitropyrene in urine (adducts are related to the half-life of white-blood cells, the usual target tissue, i.e., a few months). There are similar expectations from methylation assays, although evidence on the persistence of methylation changes is currently available only for tobacco smoking (Demetriou & Vineis, 2015).

The development of high-resolution and high-throughput technologies interrogating -omics (such as epigenomics, transcriptomics, proteomics, and metabolomics) has yielded an unprecedented perspective in air pollution science, and it can be used to identify even more biomarkers of exposure and carcinogenicity. Incorporation of validated in vitro markers in population studies will strengthen causal inferences by offering multilevel evidence for the carcinogenicity of air pollution, while highlighting the importance of exposure timing, duration, intensity, reversibility of changes, and individual susceptibility (Demetriou & Vineis, 2015).

The exposome approach

The concept of the "exposome" was proposed, initially by Wild (2005), with more recent detailed development in relation to its application to population-based studies (Wild, 2012). The original concept was expanded particularly by Rappaport and Smith (2010) who made the exposome practical in terms of chemicals detectable in bio-specimens. The exposome concept refers to the totality of exposures from a variety of external and internal sources including, but not limited to, chemical agents, biological agents, radiation, and psychosocial components from conception onward, over a complete lifetime (Wild, 2005). Under this definition, the exposome comprises biologically active chemicals in response to external environmental stimuli as well as the internal chemical environment, and offers a conceptual leap in studying the role of the environment in human disease (Rappaport & Smith, 2010; Vineis, Chadeau-Hyam, et al., 2017; Wild, 2012).

EWAS studies: Taking stock of progress

Along with the updated characterization of the exposome, the idea of Exposome-wide Association Studies (EWAS) was proposed as an environmental parallel of Genome-wide Association Studies (GWAS) (Rappaport & Smith, 2010). EWAS studies were conceptualized as studies where signatures and markers of the exposome would be measured and compared between humans with specific health outcomes. Once disease-relevant exposures were identified, additional targeted investigations would be used to tie them to specific exposures which can then be managed (e.g., air mixtures, TRAP) (Rappaport, Barupal, Wishart, Vineis, & Scalbert, 2014; Rappaport & Smith, 2010).

The practicality of EWAS studies was quickly embraced by the scientific community (Patel, 2017; Rappaport, 2012; Smith, de la Rosa, & Daniels, 2015) since by integrating exposure science and the exposome, it was viewed as an opportunity for coherence in the environmental health sciences (Lioy & Rappaport, 2011; Smith et al., 2015) and as a valuable tool in causality assessments (Rappaport, 2011; Vineis, Illari, & Russo, 2017). First, as discussed by Rappaport et al. (2014), only ~50% of noncommunicable disease burden is mapped to known exposures and the range of possible human exposures (across difference sources and classes) is extensive. Therefore, targeted, knowledge-driven studies are limited in their contribution to understanding environmental causes of disease. Instead, EWAS studies enable a streamlining of those specific parts of the exposome (chemical candidates) that can meaningfully discriminate between health outcomes (Patel, 2016; Rappaport et al., 2014). Second, interrogating all possible exposures in the exposome avoids biases relating to preconceived notions of which specific exposures or even classes of chemicals (i.e., drugs, foods, pollutants, endogenous) may contribute most to disease (Rappaport, 2011; Rappaport et al., 2014).

Since the introduction of EWAS studies and the seminal paper by Rappaport and Smith in 2010 (Rappaport & Smith, 2010), more than 300 articles with direct mentions of the human exposome have been published, but these are theoretical/perspective pieces in their majority. In order to investigate research progress since the 2010 publication, we reviewed all studies published with direct mentions of the exposome or EWAS. The goal was to identify those studies that, as per Rappaport and Smith's suggestion, interrogated the exposome agnostically through biological signatures. The studies that met our criteria are few and are discussed in the following paragraphs. EWAS-type studies were being performed even before 2010, such as studies agnostically investigating the metabolome of case and control subjects in neurologic, cardiovascular, and cancer studies (summarized in Rappaport, 2012). Following 2010, several more studies that have investigated profiles of biological molecules/chemicals with untargeted approaches in relation to particular disease outcomes have been published. However, few make direct mentions to the exposome or EWAS (Chung, Buck Louis, Kannan, & Patel, 2019; Lai et al., 2019; Maitre et al., 2014; Patel et al., 2014; Yang et al., 2015) (Table 11.1). These studies have mostly concentrated on specific parts of the exposome in relation to specific diseases (not permitting evaluation of replication of findings), and only a few of the identified exposome markers were validated either in independent studies or with different methodologies (Table 11.1).

Most of the studies identified above investigate reproduction-related outcomes (Chung et al., 2019; Maitre et al., 2014; Patel et al., 2014; Yang et al., 2015), and exposure to air pollution, including TRAP, is evidenced to be significantly associated with higher risk for preterm birth (PTB), low birth weight (LBW), and small for gestational age (SGA) births (Klepac, Locatelli, Korošec, Künzli, & Kukec, 2018; Shah & Balkhair, 2011). Such exposomic investigations into TRAP-related disease outcomes can yield useful insights into the molecular pathways via which TRAP contributes to these outcomes and can strengthen causality arguments. For example, in the studies above, urinary 1-hyproxypyrene (a metabolite associated with TRAP) was found to be significantly associated with preterm birth (Patel et al., 2014), PAHs, also associated with TRAP exposure, were associated significantly with PCOS and heavy metals, contained in TRAP, were associated with semen quality. A more comprehensive approach to exposome investigation was taken by researchers working on the National Health and Nutrition Examination Survey (NHANES) (Patel, Bhattacharya, & Butte, 2010; Patel, Cullen, Ioannidis, & Butte, 2012; Patel, Manrai, Corona, & Kohane, 2017;

Table 11.1 Studies investigating the exposome (untargeted interrogation of biological signatures) in relation to health outcomes.

Author, year	Context	Subjects, *n*	Health outcome	Exposome investigated (method)	Tissue under investigation	Validation
Patel et al. (2014)	NHANES, Cross-sectional comparison, USA	Women with a history of preterm birth ($n=72$) Women without ($n=718$)	Preterm birth	201 biological markers of external environmental factors	Serum and urine (depending on marker)	Validation of top finding in an independent study of pregnant women (16 cases and 21 controls)
Maitre et al. (2014)	Case-control study, nested in the Rhea cohort	Urine samples collected in the 1st trimester ($n=438$)	Preterm birth and Fetal growth restriction	34 major metabolites (NMR spectroscopy)	Urine	None mentioned
Yang et al. (2015)	Case-control study, Peking, Northern China	Patients with PCOS ($n=50$) Healthy controls ($n=30$)	PCOS	Organic pollutants, PAHs, phenolic compounds (gas chromatographic MS)	Serum	None mentioned
Chung et al. (2019)	LIFE Prospective cohort study, cross-sectional comparison, USA	Men ($n=473$)	Semen quality	128 EDCs concentrations	Serum and urine (depending on chemical)	None mentioned
Lai et al. (2019)	Cross-sectional investigation, USA	Active Crohn's disease samples ($n=10$), inactive Crohn's disease samples ($n=10$), Healthy non-IBD samples ($n=10$)	Crohn's disease	Metabolic profile (LC, high-resolution MS)	Serum	None mentioned

Abbreviations: *EDCs*, endocrine disrupting chemicals; *LC*, liquid chromatography; *MS*, mass spectrometry; *NHANES*, National Health and Nutrition Examination Survey; *NMR*, nuclear magnetic resonance; *PCOS*, polycystic ovary syndrome.

Patel et al., 2013). These studies have investigated a multitude of environmental and behavioral factors as well as metabolites of environmental and behavioral exposures in relation to type 2 diabetes (Patel et al., 2010), serum lipid levels (Patel et al., 2012), all-cause mortality (Patel et al., 2013), and leukocyte telomere length (Patel et al., 2017) (Table 11.2). Nevertheless, even though the cross-sectional analyses performed were in a sense agnostic (hypothesis free), the exposure factors investigated were dependent on the information available and not on untargeted interrogations of all chemicals in the internal exposome.

Even though TRAP was not directly investigated as an exposure in these studies, metabolites of chemicals found in air pollution such as hydrocarbons, and serum lead were some of the markers that were significantly associated with the investigated disease outcomes. What is more important, is that the majority of health outcomes investigated have been linked to exposure to TRAP in epidemiological studies. TRAP mixture constituents have been repeatedly linked to telomere length (Miri et al., 2019), leucocyte count (Steenhof et al., 2014), all-cause mortality (Achilleos et al., 2017; Sanyal, Rochereau, Maesano, Com-Ruelle, & Annesi-Maesano, 2018; Villeneuve et al., 2015), and type 2 diabetes (Balti, Echouffo-Tcheugui, Yako, & Kengne, 2014; Esposito, Petrizzo, Maiorino, Bellastella, & Giugliano, 2016). Therefore, the results of these studies could highlight plausible mechanisms through which TRAP exposure contributes to these outcomes.

Reviewing all studies published with direct mention of the exposome or EWAS, more commonly researchers agnostically investigated the exposome in relation to the external environment, in an effort to identify biological signatures that can serve as biomarkers of external exposures (Baker, Simpson, Lin, Shireman, & Seixas, 2017; Ellis et al., 2012; Gil, Duarte, Pinto, & Barros, 2018; Huang et al., 2018; Janssen et al., 2015; Maitre, Robinson, et al., 2018; Pino et al., 2017; Schisler et al., 2015; van Breda et al., 2015; van Veldhoven et al., 2018; Walker et al., 2016, 2018) (Table 11.3). The external exposures investigated include occupational exposure to trichloroethylene (Walker et al., 2016), environmental or occupational exposure to metals, some of which are included in TRAP (Baker et al., 2017; Ellis et al., 2012; Pino et al., 2017), unspecified environmental pollutants (Gil et al., 2018; Maitre, Robinson, et al., 2018), exposure to near-highway ultrafine particles (UFPs) (Walker et al., 2018) or other TRAP-related sources (Janssen et al., 2015; Schisler et al., 2015), smoking (Huang et al., 2018), dietary exposures (van Breda et al., 2015), and water disinfection by-products (van Veldhoven

Table 11.2 Studies investigating the exposome (environmental exposures and biochemical markers of these exposures) in relation to health outcomes.

Author, year	Context	Subjects, *n*	Health outcome	Exposome investigated	Tissue under investigation	Validation
Patel et al. (2010)	NHANES, cross-sectional comparison, USA (4/4 surveys)	Survey 1999–2000: T2D cases ($n=197$) and controls ($n=3070$); Survey 2001–02: T2D cases ($n=251$) and controls ($n=3415$); Survey 2003–04: T2D cases ($n=228$) and controls ($n=3118$); Survey 2005–06: T2D cases ($n=234$) and controls ($n=3118$)	Type 2 diabetes	226 environmental, physiological and behavioral factors, including biological markers of the environmental factors	Serum and urine (depending on marker)	No external validation, but sensitivity analyses within the study population
Patel et al. (2012)	NHANES, cross-sectional comparison, USA (3/4 surveys)	Survey 1999–2000: healthy participants ($n=3002$); Survey 2001–02: healthy participants ($n=3610$); Survey 2005–06: healthy participants ($n=2912$)	Serum lipid levels	188 environmental, physiological and behavioral factors, including biological markers of the environmental factors	Serum and urine (depending on marker)	Independent validation in the 2003–04 survey: healthy participants ($n=3449$)

Patel et al. (2013)	NHANES, cross-sectional comparison, USA (2/4 surveys)	Surveys 1999–2002: survivors ($n=5353$) and deceased ($n=655$)	All-cause mortality	249 environmental, physiological and behavioral factors, including biological markers of the environmental factors		Independent validation in the 2003–04 survey: survivors ($n=3059$) and deceased ($n=203$)
Patel et al. (2017)	NHANES, cross-sectional comparison, USA (2/4 surveys)	Surveys 1999–2002: no. of participants dependent on variable ($n=149–7118$)	Leukocyte telomere length	461 environmental, physiological and behavioral factors, including biological markers of the environmental factors and Gene expression of selected genes	Serum and urine (depending on marker)	No external validation, but sensitivity analyses within the study population

Abbreviations: *T2D*, type 2 diabetes.

Table 11.3 Studies investigating the exposome (untargeted interrogation of biological signatures) in relation to specific external environmental exposures.

Author, year	Context	Subjects, *n*	External environmental exposure	Exposome investigated (method)	Tissue under investigation	Validation
Ellis et al. (2012)	Cross-sectional investigation	Healthy human volunteers ($n=178$)	Cadmium	Metabolic profiles (NMR spectroscopy)	Urine	Validation on a subset of samples
Janssen et al. (2015)	Cross-sectional investigation, ENVIRONAGE birth cohort study	Mother-newborn pairs ($n=381$)	Mother's $PM_{2.5}$ exposure based on home address	Mitochondrial methylation (bisulfite pyrosequencing)	Placental tissue	None mentioned
Schisler et al. (2015)	Experimental investigation	Healthy human subjects ($n=6$) Samples collected before, immediately post and 24h after exposure	Exposure to 100 μg/m^3 diesel exhaust or filtered air for 2h	Transcriptomic signatures (Gene expression microarrays)	Coronary artery endothelial cells obtained from plasma	None mentioned
van Breda et al. (2015)	Interventional study	Healthy nonsmoking volunteers aged 18–45 years ($n=168$). Samples collected before and after the intervention	Consumption of 1L of a custom-made blueberry-apple juice mixture per day for 4 weeks	Transcriptomic signatures (Agilent 4x44K whole human genome oligo arrays)	Blood lymphocytes	None mentioned

Walker et al. (2016)	Cross-sectional investigation, China	TCE exposed workers ($n=80$) and matched controls ($n=95$)	Occupational exposure to TCE	Metabolic signatures (ultrahigh-resolution MS)	Plasma	None mentioned
Pino et al. (2017)	PROBE survey, Italy	Adolescents aged 13–15 years ($n=453$)	Sociodemographic determinants, environmental parameters, and geo-coding of residential address	Concentration of 19 metals	Blood	No external validation, but analysis performed with two independent method-ologies: EWAS and GLM capturing co-exposures
Baker et al. (2017)	Cross-sectional investigation	Manganese exposed workers ($n=12$) and nonexposed workers ($n=10$)	Manganese	Metabolic profiles (HPLC MS)	Urine	Validation in 8 exposed and 7 unexposed workers

Continued

Table 11.3 Studies investigating the exposome (untargeted interrogation of biological signatures) in relation to specific external environmental exposures—cont'd

Author, year	Context	Subjects, *n*	External environmental exposure	Exposome investigated (method)	Tissue under investigation	Validation
Maitre, Robinson, et al. (2018)	INMA birth cohort, cross-sectional investigation, Spain	Pregnant women ($n=750$)	Environmental pollutants some of which were quantified in samples: organochlorine pesticides, PCBs, PFAS in first trimester blood samples; mercury in cord blood; metals, phthalates, and bisphenol A in urine samples at 12 and 32 weeks of pregnancy	Metabolic signatures (NMR spectroscopy for urine, LC MS for serum)	First trimester blood samples, cord blood samples, urine samples at 12 and 32 weeks of pregnancy	None mentioned
Walker et al. (2018)	CAFEH study, cross-sectional investigation, USA	Healthy participants ($n=59$)	Near high-way UFPs	Metabolic signatures (LC, high-resolution MS)	Plasma	None mentioned
Huang et al. (2018)	Cross-sectional investigation, USA	Male twins ($n=180$)	Smoking	Metabolic signatures (LC, high-resolution MS)	Plasma	None mentioned
			Smoking-related metabolites	Epigenomics	Genomic DNA from PBL	

van Veldhoven et al. (2018)	PISCINA II Study, Spain	Volunteer swimmers at indoor pool ($n = 60$)	Disinfection by-products in swimming pool water and in volunteers' exhaled breath	Metabolic profile before and 2 h after swimming (LC, high-resolution MS)	Serum	None mentioned
Lau et al. (2018)	HELIX study, 6 European birth cohorts	Children ($n = 1192$)	BMI and dietary intakes frequency	Metabolic profiles (NMR spectroscopy for urine, LC MS for serum)	Serum and urine	None mentioned
Gil et al. (2018)	Cross-sectional investigation, Portugal	Pregnant women ($n = 107$)	Two different central Portugal exposomes, one of which comprised an important source of pollutants (the Estarreja Chemical Complex, ECC)	Metabolic profiles (NMR spectroscopy)	Urine	None mentioned

Abbreviations: *BMI*, body mass index; *CAFEH*, community assessment of freeway exposure and health; *EWAS*, environment-wide association study; *GLM*, generalized linear model; *HELIX*, human early-life exposome; *HPLC*, high-performance liquid chromatography; *INMA*, infancia y medio ambiente; *LC*, liquid chromatography; *MS*, mass spectrometry; *NMR*, nuclear magnetic resonance; *PBL*, peripheral blood leukocytes; *PROBE*, PROgramme for Biomonitoring general population Exposure; *TCE*, trichloroethylene; *UFPs*, ultrafine particles.

et al., 2018) (Table 11.3). In parallel, one study attempted to characterize the major determinants of the child metabolome by obtaining metabolic phenotypes of matched urine and serum samples from birth cohort recruited children (Lau et al., 2018). The list of exposures investigate demonstrates the popularity of air pollution, including TRAP, in exposomic investigations, highlighting the promise of valuable results as these studies become more streamlined and increase in number.

The amalgamation of the aforementioned studies indicates that despite the popularity of the "exposome" and the value of its incorporation into epidemiological assessments, risk assessments and assessments of causality, research efforts, especially with respect to TRAP exposures, remain fragmented, with little overlap and little validation. The original approach proposed by Rappaport and Smith (2010), which would include an untargeted investigation of all prominent classes of chemicals/metabolites in relation to disease and then the matching of disease differentiating chemicals/metabolites with specific external or internal environmental exposures, has not been comprehensively performed for any health outcome. In fact, there has been little effort so far to examine the totality of stressors in the exposome. Thankfully, TRAP-related exposomics is a research field where much progress is anticipated in the next few years, as highlighted by the research findings presented in the following section.

Exposome findings on air pollution: Impact on inflammation and the immune system

Here we report on one of the very few examples of TRAP-related exposome research, provided by a large European consortium, funded by the European Commission, Exposomics (Vineis, Chadeau-Hyam, et al., 2017).

Prior to the formation of the Exposomics consortium, air pollution, including TRAP, was mainly investigated in reference to biomarkers, considered as an outcome of exposure, or in relation to long-term health effects.

These investigations did not follow the "**meet-in-the-middle**" (MITM) approach we had previously proposed (Fig. 11.1) (Chadeau-Hyam et al., 2011). The MITM approach consists of measuring intermediate biomarkers (often with an agnostic omic investigation) and relating them (a) retrospectively to measurements of external exposure and (b) prospectively to a health outcome (disease, or aging, or other outcomes). Robust association of a set of markers with both ends of the exposure-to-disease continuum would serve to *validate a causal hypothesis based on the paradigm of pathway perturbation.*

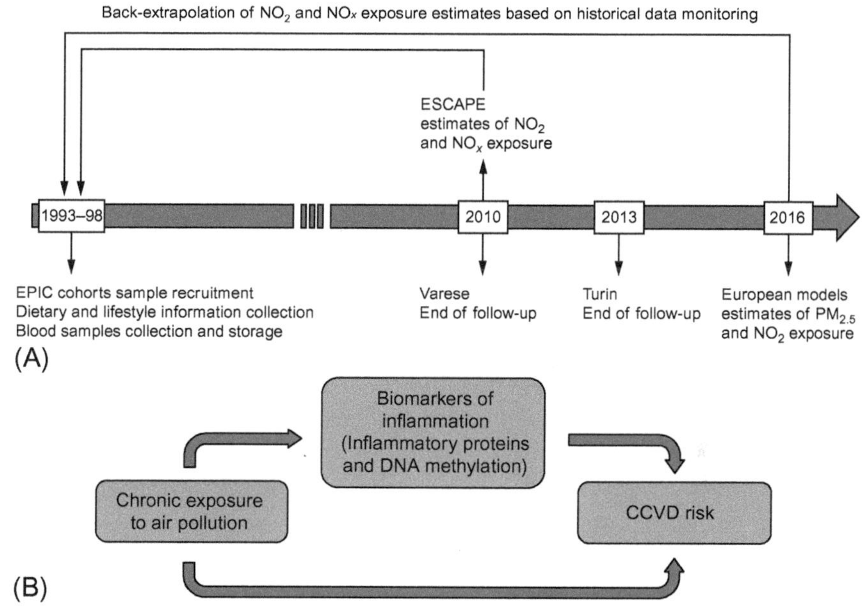

Fig. 11.1 Oxidative stress and inflammation mediate the effect of air pollution on cardio- and cerebrovascular diseases: Meet-in-the-middle approach. (A) Timeline for collection of the relevant variables and exposure estimates in an exposomic investigation based on the meet-in-the-middle approach. (B) Study hypothesis. *(Reproduced with permission from Fiorito, G., Vlaanderen, J., Polidoro, S., Gulliver, J., Galassi, C., Ranzi, A., et al. (2018). Oxidative stress and inflammation mediate the effect of air pollution on cardio- and cerebrovascular disease: A prospective study in nonsmokers. Environmental and Molecular Mutagenesis, 59, 234–246, First published: 08 November 2017. https://doi.org/10.1002/em.22153.)*

The MITM approach aims to address biological plausibility, one of Bradford Hill's guidelines for causality assessment in epidemiology. The applicability, usefulness, and timeliness of this approach, especially with respect to TRAP, rest on the recent technological developments in omics and external measurements, as well as on the existence of long-term longitudinal population cohorts with biological samples stored for many years.

As an example of the application of the MITM approach using exposome data, we considered how proteins, DNA methylation, and metabolome markers may mediate the relationship between TRAP and coronary heart disease and asthma. These disease outcomes were selected based on evidence that related adverse health effects, including mortality and morbidity due to respiratory diseases and coronary heart disease (CHD), associate significantly with exposure to ambient air pollution (Newby et al., 2015; Uzoigwe, Prum, Bresnahan, & Garelnabi, 2013; Wolf et al., 2015).

Despite previous proposal of mechanistic hypotheses aiming to explain the effect of air pollution on the cardiovascular and respiratory system, particularly oxidative stress and inflammation, this was the first application of the MITM approach.

Participants of the Italian component (Turin and Varese Centers, 18,982 individuals) of the European Prospective Investigation into Cancer and Nutrition (EPIC) cohort were included in the CCVD (cerebro-cardiovascular disease) study. Newly diagnosed cases of CCVD and revascularization, arising during a mean follow-up time of 12.2 were ascertained from hospital discharge records. Prospectively collected and archived blood was used for biomarker analyses (inflammatory proteins, genome-wide DNA methylation, metabolites) in a specially designed case-control study nested in the cohort, using the incident density sampling method (Fiorito et al., 2018). Enrichment of altered DNA methylation in "ROS/Glutathione/Cytotoxic granules" and "Cytokine signaling" pathways-related genes was identified to be associated with both TRAP and with CCVD risk. Interleukin-17 was evidenced to associate with higher exposure to NO_2, NO_x, as well as with CCVD risk ($OR_{\text{highest vs lowest tertile}} = 1.79$; CI 1.04–3.11) (Fiorito et al., 2018). Perturbation of the linoleate metabolism pathway was associated with both TRAP exposure and two disease outcomes, namely CCVD and asthma, in a subsequent metabolomics investigation (Jeong et al., 2018). Linoleate is involved in the modulation of IL-8, and thus also in the immune response. These results add credibility to previously proposed pathophysiological mechanisms and suggest that immune proteins and DNA methylation perturbations related to air pollution can be detected several years before CCVD diagnosis in blood samples.

Exposome research as an approach to investigate mixtures of air pollutants

The focus of the abovementioned studies was broadly characterized air pollutants (i.e., $PM_{2.5}$, PM_{10}, NO_2, NO_x, ultrafine particles). Yet, both within and across environmental compartments, people are usually exposed to complex chemical mixtures (air, water, food; air pollution mixtures) rather than singular exposures. This observation raises a practical research question, namely whether components in a mixture act separately via different metabolic or molecular pathways or whether they impact on common pathways. Omics investigations under the Exposomics initiative, evidenced that both cases are realistic depending on the mixture investigated.

When considering air pollutants from the Oxford Street randomized cross-over trial, the hypothesis of separate effects is supported (Sinharay et al., 2018). In this trial, volunteers were exposed to high (Oxford Street, London) or low (Hyde Park) TRAP levels. Investigation of RNA (mRNA and miRNA) and metabolic profiles, showed that the different components of air pollution gave rise to signals that did not overlap, suggesting that each pollutant may follow a different metabolic or molecular pathway to exert its effects (Espín-Pérez et al., 2018; Krauskopf et al., 2019; van Veldhoven et al., 2019).

On the other hand, investigations of other exposures, such as pool water contaminated by chlorinated or brominated disinfection by-products, in the Piscina experimental study (van Veldhoven et al., 2018), showed an overlap between miRNA and metabolomic signals across five different disinfection by-products.

Toward improved causal inference: Meaningful biological pathways affected by urban air pollution

The most promising signals associated with TRAP that were replicated across studies in Exposomics were IL-8 in three studies (including linoleate that is involved in IL-8 activation), carnitine shuttle in two studies, and phosphatydilcholines in two studies. Therefore, pathways commonly perturbed after exposure to air pollution most likely involve immunity, inflammation, and oxidative stress. On the contrary, DNA methylation results did not replicate across studies. Consistent hypo- or hypermethylated genes in relation to environmental exposures were only found for tobacco smoke (notably *AHRR*; Guida et al., 2015; Joehanes et al., 2016). For all other exposures only a few genes have been found so far, with little replication. In the attempt to understand better how exposures (individual or mixtures) impact on disease, a research program in the exposome community is aiming to reconcile the exposure-associated pathways with "hallmarks" of disease or toxicants. López-Otín, Blasco, Partridge, Serrano, and Kroemer (2013) recently proposed nine "hallmarks of aging" which partly overlap with the more well-known and previously published, hallmarks of cancer (Rappaport, 2011). Independently, similar "key characteristics of carcinogens" have been proposed (Fig. 11.2) (Hanahan & Weinberg, 2011). In Exposomics when signals corresponding to hallmarks of cancer or key characteristics of carcinogens, were investigated, significant perturbation of the following categories were detected: miRNA expression, cytokine signaling, reactive oxygen species (oxidative stress). glycosphingolipids (apoptosis, cell

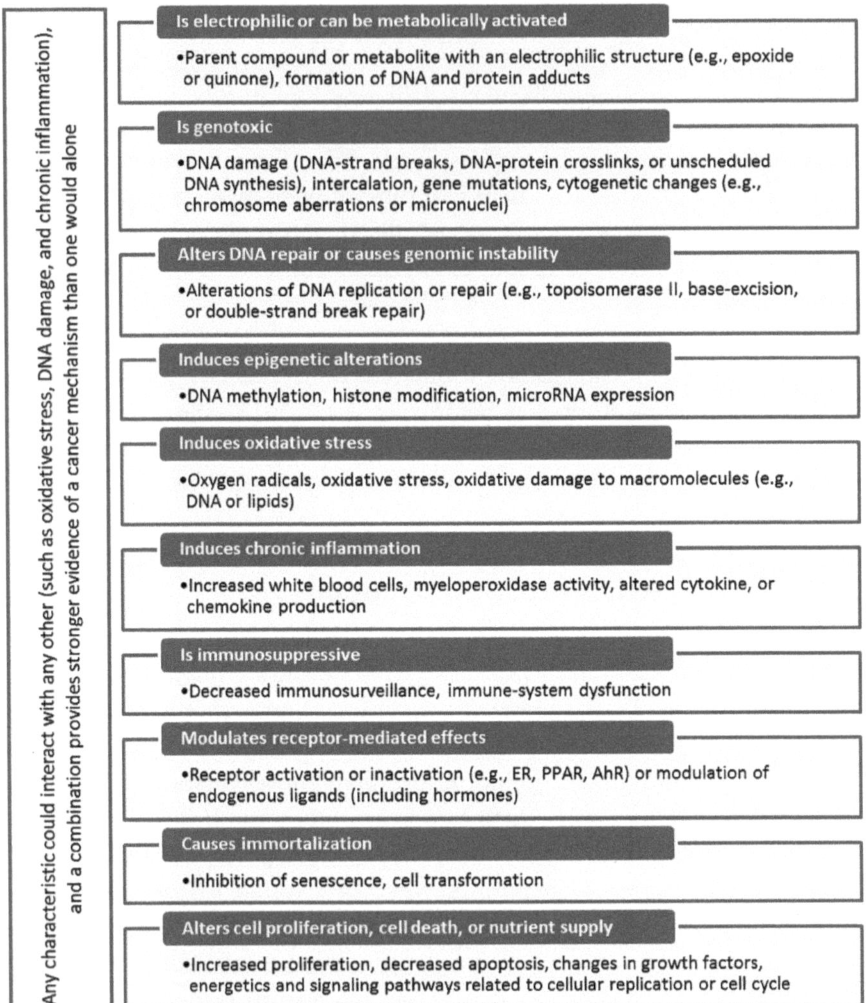

Fig. 11.2 Key characteristics of carcinogens. *(Reproduced with permission from Smith, M. T., Guyton, K. Z., Gibbons, C. F., Fritz, J. M., Portier, C. J., Rusyn, I., et al. (2016). Key characteristics of carcinogens as a basis for organizing data on mechanisms of carcinogenesis. Environmental Health Perspectives, 124, 713–721.)*

growth, senescence, cell cycle control), NOTCH1 (cell cycle control). carnitine shuttle and acylcarnitines (oxidative stress). Immunity, inflammation, oxidative stress, and miRNA have also been shown to be involved in respiratory and cardiovascular diseases. These pathways are identical to the pathways observed to be significantly and consistently perturbed in response to air pollutants as measured in filters (Gulliver et al., 2018) as well as in our previously discussed Exposomics investigations.

Moving forward

There are now several more comprehensive efforts underway to characterize aspects of the exposome by linking exposures in the external environment, including TRAP, with biomarkers of the internal environment as well as by linking biomarkers of the internal environment with health outcomes (Balshaw, Collman, Gray, & Thompson, 2017; Maitre, de Bont, et al., 2018; Neveu et al., 2017; Steckling et al., 2018; Vineis, Chadeau-Hyam, et al., 2017). The Children's Health Exposure Analysis Resource (CHEAR) is a new endeavor aiming to facilitate the inclusion of the concept of the exposome in research pertaining to children's health. More specifically, through CHEAR, the National Institute of Health provides to the children's health research community access to an infrastructure for laboratory and data analyses, which will "support exposure analysis through traditional targeted biomonitoring methods and will also support the untargeted analysis of the exposome" (Balshaw et al., 2017). Furthermore, the International Agency for Research on Cancer (IARC) collected data from peer-reviewed publications and organized it into the first database dedicated to biomarkers of exposure to environmental risk factors: the Exposome Explorer (Neveu et al., 2017). The database contains information on the nature (i.e., structure) of biomarkers, their expected concentrations in different human biospecimens, the populations in which these markers were studied, links to the studies that investigated these markers, as well as correlations of these markers with external exposure measurements. In addition, the Human Early Life Exposome (HELIX) study aims to quantify and describe numerous environmental exposures during pregnancy and childhood (early life) in the context of a prospective cohort study and to investigate these exposures in relation to molecular signatures (-omics) and health outcomes during childhood (Maitre, de Bont, et al., 2018). The Health and Environment-wide Associations based on Large Population Surveys (HEALS) project (Steckling et al., 2018) reviewed published evidence to summarize available biomarkers of exposure for a broad range of environmental stressors and exposure determinants. In the resulting comprehensive publication, 13 stressor categories are considered and for each stressor/biomarker pair, reference and exposure limit values as well as biomonitoring equivalents are also reported, where available. Air pollution is one of the investigated stressor categories, and TRAP-related exposures such as diesel exhaust, NO_x, PM, and UFPs, were included within the category. However, only for diesel exhaust and NO_x is there a reported biomarker of exposure (1-HP and NO_x, respectively) (Steckling et al., 2018). Some additional TRAP-related exposures, such as metals, PAHs, and benzene are also reported but under different stressor categories (relevant biomarkers of exposure: 1-HP, the metal itself, and benzene, respectively).

Existing challenges in exposome investigations

Despite the progress evident in the design of these comprehensive efforts, linking the exposome to disease and specific exposures continues to be hindered by methodological challenges inherent in exposome investigations. First, complex statistical methodologies are needed in these investigations and the appropriateness of different methods has been compared and extensively discussed (Agier et al., 2016; Barrera-Gómez et al., 2017; Chadeau-Hyam et al., 2013; Jain et al., 2018; Manrai et al., 2017). However, despite advances in the available statistical methods for exposome investigations, this continues to be a complicated and often daunting issue especially when considering mixtures of exposures and multilevel "omics" (Patel et al., 2017; Taylor et al., 2016). Such complex investigations will require scientists who can "bridge public health, genomics, and biomedicine in informatics and statistics" (Manrai et al., 2017).

Second, as already mentioned, exposures often occur in mixtures, and individual exposure components (whether external or internal) are often correlated (Patel, 2016; Patel & Manrai, 2015; Rappaport et al., 2014). As described before, given the "exposure-wide" nature of the exposome, there is an expansive number of exposure combinations possible, from which there is a need to dissect prevalent co-occurring combinations (Patel, 2016). Additionally, when considering mixtures, there is a challenge in deciphering whether correlation is simply co-occurrence or whether there is also interaction of exposure components to influence health (i.e., one exposure might act as a trigger in the presence of another exposure to influence phenotype) (Braun, Gennings, Hauser, & Webster, 2016; Patel, 2017). In order to ascertain whether and how mixtures are associated with disease, new analytical methods and large sample sizes are required (Patel, 2016).

Third, when interrogating the exposome, there is a "mingling of causal and reactive -omes," i.e., it is not always clear whether a change (e.g., in methylation) represents a defensive response of the organism, or it is a pathological signature. This makes it difficult to differentiate those markers that are exposure biomarkers (irrespective of disease) and those that are biomarkers of disease, irrespective of exposure (Rappaport, 2012).

Fourth, assessment of cumulative risks requires a clear, or at least a clearer, understanding on how the totality of stressors (including stressors from the natural and social environments) at several time-points throughout life interact to affect individual health (Miller & Jones, 2014; Smith et al., 2015). To take this further, incorporating natural, social, built, and policy environments with the internal exposome, can also serve in the assessment of risk at the community level, termed the public health exposome (Juarez et al., 2014).

Fifth, current analytical platforms cannot reliably measure concentrations of chemicals below a certain threshold and a large number of chemicals/metabolites are found in the human body in smaller than the threshold concentrations (Rappaport et al., 2014). From the thousands of compounds detectable in bio-specimens using current analytical methodologies, only a small fraction can be meaningfully identified, leaving the largest fraction of detectable substances virtually unexploitable (Escher et al., 2017). Also, a characterization of the inherent variability of compounds detectable in bio-specimens is necessary and such results are now becoming available (Casas et al., 2018; Maitre et al., 2017). Moreover, the longitudinal investigations necessary for a characterization of the exposome throughout the lifespan require vigorous and challenging, not to mention expensive, study methodologies (Escher et al., 2017).

Lastly, there is also the challenge of "where to look." Blood has been argued to be the most appropriate medium for exposome investigations since it transports chemicals around the body and is also a reservoir for many chemicals, both exogenous and endogenous (Escher et al., 2017; Nicholson et al., 2012; Rappaport et al., 2014). It can also be obtained via minimally invasive procedures and has been collected for years in population studies (Escher et al., 2017). However, other tissues such as saliva (Bessonneau, Pawliszyn, & Rappaport, 2017), urine (Lau et al., 2018; Maitre, Robinson, et al., 2018), and fecal samples (Wu et al., 2019) are now emerging as valuable platforms for exposomic investigations.

Summary and conclusions

Though still preliminary, the exposome approach to air pollution/TRAP looks like a promising avenue to make sense of the multitude of signals emerging from omics investigations. Further, it can lead to the identification of valid biomarkers of exposure or biological effect, which can be incorporated into epidemiological studies in order to strengthen causal inferences, since looking at pathways and pathway perturbation is a way to confer biological plausibility to observations and to increase statistical power. In this context, the next generation of omics research should aim to fill the gap between the "hallmarks" of disease or "key characteristics" of carcinogens and the empirical results of exposome studies.

Acknowledgment

This work was supported by the Exposomics (European Commission FP7 Grant number 308610) grant to PV.

Conflict of interest

We have no conflicts of interest to declare.

References

Achilleos, S., Kioumourtzoglou, M.-A., Wu, C.-D., Schwartz, J. D., Koutrakis, P., & Papatheodorou, S. I. (2017). Acute effects of fine particulate matter constituents on mortality: A systematic review and meta-regression analysis. *Environment International, 109*, 89–100.

Agier, L., Portengen, L., Chadeau-Hyam, M., Basagaña, X., Giorgis-Allemand, L., Siroux, V., et al. (2016). A systematic comparison of linear regression-based statistical methods to assess exposome-health associations. *Environmental Health Perspectives, 124*, 1848–1856.

Baker, M. G., Simpson, C. D., Lin, Y. S., Shireman, L. M., & Seixas, N. (2017). The use of metabolomics to identify biological signatures of manganese exposure. *Annals of Work Exposures and Health, 61*, 406–415.

Balshaw, D. M., Collman, G. W., Gray, K. A., & Thompson, C. L. (2017). The Children's Health Exposure Analysis Resource (CHEAR): Enabling research into the environmental influences on children's health outcomes. *Current Opinion in Pediatrics, 29*, 385–389.

Balti, E. V., Echouffo-Tcheugui, J. B., Yako, Y. Y., & Kengne, A. P. (2014). Air pollution and risk of type 2 diabetes mellitus: A systematic review and meta-analysis. *Diabetes Research and Clinical Practice, 106*, 161–172.

Barrera-Gómez, J., Agier, L., Portengen, L., Chadeau-Hyam, M., Giorgis-Allemand, L., Siroux, V., et al. (2017). A systematic comparison of statistical methods to detect interactions in exposome-health associations. *Environmental Health, 16*, 74.

Bessonneau, V., Pawliszyn, J., & Rappaport, S. M. (2017). The saliva exposome for monitoring of individuals' health trajectories. *Environmental Health Perspectives, 125*, 077014.

Braun, J. M., Gennings, C., Hauser, R., & Webster, T. F. (2016). What can epidemiological studies tell us about the impact of chemical mixtures on human health? *Environmental Health Perspectives, 124*, A6–A9.

Burnett, R., Chen, H., Szyszkowicz, M., Fann, N., Hubbell, B., Pope, C. A., et al. (2018). Global estimates of mortality associated with long-term exposure to outdoor fine particulate matter. *PNAS, 115*, 9592–9597.

Casas, M., Basagaña, X., Sakhi, A. K., Haug, L. S., Philippat, C., Granum, B., et al. (2018). Variability of urinary concentrations of non-persistent chemicals in pregnant women and school-aged children. *Environment International, 121*, 561–573.

Chadeau-Hyam, M., Athersuch, T. J., Keun, H. C., Iorio, M. D., Ebbels, T. M. D., Jenab, M., et al. (2011). Meeting-in-the-middle using metabolic profiling—A strategy for the identification of intermediate biomarkers in cohort studies. *Biomarkers, 16*, 83–88.

Chadeau-Hyam, M., Campanella, G., Jombart, T., Bottolo, L., Portengen, L., Vineis, P., et al. (2013). Deciphering the complex: Methodological overview of statistical models to derive OMICS-based biomarkers. *Environmental and Molecular Mutagenesis, 54*, 542–557.

Chung, M. K., Buck Louis, G. M., Kannan, K., & Patel, C. J. (2019). Exposome-wide association study of semen quality: Systematic discovery of endocrine disrupting chemical biomarkers in fertility require large sample sizes. *Environment International, 125*, 505–514.

Cohen, A. J., Brauer, M., Burnett, R., Anderson, H. R., Frostad, J., Estep, K., et al. (2017). Estimates and 25-year trends of the global burden of disease attributable to ambient air pollution: An analysis of data from the Global Burden of Diseases Study 2015. *The Lancet, 389*, 1907–1918.

Cohen, G., Levy, I., Yuval, Kark, J. D., Levin, N., Broday, D. M., et al. (2017). Long-term exposure to traffic-related air pollution and cancer among survivors of myocardial infarction: A 20-year follow-up study. *European Journal of Preventive Cardiology, 24*, 92–102.

Cohen, G., Levy, I., Yuval, n., Kark, J. D., Levin, N., Witberg, G., et al. (2018). Chronic exposure to traffic-related air pollution and cancer incidence among 10,000 patients undergoing percutaneous coronary interventions: A historical prospective study. *European Journal of Preventive Cardiology, 25*, 659–670.

Cui, P., Huang, Y., Han, J., Song, F., & Chen, K. (2015). Ambient particulate matter and lung cancer incidence and mortality: A meta-analysis of prospective studies. *European Journal of Public Health, 25*, 324–329.

Demetriou, C. A., Raaschou-Nielsen, O., Loft, S., Møller, P., Vermeulen, R., Palli, D., et al. (2012). Biomarkers of ambient air pollution and lung cancer: A systematic review. *Occupational and Environmental Medicine, 69*, 619–627.

Demetriou, C. A., & Vineis, P. (2015). Carcinogenicity of ambient air pollution: Use of biomarkers, lessons learnt and future directions. *Journal of Thoracic Disease, 7*, 67–95.

Ellis, J. K., Athersuch, T. J., Thomas, L. D. K., Teichert, F., Pérez-Trujillo, M., Svendsen, C., et al. (2012). Metabolic profiling detects early effects of environmental and lifestyle exposure to cadmium in a human population. *BMC Medicine, 10*, 61.

Escher, B. I., Hackermüller, J., Polte, T., Scholz, S., Aigner, A., Altenburger, R., et al. (2017). From the exposome to mechanistic understanding of chemical-induced adverse effects. *Environment International, 99*, 97–106.

Espín-Pérez, A., Font-Ribera, L., van Veldhoven, K., Krauskopf, J., Portengen, L., Chadeau-Hyam, M., et al. (2018). Blood transcriptional and microRNA responses to short-term exposure to disinfection by-products in a swimming pool. *Environment International, 110*, 42–50.

Esposito, K., Petrizzo, M., Maiorino, M. I., Bellastella, G., & Giugliano, D. (2016). Particulate matter pollutants and risk of type 2 diabetes: A time for concern? *Endocrine, 51*, 32–37.

Fiorito, G., Vlaanderen, J., Polidoro, S., Gulliver, J., Galassi, C., Ranzi, A., et al. (2018). Oxidative stress and inflammation mediate the effect of air pollution on cardio- and cerebrovascular disease: A prospective study in nonsmokers. *Environmental and Molecular Mutagenesis, 59*, 234–246.

Fischer, P. H., Marra, M., Ameling, C. B., Hoek, G., Beelen, R., de Hoogh, K., et al. (2015). Air pollution and mortality in seven million adults: The Dutch Environmental Longitudinal Study (DUELS). *Environmental Health Perspectives, 123*, 697–704.

Gandini, M., Scarinzi, C., Bande, S., Berti, G., Carnà, P., Ciancarella, L., et al. (2018). Long term effect of air pollution on incident hospital admissions: Results from the Italian Longitudinal Study within LIFE MED HISS project. *Environment International, 121*, 1087–1097.

Gil, A. M., Duarte, D., Pinto, J., & Barros, A. S. (2018). Assessing exposome effects on pregnancy through urine metabolomics of a Portuguese (Estarreja) cohort. *Journal of Proteome Research, 17*, 1278–1289.

Guida, F., Sandanger, T. M., Castagné, R., Campanella, G., Polidoro, S., Palli, D., et al. (2015). Dynamics of smoking-induced genome-wide methylation changes with time since smoking cessation. *Human Molecular Genetics, 24*, 2349.

Gulliver, J., Morley, D., Dunster, C., McCrea, A., van Nunen, E., Tsai, M.-Y., et al. (2018). Land use regression models for the oxidative potential of fine particles (PM2.5) in five European areas. *Environmental Research, 160*, 247–255.

Hanahan, D., & Weinberg, R. A. (2011). Hallmarks of cancer: The next generation. *Cell, 144*, 646–674.

Huang, Y., Hui, Q., Walker, D. I., Uppal, K., Goldberg, J., Jones, D. P., et al. (2018). Untargeted metabolomics reveals multiple metabolites influencing smoking-related DNA methylation. *Epigenomics, 10*, 379–393.

Jain, P., Vineis, P., Liquet, B., Vlaanderen, J., Bodinier, B., van Veldhoven, K., et al. (2018). A multivariate approach to investigate the combined biological effects of multiple exposures. *Journal of Epidemiology and Community Health, 72*, 564–571.

Janssen, B. G., Byun, H.-M., Gyselaers, W., Lefebvre, W., Baccarelli, A. A., & Nawrot, T. S. (2015). Placental mitochondrial methylation and exposure to airborne particulate matter in the early life environment: An ENVIRONAGE birth cohort study. *Epigenetics, 10*, 536–544.

Jeong, A., Fiorito, G., Keski-Rahkonen, P., Imboden, M., Kiss, A., Robinot, N., et al. (2018). Perturbation of metabolic pathways mediates the association of air pollutants with asthma and cardiovascular diseases. *Environment International, 119*, 334–345.

Joehanes, R., Just, A. C., Marioni, R. E., Pilling, L. C., Reynolds, L. M., Mandaviya, P. R., et al. (2016). Epigenetic signatures of cigarette smoking. *Circulation. Cardiovascular Genetics, 9*, 436.

Juarez, P. D., Matthews-Juarez, P., Hood, D. B., Im, W., Levine, R. S., Kilbourne, B. J., et al. (2014). The public health exposome: A population-based, exposure science approach to health disparities research. *International Journal of Environmental Research and Public Health, 11*, 12866–12895.

Klepac, P., Locatelli, I., Korošec, S., Künzli, N., & Kukec, A. (2018). Ambient air pollution and pregnancy outcomes: A comprehensive review and identification of environmental public health challenges. *Environmental Research, 167*, 144–159.

Krauskopf, J., van Veldhoven, K., Chadeau-Hyam, M., Vermeulen, R., Carrasco-Turigas, G., Nieuwenhuijsen, M., et al. (2019). Short-term exposure to traffic-related air pollution reveals a compound-specific circulating miRNA profile indicating multiple disease risks. *Environment International, 128*, 193–200.

Lai, Y., Xue, J., Liu, C.-W., Gao, B., Chi, L., Tu, P., et al. (2019). Serum metabolomics identifies altered bioenergetics, signaling cascades in parallel with exposome markers in Crohn's disease. *Molecules, 24*. https://doi.org/10.3390/molecules24030449.

Lau, C.-H. E., Siskos, A. P., Maitre, L., Robinson, O., Athersuch, T. J., Want, E. J., et al. (2018). Determinants of the urinary and serum metabolome in children from six European populations. *BMC Medicine, 16*, 202.

Lewis, S. L., & Maslin, M. A. (2015). Defining the anthropocene. *Nature, 519*, 171–180.

Lioy, P. J., & Rappaport, S. M. (2011). Exposure science and the exposome: An opportunity for coherence in the environmental health sciences. *Environmental Health Perspectives, 119*, A466–A467.

López-Otín, C., Blasco, M. A., Partridge, L., Serrano, M., & Kroemer, G. (2013). The hallmarks of aging. *Cell, 153*, 1194–1217.

Maitre, L., de Bont, J., Casas, M., Robinson, O., Aasvang, G. M., Agier, L., et al. (2018). Human early life exposome (HELIX) study: A European population-based exposome cohort. *BMJ Open, 8*, e021311.

Maitre, L., Fthenou, E., Athersuch, T., Coen, M., Toledano, M. B., Holmes, E., et al. (2014). Urinary metabolic profiles in early pregnancy are associated with preterm birth and fetal growth restriction in the Rhea mother-child cohort study. *BMC Medicine, 12*, 110.

Maitre, L., Lau, C.-H. E., Vizcaino, E., Robinson, O., Casas, M., Siskos, A. P., et al. (2017). Assessment of metabolic phenotypic variability in children's urine using 1H NMR spectroscopy. *Scientific Reports*, 7, 46082.

Maitre, L., Robinson, O., Martinez, D., Toledano, M. B., Ibarluzea, J., Marina, L. S., et al. (2018). Urine metabolic signatures of multiple environmental pollutants in pregnant women: An exposome approach. *Environmental Science & Technology*, *52*, 13469–13480.

Manrai, A. K., Cui, Y., Bushel, P. R., Hall, M., Karakitsios, S., Mattingly, C. J., et al. (2017). Informatics and data analytics to support exposome-based discovery for public health. *Annual Review of Public Health*, *38*, 279–294.

Miller, G. W., & Jones, D. P. (2014). The nature of nurture: Refining the definition of the exposome. *Toxicological Sciences*, *137*, 1–2.

Miri, M., Nazarzadeh, M., Alahabadi, A., Ehrampoush, M. H., Rad, A., Lotfi, M. H., et al. (2019). Air pollution and telomere length in adults: A systematic review and meta-analysis of observational studies. *Environmental Pollution*, *244*, 636–647.

Monastersky, R. (2015). Anthropocene: The human age. *Nature*, *519*, 144–147.

Neveu, V., Moussy, A., Rouaix, H., Wedekind, R., Pon, A., Knox, C., et al. (2017). Exposome-Explorer: A manually-curated database on biomarkers of exposure to dietary and environmental factors. *Nucleic Acids Research*, *45*, D979–D984.

Newby, D. E., Mannucci, P. M., Tell, G. S., Baccarelli, A. A., Brook, R. D., Donaldson, K., et al. (2015). Expert position paper on air pollution and cardiovascular disease. *European Heart Journal*, *36*, 83–93.

Nicholson, J. K., Holmes, E., Kinross, J. M., Darzi, A. W., Takats, Z., & Lindon, J. C. (2012). Metabolic phenotyping in clinical and surgical environments. *Nature*, *491*, 384–392.

Patel, C. J. (2016). Analytical complexity in detection of gene variant-by-environment exposure interactions in high-throughput genomic and exposomic research. *Current Environmental Health Reports*, *3*, 64–72.

Patel, C. J. (2017). Analytic complexity and challenges in identifying mixtures of exposures associated with phenotypes in the exposome era. *Current Epidemiology Reports*, *4*, 22–30.

Patel, C. J., Bhattacharya, J., & Butte, A. J. (2010). An environment-wide association study (EWAS) on type 2 diabetes mellitus. *PLoS One*, *5*, e10746.

Patel, C. J., Cullen, M. R., Ioannidis, J. P. A., & Butte, A. J. (2012). Systematic evaluation of environmental factors: Persistent pollutants and nutrients correlated with serum lipid levels. *International Journal of Epidemiology*, *41*, 828–843.

Patel, C. J., & Manrai, A. K. (2015). Development of exposome correlation globes to map out environment-wide associations. *Pacific Symposium on Biocomputing*, 231–242.

Patel, C. J., Manrai, A. K., Corona, E., & Kohane, I. S. (2017). Systematic correlation of environmental exposure and physiological and self-reported behaviour factors with leukocyte telomere length. *International Journal of Epidemiology*, *46*, 44–56.

Patel, C. J., Rehkopf, D. H., Leppert, J. T., Bortz, W. M., Cullen, M. R., Chertow, G. M., et al. (2013). Systematic evaluation of environmental and behavioural factors associated with all-cause mortality in the United States national health and nutrition examination survey. *International Journal of Epidemiology*, *42*, 1795–1810.

Patel, C. J., Yang, T., Hu, Z., Wen, Q., Sung, J., El-Sayed, Y. Y., et al. (2014). Investigation of maternal environmental exposures in association with self-reported preterm birth. *Reproductive Toxicology*, *45*, 1–7.

Pino, A., Chiarotti, F., Calamandrei, G., Gotti, A., Karakitsios, S., Handakas, E., et al. (2017). Human biomonitoring data analysis for metals in an Italian adolescents cohort: An exposome approach. *Environmental Research*, *159*, 344–354.

Puett, R. C., Hart, J. E., Yanosky, J. D., Spiegelman, D., Wang, M., Fisher, J. A., et al. (2014). Particulate matter air pollution exposure, distance to road, and incident lung cancer in the nurses' health study cohort. *Environmental Health Perspectives*, *122*, 926–932.

Raaschou-Nielsen, O., Andersen, Z. J., Beelen, R., Samoli, E., Stafoggia, M., Weinmayr, G., et al. (2013). Air pollution and lung cancer incidence in 17 European cohorts: Prospective analyses from the European Study of Cohorts for Air Pollution Effects (ESCAPE). *The Lancet Oncology*, *14*, 813–822.

Raaschou-Nielsen, O., Beelen, R., Wang, M., Hoek, G., Andersen, Z. J., Hoffmann, B., et al. (2016). Particulate matter air pollution components and risk for lung cancer. *Environment International*, *87*, 66–73.

Rappaport, S. M. (2011). Implications of the exposome for exposure science. *Journal of Exposure Science & Environmental Epidemiology*, *21*, 5–9.

Rappaport, S. M. (2012). Biomarkers intersect with the exposome. *Biomarkers*, *17*, 483–489.

Rappaport, S. M., Barupal, D. K., Wishart, D., Vineis, P., & Scalbert, A. (2014). The blood exposome and its role in discovering causes of disease. *Environmental Health Perspectives*, *122*, 769–774.

Rappaport, S. M., & Smith, M. T. (2010). Environment and disease risks. *Science (New York, N.Y.)*, *330*, 460.

Sanyal, S., Rochereau, T., Maesano, C. N., Com-Ruelle, L., & Annesi-Maesano, I. (2018). Long-term effect of outdoor air pollution on mortality and morbidity: A 12-year follow-up study for metropolitan France. *International Journal of Environmental Research and Public Health*, *15*. https://doi.org/10.3390/ijerph15112487.

Schisler, J. C., Ronnebaum, S. M., Madden, M., Channell, M., Campen, M., & Willis, M. S. (2015). Endothelial inflammatory transcriptional responses to an altered plasma exposome following inhalation of diesel emissions. *Inhalation Toxicology*, *27*, 272–280.

Shah, P. S., & Balkhair, T. (2011). Knowledge synthesis group on determinants of preterm/LBW births, air pollution and birth outcomes: A systematic review. *Environment International*, *37*, 498–516.

Sinharay, R., Gong, J., Barratt, B., Ohman-Strickland, P., Ernst, S., Kelly, F. J., et al. (2018). Respiratory and cardiovascular responses to walking down a traffic-polluted road compared with walking in a traffic-free area in participants aged 60 years and older with chronic lung or heart disease and age-matched healthy controls: A randomised, crossover study. *The Lancet*, *391*, 339–349.

Smith, M. T., de la Rosa, R., & Daniels, S. I. (2015). Using exposomics to assess cumulative risks and promote health. *Environmental and Molecular Mutagenesis*, *56*, 715–723.

Snyder, E. G., Watkins, T. H., Solomon, P. A., Thoma, E. D., Williams, R. W., Hagler, G. S. W., et al. (2013). The changing paradigm of air pollution monitoring. *Environmental Science & Technology*, *47*, 11369–11377.

Steckling, N., Gotti, A., Bose-O'Reilly, S., Chapizanis, D., Costopoulou, D., De Vocht, F., et al. (2018). Biomarkers of exposure in environment-wide association studies—Opportunities to decode the exposome using human biomonitoring data. *Environmental Research*, *164*, 597–624.

Steenhof, M., Janssen, N. A. H., Strak, M., Hoek, G., Gosens, I., Mudway, I. S., et al. (2014). Air pollution exposure affects circulating white blood cell counts in healthy subjects: The role of particle composition, oxidative potential and gaseous pollutants—The RAPTES project. *Inhalation Toxicology*, *26*, 141–165.

Taylor, K. W., Joubert, B. R., Braun, J. M., Dilworth, C., Gennings, C., Hauser, R., et al. (2016). Statistical approaches for assessing health effects of environmental chemical mixtures in epidemiology: Lessons from an innovative workshop. *Environmental Health Perspectives*, *124*, A227–A229.

Tomczak, A., Miller, A. B., Weichenthal, S. A., To, T., Wall, C., van Donkelaar, A., et al. (2016). Long-term exposure to fine particulate matter air pollution and the risk of lung cancer among participants of the Canadian National Breast Screening Study. *International Journal of Cancer*, *139*, 1958–1966.

Uzoigwe, J. C., Prum, T., Bresnahan, E., & Garelnabi, M. (2013). The emerging role of outdoor and indoor air pollution in cardiovascular disease. *North American Journal of Medical Sciences*, *5*, 445.

van Breda, S. G. J., Wilms, L. C., Gaj, S., Jennen, D. G. J., Briedé, J. J., Kleinjans, J. C. S., et al. (2015). The exposome concept in a human nutrigenomics study: Evaluating the impact of exposure to a complex mixture of phytochemicals using transcriptomics signatures. *Mutagenesis*, *30*, 723–731.

van Veldhoven, K., Keski-Rahkonen, P., Barupal, D. K., Villanueva, C. M., Font-Ribera, L., Scalbert, A., et al. (2018). Effects of exposure to water disinfection by-products in a swimming pool: A metabolome-wide association study. *Environment International*, *111*, 60–70.

van Veldhoven, K., Kiss, A., Keski-Rahkonen, P., Robinot, N., Scalbert, A., Cullinan, P., et al. (2019). Impact of short-term traffic-related air pollution on the metabolome—Results from two metabolome-wide experimental studies. *Environment International*, *123*, 124–131.

Villeneuve, P. J., Weichenthal, S. A., Crouse, D., Miller, A. B., To, T., Martin, R. V., et al. (2015). Long-term exposure to fine particulate matter air pollution and mortality among Canadian women. *Epidemiology*, *26*, 536–545.

Vineis, P., Chadeau-Hyam, M., Gmuender, H., Gulliver, J., Herceg, Z., Kleinjans, J., et al. (2017). The exposome in practice: Design of the EXPOsOMICS project. *International Journal of Hygiene and Environmental Health*, *220*, 142–151.

Vineis, P., Illari, P., & Russo, F. (2017). Causality in cancer research: A journey through models in molecular epidemiology and their philosophical interpretation. *Emerging Themes in Epidemiology*, *14*, 7.

Walker, D. I., Lane, K. J., Liu, K., Uppal, K., Patton, A. P., Durant, J. L., et al. (2018). Metabolomic assessment of exposure to near-highway ultrafine particles. *Journal of Exposure Science & Environmental Epidemiology*. https://doi.org/10.1038/s41370-018-0102-5.

Walker, D. I., Uppal, K., Zhang, L., Vermeulen, R., Smith, M., Hu, W., et al. (2016). High-resolution metabolomics of occupational exposure to trichloroethylene. *International Journal of Epidemiology*, *45*, 1517–1527.

Wild, C. P. (2005). Complementing the genome with an "exposome": The outstanding challenge of environmental exposure measurement in molecular epidemiology. *Cancer Epidemiology, Biomarkers & Prevention*, *14*, 1847–1850.

Wild, C. P. (2012). The exposome: From concept to utility. *International Journal of Epidemiology*, *41*, 24–32.

Wolf, K., Stafoggia, M., Cesaroni, G., Andersen, Z., Beelen, R., Galassi, C., et al. (2015). Long-term exposure to particulate matter constituents and the incidence of coronary events in 11 European cohorts. *Epidemiology*, *26*, 565–574.

World Health Organization. (2009). *Global health risks: Mortality and burden of disease attributable to selected major risks*. Geneva: World Health Organization.

Wu, W.-K., Chen, C.-C., Panyod, S., Chen, R.-A., Wu, M.-S., Sheen, L.-Y., et al. (2019). Optimization of fecal sample processing for microbiome study—The journey from bathroom to bench. *Journal of the Formosan Medical Association*, *118*, 545–555.

Yang, Q., Zhao, Y., Qiu, X., Zhang, C., Li, R., & Qiao, J. (2015). Association of serum levels of typical organic pollutants with polycystic ovary syndrome (PCOS): A case-control study. *Human Reproduction*, *30*, 1964–1973.

Yin, P., Brauer, M., Cohen, A., Burnett, R. T., Liu, J., Liu, Y., et al. (2017). Long-term fine particulate matter exposure and nonaccidental and cause-specific mortality in a large national cohort of Chinese men. *Environmental Health Perspectives*, *125*, 117002.

CHAPTER 12

Qualitative health impact assessment

Jennifer S. Mindell[a] **and Martin Birley**[b]
[a]Research Department of Epidemiology & Public Health, UCL (University College London), London, United Kingdom
[b]BirleyHIA, London, United Kingdom

Abbreviations

CO_2	Carbon dioxide
DMRB	Design Manual for Roads and Bridges
EC	European Commission
EHIA	environmental health impact assessment
EIA	environmental impact assessment
ESHIA	environment social and health impact assessment
HIA	health impact assessment
HiAP	health in all policies
HNA	health needs assessment
IAIA	International Association for Impact Assessment
IIA	integrated impact assessment
NO_x	nitrous oxides
SEA	strategic environmental assessment
SIA	social impact assessment
UK	United Kingdom
WHO	World Health Organization

Introduction

Health impact assessment (HIA) is a systematic approach to assessing potential health impacts—both beneficial and adverse—of a proposal, in order to make recommendations to decision-makers to mitigate or prevent any negative impacts and enhance the positive impacts, while also reducing any health inequalities that may ensue. Many summary journal articles provide a short introduction (Huang, 2012; Joffe & Mindell, 2005; Lock, 2000; Scott-Samuel, 1998), while books are available for those seeking more detailed descriptions and discussion of different aspects of HIA (Asian Development Bank, 2018; Birley, 2011; Kemm, Parry, & Palmer, 2004; Scott-Samuel, Ardern, & Birley, 2001).

Traffic-Related Air Pollution
https://doi.org/10.1016/B978-0-12-818122-5.00012-0

This chapter discusses the development of HIA over the past 30 years; the underlying paradigms of health and inequalities; the key stages in HIA; and some additional benefits of and problems with HIA. It ends with a summary of ongoing controversies within the field of HIA. Advances in HIA in the past decade include improved quantification, understanding of the planning process, improved analytic tools, risk perception, consideration of cumulative impacts, and consequences of mainstreaming HIA on the global stage. Quantification is not addressed in this chapter as it is described in detail in the following chapter.

What is health impact assessment?

Definition

The World Health Organization (WHO) and International Association for Impact Assessment (IAIA) define Health Impact Assessment (HIA) as a combination of procedures, methods, and tools that systematically judges the potential effects of a policy, plan, program or project, upon the health of a population; and the distribution of those effects within the population. HIA identifies appropriate actions to manage those effects (Quigley et al., 2006).

The main values underpinning HIA are promoting health (in a broad sense); democracy; equity; sustainability; equality of all stakeholders, especially the affected communities; and the ethical use of evidence. A proposal is sustainable, as used here, if it safeguards human health.

Brief history of HIA

There are many different origins of HIA. One of them was as an extension to project EIAs. For example, assessing before the implementation of how the design of tropical irrigation systems could affect environmental determinants of health associated with malaria allowed mitigation measures to be included (Birley, 1991).

HIA was first recommended at a policy level in British Columbia, in 1991 (National Collaborating Centre for Healthy Public Policy, n.d.; Mindell & Joffe, 2003). The application of HIA to a broader set of project types and then, around the turn of the millennium, to policy proposals, coincided with a shift in perspective to a social view of health, shaped by the circumstances in which people lived. These circumstances are generally referred to as the social determinants of health (Robert Wood Johnson Foundation et al., 2010; Wilkinson & Marmot, 2003).

The other major change was a move from focusing on assessing and mitigating potential adverse impacts on health to also considering what beneficial impacts a project or policy could have and how that could be amplified, as well as how inequalities could be reduced. HIA did not appear in the United States until the early 2000s (Dannenberg et al., 2006). In the United States, Social Impact Assessment (SIA) was conducted, possibly delaying the focus on health (Mindell & Joffe, 2003).

The evolution of HIA is discussed here in terms of policy, procedure, and method. The policy provides some of the drivers, explaining why an entity would commission an HIA. Procedure or process provides the framework: the distinct steps required to initiate and complete an HIA, such as screening and scoping. The method provides the analytic tool for justifying statements that there is a likelihood of a health condition changing significantly as a result of a policy, strategy, program, plan, or project (collectively referred to here as proposals).

HIA is not a single entity to be conducted in a single way but involves a range of approaches. Project-level HIAs are often conducted as part of an Environmental Impact Assessment (EIA). In some cases, these are referred to as Environmental Health Impact Assessments (EHIA), or Environmental, Social, and Health Impact Assessments (ESHIA). At the simplest, a health chapter is added to an EIA. EIAs are required by law in most countries in the world for some categories of proposal. Although there are successful examples of integrating health into EIA, most EIAs that have been studied did not include health in any comprehensive manner, even where an assessment of health impacts are required (Bhatia & Wernham, 2008; Birley, 2011). By contrast, HIAs are mandatory in a few countries (Asian Development Bank, 2018; National Health Commission Office, 2016). Within Europe, HIA is legally required as part of a Strategic Environment Assessment (SEA) and the revised EC Directive on EIA (EC, 2014).

Drivers

Several distinct drivers promote the use of HIA, including the following.

Health sector requirements

In 1997, the WHO Jakarta Declaration recommended equity-focused HIA when developing both private and public sector policies (WHO, 1997). By 2006, HIA was highlighted in the 2006 Bangkok Charter on Health Promotion in a Globalized World as a key decision-making aid (Bos, 2006).

Human rights legislation

There are many links between health and human rights (United Nations, n.d.). Some relate directly to the Human Right to Health, others link health and human rights through health determinants, such as the rights to an adequate standard of living, which include the human rights to safe drinking water and sanitation, food, housing, and clothing.

The Universal Declaration on Human Rights has two covenants: the Covenant on Political and Civil Rights, and the Covenant on Economic, Social, and Cultural Rights (United Nations, n.d.). The second covenant includes the Right to Health. The covenants are intergovernmental; the second Covenant has been signed and ratified by many countries.

Under the UN Guiding Principles on Business and Human Rights, governments, multinational corporations, and financial institutions have the responsibility to promote respect for human rights (United Nations, 2011). Several of the human rights criteria overlap with the indicators to measure progress toward achieving the Sustainable Development Goals (UNDP, n.d.). HIA is part of the due diligence process for ensuring that the Right to Health has been protected.

Business case

Many institutions undertake some form of HIA of their new proposals in order to protect their reputation and their social license to operate. Examples include the extractive sectors—oil and gas and mining and minerals. These sectors have trade associations which have issued guidance on HIA (ICMM, n.d.; IPIECA, 2016).

Lenders' requirements

The multilateral development institutions, the World Banks, have established performance standards. Under these standards, they should lend money only for proposals that safeguard human health (IFC, 2012). Most of the major commercial banks in the world have adopted similar standards under the Equator Principles (Equator Principles, n.d.).

The revised EC Directive on EIA

In the European Union, there has long been a legally binding Directive that determines the minimum conditions under which an EIA is necessary. During 2014, the Directive was revised (EC, 2014). The revision included an explicit requirement that health must be safeguarded. The revised directive was transcribed into English law during 2017.

The enthusiasm of charismatic leaders

Even where there is no legal requirement for HIA, there has often been a demand promoted by charismatic leaders with a personal interest in safeguarding human health. These leaders arise at every level from the local authority to the multilateral institution. We have observed that when these leaders are promoted or change their roles, the demand may wane again as it has not been embodied in the institution.

Determinants of health

Although people's health depends partly on fixed or unmodifiable aspects such as their genes, age, and sex, their physical and mental health and well-being are affected predominantly by their income, education, occupation, housing, transport modes, opportunities, social connectedness, the built, and fiscal environment in which they live, and the natural environment. These affect their lifestyle and the choices actually available to them, and thus their exposures. For example, a change in air quality is an environmental determinant of health.

The HIA method is based on a consideration of changes in the determinants of health (Fig. 12.1). Inequalities in these determinants are the underlying causes of most inequalities in health (Marmot, 2008).

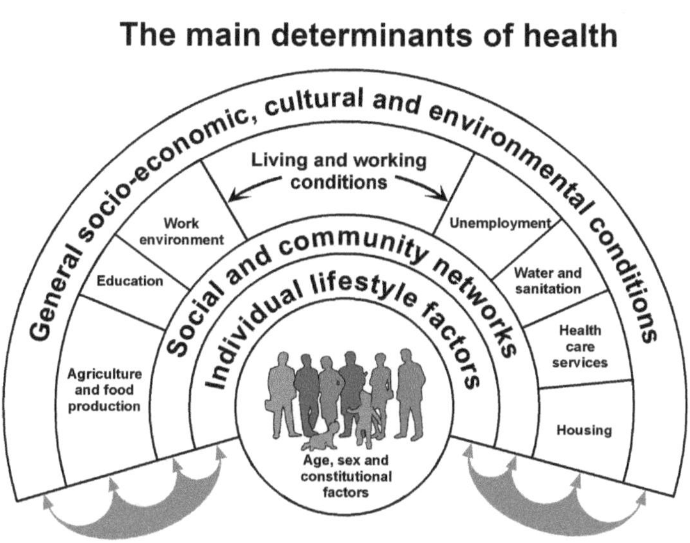

Fig. 12.1 Determinants of health and well-being (Dahlgren & Whitehead, 1993).

Types

Many terms are used in HIA, which have been described in a glossary of HIA (Mindell, Ison, & Joffe, 2003). Several categorizations have been used, for example, differentiating between rapid and comprehensive HIA, prospective and retrospective, participatory and "expert," and HIAs of projects or policies. Harris-Roxas and Harris described a typology of four forms of HIA: mandated, decision-support, advocacy, and community-led (Harris-Roxas & Harris, 2011). Whether a simple dichotomy or the four categories of HIA, the different forms of HIA aim to achieve different purposes and enable HIA to be used to address a wide range of issues, circumstances, and concerns.

HIA is intended to be a holistic tool. However, several specialist HIA subtypes have evolved to meet specific needs. Examples include mental well-being impact assessment (Cooke et al., 2011) and health inequalities impact assessment (Heller et al., 2014; Simpson, Mahoney, Harris, Aldrich, & Stewart-Williams, 2005).

Techniques used in HIA

By 2008, there were many different tools or guides to conducting general HIA, as well as guidance on specific elements of an HIA or guidance restricted to a single topic (Mindell, Boltong, & Forde, 2008). Since then, the World Health Organization has developed an online repository that includes a list of short introductory guides from around the world (WHO, n.d.-a) and more detailed toolkits and guides. Tools advising local public health teams on when and how to conduct an HIA have proliferated (e.g., Cave, Fothergill, Gibson, & Pyper, 2017). Tools have also been produced internationally by financial institutions and trade organizations (ICMM, n.d.; Asian Development Bank, 2018; IFC, 2009; IPIECA, 2016).

Many techniques used in HIA, such as community consultation and engagement, or community profiling, will be familiar to public health practitioners and others from similar community health approaches. However, HIA of a proposal differs in purpose, starting point, and relationship to interventions.

The HIA procedure

Although many different tools and sets of guidelines exist (Mindell et al., 2008), almost all incorporate the same procedural steps: screening, scoping, analysis, formulation of recommendations, and submitting recommendations to decision-makers (Mindell et al., 2003). Ideally, evaluation and

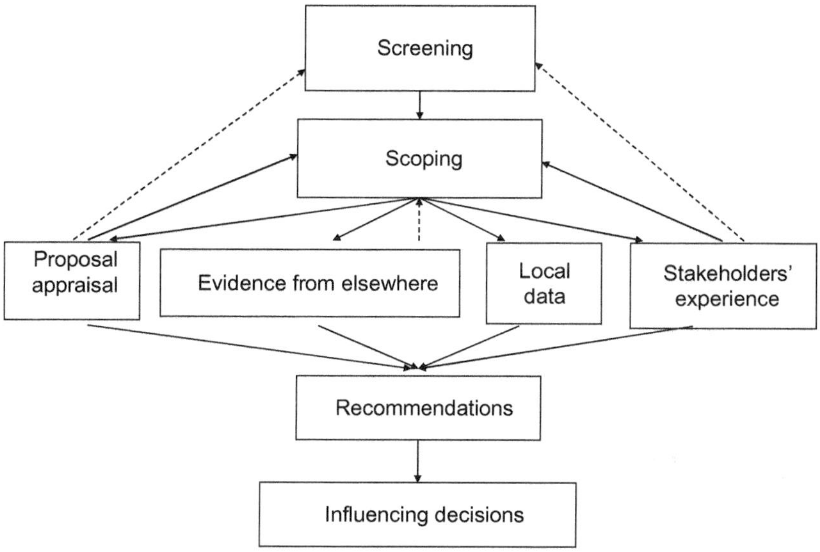

Fig. 12.2 The stages of health impact assessment. *(Reprinted with permission from Mindell, J., Boaz, A., Joffe, M., Curtis, S., Birley, M. (2004). Enhancing the evidence base for health impact assessment. Journal of Epidemiology & Community Health 58, 546–551. https://doi.org/10.1136/jech.2003.012401.)*

monitoring are final steps, but these are seldom undertaken, for reasons described in the following (Fig. 12.2).

Screening

The initial stage is usually screening, to determine whether a proposal merits an HIA. The two criteria generally used for screening are:

1. whether the proposal is likely to affect the determinants of health,
2. whether there is likely to be an opportunity to make recommendations that could change the proposal.

An HIA may still be conducted even where changes are not possible, for political or timing reasons, if it is considered that adverse health consequences might affect health care needs and therefore planning for enhanced provision. For example, an influx of construction workers can overwhelm rural health care facilities. An HIA can recommend additional provisions.

Scoping

Scoping sets a boundary about what is to be included and not included in the HIA. Some components follow.

The determinants of health and/or the health impacts

The scope should include both the social and environmental determinants of health. Some specific determinants may be identified at this stage, such as air pollution. Others will emerge later during the analysis stage.

The geographical boundary

The scope will define the distance from the geographical focus of the proposal that should be included, e.g., the airshed.[a]

The timeframe(s)

There may be short-, mid-, or long-term impacts. For example, the health impacts of infrastructure development will usually vary between the construction, operation, and decommissioning stages. A generally pro-health development may cause adverse impacts during the development phase as a result of noise and air pollution, and injury risk from heavy vehicle traffic.

The population(s) to be considered

The populations, or communities, affected by a proposal include those who currently inhabit the geographical scope and of those who are likely to migrate there. They are usually disaggregated by age, gender, ethnicity, income, health status, or other categories of concern for health inequalities or the unequal impact of the proposal.

Excluding other assessments

The scope usually excludes the occupational health and safety assessment of project associated workforces as this is included in other management processes, such as Health Risk Assessment. In addition, the HIA is not a replacement for Health Needs Assessment (HNA). HNA focuses on the existing health needs of the community or population independently of the proposal.

There are usually complex overlaps between the scope of an HIA and EIA or SIA. These assessments are the responsibility of other specialists and a pragmatic approach is often required for deciding who does what.

Analysis

This is the most complicated and time-consuming stage of an HIA. While rapid assessments tend to focus primarily on published scientific literature (referred to below as "evidence from elsewhere"), amplified by some local

[a] An airshed is a geographical area within which the air frequently is confined or channeled, with all parts of the area thus being subject to similar conditions of air pollution.

data, a comprehensive HIA would include and give weight to the published evidence, a more detailed analysis of local data, and the experiences of local stakeholders, both professionals, and community members.

There are several ways in which stakeholders' experiences can be obtained, including interviews with key individuals, purposive focus groups, and broader meetings. If the last approach is used, the meeting will often begin with short presentations summarizing the proposal and its appraisal, the published literature, and the relevant local data before participants are invited to move to small groups to discuss specific, pre-identified topics or express their views more broadly.

Proposal content

A good understanding of the proposal is required. This can include the study of planning documents and maps, as well as a visit to the project site and surrounds (if there is one). Familiarity with the sector is an advantage. For example, this might require the experience of upstream and downstream processes in the oil sector or knowledge of modal shift strategies in the road transport sector.

In the HIAs of policy, the "environment" is used to refer to a variety of factors that influence people's opportunities and decisions for health. It refers to physical exposures, such as air pollution, but also to the fiscal, legislative, social, and cultural environments in which people live and make their decisions (Huang, 2012). For HIAs of projects, it is generally used to refer to the physical (built and natural) environment.

Evidence from elsewhere

Many specific difficulties have been identified relating to reviewing published evidence for use in an HIA, including whether there is sufficient evidence to consider relationships causal (Joffe & Mindell, 2002); limited time for the review; a diverse evidence base with widely differing study methods, and the limited evidence on the reversibility[b] of health impacts of adverse exposures and unequal effects; the wide range of stakeholders, the need for definite recommendations; and lack of capacity in practitioners in undertaking literature searches, critical appraisal, and synthesis of findings

[b] *Reversibility*: Even where there is good evidence that A causes B, there may not be any information on whether a reduction in B results in a reduction in A, i.e., whether the effects of the exposure are *reversible*. For example, stopping smoking stops further deterioration of respiratory function due to smoking but does not reverse the existing damage to improve lung function to what it would have been in a nonsmoker of that age.

(Mindell, Boaz, Joffe, Curtis, & Birley, 2004). Even where there is good quality evidence, not all studies are transferable to the context of the particular proposal being assessed.

A multidisciplinary group of researchers and HIA practitioners developed and tested a "Guide to reviewing evidence for use in health impact assessment" (Mindell et al., 2006). It guides conducting high-quality reviews and gives minimum standards that should be observed when resources do not permit more.

Local data

Local data will generally be "routine" data, i.e., data that are collected routinely for administrative or other purposes. Such data may not be accurate. Demographic data is almost always available, such as the number or proportion of the population in each age- and sex group. It may include data from a Census (e.g., education levels, income, occupation, and general health); mortality data (ideally including age, sex, and major causes); and morbidity data (information about poor health, including hospital admissions, primary care visits, or other health care use).

In some cases, there may be bespoke data available, from a health survey, for example. Regional or even national data can still be useful, particularly when interpreted with local knowledge. For an HIA of regional or national policy, the appropriate level of data is likely to be available, if it exists. However, unless a local level has been undertaken, it is unlikely that data from a national survey could be released for a local HIA because of data confidentiality or sample size issues.

In most cases, this would be health interview data but there may be health examination data available. This has the advantage of identifying undiagnosed disease (such as raised blood pressure or diabetes), which is not obtained from routine health care data (Mindell et al., 2017; Tolonen et al., 2018). Occasionally, time and funding might permit the collection of health-related data specifically for the HIA, but this is uncommon and subject to the rules of medical ethics.

Stakeholders' experience

Community engagement

For an HIA to be comprehensive, the inclusion of members of the affected population(s) is crucial. For local policies or projects, the population groups and subgroups likely to be affected (whether positively or negatively) would have been identified during the scoping stage. For national policies, leaders of organizations representing groups that may be affected would generally be included. For regional policies, representatives from community groups or organizations would normally be invited but it is sometimes possible to identify local communities to include. The extent of intended or actual stakeholder engagement varies widely, depending on the timescale of the HIA and the commitment of those running or commissioning the HIA.

Perceptions of risk

Communities who are the involuntary recipients of a proposal often have a different perception of the health risks from that of the proponents and the scientific community. Some members of the scientific community may believe that their perception of risk, based on sound epidemiological data, modeling, and understanding of causality, is superior. They may believe that the community perception of risk can be changed by providing more information. In the social sciences, this is often referred to as the information deficit model and it is rarely successful.

Community perception of risk is influenced by many factors, including social media. There is both social amplification and social attenuation of risk perceptions. These perceptions not only give rise to anxiety and stress, they also create obstacles to project delivery through anger, demonstrations, and legal challenges. Risk perception is a well-developed field of sociology, psychology, economics, and anthropology (Slovic, 2010; Boholm, 2015; Gigerenzer, 2015; Kahneman, 2012). The following list summarizes some of the conclusions:

- A formal process of information, education, and communication associated with a project is unlikely to change the risk perceptions of a community.
- Risk perception is subjective and based on earlier experiences and trusted informants.
- Perceived risks can give rise to anxiety and anger. These can lead to changes in mental health and well-being.
- There will always be substantial differences in the risk perceptions made by project proponents and by some of the project-affected communities. The proponents are voluntary risk takers; most of the project-affected communities are involuntary risk receivers. Many people accept high voluntary risks but reject low involuntary risks.
- Whether a project is perceived as a hazard or an opportunity depends on the observer. Individuals who believe themselves powerful and in control of their destiny rank hazards as being of lower importance. Individuals

who believe themselves to be underprivileged or powerless rank those same hazards much higher.

- We all evaluate risks using criteria such as: voluntary or involuntary; affecting children or affecting adults; familiar or unfamiliar; risk of killing people one-by-one or in groups.

Professional stakeholders

Professional stakeholders are often referred to as key informants and their knowledge and experience are crucial. A list of key informants is usually identified, and they are interviewed either face-to-face or by telephone. The objective of the interview is to elicit their concerns about how the proposal could affect human health, identify unpublished sources of data, identify additional key informants, and learn from them about actions that they would recommend.

Prioritization

The conclusion of the analysis is usually a set of positive and negative health concerns that are ranked according to their likelihood and severity and these are combined in a Risk Assessment Matrix. The ranking process usually draws on group discussions and preferences. It is a judgment that draws on the evidence that has been obtained. Potential health impacts that have high likelihood and high severity are judged to be of high significance and recommendations will be required for managing them.

Formulating evidence-based recommendations

Recommendations should be evidence-based and should bring together different information obtained during the assessment stage (WHO, n.d.-b). Although people may value and interpret the various types of evidence differently, if the recommendations, and the evidence underpinning them, is provided in a clear manner, the decision-makers can judge this for themselves without those conducting the HIA making their value judgments. Recommendations are made for mitigating significant negative health impacts and enhancing significant health benefits.

Mitigation

Mitigation measures are arranged in a hierarchy (Table 12.1). At the top of the hierarchy are actions taken by the project proponent to safeguard the health and safety of the community through healthy engineering design and social measures. Lower down in the hierarchy are changes in behavior

Table 12.1 The health mitigation hierarchy with some examples.

Hierarchy	How	Examples concerning air pollution
Avoid	Design out by healthy engineering design and social measures	Encourage modal shift from roads to rail
Abate at site	Add to design by healthy engineering design and social measures	Site new major highways away from residential areas and schools
Abate at receptor	Personal protection	Supply gas masks to population
Repair	Medical treatment	Rapid diagnosis, surveillance, and effective treatment
Compensate	Finance and insurance	Provide funding to the medical sector to cope with additional medical demands

required of the community as individuals. The dominant principle is that people should not be required to change their behavior until all higher measures in the hierarchy have been implemented. At the bottom are the consequences of a failure to safeguard. These often include medical treatment. This cost of treatment may be transferred from the project to the health sector. The health sector should be forewarned that this may occur so that it can work with the proposal for any additional medical services that may be required. Whatever mitigation measures are chosen should be sustainable.

Enhancement

Enhancement is about maximizing the health benefits of the project and identifying opportunities for improving the health of affected communities. For example:

- using social investment to further improve the quality of life of affected communities or to support existing proposals by the public sector;
- providing vocational training to increase the employability of local workers;
- improving access to social services;
- promoting secondary industry;
- improving air quality;
- improving housing infrastructure.

There is no universal health benefit hierarchy equivalent to the mitigation hierarchy. A suggested hierarchy is presented in Table 12.2 (Asian Development Bank, 2018).

Table 12.2 The health benefit hierarchy with some examples.

Hierarchy	Examples concerning air pollution
Permanent modifications	• Benefits that last the length of the project, such as roads with space for nonmotorized users • Parks and recreational facilities included in urban housing projects
Enhancing equity	• Ensuring that disadvantaged groups benefit, such as reducing transport-related air pollution and therefore exposure • Creating healthy choices, such as making public spaces attractive to users
Repeated actions	• Maintenance and repair of project structures
Promoting healthy behavior	• Multifaceted actions to make active travel an easier choice
Enhancing medical care	• Sustainable improvements to public clinics, that continue to function when project-related finance is removed

Submitting recommendations

Care needs to be taken over how the recommendations are submitted to the decision-makers (Elliott & Francis, 2005). Decisions are needed not only on how detailed or brief the recommendations and documentation are but also the format in which they are presented. This is something the steering group for the HIA should start to consider early on, in the scoping stage.

Throughout an HIA, it should be accepted by all those conducting or participating that the role of an HIA is not to make decisions but to try to influence decisions. It should always be borne in mind that health (and health inequalities) is only one area of outcomes that decision-makers may consider. Other issues—political, financial, legislative, or other—may appear to them to be more urgent and/or important.

Evaluation

Evaluation can take three forms: process, impact, and outcome. The easiest form is process evaluation, which assesses how the HIA was conducted, providing both quality assurance for the HIA and as a source of learning for the future. Impact evaluation examines what happened in the short term, namely the response of the decision-makers to the HIA recommendations and whether (and if so, how) any adjustment was made to the proposal. Outcome evaluation assesses what happened to health outcomes and intermediate indicators of current or future health after the implementation of the proposal.

There are three main problems with evaluation. In most cases, the limited resources (whether people's time or other resources) cease as soon as the recommendations are submitted so evaluation is seldom undertaken (Waheed et al., 2018). Additionally, a "successful" HIA may have influenced the final proposal that was implemented, such that the adverse outcomes identified in the assessment did not occur because of changes to the proposal. This is often referred to as the counterfactual argument. Finally, even if changes are seen between the versions of the proposal before and after the HIA was conducted, attributing causality can be problematic. In the case of the London Mayor's initial Transport Strategy, both the Mayor and the senior transport official mentioned in a public forum that the strategy was amended because of the HIA (Mindell, Sheridan, Joffe, Samson-Barry, & Atkinson, 2004), but this information is seldom available.

Monitoring

As with evaluation, a lack of resources often limits the amount of monitoring undertaken. It is good practice to enable communities to take control of the risks that affect them. This includes consulting on and involving communities in development proposals as well as involving them by providing monitoring reports; having independent monitoring by trusted third parties; or enabling communities to monitor risk and impacts themselves by providing them with equipment.

Engagement

An HIA should generally be multidisciplinary and involve people from a range of organizations, as well as those potentially affected by the proposal. The lead person may be a public health practitioner, an independent HIA practitioner, or someone skilled in community engagement. Even if led by a single organization, a steering group of key stakeholders from a range of organizations and local community groups is recommended. Regardless of the level of the proposal and HIA, involving decision-makers in the HIA process is likely to enhance the effectiveness of the recommendation-making process and the extent to which the HIA recommendations influence the decisions.

Strengths and limitations of HIAs and their use in decision-making

Strengths of HIA

A major strength of a well-conducted HIA is the multidisciplinary approach, including working across organizations. This not only strengthens the outputs from the HIA itself but also enables HIA to be a mechanism to bring together practitioners from different disciplines and help them appreciate the interrelationship of their areas of expertise, e.g., environment and health (Negev, Levine, Davidovitch, Bhatia, & Mindell, 2012). Involvement in an HIA can be educational, helping practitioners, policy-makers, and civil society to understand the social and environmental determinants of health and the importance of their decisions in influencing the health of the population they serve. Teaching across disciplines can also be facilitated using HIA (Chinchilla & Arcaya, 2017). An HIA can help build consensus.

A comprehensive HIA is also participatory. An HIA systematically enquires about current health and inequalities and potential positive and negative impacts on these, in a holistic way. The health impact of air pollution is simply one concern. HIA can, therefore, formulate evidence-based recommendations to enhance the positive and mitigate or prevent the potential negative impacts.

HIA can be conducted on proposals at any level of intervention or geographical reach. Consideration of equity or inequalities (social justice, namely the distribution of effects) is an integral part of HIA. Another strength of HIA, particularly where a number of these are conducted covering the same geographical area or population(s) is in identifying synergies (and barriers) between different policies (Mindell, Bowen, Herriot, Findlay, & Atkinson, 2010).

Limitations

Limitations of HIA—or more often, barriers to conducting the best HIA—include time pressures and lack of resources (funding, staff time, and staff expertise). Limited evidence (Lock, 2000)—and limited time to search systematically—is often a problem. There is not the luxury of 18 months to conduct a systematic review of the published literature; much relevant information may be in hard-to-access "grey" literature; and there is often lack of evidence of reversibility.[b] Lack of evaluation or follow-up is the norm (Waheed et al., 2018). Finally, there is often misunderstanding by participants of the decision-making process and a lack of acceptance that health is not the only factor driving decisions. HIA is a management, not a scientific, tool. The objective of HIA is not to uncover absolute truths about the universe, or to advance scientific knowledge, but to nudge proposals toward being healthier. HIAs do not generally provide predictions, but they provide assessments.

Differential health impacts of traffic-related air pollution in sensitive subpopulations

It is well-recognized that air pollution affects disproportionately those with preexisting cardiorespiratory disease, the very young, and the frail, as has been described in Chapter 9. More vulnerable groups are generally more likely to be living in poverty or in deprived areas, which are also the areas more likely to be exposed to air pollution. Thus, changes in air pollution as a result of a proposal are likely to change health inequalities. Improvements in air quality would generally reduce health inequalities, while increased emissions would worsen both ill-health and health inequalities.

Examples of HIAs of road transport

A recent systematic review examined transportation HIAs (Waheed et al., 2018). The authors reported that project proponents' EIAs and other assessments were more likely to include those exposures that were potentially quantifiable, such as changes in air pollution, than other impacts, such as changes to access to employment or goods, or social contacts. When examining the 158 HIAs reported in peer-reviewed journals or the grey literature, Waheed et al. found that most of the HIAs had taken a qualitative approach to health impact, and almost all had considered vulnerable populations or equity issues. They identified two key areas. First, the existence and extent of a scoping stage were related to quality. Second, almost all HIAs that included decision-makers had at least some impact on the final proposal; those that did not either had no impact or did not evaluate or report on this.

Road transport projects

The development of road transport infrastructure is often the responsibility of a highways agency. The objective of the agency is generally to build as many highways and bridges as possible, as practitioners and policy-makers tend to focus on moving vehicles not people. They often use economic cost-benefit analysis to decide their priorities. The economic benefit may be derived from the number of minutes that a new highway is expected to reduce journey times. The economic costs include design and construction costs. In England, design considerations are articulated in the Design Manual for Roads and Bridges (DMRB) (Highways Agency, n.d.).

The environmental impacts of the design are assessed directly (EIA), but many of the health impacts are not assessed. Typical chapters in the EIA required by the DMRB are air quality, cultural heritage, landscape, ecology and nature conservation, geology and soils, materials (construction), noise and vibration, effects on all travelers, community and private assets, road drainage and the water environment, traffic and transportation, and lighting. Most of these chapters are determinants of health but the link to health is generally not made. The economic cost of air pollution is sometimes quantified and may include the health cost (see Chapter 18). Mitigation measures reduce the negative environmental impacts, and these are input to the design and construction plan.

There are many checklists available that help to identify the potential health impacts of road transport projects (GLA, 2014; Health Scotland, 2007; THSG, 2000, 2011; WHO, 2000). When these are compiled, they form 14 subsections and over 100 questions. The questions in two contrasting subsections are illustrated in Tables 12.3 and 12.4. The other subsections are active travel and physical activity, community stress, crime, ecology and nature conservation, economic health, other hazards, equality, housing + land-take + resettlement, landscape and lighting, noise and vibration, road drainage and water environment, and safety.

One personal observation (MB) of a highways project was as follows. The project was the dualling of a highway that started about 3 km from a major city, at a major intersection. A segregated cycle lane was included parallel to the highway. The scope of the impact assessment was from the beginning to the end of the highway and some hundreds of meters on either side. The concentration of air pollutants was modeled around the highway as part of the EIA and no exceedances were expected. The HIA observed that the segregated cycle lane finished 3 km from the city. There was no provision for continuing the cycle lane from the end of the highway to the city. There was no provision for enabling cyclists to cross the major intersection safely. The scope of the HIA needed to be greater than the scope of the EIA.

Road transport policies: The London example

The London Mayor committed to having HIAs conducted of each Mayoral strategy. The first was the Transport Strategy. The process, recommendations, and impacts of the HIA have been summarized elsewhere (Negev et al., 2012) and reported in detail (Mindell, Sheridan, et al., 2004). Despite the very short timeframe, brief reviews of the evidence and a half-day workshop for stakeholders across London supported most of the proposals in the

Table 12.3 Checklist questions about access and accessibility in a road transport project.

Does the proposal improve, change, or disrupt access to schools and other local services by active transport?
Does the proposal improve the quality and accessibility of public transport, local services, and facilities?
Does the proposal allow people with mobility problems or a disability to access buildings and places?
Does the proposal provide access to local employment and training opportunities, including temporary construction and permanent "end-use" jobs?
Does the proposal include opportunities for work for local people via local procurement arrangements?
Does the scheme improve access and accessibility to income, employment, housing, education, services, amenities, facilities, and social networks crucial to maintaining a healthy vibrant and cohesive community?
Does the proposal affect access from communities to residential, commercial, and community lands?
Does the proposal identify all the intersecting paths, cycleways, and desire lines and make provision to avoid severance?
Does the proposal connect with existing communities, i.e., layout and movement which avoids physical barriers and severance and land uses and spaces which encourage social interaction?
Does the proposal affect access to green or blue space?
Does the design of pedestrian amenities create a barrier for the elderly or disabled? For example, is there a risk—or perceived risk—of trips, slips, or physical strain?

Table 12.4 Checklist questions about air quality in a road transport project.

Does the project consider the differential exposure of disadvantaged groups to air pollution?
Does the air quality chapter of the EIA summarize the health effects of air pollutants NO_2 and PM_{10}, associated legislation, Air Quality Objectives, and EU Limit Values?
Will the scheme result in nuisance dust during construction?
Will the scheme improve air quality including areas currently covered by Air Quality Management Areas?
Will the scheme result in any exceedances in annual NO_2 and PM_{10} concentrations?
Will the scheme result in any changes to regional emissions?
Will the project reduce the net emissions of Green House Gasses from transport?
Does the Air Quality chapter assess the exposure to air pollution of cyclists and pedestrians adjacent to the road alignment?

Strategy, which focused on reducing private car use and increasing public transport infrastructure and services. The HIA endorsed these and also recommended an increased consideration of promoting walking and cycling. Almost all the recommendations were incorporated into the final Strategy.

Since then two further Mayoral Transport Strategies have been developed, in 2009 and in 2016/2017 (Transport for London, 2009, 2017). Each had an integrated impact assessment (see below) that included consideration of health and health inequalities.

Further developments

Integrated impact assessment

London, one of the first places to institutionalize HIA for policies and strategies, has moved in the past decade toward using integrated impact assessment. This includes elements covering sustainability appraisal, SEA, and equality impact assessment (Mindell et al., 2010). Integrated impact assessments are also recommended in the extractive sector, where they are referred to as ESHIA.

Health in All Policies (HiAP) is based on the same understanding of health as HIA, namely the wide range of factors outside the health care services that determine people's health. It was formalized in the Helsinki Statement on Health in All Policies (WHO, 2013). HiAP aims to be systematic in considering the health impacts of decisions on public policy to avoid harmful impacts and to maximize benefits through synergies between policies. This approach includes a focus on communities and formal collaborations across organizations as key features. HIA is thus a natural tool to use as a key element of HiAP planning (Public Health New Zealand, n.d.).

Ongoing debates regarding HIA

Some HIA practitioners have urged the introduction of a legal requirement for HIA. They point out that this is the only way that HIA will be conducted routinely on the wide range of proposals that would benefit from an HIA. It would also require resources to be made available for the mandatory HIAs—and lack of resources (including staff time) is the biggest obstacle to conducting an HIA. In private sector development and development supported by major lending agencies, these resources are available.

The European Union's Directive on Strategic Environmental Assessment requires consideration of health impacts as part of the SEA. In the United Kingdom, EIA is supposed to include an assessment of the impacts on human health as well as on the environment. Nevertheless, HIA is often missing. Thus, mandating HIA does not necessarily result in improved design or operation of proposals.

The final strand of arguments relates to the capacity and competence of practitioners. While HIA is a limited field, practitioners generally know each other and their skill sets. If there is no qualification nor a requirement for one, there is concern over the validity of HIAs conducted by anyone who chooses to set themselves up as an HIA practitioner.

Many of the large environmental consultancy companies undertake commercial health impact assessment. Unfortunately, their staff rarely have training and experience in public health and related disciplines. The staff is usually bright generalists who can pick up new ideas quickly from reading a textbook. Consequently, the standard of the HIAs is very low and old ideas are reinvented. At the same time, those who commission the HIAs also do not know the appropriate competencies to expect from their consultants. The solution is to introduce a program of certification, as is already commonplace in EIA. Attempts to introduce certification have so far proved unsuccessful. The reasons include insurance, the small size of the profession, lack of agreement on standards, and lack of an institution willing to accept the task.

Cumulative impacts

Cumulative health impacts occur at local, national, and global scales (Asian Development Bank, 2018). Local cumulative health impacts occur when different development projects are colocated. The impacts of one project may be insignificant but the impact of many projects may be significant. For example, a point source of air pollution may be insignificant, but multiple point sources may change an airshed significantly. The management of local cumulative impacts is strengthened by undertaking a Strategic Assessment that includes human health.

Cumulative impacts should be distinguished from "in-combination" impacts. In-combination impacts occur when two or more health determinants work together in a single project, for example, the in-combination impacts of changes in noise and air pollution on cardiovascular disease. The two actings together (in-combination) are likely to be worse than each effect acting singly.

National cumulative impacts are most evident in industrial sectors such as oil extraction (Schrecker, Birn, & Aguilera, 2018). When a country relies on a small number of natural resources for a large part of its income, there may be economic destabilization and a deterioration in community health status. One sector of the economy booms and this causes price inflation that excludes other sectors of the economy and the communities that work in them. The causal pathways are complex and are likely to include the effect of inequality.

Global cumulative impacts are linked to a project's contribution to global changes, such as CO_2 concentrations. Major projects, such as highway networks, that consume fossil fuels contribute to rising CO_2 concentrations. Rising CO_2 concentrations contribute to climate change, which has been termed a public health emergency by the WHO (Watts et al., 2017). Atmospheric CO_2 concentrations have already at a historic high; temperature extremes and other climatic events, such as typhoons and reducing river flows, have already creating health impacts. Food insecurity, changing monsoons, and coastal flooding are expected to affect the health and safety of many millions of people (Asian Development Bank, 2011; Bush et al., 2011; Kumaresan, Narain, & Sathiakumar, 2011; WHO, n.d.-c) The economic costs of climate change are very high. For example, an initiative to mitigate the health impacts of climate change estimated that it had saved US$5.7 billion between 2009 and 2017 (MSCCH, n.d.).

Global cumulative health impacts are hard to mitigate at a project level. An HIA, therefore, refers to the national and international policies and strategies with which the project must comply. It is unlikely that infrastructure can be adapted to withstand the expected effect of food shortages and sea level rises. Projects that reduce national CO_2 emissions are likely to help mitigate climate change and may have positive impacts on health and inequalities within and between countries (Brown & Spickett, 2014). For example, in the urban transport sector, the co-benefits of change are widely acknowledged (Hosking, Mudu, & Dora, 2011; The Lancet Planetary Health, 2017). Fossil fuel consumption not only contributes to climate change but also produces black carbon and other air pollutants that contribute to local health impacts. Policies that promote a modal shift from private car use to active travel not only reduce emissions of CO_2 and other air pollutants but also increase mental and physical health through an increase in physical activity, lower transport congestion, and reduced inhalation of pollutants (Davis, 2014; Gouldson, Sudmant, Khreis, & Papargyropoulou, 2018; Mindell, Cohen, Watkins, & Tyler, 2011).

Summary and conclusions

Qualitative health impact assessment (HIA) emerged in the 1990s and has evolved through the experience of early adopters to create policies, procedures, methods, and tools, together with capacity-building measures. It is defined as "a combination of procedures, methods, and tools by which a policy, program, or project may be judged as to its potential effects on the health of a population, and the distribution of those effects within the population." HIA identifies appropriate actions to manage those effects. Weaknesses of HIA include a lack of good quality, transferable, relevant evidence for some cases; a reluctance to engage stakeholders; a lack of legal requirement for HIAs, a lack of qualified practitioners; poor integration with other impact assessment; and limited resources of time, money, and people. Even where HIA is mandated, this has often been ignored or rudimentary assessments conducted. The main strengths of HIAs are their consideration of broad determinants of health; their use of evidence; involvement of stakeholders; and the more systematic assessment of beneficial and adverse impacts of a proposal on health and health inequalities. Recommendations can thus be presented to decision-makers to enhance positive health impacts and mitigate or prevent negative impacts before proposals are implemented.

Further information

Most of the recent HIAs use a quantitative approach, which is the subject of the following chapter.

Collections of HIA resources

https://www.who.int/hia/en/
https://www.health.govt.nz/our-work/health-impact-assessment/resources-health-impact-assessment
https://www.gov.uk/search?q=health+impact+assessment

References

Asian Development Bank. (2011). *Accounting for health impacts of climate change*. Asian Development Bank.

Asian Development Bank. (2018). *Health impact assessment: A good practice sourcebook*. Asian Development Bank. https://doi.org/10.22617/TIM189515-2.

Bhatia, R., & Wernham, A. (2008). Integrating human health into environmental impact assessment: An unrealized opportunity for environmental health and justice. *Environmental Health Perspectives, 116*, 991–1000. https://doi.org/10.1289/ehp.11132.

Birley, M. (1991). *PEEM guidelines for forecasting the vector-borne disease implications of water resources development (No. 2)*. Panel of Experts on Environmental Management.

Birley, M. (2011). *Health impact assessment, principles and practice*. Earthscan/Routledge.

Boholm, A. (2015). *Anthropology and risk, Earthscan*. Routledge.

Bos, R. (2006). Health impact assessment and health promotion. *Bulletin of the World Health Organization, 84*, 914–915.

Brown, H., & Spickett, J. (2014). Health consequence scales for use in health impact assessments of climate change. *International Journal of Environmental Research and Public Health, 11*, 9607–9620. https://doi.org/10.3390/ijerph110909607.

Bush, K. F., Luber, G., Kotha, S. R., Dhaliwal, R. S., Kapil, V., Pascual, M., et al. (2011). Impacts of climate change on public health in India: Future research directions. *Environmental Health Perspectives, 119*, 765–770.

Cave, B., Fothergill, J., Gibson, G., & Pyper, R. (2017). *Health and environmental impact assessment: A briefing for public health teams in England*. Public Health England.

Chinchilla, M., & Arcaya, M. C. (2017). Using health impact assessment as an interdisciplinary teaching tool. *International Journal of Environmental Research and Public Health, 14*, 744. https://doi.org/10.3390/ijerph14070744.

Cooke, A., Friedli, L., Coggins, T., Edmonds, N., Michaelson, J., O'Hara, K., et al. (2011). *Mental wellbeing impact assessment: A toolkit for wellbeing*. London: National MWIA Collaborative.

Dahlgren, G., & Whitehead, M. (1993). *Tackling inequalities in health: What can we learn from what has been tried?* In: *Working paper prepared for the King's Fund international seminar on tackling inequalities in health, September 1993, Ditchley Park, Oxfordshire*. London, King's Fund, accessible in: Dahlgren G, Whitehead M. (2007) European strategies for tackling social inequities in health: Levelling up Part 2. Copenhagen: WHO Regional office for Europe. http://www.euro.who.int/__data/assets/pdf_file/0018/103824/E89384.pdf.

Dannenberg, A. L., Bhatia, R., Cole, B. L., Dora, C., Fielding, J. E., Kraft, K., et al. (2006). Growing the field of health impact assessment in the United States: An agenda for research and practice. *American Journal of Public Health, 96*, 262–270. https://doi.org/10.2105/AJPH.2005.069880.

Davis, A. (2014). *Claiming the health dividend: A summary and discussion of value for money estimates from studies of investment in walking and cycling*. Department for Transport (UK).

EC. (2014). *Directive 2014/52/EU* (p. 124).

Elliott, E., & Francis, S. (2005). Making effective links to decision-making: Key challenges for health impact assessment. *Environmental Impact Assessment Review, 25*, 747–757. https://doi.org/10.1016/j.eiar.2005.07.007.

Equator Principles n.d. WWW document, http://www.equator-principles.com/ (Accessed 11.9.16).

Gigerenzer, G. (2015). *Risk Savvy: How to make good decisions*. Penguin.

GLA. (2014). *Transport and health in London: The main impacts of London road transport on health*. Greater London Authority.

Gouldson, A., Sudmant, A., Khreis, H., & Papargyropoulou, E. (2018). *The economic and social benefits of low-carbon cities: A systematic review of the evidence*. London and Washington DC: Coalition for Urban Transitions. http://newclimateeconomy.net/content/cities-working-papers.

Harris-Roxas, B., & Harris, E. (2011). Differing forms, differing purposes: A typology of health impact assessment. *Environmental Impact Assessment Review, Health Impact Assessment in the Asia Pacific, 31*, 396–403. https://doi.org/10.1016/j.eiar.2010.03.003.

Health Scotland. (2007). *Health impact assessment of transport initiatives: A guide*.

Heller, J., Givens, M. L., Yuen, T. K., Gould, S., Jandu, M. B., Bourcier, E., et al. (2014). Advancing efforts to achieve health equity: Equity metrics for health impact assessment practice. *International Journal of Environmental Research and Public Health, 11*, 11054–11064. https://doi.org/10.3390/ijerph111111054.

Highways Agency, n.d. Design Manual for Roads and Bridges (DMRB) | Standards for Highways [WWW document]. http://www.standardsforhighways.co.uk/ha/standards/dmrb/ (Accessed 03.14.19).

Hosking, J., Mudu, P., & Dora, C. (2011). *WHO | Health co-benefits of climate change mitigation—Transport sector [WWW Document]*. WHO. http://www.who.int/hia/green_economy/transport_sector_health_co-benefits_climate_change_mitigation/en/ (Accessed December 11, 2015).

Huang, Z. (2012). Health impact assessment in China: Emergence, progress and challenges. *Environmental Impact Assessment Review, 32*, 45–49. https://doi.org/10.1016/j.eiar.2011.03.001.

ICMM, n.d. New guidance on health risk assessment.

IFC. (2009). *Introduction to health impact assessment.* International Finance Corporation.

IFC. (2012). *Performance standards on environmental and social sustainability.* Washington: International Finance Corporation.

IPIECA. (2016). *Health impact assessments.* International Association of Oil and Gas Producers.

Joffe, M., & Mindell, J. (2002). A framework for the evidence base to support health impact assessment. *Journal of Epidemiology & Community Health, 56*, 132–138. https://doi.org/10.1136/jech.56.2.132.

Joffe, M., & Mindell, J. (2005). Health impact assessment. *Occupational and Environmental Medicine, 62*, 907–912. https://doi.org/10.1136/oem.2004.014969.

Kahneman, D. (2012). *Thinking, fast and slow.* London: Penguin Books.

Kemm, J., Parry, J., & Palmer, S. (Eds.), (2004). *Health impact assessment.* Oxford University Press.

Kumaresan, J., Narain, J. P., & Sathiakumar, N. (2011). Climate change and health in South East Asia. *International Journal of Climate Change Strategies and Management, 3*, 200–208. https://doi.org/10.1108/17568691111129020.

Lock, K. (2000). Health impact assessment. *BMJ, 320*, 1395–1398. https://doi.org/10.1136/bmj.320.7246.1395.

Marmot, M. (2008). *WHO Commission on Social Determinants of Health: Closing the gap in a generation.* Geneva: World Health Organization.

Mindell, J., Biddulph, J., Boaz, A., Curtis, S., Joffe, M., Lock, K., et al. (2006). *A guide to reviewing the evidence for use in HIA.* London Health Observatory.

Mindell, J., Boaz, A., Joffe, M., Curtis, S., & Birley, M. (2004). Enhancing the evidence base for health impact assessment. *Journal of Epidemiology & Community Health, 58*, 546–551. https://doi.org/10.1136/jech.2003.012401.

Mindell, J. S., Boltong, A., & Forde, I. (2008). A review of health impact assessment frameworks. *Public Health, 122*, 1177–1187. https://doi.org/10.1016/j.puhe.2008.03.014.

Mindell, J. S., Bowen, C., Herriot, N., Findlay, G., & Atkinson, S. (2010). Institutionalizing health impact assessment in London as a public health tool for increasing synergy between policies in other areas. *Public Health, 124*, 107–114. https://doi.org/10.1016/j.puhe.2010.01.007.

Mindell, J. S., Cohen, J. M., Watkins, S., & Tyler, N. (2011). Synergies between low-carbon and healthy transport policies. *Proceedings of the Institution of Civil Engineers Transport, 164*, 127–139. https://doi.org/10.1680/tran.2011.164.3.127.

Mindell, J., Ison, E., & Joffe, M. (2003). A glossary for health impact assessment. *Journal of Epidemiology and Community Health, 57*, 647–651. https://doi.org/10.1136/jech.57.9.647.

Mindell, J., & Joffe, M. (2003). Health impact assessment in relation to other forms of impact assessment. *Journal of Public Health (Oxford, England), 25*, 107–112. https://doi.org/10.1093/pubmed/fdg024.

Mindell, J. S., Moody, A., Alvarado, A., Vecino-Ortiz, A. I., Triana, C. A., de Oliveira, C., et al. (2017). Comparison of health examination survey methods in Brazil, Chile, Colombia, Mexico, England, Scotland, and the United States. *American Journal of Epidemiology, 186*, 648–658. https://doi.org/10.1093/aje/kwx045.

Mindell, J., Sheridan, L., Joffe, M., Samson-Barry, H., & Atkinson, S. (2004). Health impact assessment as an agent of policy change: Improving the health impacts of the mayor of London's draft transport strategy. *Journal of Epidemiology and Community Health, 58*, 169. https://doi.org/10.1136/jech.2003.012385.

MSCCH, n.d. Medical alert! Climate change is harming our health [WWW document]. The Medical Society Consortium on Climate and Health (MSCCH). https://medsocietiesforclimatehealth.org/reports/medical-alert/ (Accessed 03.27.19).

National Collaborating Centre for Healthy Public Policy, n.d. Health impact assessment [WWW document]. http://www.ncchpp.ca/54/health_impact_assessment.ccnpps (Accessed 3.25.19).
National Health Commission Office. (2016). *Thailand's rules and procedures for the health impact assessment of public policies*. NHCO. https://en.nationalhealth.or.th/knowledge-base/thailands-rules-procedures-health-impact-assessment-public-policies-2016/ (Accessed July 5, 2019).
Negev, M., Levine, H., Davidovitch, N., Bhatia, R., & Mindell, J. (2012). Integration of health and environment through health impact assessment: Cases from three continents. *Environmental Research, 114*, 60–67. https://doi.org/10.1016/j.envres.2012.02.003.
Public Health New Zealand, n.d. Health in all policies | Toi Te Ora Public Health [WWW document] https://www.toiteora.govt.nz/health_in_all_policies (Accessed 03.26.19).
Quigley, R., den Broeder, L., Furu, P., Bond, A., Cave, B., & Bos, R. (2006). *Health impact assessment international best practice principles (No. 5)*. Fargo: International Association for Impact Assessment.
Robert Wood Johnson Foundation, Carger, E., & Westen, D. (2010). *A new way to talk about the social determinants of health [WWW document]*. RWJF. https://www.rwjf.org/en/library/research/2010/01/a-new-way-to-talk-about-the-social-determinants-of-health.html (Accessed March 26, 2019).
Schrecker, T., Birn, A.-E., & Aguilera, M. (2018). How extractive industries affect health: Political economy underpinnings and pathways. *Health & Place, 52*, 135–147. https://doi.org/10.1016/j.healthplace.2018.05.005.
Scott-Samuel, A. (1998). Health impact assessment—Theory into practice. *Journal of Epidemiology and Community Health, 52*, 704–705.
Scott-Samuel, A., Ardern, K., & Birley, M. (2001). *The Merseyside guidelines for health impact assessment* (2nd ed.). International Health Impact Assessment Consortium.
Simpson, S., Mahoney, M., Harris, E., Aldrich, R., & Stewart-Williams, J. (2005). Equity-focused health impact assessment: A tool to assist policy makers in addressing health inequalities. *Environmental Impact Assessment Review, Health Impact Assessment, 25*, 772–782. https://doi.org/10.1016/j.eiar.2005.07.010.
Slovic, P. (2010). *The feeling of risk: New perspectives on risk perception*. London: Routledge. https://doi.org/10.4324/9781849776677. eBook ISBN: 9781849776677.
The Lancet Planetary Health. (2017). *Health co-benefits of climate action [WWW document]*. https://www.thelancet.com/journals/lanplh/article/PIIS2542-5196(17)30003-7/fulltext (Accessed 18 March 2019).
THSG. (2000). *Carrying out a health impact assessment of a transport policy—Guidance from the Transport & Health Study Group*.
THSG. (2011). *Health on the Move 2: Policies for health promoting transport*.
Tolonen, H., Koponen, P., Al-kerwi, A., Capkova, N., Giampaoli, S., Mindell, J., et al. (2018). European health examination surveys—A tool for collecting objective information about the health of the population. *Archives of Public Health, 76*, 38. https://doi.org/10.1186/s13690-018-0282-4.
Transport for London. (2009). *Mayor's transport strategy integrated impact assessment: Scoping report*. London.
Transport for London. (2017). *Integrated impact assessment of the Mayor's transport strategy 3, scoping report*. London.
UNDP, n.d. Sustainable development goals [WWW document]. UNDP. https://www.undp.org/content/undp/en/home/sustainable-development-goals.html (Accessed 03.26.19).
United Nations. (2011). *UN "Protect, Respect and Remedy" framework and guiding principles [WWW Document]*. Business & Human Rights Resource Centre. https://www.business-humanrights.org/en/un-secretary-generals-special-representative-on-business-human-rights/un-protect-respect-and-remedy-framework-and-guiding-principles (Accessed 25 March 2019).

United Nations, n.d. International covenant on economic, social and cultural rights [WWW document]. Human Rights, Office of the High Commissioner. http://www.ohchr.org/EN/ProfessionalInterest/Pages/CESCR.aspx (Accessed 01.06.17).

Waheed, F., Ferguson, G. M., Ollson, C. A., MacLellan, J. I., McCallum, L. C., & Cole, D. C. (2018). Health impact assessment of transportation projects, plans and policies: A scoping review. *Environmental Impact Assessment Review, 71*, 17–25. https://doi.org/10.1016/j.eiar.2017.12.002.

Watts, N., Amann, M., Ayeb-Karlsson, S., Belesova, K., Bouley, T., Boykoff, M., et al. (2017). The Lancet Countdown on health and climate change: From 25 years of inaction to a global transformation for public health. *The Lancet*, https://doi.org/10.1016/S0140-6736(17)32464-9.

WHO. (1997). *Jakarta declaration on leading health promotion into the 21st century [WWW document]*. WHO. https://www.who.int/healthpromotion/conferences/previous/jakarta/declaration/en/ (Accessed 25 March 2019).

WHO. (2000). *Transport, environment and health*.

WHO. (2013). *The 8th global conference on health promotion, Helsinki [WWW document]*. WHO. https://www.who.int/healthpromotion/conferences/8gchp/en/ (Accessed 26 March 2019).

WHO, n.d.-a Health impact assessment (HIA). Short guides [WWW document]. WHO. https://www.who.int/hia/about/guides/en/ (Accessed 03.26.19).

WHO, n.d.-b Using evidence within HIA [WWW document]. WHO. https://www.who.int/hia/tools/evidenceuse/en/index6.html (Accessed 03.26.19).

WHO, n.d.-c Climate change and health [WWW document]. WHO. http://www.who.int/mediacentre/factsheets/fs266/en/ (Accessed 05.10.17).

Wilkinson, R. & Marmot, M. (Eds.), (2003). *Social determinants of health: The solid facts* (2nd ed). Geneva: World Health Organization European Regional Office.

CHAPTER 13

Quantitative health impact and burden of disease assessment of traffic-related air pollution

Natalie Mueller[a,b,c], **Mark J. Nieuwenhuijsen**[a,b,c], **and David Rojas-Rueda**[a,b,c,d]
[a]Barcelona Institute for Global Health (ISGlobal), Barcelona, Spain
[b]Universitat Pompeu Fabra (UPF), Barcelona, Spain
[c]CIBER Epidemiologia y Salud Pública (CIBERESP), Madrid, Spain
[d]Department of Environmental and Radiological Health Sciences, Colorado State University, Fort Collins, CO, United States

Abbreviations

AB	attributable burden
BC	black carbon
CBA	cost-benefit analysis
CUA	cost-utility analysis
DALY	disability-adjusted life year
EC	European Commission
ERF	exposure response function
GBD	global burden of disease
HALY	health-adjusted life year
HIA	health impact assessment
NO_2	nitrogen dioxide
PAF	population attributable fraction
$PM_{2.5}$	particular matter with diameter $\leq 2.5\,\mu m$
PM_{10}	particular matter with diameter $\leq 10\,\mu m$
QALY	quality-adjusted life year
RR	relative risk
TB	total burden
YLD	years lived with disability
YLL	years of life lost
WHO	World Health Organization

Introduction

Environmental, economic, social, and other public policies are often so closely related that a decision in one sector affects the objectives and performance of another sector. To acknowledge this, in the past, public policies

Traffic-Related Air Pollution
https://doi.org/10.1016/B978-0-12-818122-5.00013-2

have been assessed based on their environmental, economic, or social impacts. Considering explicitly the impacts of public policies on population health is a relatively new field.

Health impact assessment (HIA) is an important tool to integrate health evidence into the policy decision-making process, and introduce and protect public health in all policies. HIA is a tool that can help policymakers foresee how different interventions, policies, and programs outside the health sphere, will affect health and well-being by informing on the associated health benefits and risks (Ståhl et al., 2006). HIA combines different methods and procedures to systematically identify and judge all relevant, whether intended or unintended, exposure pathways and health effects the proposed intervention, policy, or program might have on the health of a population and the distribution of those effects within this population (Mindell, Ison, & Joffe, 2003; WHO, 1999). Most HIAs are carried out prospectively and aim to give an outlook on the expected health consequences in the future (Wismar et al., 2007). Because *health* is usually a strong-weighting factor to the public, outcomes of HIAs can raise health awareness and advocacy and can influence the decision-making process (Burstein, 1998; Jacobs & Shapiro, 1994). Therefore, outcomes of HIAs can have a strong bearing on public policy proposals (Wismar et al., 2007).

The aim of this chapter is to provide an overview of what quantitative HIA is, how it is commonly being conducted, provide examples of quantitative HIA studies on traffic-related air pollution, and to discuss uncertainties, utilities, and further considerations of quantitative HIA.

Health impact assessment of transport policies and traffic-related air pollution

High levels of air pollution have become a global crisis and are a popular matter for societal concern. The Global burden of disease (GBD) Study classifies air pollution as a leading risk factor for premature mortality and morbidity. In 2016, 6.1 million premature deaths worldwide were caused by air pollution, whereas 4.1 million premature deaths were in particular attributable to ambient air pollution (GBD 2016 Risk Factors Collaborators, 2017). Many cities worldwide regularly exceed the established air quality health thresholds by the World Health organization (WHO), which is estimated to result in a significant health burden (European Environment Agency, 2019). In urban areas, ambient air pollution is often traffic related, with the traffic contribution to PM_{10}, $PM_{2.5}$, and NO_2 in European cities being on average 39% (range 9%–53%), 43% (9%–66%), and up to 80%, respectively (Sundvor et al., 2012). Most recent systematic reviews and meta-analyses established significant associations between long-term exposure to ambient air pollution and traffic markers with various health end points, including mortality, various cardiovascular and respiratory disease end points, cerebrovascular disease, lung cancer, type 2 diabetes, preterm birth, low birth weight, reduced fertility as well as working days lost and restricted-activity days (Atkinson et al., 2018; Cesaroni et al., 2014; Dadvand et al., 2014; Eze et al., 2015; Héroux et al., 2015; Nieuwenhuijsen et al., 2014; Pedersen et al., 2013; Stafoggia et al., 2014; WHO, 2014). Supporting toxicological and clinical evidence suggests that there are documented biological mechanisms by which air pollution leads to premature mortality and a wide range of morbidity end points.

Transport planning is a public policy sector that is increasingly recognized for its significant effects on human health and well-being through multiple exposure pathways such as air pollution, noise and heat exposure, traffic accidents, missed opportunities for physical activity, and the road network and parking infrastructure taking up large amounts of land, which could be used for other purposes such as parks and green spaces with many documented health benefits (Nieuwenhuijsen & Khreis, 2016). Currently, transport planning evaluation is driven by cost appraisal schemes such as cost-benefit analysis (CBA) that compare economic costs with economic benefits. CBA attempts to monetize the expected effects of transport projects. Monetized items that so far have gone into the calculations include changes in travel time, providers' revenues and costs, consumers' surplus, changes in business and employment activities, infrastructure costs, traffic accidents, carbon emissions, congestion, habitat damage, air pollution, and noise impacts (CE Delft, 2019; Geurs, Boon, & Van Wee, 2009). Nevertheless, it is often difficult to understand what the monetized values of, e.g., air pollution and noise in CBAs exactly include and how these values were defined. For example, to define the external costs of transport-related air pollution, the European Commission's (EC's) handbook on the external costs of transport (2019) provides a combined monetary value for the four impact domains of health effects, crop losses, material and building damage, and biodiversity loss (CE Delft, 2019). While some specific mortality, respiratory and cardiovascular disease health end points are being addressed, the full range of air pollution-related health impacts is not comprehensively assessed and included, potentially leading to an underestimation of the true health costs of transport projects.

Qualitative versus quantitative health impact assessment

HIA models can be qualitative or quantitative in nature. Both methods can coexist and can profit from each other (Nieuwenhuijsen et al., 2017). While qualitative HIA can be carried out in a timelier manner and is less resource intensive, it does not provide objective information on the magnitude and direction (i.e., positive or negative) of the expected health impacts. Therefore, qualitative HIA has a higher risk of subjective and selective assessment and therefore biased results, conclusions, and recommendations. A qualitative HIA most likely draws on information and evidence that is already available and is easily accessible (Mindell et al., 2003). In transport planning, proposed interventions or projects have commonly been assessed more qualitatively instead of offering more powerful, quantitative estimates to stakeholders through quantitative approaches (Mindell et al., 2010; Shafie, Omar, & Karuppannan, 2013). This has led to the criticism of an existing lack of rigor in collecting and analyzing evidence quantitatively (Mindell et al., 2010).

Quantitative HIA, on the other hand, informs numerically on the expected health benefits and risks of the proposed intervention, policy, or program and therefore provides a direction *and* magnitude of expected health impacts. Quantitative HIA can be especially important for influencing policy proposals, because decision-makers may give more weight to outcomes that are measurable (i.e., quantifiable) (Joffe & Mindell, 2005). Having indications on the effect size of expected impacts can help distinguish between main issues and details and can provide clarity on the tradeoffs (Veerman, Barendregt, & Mackenbach, 2005). Quantitative HIA, unlike qualitative HIA, is less sensitive to societal opinions and perceptions. Instead, quantitative HIA draws on the best available epidemiological evidence. Consequently, quantitatively estimated health impacts might better reflect the real situation and true impact to be expected, which can help decision-makers understand to what extent their actions can be targeted effectively to benefit health and prioritize the best policy option available. However, despite the outlined advantages of quantitative HIA, there is an apparent lack of communication between research and practice and it has been recognized that the existing tools and models, despite being scientifically sound, often do not meet the needs of end users (Mindell et al., 2010).

Comparative risk assessment framework

A quantitative HIA approach commonly follows the comparative risk assessment framework, where the burden of mortality or disease due to an observed exposure distribution (e.g., air pollution) in a population

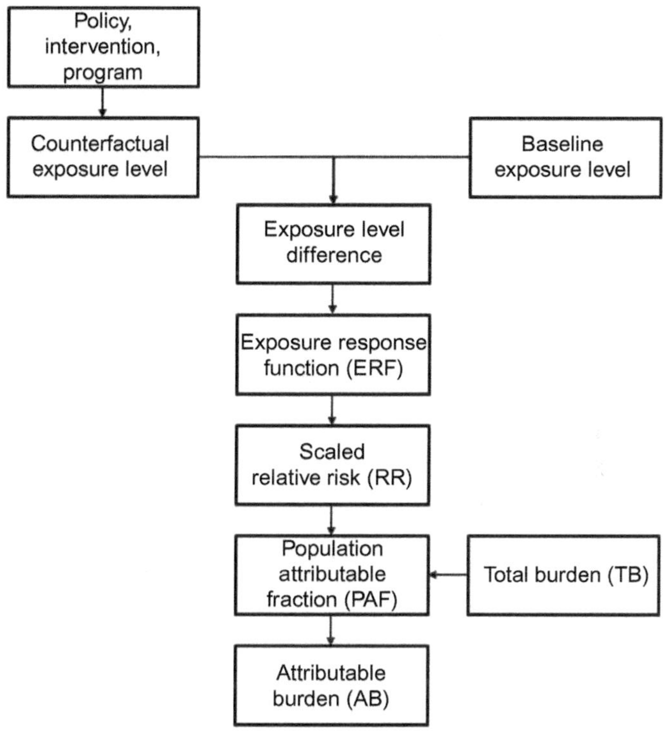

Fig. 13.1 Comparative risk assessment framework.

(i.e., baseline situation/business-as-usual) is compared with the burden of mortality or disease from a hypothetical exposure distribution or a series of distributions (i.e., counterfactual scenario(s)) associated with the proposed intervention, policy, or program (Murray et al., 2004) (Fig. 13.1). The aim is to provide quantitative estimates of the expected health impacts (e.g., changes in premature deaths, cases of disease, disability-adjusted life years (DALYs), etc.) and the distribution thereof among the population.

In quantitative HIA the steps are usually as follows:

1. *Scenario(s) definition*: A counterfactual scenario (or a set of scenarios) is defined which describes the hypothetical exposure level that is aimed at with the proposed intervention, policy, or program. Various criteria may determine the choice of the counterfactual exposure distribution. Ideally, counterfactuals should represent intervention, policy, or program actions that actually can be implemented (e.g., motorized traffic reductions to reduce air pollution levels in cities), rather than counterfactuals that are not achievable (e.g., zero air pollution). Murray and Lopez (1999) looked into different counterfactual scenario categories and identified

exposure distributions corresponding to theoretical minimum risk (i.e., exposure distribution with lowest population risk), plausible minimum risk (i.e., imaginable exposure distribution), feasible minimum risk (i.e., exposure distribution observed in some populations), and cost-effective minimum risk (i.e., considers costs of exposure reduction).

2. *Hazard identification*: The HIA addressed exposure (or multitude of exposures) needs to be specified (e.g., air pollution).
3. *Exposure assessment*: The baseline and counterfactual exposure level distributions of the population under study need to be quantified. The exposure level at baseline is compared with the aimed at exposure level of the counterfactual scenario and the resulting difference in exposure level is quantified.
4. *Health outcome definition*: The health outcome (or a multitude of health outcomes) of interest needs to be defined and should have been shown by previous epidemiological studies to be associated with the exposure. A causal relationship between exposure and health outcome of interest is a prerequisite.
5. *Health data identification*: The total burden (TB) of the health outcome of interest needs to be available for the population under study. Incidence rates are preferred over prevalence proportions, because only new cases expected over a given time (i.e., incident cases) are preventable under the assumption that exposure levels will change.
6. *Exposure response relationship*: An exposure response function (ERF) that quantifies the strength of association between the exposure and the health outcome needs to be available from the literature or available guidance. The ERF needs to come from the best available evidence. In an ideal case, the ERF would be estimated particularly for the population under study, reflecting most accurately the level of risk studied. However, in many cases, a particular ERF for the population under study is not available; therefore, the ERF should preferably be a pooled, generalized estimate of the overall effect coming from a meta-analysis or large longitudinal study.
7. *Risk characterization*: The risk estimate obtained from the ERF that most epidemiological studies report in terms of relative risk (RR), which is the ratio of incidence observed at two different exposure levels, quantifies the strength of association between the exposure and health outcome. The risk estimate is scaled to the difference in exposure level resulting of the comparison of the baseline exposure level with the counterfactual exposure level.

The scaling is done as follows:

n is the number of exposure categories (e.g., census tracts, neighborhoods, districts, etc.),

i is an element of the set $i \in \{1,\ldots,n\}$,

$$RR_i = e^{\left(\frac{\ln(RR_E)}{E} \times ED_i\right)}$$

where RR_E is the relative risk obtained from the ERF, E is the exposure unit that corresponds to the RR_E obtained from the ERF, and ED_i is the exposure level difference, which is the difference in the exposure level resulting of the comparison of the baseline exposure level with the counterfactual exposure level.

The population attributable fraction (PAF) defines the proportional health burden of the health outcome of interest that is attributable to exposure level difference ED_i.

The PAF is calculated as follows:

$$PAF_i = \frac{P_i \times RR_i - 1}{P_i \times RR_i}$$

where P_i is the proportion of the population exposed andRR_i is the previously scaled relative risk.

The attributable burden (AB) describes the total burden of the health outcome of interest that is attributable to the exposure level difference ED_i.

The AB is calculated as follows:

$$AB = \sum_{i=1}^{n} TB_i \times PAF_i$$

where TB_i is the total burden of the health outcome of interest and PAF_i is the previously calculated proportional health burden of the health outcome of interest that is attributable to the exposure level difference ED_i.

Burden of disease assessment

The global burden of disease concept emerged in 1992 when the World Bank commissioned the first GBD study to comprehensively assess the disease burden in 1990. The idea was to describe the mortality and ill-health burden due to disease, injury, and risk factors for different geographic regions of the world to assess the comparative importance of risk factors to health and their outcomes (Lopez et al., 2006). The burden of a particular health outcome is estimated by summing (1) the years of life a person loses because of premature death due to that particular health outcome (i.e., years of life lost (YLLs)) and (2) the years of life a person lives with disability caused by that particular health outcome (i.e., years lived with disability (YLDs)). Adding up the YLLs and YLDs provides an estimate of the burden of that particular health outcome, expressed as DALYs. A DALY is defined as the loss of 1 year of life lived in full health (Lopez et al., 2006). Besides the DALY, several other measures have emerged, including the quality-adjusted life year (QALY) and the health-adjusted life year (HALY). QALYs combine the effects of health interventions on mortality and morbidity and are commonly used in cost-utility analysis (CUA) (Whitehead & Ali, 2010). HALYs are a summary measure of population health that integrates mortality and quality of life (Gold, Stevenson, & Fryback, 2002). Nevertheless, DALYs are the most commonly used measure in burden of disease assessments such as the GBD studies and make different health outcomes and their disabilities comparable.

DALYs are estimated as

$$DALYs = YLLs + YLDs$$

The YLLs correspond to the number of deaths due to a particular health outcome multiplied by the number of remaining years to standard life expectancy at the age of death (Hänninen & Knol, 2011).

YLLs are estimated as follows:

$$YLLs = N \times L$$

where N is the number of deaths and L is the number of remaining years to standard life expectancy at the age of death.

To estimate YLDs, the number of incident disease cases (of that particular health outcome) is multiplied by the average duration of the disease and a disease disability weight, which reflects the severity of the disease on a scale from 0 (i.e., perfect health) to 1 (i.e., death). YLDs are estimated as

$$YLD = n \times L \times DW$$

where n is the number of incident cases, L is the average duration of the disease in years, and DW is the disability weight of the disease.

YLLs, YLDs, and DALYs can be selected as the health outcomes of interest in quantitative HIA studies. The burden of disease units are especially useful in studies where multiple health outcomes or different geographical settings are studied simultaneously (e.g., the GBD studies). The burden of disease approach provides outcome units that summarize mortality and morbidity effects of health outcomes and therefore make them comparable even though they carry different severities and produce different disability weights.

Examples of quantitative HIA studies assessing traffic-related air pollution impacts

In the last two decades, quantitative HIA studies of transport projects and traffic-related air pollution have seen a large increase, especially in the academic literature. These HIA studies relied on modeling methods and tools that are still largely restricted to research, but if developed further and improved, they have the potential to be used in practice. This, as mentioned above, reflects a communication gap between the different sectors. While researchers have the know-how to develop and apply quantitative HIA models, they lack the sense and experience of knowing whether their counterfactual scenarios are realistic and the developed models and tools are often not user friendly. On the other hand, practitioners, policy and decision-makers lack the motivation, knowledge, methods, and tools (i.e., resources) to routinely carry out these health impact modeling exercises to inform their decision-making process (Mindell et al., 2010; Nieuwenhuijsen et al., 2017). This section will provide an overview of the existing studies modeling traffic-related air pollution health impacts.

McKinley et al. (2005) quantified health benefits and costs of a combination of air pollution control measures for Mexico City, including the city's taxi fleet renovation, metro network expansions, and the introduction of hybrid buses. Control measures were estimated to reduce the city's PM_{10} concentrations by 1% and O_3 by 3%. Health benefits were substantial and were estimated to overall outweigh investment costs, with benefit-cost ratios being 3.3, 0.7, and 1.3 for the taxi fleet renovation, metro expansion, and hybrid bus introduction, respectively. Woodcock et al. (2009) looked into different transport scenarios for London and Delhi and found that a combination of low-carbon-emission vehicles and the uptake of active transport would reduce $PM_{2.5}$ levels most effectively from 10.1 to 7.4 $\mu g/m^3$ for London and from 88.7 to 72.3 $\mu g/m^3$ for Delhi, saving annually 319 DALYs and 2749 DALYs, respectively. Creutzig, Mühlhoff, and Römer (2012) studied the co-benefits of reducing greenhouse gas emissions for Barcelona, Malmö, Sofia, and Freiburg. Their most ambitious scenario, which included congestion charging, aggressive land-use policies, and car-free areas with active and public transport development and construction restrictions, reduced air pollution levels by 58%–73%, depending on the city, resulting in the largest health co-benefits. Grabow et al. (2012) modeled the health impacts of replacing 50% of shorter car trips with the bicycle in 11 metropolitan regions in the US in order to improve air quality. They estimated that the annual average urban $PM_{2.5}$ concentration would decline by 0.1 μg/m^3 and mortality would decrease by 1295 deaths per year. Xia et al. (2015) hypothetically shifted 40% of vehicle kilometers traveled to alternative modes of transport in Adelaide, Australia. The associated 26% reduction of $PM_{2.5}$ levels was estimated to result in 13 preventable deaths and 118 averted DALYs, besides further health benefits being related to increased physical activity levels among the population and reduced traffic accidents. Tainio (2015) quantified the burden of disease caused by local transport in Warsaw and estimated that transport caused around 58,000 DALYs annually of which 44% were attributable to traffic-related air pollution. Stevenson et al. (2016) estimated health impacts under a compact city scenario, compromising a city of short distances, increased residential density, mixed land use, and the promotion of active travel for Melbourne, Boston, London, Copenhagen, Sao Paulo, and Delhi. All cities benefited from the compact city scenario and the scenario resulted in 3%–12% reduced transport-related PM emissions and 393–827 averted DALYs. Mueller et al. (2017a, 2017b) attributed almost 660 premature deaths and almost 10,000 DALYs to breaching WHO $PM_{2.5}$ air-quality guidelines in Barcelona, recognizing that motorized traffic is the most important contributor to local air pollution levels. Khreis et al. (2018) studied the total traffic-related air pollution childhood asthma burden in Bradford, UK and attributed between 279 and 612 cases of asthma in children to local $PM_{2.5}$, PM_{10}, and black carbon (BC) concentrations, representing 15%–33% of all asthma cases in the city. As a further step, Khreis et al. (2019) looked into the air pollution childhood asthma burden in 18 European countries with more than 64 million children and estimated that if minimum pollution levels recorded for NO_2, $PM_{2.5}$, and BC could be complied more than 135,000, 190,000, and 89,000 incident cases of asthma could be prevented, respectively. A recent study assessed the Barcelona Superblock model, an innovative land use intervention with the aim to reclaim space for people, reduce motorized transport, promote active transport, increase urban greening, and mitigate effects of climate change found that with full implementation of the model, current city-wide annual mean NO_2 concentrations of 47.2 μg/m^3 could be reduced to 35.7 μg/m^3, which in return would result in 291 preventable deaths and an increase of average life expectancy by 129 days (Mueller et al., 2019).

Uncertainty in quantitative health impact assessment

Like any model, quantitative HIA modeling processes have limitations that need to be acknowledged. Quantitative HIA is rather an indicative than empirical research tool and carries uncertainties in the estimation of health impacts (Parry & Stevens, 2001). Currently, there is a lack of standards and best practices. The predictive validity and plausibility of quantified health impacts depends on the application and interpretation of the available supportive evidence (Thomson et al., 2008). Uncertainty is introduced where epidemiological evidence is lacking or causality has not been established. HIA draws on assumptions and extrapolations, where gaps in knowledge on true properties of parameters exist.

Most often HIAs make use of counterfactual scenarios that describe ideal situations. However, these ideal situations are often too optimistic, hardly attainable, and bear little relevance to authorities and policy-makers (Nieuwenhuijsen et al., 2017). HIAs are fundamentally different from evaluation studies and pre-post intervention studies. It remains uncertain whether health impacts will truly occur as estimated. Therefore, outputs of HIA studies can only be interpreted as an indication of the magnitude of expected health impacts under the counterfactual scenario. Health impacts are sensitive to contextual settings and underlying population parameters and depend largely on the baseline exposure level, the exposure's distribution among the population as well as the general health status of that population. Therefore, varying results can be expected for varying settings and contexts, which can also be considered as a strength of such assessments as they speak to local contexts. Furthermore, health impact estimations are incapable of reflecting and capturing personal choices and intrinsic motivations for behavior change that are implied by the counterfactual scenario (e.g., choosing the bicycle over the car to help reduce air pollution levels in the city).

Quantitative HIA draws on exposure proxies (e.g., annual mean air pollution concentration) that are unable to fully capture the variability of exposure levels within the population under study. This results in uncertainty in exposure characterization and may lead to exposure misclassification. For instance, many HIAs use $PM_{2.5}$ as a proxy to reflect the overall air pollution burden in a city; however, $PM_{2.5}$ might vary from place to place in its chemical composition and associated toxicity and disentangling the contribution of different sources (e.g., traffic versus industry) is challenging and often underexplored.

Local ERFs that quantify the strength of association between exposure (i.e., air pollution) and the health outcome of interest in the population under study are often not available. Therefore, risk estimates are obtained from elsewhere and are extrapolated to the population under study. This might limit the comparability of results of different HIA studies because different risk estimates for the same exposure-health outcome combination may have been used. Moreover, extrapolating risk estimates from one population to another implies that causality of the health effect in the population under study is simply assumed. While estimates that come from meta-analyses are preferred, they are limited in the sense that they represent a pooled, generalized estimate. Differences in risk between different studies and different populations may actually reflect the true variability in the association between exposure and health outcome and therefore different susceptibilities of populations, especially if the underlying population parameters vary a lot. Some exposure-health outcome combinations are easier to quantify than others. The attributable burden can only be quantified for those health outcomes that previous epidemiological studies established causal effects for. The more holistically an exposure is assessed and the stronger the` exposure-risk gradient, the larger the contribution of this exposure to the health burden will be (Stansfeld, 2015). Air pollution epidemiology has a long tradition and the attributable health burden has been assessed comprehensively. Other exposures, that are highly correlated, such as noise exposure in urban areas, have been given less attention to until recently and the full health burden related to noise exposure is currently being assessed.

Besides exposure proxies, health outcomes might also be approximated if no detailed data is available for the population under study. While for instance an all-cause mortality rate might be available at the city level, this mortality rate will most likely vary among the population, e.g., by age, sex, or socioeconomic status. Disease incidence rates might be more difficult to obtain due to different diagnosis, recording, and reporting practices. Other health outcomes, such as burden of disease outcomes, including YLLs, YLDs, and DALYs, are most likely not available for the population under study. In this case, national estimates have to be used that can be obtained through the GBD study and the GBD Results Tool (http://ghdx.healthdata.org/gbd-results-tool) and have to be scaled down to the population under study. This scaling, however, despite possibly being sensitive to age and sex variations, is most likely insensitive to other influential population parameters, such as differences in health status between urban and rural populations, ethnicity, socioeconomic status, etc. This may lead to misclassification of the true baseline health burden for the population under study. In addition, results of HIAs can be controversial when comparing the health burden of less severe but common health outcomes to severe but rare outcomes (Hänninen & Knol, 2011). Tainio (2015) and Mueller et al. (2017a, 2017b) compared the health burden of traffic-related air pollution with road traffic noise in Warsaw and Barcelona, respectively. While both studies attributed a larger mortality burden to traffic-related air pollution, both studies attributed a larger total burden of disease to road traffic noise, especially with noise annoyance and noise-related sleep disturbance contributing strongly, due to a high prevalence in both populations, despite small disability weights.

Further uncertainty in quantitative HIA is introduced when the health effects of one exposure (e.g., traffic-related air pollution) might possibly depend on another exposure (e.g., road traffic noise). While review studies suggest physiological mechanisms of air pollution and noise to vary and therefore suggest to treat both risk factors as independent in HIA studies (Stansfeld, 2015; Tétreault, Perron, & Smargiassi, 2013), one might run the risk of potentially double-counting the attributable health impacts (i.e., attributable cases), if the independence of health effects is not true and synergies between the exposures occur. Moreover, the applied time horizon and possible time lags in benefit buildup are important to consider. Under the counterfactual scenario it will take time for benefits to build up and they will not occur immediately (unless acute effects rather than chronic effects are studied). Hence, long-term benefits (e.g., changes in mortality attributable to reductions in traffic-related air pollution) will occur with delay. To policy-makers and the public, benefits that occur in the future are less valuable than benefits that occur immediately. Comparing health benefits in the future (e.g., traffic reduction leading to cleaner air and possibly preventable premature mortality in the future) with immediate risks (e.g., not being able to use the car today to go to work) can distort the overall benefit-risk tradeoff.

Finally, the interpretability of the health outcome needs to be discussed. The choice of health outcome, for which changes are estimated under the counterfactual scenario, depends on the target audience the HIA is conducted for. Estimating changes in premature mortality (i.e., number of preventable premature deaths under a cleaner air scenario) can be misleading as mortality as such is not preventable. The right interpretation is that a risk factor (i.e., air pollution) can cause death to occur earlier than expected, as determined by life expectancy. However, outside the public health sector, this concept is more difficult to understand. Using premature mortality as the health outcome of interest implies the risk of underestimation of the overall health effect because mortality is an extreme event and only "the tip of the iceberg" (Künzli, Perez, & Rapp, 2010). Nonfatal outcomes, such as doctor visits, medication use, hospital admissions, activity restrictions, etc., will occur before, however, some of these health outcomes might be more difficult to quantify as no ERFs might be available for them. Also, the concept of life expectancy can be challenging, as it represents a population average, and can be ambiguous if a large number of persons lives past the average life expectancy in a good quality of life (Thacker et al., 2006). Using morbidity outcomes is useful to demonstrate the range of health effects the exposure has been associated with. Nevertheless, comparing the different disabilities these health outcomes produce is challenging (e.g., comparing the case of hypertension with a case of lung cancer) because these health outcomes produce different disability weights (Lopez et al., 2006). Using burden of disease concepts such as YLLs, YLDs, and DALYs are useful to compare different health conditions (National Research Council of the National Academies, 2011), however, the comprehensibility and interpretability of these concepts by non-health care professionals (e.g., policymakers, transport planners, the public, etc.), for which the HIA is conducted for, may be limited. Monetization of health impacts can be useful and might be a preferred and understandable outcome for the HIA target audience; however, monetary values in CBAs can have different underlying concepts (e.g., direct costs versus indirect costs, willingness to pay, etc.), input parameters, and are setting-specific. Moreover, monetization fails to holistically reflect changes in health-related quality of life (e.g., personal disability perception, psychological distress, etc.) (Thacker et al., 2006). Comprehensively considering and quantifying all direct costs (e.g., burden on the health-care system) and indirect costs (e.g., work absence, productivity loss, personal pain, and suffering, etc.) a health condition produces is nearly impossible.

Utility of quantitative health impact assessment

Despite the many uncertainties implied, the utility of quantitative HIA needs to be emphasized. HIA is increasingly recognized as a tool for disease prevention and health promotion. In line with the Health in All Policies principle, quantitative HIA provides opportunities to place *health* on the agenda of sectors outside the health sphere and provides quantitative estimates of expected health benefits and risks (Ståhl et al., 2006). Quantitative HIA provides objective evidence based on the best available epidemiological data and thereby advocates for healthy public policy-making. Quantitative HIA adds to the comprehension of causes of illness, the strength of association between an exposure and a health outcome, and the role of policy in shaping and determining health outcomes. HIA can help overcome misperceptions that health inequalities are exclusively due to lifestyle choices, genetic predispositions, and access to health-care services (National Research Council of the National Academies, 2011). In contrast to risk assessment that rather focuses on environmental stressors, agents, and events, HIA assesses the health impacts of interventions, policies, and programs outside the health sector (Briggs, 2008). In addition, HIA provides an alternative paradigm by acknowledging that the environment is not just a hazard and exposure implies risk, but that the environment also provides health benefits through, for instance, ecological services (e.g., habitat, nutrition, access to nature and green spaces, etc.) (Briggs, 2008).

Quantitative HIA tries to disentangle the complexity of the multitude of risk/protective factors operating simultaneously and tries to comprehensively and systematically assess their repercussions, causal health pathways, synergetic operations, and associated health effects. In this sense, quantitative HIA sheds light on the magnitude and direction of each exposure. It also allows the comparison of severities and ramifications across the exposures that with other epidemiological study designs would not be possible. The overall outcome of a quantitative HIA study is an indication of the expected health benefit-risk tradeoff of the assessed intervention, policy, or program. Failure to consider the health impacts of public policies can result in health harms among the population and lost opportunities for health protection and promotion (National Research Council of the National Academies, 2011). Comprehensively assessing all relevant risks and benefits that a public policy proposal implies allows policy-makers to take informed and proactive decisions and actions. The public usually has a strong self-interest in health preservation (Burstein, 1998; Jacobs & Shapiro, 1994), thus, demonstrating that all intended and unintended health impacts can be a convincing and defensible argument in the policy decision-making process and ensures that the proposed action is practicable and supported by the communities affected (Wismar et al., 2007).

Outcomes of quantitative HIA studies can be compared with established numerical criteria or threshold levels that determine the significance of the health effect (e.g., comparison to WHO air quality standards to understand the health implication of complying with or exceeding those established standards). Quantitative estimates allow the comparison of alternatives (i.e., different counterfactual scenarios) and provide insight into the best available choice in terms of maximizing health benefits and mitigating harms. Furthermore, since quantitative HIA is most commonly prospectively applied and estimated impacts provide an outlook into the future, HIAs can be useful for economic evaluations, planning purposes, budgeting, and resource allocation. Quantitative HIA not only looks into the direction and magnitude of expected health impacts, but also into the distribution thereof among the population. Hence, quantitative HIA can help identify subpopulations that would disproportionally be affected or disadvantaged by the intervention, policy, or program under study and therefore can provide valuable information on unjust differences and possible gradients of susceptibility and health status of the population. Therefore, quantitative HIA can contribute to identifying and eliminating health inequalities and instead promote health equity and justice (Collins & Koplan, 2009; Forsyth, Schively Slotterback, & Krizek, 2010).

HIA is an independent discipline that lies at the intersection of science, policy, community engagement, and outreach. HIA can offer a way to engage different policy sectors, stakeholders, and interest groups that usually do not work together because of different priorities, authorities, and objectives (National Research Council of the National Academies, 2011). Hence, HIA can contribute to interdisciplinarity, ensuring that many viewpoints are considered and can help overcome isolated silos of practice. Quantitative HIA provides a platform for scientists to get directly involved in the application of their science to improve public health, while at the same time, HIA outputs can help policy-makers to better understand the health implications and repercussions of their policies. Active engagement of the communities affected by the policy proposals helps to gain a better understanding of their needs and helps to gain their trust and support. Democratically considering the health impacts of proposed policies and adjusting them accordingly ensures that outputs are realistic, practicable, and acceptable by the communities affected by them.

Further considerations

Until now, most quantitative HIA studies have been static and modeled expected health impacts linearly, where a given change in exposure level leads to a defined change in the health outcome. However, reality is usually not that simple, but more complex and dynamic. A few exceptions emerge in the literature, such as the studies by Macmillan et al. (2014) and Macmillan and Woodcock (2017), which used system dynamics modeling (SDM) to model causal loop diagrams of policy changes that promote cycling in Auckland, London, and Nijmegen, incorporating feedback effects, nonlinear relationships, and time delays. More HIA modeling approaches are needed that consider the complexities and nonlinearity of reality and incorporate in their models feedback loops, allow for varying population parameters over time, time delays, and dynamic relationships.

In addition, there is a lack of integrated, full-chain, participatory HIA modeling approaches (Nieuwenhuijsen et al., 2017) (Fig. 13.2). Until now, barely any HIA studies that modeled health impacts of traffic-related air pollution underwent the full modeling process from air pollution source to health outcome (Khreis, de Hoogh, & Nieuwenhuijsen, 2018). Expertise lies within the different sectors and exchange and communication between

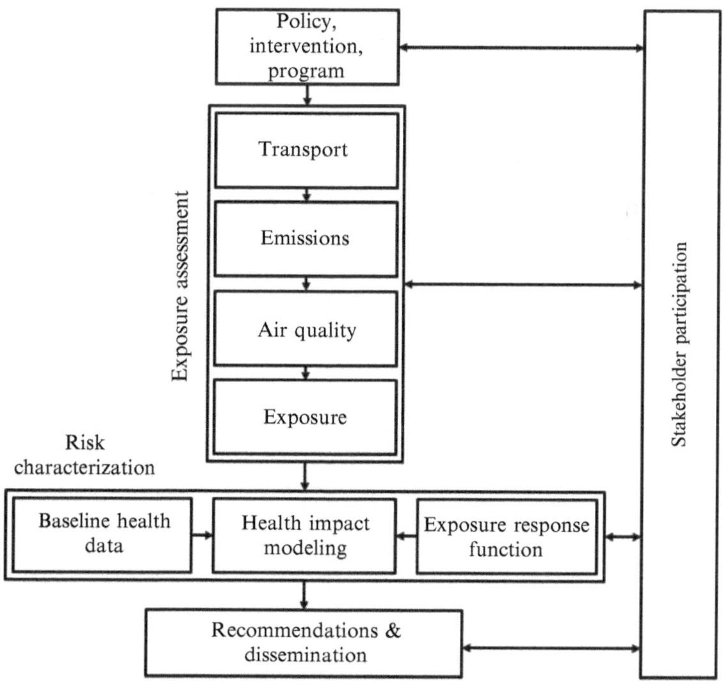

Fig. 13.2 Integrated, full-chain, participatory HIA approach. *Adapted from Nieuwenhuijsen, M.J., et al. (2017). Participatory quantitative health impact assessment of urban and transport planning in cities: A review and research needs. Environment International 103, 61–72.*

these sectors is often nonexistent. Whereas civil and transport engineers and planners might be familiar with models of traffic activities and traffic emissions, considering variables such as car densities, vehicle fleet makeup, traffic conditions, and street design, it might be more the environmental and computer scientists developing air dispersion models to predict air quality, while public health professionals and epidemiologists understand how to model and quantify associated health impacts. A better integration and understanding of these models between the different sectors is desirable and will help to understand and disentangle the traffic contribution to the overall health burden. Moreover, citizens and other stakeholders need to be involved more routinely. Involvement of citizens and diverse stakeholders ensures that contextual factors are considered, equity impacts are brought to the table and research, policy, and practice are made more democratic and inclusive (Bach et al., 2017). Carrying out evaluations and taking decisions without consultation can lead to pushback and refusal, thus participatory approaches allow more viewpoints in the interpretation of complex issues and shed light on practical issues related to implementation and the acceptance of implemented schemes.

Conclusions

Health impact assessment is a valuable tool to assess the anticipated health benefits and risks of a policy proposal outside the health sphere, such as transport planning. Despite uncertainties mainly related to causal inferences, quantitative HIA systematically assesses all relevant risks (and benefits) a policy proposal implies and can help advice decision-makers to take informed and proactive decisions and actions. By looking into the direction, magnitude, and distribution of health effects among the population, quantitative HIA can uncover unjust differences and help in eliminating health inequalities. Quantitative HIA is an emerging field which until now has been restricted to research and academia. A better exchange and communication between research and practice is desirable, because practitioners and decision-makers need more user-friendly models and tools that allow them to more routinely quantify health consequences of their policy proposals. More dynamic, participatory, and full-chain HIA approaches are needed, which need to include citizen and stakeholder viewpoints, to ensure that evaluated scenarios are desired and supported by the affected communities and that the complexities of reality are more reasonably reflected.

References

Atkinson, R. W., et al. (2018). Long-term concentrations of nitrogen dioxide and mortality. *Epidemiology, 29*(4), 460–472.

Bach, M., et al. (2017). Participatory epidemiology: The contribution of participatory research to epidemiology. *Emerging Themes in Epidemiology, 14*(2), 1–15.

Briggs, D. J. (2008). A framework for integrated environmental health impact assessment of systemic risks. *Environmental Health,* 7(61), 1–17.

Burstein, P. (1998). Bringing the public back in: Should sociologists consider the impact of public opinion on public policy? *Social Forces,* 77(1), 27–62.

CE Delft. (2019). *Handbook on the external costs of transport: Version 2019.* Available at: https://ec.europa.eu/transport/sites/transport/files/studies/internalisation-handbook-isbn-978-92-79-96917-1.pdf.

Cesaroni, G., et al. (2014). Long term exposure to ambient air pollution and incidence of acute coronary events: Prospective cohort study and meta-analysis in 11 European cohorts from the ESCAPE Project. *BMJ, 348*(f412), 1–16.

Collins, J., & Koplan, J. (2009). Health impact assessment: A step toward health in all policies. *JAMA, 302*(3), 315–317.

Creutzig, F., Mühlhoff, R., & Römer, J. (2012). Decarbonizing urban transport in European cities: Four cases show possibly high co-benefits. *Environmental Research Letters,* 7(4), 044042.

Dadvand, P., et al. (2014). Residential proximity to major roads and term low birth weight: The roles of air pollution, heat, noise, and road-adjacent trees. *Epidemiology (Cambridge, Mass.), 25*(4), 518–525.

European Environment Agency. (2019). *Air quality in Europe.* Available at: http://www.eea.europa.eu/publications/air-quality-in-europe-2012.

Eze, I. C., et al. (2015). Association between ambient air pollution and diabetes mellitus in Europe and North America: Systematic review and meta-analysis. *Environmental Health Perspectives, 123*(5), 381–389.

Forsyth, A., Schively Slotterback, C., & Krizek, K. (2010). Health impact assessment (HIA) for planners: What tools are useful? *Journal of Planning Literature, 24*(3), 231–245.

GBD 2016 Risk Factors Collaborators. (2017). Global, regional, and national comparative risk assessment of 84 behavioural, environmental and occupational, and metabolic risks or clusters of risks, 1990–2016: A systematic analysis for the Global Burden of Disease Study 2016. *The Lancet, 390,* 1345–1422.

Geurs, K. T., Boon, W., & Van Wee, B. (2009). Social impacts of transport: Literature review and the State of the practice of transport appraisal in the Netherlands and the United Kingdom. *Transport Reviews, 29*(1), 69–90.

Gold, M. R., Stevenson, D., & Fryback, D. G. (2002). HALYs and QALYs and DALYs, Oh My: Similarities and differences in summary measures of population health. *Annual Review of Public Health, 23*(1), 115–134.

Grabow, M. L., et al. (2012). Air quality and exercise-related health benefits from reduced car travel in the midwestern United States. *Environmental Health Perspectives, 120*(1), 68–76.

Hänninen, O., & Knol, A. (2011). *European perspective on environmental burden of disease. Estimates for nine stressors in six European countries.* Helsinki.

Héroux, M., et al. (2015). Quantifying the health impacts of ambient air pollutants: Recommendations of a WHO/Europe project. *International Journal of Public Health, 60,* 619–627.

Jacobs, L. R., & Shapiro, R. Y. (1994). Questioning the conventional wisdom on public opinion toward health reform. *Political Science & Politics,* 27(2), 208–214.

Joffe, M., & Mindell, J. (2005). Health impact assessment. *Occupational and Environmental Medicine, 62*(12), 907–912.

Khreis, H., de Hoogh, K., & Nieuwenhuijsen, M. J. (2018). Full-chain health impact assessment of traffic-related air pollution and childhood asthma. *Environment International, 114*, 365–375.

Khreis, H., et al. (2018). Traffic-related air pollution and the local burden of childhood asthma in Bradford, UK. *International Journal of Transportation Science and Technology, 8*(2), 116–128.

Khreis, H., et al. (2019). Outdoor air pollution and the burden of childhood asthma across Europe. *European Respiratory Journal, 54*.

Künzli, N., Perez, L., & Rapp, R. (2010). *Air quality and health* (pp. 1–66). European Resipratory Society.

Lopez, A. D., et al. (2006). *Global burden of disease and risk factors*. Oxford University Press and the World Bank.

Macmillan, A., & Woodcock, J. (2017). Understanding bicycling in cities using system dynamics modelling. *Journal of Transport and Health*, 7(August), 269–279.

Macmillan, A., et al. (2014). The societal costs and benefits of commuter bicycling: Simulating the effects of specific policies using system dynamics modeling. *Environmental Health Perspectives, 122*(4), 335–344.

McKinley, G., et al. (2005). Quantification of local and global benifits from air pollution control in Mexico City. *Environmental Science and Technology, 39*(7), 1954–1961.

Mindell, J., Ison, E., & Joffe, M. (2003). A glossary for health impact assessment. *Journal of Epidemiology and Community Health, 57*(9), 647–651.

Mindell, J., et al. (2010). Improving the use of evidence in health impact assessment. *Bulletin of the World Health Organization, 88*(7), 543–550.

Mueller, N., et al. (2017a). Health impacts related to urban and transport planning: A burden of disease assessment. *Environment International, 107*, 243–257.

Mueller, N., et al. (2017b). Urban and transport planning related exposures and mortality: A health impact assessment for cities. *Environmental Health Perspectives, 125*(1), 89–96.

Mueller, N., et al. (2019). Changing the urban design of cities for health: The superblock model. *Environment International*, 105132. https://doi.org/10.1016/j.envint.2019.105132.

Murray, C., & Lopez, A. (1999). On the comparable quantification of health risks: Lessons from the global burden of disease study. *Epidemiology, 10*, 594–605.

Murray, C., et al. (2004). *Comparative quantification of health risks: Conceptual framework and methodological issues. Vol. 1*. Geneva: World Health Organization.

National Research Council of the National Academies. (2011). *Improving health in the United States: The role of health impact assessment*. Washington, DC: The National Academies Press.

Nieuwenhuijsen, M. J., & Khreis, H. (2016). Car free cities: Pathway to healthy urban living. *Environment International, 94*, 251–262.

Nieuwenhuijsen, M. J., et al. (2014). Air pollution and human fertility rates. *Environment International, 70*, 9–14.

Nieuwenhuijsen, M. J., et al. (2017). Participatory quantitative health impact assessment of urban and transport planning in cities: A review and research needs. *Environment International, 103*, 61–72.

Parry, J., & Stevens, A. (2001). Prospective health impact assessment: Pitfalls, problems, and possible ways forward. *BMJ, 323*(7322), 1177–1182.

Pedersen, M., et al. (2013). Ambient air pollution and low birthweight : A European cohort study (ESCAPE). *The Lancet Respiratory Medicine, 1*(13), 695–704.

Shafie, F. A., Omar, D., & Karuppannan, S. (2013). Environmental health impact assessment and urban planning. *Procedia - Social and Behavioral Sciences, 85*, 82–91.

Stafoggia, M., et al. (2014). Long-term exposure to ambient air pollution and incidence of cerebrovascular events: Results from 11 European Cohorts within the ESCAPE Project. *Environmental Health Perspectives, 122*(9), 919–925.

Ståhl, T., et al. (2006). *Health in All Policies. Prospects and Potentials*. Available at: http://www.euro.who.int/__data/assets/pdf_file/0003/109146/E89260.pdf.

Stansfeld, S. A. (2015). Noise effects on health in the context of air pollution exposure. *International Journal of Environmental Research and Public Health, 12*(10), 12735–12760.

Stevenson, M., et al. (2016). Land use, transport, and population health: Estimating the health benefits of compact cities. *The Lancet, 388*, 2925–2935.

Sundvor, I., et al. (2012). *Road traffic's contribution to air quality in European cities*. ETC/ACM Technical Paper 2012/1.

Tainio, M. (2015). Burden of disease caused by local transport in Warsaw, Poland. *Journal of Transport & Health, 2*(3), 423–433.

Tétreault, L., Perron, S., & Smargiassi, A. (2013). Cardiovascular health, traffic-related air pollution and noise: Are associations mutually confounded? A systematic review. *International Journal of Public Health, 58*(5), 649–666.

Thacker, S., et al. (2006). Measuring the Public's health. *Public Health Reports, 121*(1), 14–22.

Thomson, H., et al. (2008). Assessing the unintended health impacts of road transport policies and interventions: Translating research evidence for use in policy and practice. *BMC Public Health, 8*(339), 1–13.

Veerman, J. L., Barendregt, J. J., & Mackenbach, J. P. (2005). Quantitative health impact assessment: Current practice and future directions. *Journal of Epidemiology and Community Health, 59*(5), 361–370.

Whitehead, S. J., & Ali, S. (2010). Health outcomes in economic evaluation: The QALY and utilities. *British Medical Bulletin, 96*(1), 5–21.

WHO. (1999). *Health impact assessment: Main concepts and suggested approach*. Gothenburg consensus paper, Gothenburg.

WHO. (2014). *WHO Expert Meeting: Methods and tools for assessing the health risks of air pollution at local, national and international level*. Copenhagen.

Wismar, M., et al. (2007). *The effectiveness of health impact assessment: Scope and limitations of supporting decision-making in Europe*. Copenhagen.

Woodcock, J., et al. (2009). Public health benefits of strategies to reduce greenhouse-gas emissions: Urban land transport. *The Lancet, 374*(9705), 1930–1943.

Xia, T., et al. (2015). Traffic-related air pollution and health co-benefits of alternative transport in Adelaide, South Australia. *Environment International, 74*, 281–290.

CHAPTER 14

Impacts of traffic-related air pollution on policy- and decision-making

Michael O. Rodgers[a] **and April Willis Rodgers**[b,c]
[a]School of Civil and Environmental Engineering and School of Public Policy, University System of Georgia and Georgia Institute of Technology, Atlanta, GA, United States
[b]Avatar Environmental and Technical Services, Marietta, GA, United States
[c]Samford University, Birmingham, AL, United States

Abbreviations

CDC	Centers of Disease Control and Prevention
EGR	exhaust gas recirculation
EPA	Environmental Protection Agency
HEI	Health Effects Institute
kg	kilogram
MSAT	mobile source air toxics
NAAQS	National Ambient Air Quality Standards
NHTSA	National Highway Traffic Safety Administration
NEPA	National Environmental Policy Act
PCV	Positive Crankcase Ventilatio
PM	particulate matter
SCR	selective catalytic reduction
USDHEW	United States Department of Health, Education and Welfare
ULSD	ultralow sulfur diesel
U.S.C	United States Code
VOC	volatile organic compounds

Introduction—Pollution in the city

After years of deterioration, the situation had finally become unbearable. The transportation system that had helped to build the city and on which its residents were now dependent had seemingly turned on them. The technology that was embraced so enthusiastically just a few decades ago now seemed to be a villain threatening the viability of the city itself moving into the future. On stagnant air days, that seemed all too frequent, residents resorted to covering their faces with masks or handkerchiefs that offered little respite from the heavy air that blanketed the city. While the wealthy often

Traffic-Related Air Pollution
https://doi.org/10.1016/B978-0-12-818122-5.00014-4

fled the city during the summer when the pollution was at its worst, those less fortunate were left to make do as well as they could. The health impacts of the pollution that choked the city were large and getting progressively worse with each passing year. While both public health officials and the press opined for the need for action, neither had the power to directly address the problem. Public officials finally realized the time had come for action and, belatedly, began to face the need for change.

While this scene could have been describing Los Angeles in the early 1960s, Beijing in the early 2000s or Delhi today, it is actually describing the conditions in New York and London in the 1890s. While the former issues are associated with the rise of the automobile and traffic-related air pollution, New York and London were dealing with another transportation-related pollution problem: the accumulation of horse manure on the city streets associated with the horse-drawn street railways of the era as well as countless other horse-drawn vehicles (Johnson, 2020; Kohlstedt, 2017). These street railways (horsecars) typically featured a single car drawn by a pair of horses along a railway set into a street. Horsecars offered a significant advancement in both speed and comfort compared to earlier horse-drawn carriages and began to become a dominant mode of city travel beginning in the 1860s. Each horsecar in service typically required eight horses operating in pairs on a shift basis for its operations and thus for a large city with many lines the horse population was quite large.

While it is difficult to get a precise estimate of the horse population in either city, it is clear that both cities were home to between 100,000 and 250,000 horses used primarily in the transportation industry (Turvey, 2005). Since a working horse typically produced 8–10 kg of manure per day as well as a significant quantity of urine, the problem was substantial. In 1895, driven by deteriorating conditions (Fig. 14.1) the City of New York hired one of the most prominent American sanitary engineers, George E. Waring Jr. to develop a plan to address the problem (Kohlstedt, 2017). Waring was not new to either public health or New York City. Waring had successfully developed the first large dedicated sanitary sewer system in North America for Memphis, Tennessee to address epidemics of yellow fever and cholera that periodically plagued the city (History Channel and Waring, 2018). In New York, he had worked with Frederic Olmstead to drain the wetlands that became Central Park. After his appointment, Waring took action quickly and pushed for laws that required stabling horses at night (to force their owners to deal with overnight waste issues on their own) and

MORTON STREET, CORNER OF BEDFORD, LOOKING TOWARD BLEECKER STREET, MARCH 17, 1893.

Fig. 14.1 New York city street showing accumulation of horse manure and urine (Kohlstedt, 2017).

hired crews to gather manure for sale as fertilizer. Horse corpses, often left rotting on the street, were collected and sold for glue. That which could not be sold was transported and (unfortunately) dumped elsewhere. Despite the costs, these efforts were both successful and popular and circumstances improved significantly. So much so that the city elected to hold a parade for the white-clad sanitation workers the following year (1896).

Traffic-related air pollution

Unfortunately, there are no parades for modern traffic-related air pollution. While the costs, sanitary considerations, and care requirements made residents relatively eager to substitute internal combustion engines for horses in transportation applications, no similar transportation revolution has yet occurred to displace the internal combustion engine responsible for modern air pollution concerns. Such changes may be coming, however. As Professor Dan Sperling describes in this book, *The Three Revolutions* (Sperling et al., 2018), that the simultaneous effects of vehicle electrification, shared mobility, and the introduction of autonomous and connected vehicles will likely significantly alter our transportation system over the next generation. While such changes offer additional hope for the future of transportation, current issues must be addressed within the context of current technologies and the existing vehicle fleet.

Policy- and decision-making

Decision-making is the process of selecting among alternatives. This process can involve choices *within* the same context: "Do I walk, drive or take the bus to the movies?" or *between* contexts: "Do I go to the movies or take a nap?" Decision-making *within* a context carries with it the consequences of a series of previous decisions that set the context for the current decision. For example, before buying an article of clothing in a store, you have made a whole series of decisions that lead you to being at that store at that time and in possession of the resources necessary to make the purchase. Decision-making *between* contexts can be thought as the colloquial "apples vs oranges" comparison (paradoxically, that comparison is actually a *within* decision between two fruits). While in the broadest sense there are very few true *between* decisions to be made since all decisions are made within a certain context, some decisions (e.g., do I go to college or do I go to work after high school) offer a significant enough difference in potential outcomes to effectively be *between* decisions.

Like individuals, organizations including governments need to make routine decisions in the context of their responsibilities. These decisions range from the routine "do I buy this type of paper or the other" to the strategic "how should we treat a person who has stolen goods from another." Since these latter decisions can have a broad impact on the way that the organization or society is organized and operates, these decisions are normally reserved to selected members of the group who are entrusted to make the highest level decisions. In a corporate setting, these individuals would normally be the corporate officers and the board of directors. In a governmental setting these individuals would most often be legislators and/or the highest level governmental officials (in democratic societies these officials would normally be elected).

In the public sector, the highest level decision-making involves the development of *laws* that represent the overall context *within* which other decision-making will occur. Typically, laws are not written to be absolute endorsement or prohibition of an activity but rather provide a *legal context* or authority for additional decision-making at a lower level. For example, the Clean Air Act of 1970 (42 USC §7401 et seq., 1970) did not specify the standards required for carbon monoxide in the air but rather instructed US EPA to develop such standards. These subsequent lower level decisions within a legal framework form what we refer to as either *standards* or *policies.* In most cases these policies can be thought of a "default decisions" that will be followed unless unusual conditions are present. As opposed to laws, policies generally include provisions for exceptions. Normally these exceptions can be authorized only by the highest ranking officials within organization and only under certain (normally emergency) conditions. These exceptions are designed to allow officials the flexibility to deal with conditions that are either rare or were not considered at the time that either the laws or policies were developed.

Air pollution in a historical context

Air pollution is not a new problem. In ancient Sumer, Hammurabi's code (no. 232) required compensation for goods ruined by smoke. In Egypt, evidence of air pollution-induced lung disease have been found in many ancient mummies (Mongomerie, 2013). In Roman times, the poet Horace decried "smoke-grimed statue scarce divine" (Odes and Carmen Saeculare) within Rome itself. Later, Seneca the Younger, the roman stoic philosopher and teacher of Nero, noted:

> *As soon as I escaped the heavy air of Rome and the stench of its smoky chimneys which when stirred poured forth whatever pestilent vapors and soot they held enclosed, I felt a change in my disposition (Letter to Lucilius, 61 CE).*

About the same time Pliny the Elder noted (*Naturalis Historia*) that slaves who both mined and wove asbestos often suffered from lung disorders. Interestingly, Rome did avoid many of the problems associated with horses that we described earlier as Julius Caesar banned the use of horse-drawn carts within the boundaries of Rome [*Lex Julia Municipalis* (44 BCE)]. Later the Justinian Code [*Corpus Juris Civilis* (535 CE)] noted that "the air, running water, the sea, and consequently the shores of the sea" were held in common by mankind setting the stage for environmental law.

Things did not get much better later in the middle ages. Like their Egyptian counterparts' millennia earlier, naturally mummified Viking remains often show signs of lung disease from living in long houses kept intentionally smoky for insect control. In the 12th century, the Jewish philosopher and personal physician to the Sultan Saladin, Moses Maimonides, traveled extensively in North Africa and the Middle East and found the cities that he visited to be "stuffy, smoky, polluted, obscure, and foggy" and unhealthy (Rosner, 1987).

In England, Edward I ("Long Shanks") banned the burning of sea coal in London when parliament was in session due to "smoggy" conditions (Freese, 2005). Edward's laws were ineffective, and much later William Shakespeare noted in his play *Julius Caesar* (Act 1, Scene 2) that "...I durst not laugh for fear of opening my lips and receiving the bad air." About the same time John Evelyn, the father of the English garden, wrote a proposal to parliament and the English king titled "FUMIGUGIUM or the Inconvenience of the Aer and Smoak of London Dissipated" that called for replacing sea coal with wood-based fuels for London and its environs (Evelyn, 1661).

Up until the beginning of the 19th century, most of this urban air pollution was associated with the use of fuels for cooking, for domestic heating, and for industry. With the advent of the steam engine and the rapid growth of cities fueled by the industrial revolution, the contribution of transportation to air pollution began to rise rapidly arising from the aforementioned horse sources as well as from the railroad and, in some cases, steamships.

In 1872, Robert Angus Smith, the first alkali inspector in England who also coined the term "acid rain," provided the first known technical description of air pollution in the book "Air and Rain: The Beginning of a Chemical Climatology" (Smith, 1872). In this book, he noted the pollution associated with the railroads of the day:

> *When sitting in a railway carriage with a friend ... that Gentleman observed that the particles of dust which floated in the air seemed to shine with a metallic lustre. I immediately collected some, and found that the larger class were in reality rolled plates of iron...Another smaller class were less brilliant and when looked at with considerable power showed many inequalities of surface...The dust enters the mouth and lungs and has to be taken as one of the evils of railway travelling... (p. 447).*

It may be surprising to a modern observer, but the potential health impacts of air pollution were often dismissed during this period. Smith also noted:

> *Medical men have objected to the argument that any evil can result from these effects, saying that man is formed so as to resist such influences ... but that there are evil omens is true or we have long been deceived. (p. 224).*

The beginning of the 20th century saw two changes that would dramatically impact the nature of air pollution. The first was the gradual redirection of the use of steam power from industrial and transportation applications to the generation of electricity. While the initial development of electricity was strongly impacted by the "current wars" between Thomas Edison (DC current) and George Westinghouse/Nicola Tesla (AC current). The ultimate triumph of AC current was due, in no small measure, to its ability to carry electricity over longer distances. This allowed electrical power generation facilities to move progressively greater distances from populated areas, thereby reducing direct air pollution impacts on its customers.

The second major trend impacting urban air pollution was in the widespread adoption of the internal combustion engine, especially in the transportation sector. The high power-to-weight ratio of internal combustion engines made aviation practical and strongly influenced its adoption in other sectors. As cars and trucks became more prevalent, oil production and refining increased in step. As a consequence of this increased petroleum production and refining, the use of petroleum-based products in nonfuel applications (e.g., asphalt for roads, plastics, and other petrochemicals including pesticides) increased as well.

By mid-century, motor vehicles were ubiquitous, and the design of cities were strongly impacted by the needs of this new transportation system. Movements between cities were also impacted. The National Interstate and Defense Highways Act [70 Stat. 374 (1956)] that created the US Interstate Highway System significantly altered the mobility of both goods and passengers over longer distances. In response, overall vehicle-miles-traveled began a decade-long increase that has only recently begun to slow down. Not surprisingly, this increase in travel had significant implications for air pollution. In the next section, we will examine the development of some of policies arising from these developments.

Air pollution, human health, and public policy

In looking back over the history of policy responses to air pollution, one is tempted to have such responses as "How could they not a known that it was a problem?" or "I'm sure glad that we would never make such a mistake." While there are certainly many examples of appalling mistakes or "wrong-headed" policies, it is far more common that older policies were based on well-reasoned decisions based on the different social priorities of the time. The extent to which these older concerns are no longer applicable often makes these decisions difficult to understand. This is particularly true in the case of human health concerns.

For example, when automobiles first became prevalent in cities they were replacing horses and other draft animals for human and goods mobility and were thus often marketed as being "cleaner," cheaper, and more reliable than the mode they replaced. While they certainly had emissions, they were certainly less noticeable to the individuals involved than the ones they replaced. These vehicles did, however, have an appalling safety record. For example, the highway motor vehicle death rate in 1920 was approximately *25 times* higher than current fatality rates on a per-mile basis (NHTSA, 2016). Similarly, the 1918 Spanish Flu epidemic impacted approximately one-third of the human population and was responsible for the deaths of approximately 675,000 people in the United States (CDC, 2020). Not surprisingly, public health officials at the time were more concerned with the basic safety of these new motor vehicles and their contribution to the spread of infectious disease than in their potential contribution to air pollution.

By the end of the World War II, the annual death toll from infectious disease had begun to wane and concern over the health effects of air pollution was building. The London "Killer Fog" that was responsible for the deaths of more than 4000 people during December 1952 received worldwide attention as had earlier, albeit smaller, incidents in Donora, Pennsylvania in 1948 and the Meuse Valley in Belgium in 1930. Although these cases were largely caused by the stationary sources of pollution, the influence of transportation source on air pollution was also being considered. A. J. Haagen-Smit a professor of chemical engineering at Cal Tech in the United States and Christian Junge in Germany established the chemical basis for urban photochemical smog in the 1950s. Dr. Haagen-Smit went on to become the first chairman of the California Air Resources Board (CARB) and helped to design the first emissions control device to be installed on an automobile in 1961.

While photochemical smog was often considered to be a "California Problem" much of the attention of the health care community in the 1960s was focused on what were perceived to be bigger issues. In January 1964 the Surgeon General of the United States issued the report *Smoking and Health* that for the first time in a government report directly linked cigarette smoking to lung cancer and other respiratory diseases (USDHEW, 1964). At that time, approximately 40% of the adult population smoked and second-hand smoke was prevalent in virtually all indoor environments. As a result of this, and subsequent reports, nonsmoking campaigns became a major public health effort that extends to the present day.

Other concerns were also present. The rise in highway speeds related to the new interstate highways renewed concerns regarding highway safety.

When the new MotorVehicle Safety Standard that required all new vehicles be equipped with seat belts took effect in 1968 (49USC-301); there were more than 50,000 fatalities per year on American Roadways approximately 70% more than that today from a much smaller population (NHTSA, 2016).

There was also increasing concern regarding occupational exposure to air pollutants. Building on the legacy of Dr. Alice Hamilton at Hull House in studying occupational medicine, numerous studies identified high levels of lung disease among coal miners (Black Lung Disease), cotton workers (Brown Lung Disease), and various jobs involved with silica-containing materials (Silicosis) as well as other less prevalent diseases (e.g., asbestosis) (see, e.g., American Thoracic Society, 2010).

Air pollution was, of course, not the only environmental issue of concern to decision makers of the era. Rachel Carson in her seminal work *Silent Spring* in 1962 (Carson, Darling, & Darling, 1962) pointed out the widespread impacts of the pesticide DDT on wildlife. Perhaps more visible were the impacts of water pollution on the nation's rivers and streams. Eutrophication of lakes resulted in massive fish kills that received widespread attention, especially in the area of Lake Erie that was perhaps the most polluted of the great lakes. Concerns about water pollution were further compounded on June 22, 1969 when the Cuyahoga River in Cleveland Ohio caught fire and burned for several hours. While it was not the first time that the Cuyahoga had burned (a 1952 fire was actually larger), it was much more extensively covered and provided a visible image of pollution problems for the nascent environmental movement. The same year an oil spill near Santa Barbara California was also influential in impacting public opinion regarding the need for new environmental laws. At about the same time, the Sierra Club helped to publish a book on the national problem of air pollution: *Moment in the Sun*. Written by Dr. Robert Rienow, a professor of political science in the State University of New York at Albany and his wife Leona Train Rienow, this best-selling book did much to galvanize the concern over air pollution as well (Rienow & Rienow, 1967).

Motivated into action, the US Congress passed the National Environmental Policy Act (NEPA, 83 Stat. 852) that was signed into law on New Year's Day 1970. NEPA established the legal basis for environmental impact assessment. A few months after the signing of NEPA, organizers (including several influential members of congress) promoted the first "Earth Day" on April 22, 1970. At the time, Earth Day was the largest demonstration in US history involving more than 20 million participants, about 10% of the US population at that time (Earthday, 2020).

Motivated by these concerns, 1970 saw the passage of the *Clean Air Act*, the *Clean Water Act*, the *Endangered Species Act*, and the founding of the US Environmental Protection Agency (US EPA) along with NEPA.

The clean air act and environmental regulation

The Clean Air Act of 1970 [42 USC §7401 et seq. (1970)] was the first comprehensive air pollution legislation in the United States as earlier legislation had focused mostly on voluntary programs. The new EPA was tasked with developing National Ambient Air Quality Standards (NAAQS) for six pollutants. Specifically, these were carbon monoxide, nitrogen dioxide, sulfur dioxide, lead, ozone, and particulate matter. Since then, the standards for individual pollutants have changed several times in terms of allowable levels and/or averaging time, however, the only standard that has undergone significant changes in the form of the standard is that for particulate matter. The original standard for total suspended particulates (TSP) was amended to include a limit for particulate matter with an aerodynamic diameter of less than 10 μm (PM_{10}) as a result of the 1977 amendments (91 Stat. 685) and a separate limit for "fine" particulate matter of less than 2.5 μm ($PM_{2.5}$) as a result of the 1990 amendments (104 Stat. 2468).

The Clean Air Act established two different NAAQS, a primary standard for the protection of human health and a secondary standard for "welfare" effects including impacts on wildlife or cultural resources. In practice, in virtually all cases the adopted standards are identical or effectively so. Since the NAAQS primary standard for each pollutant is established to "protect human health of sensitive populations with an adequate margin of safety," health concerns were, and are, essential in determining both the standards according to which air pollution is judged and the policies put in place to achieve these standards.

At this point it is important to distinguish between two types of pollutants: those that are emitted directly from a source (e.g., an automobile tailpipe), a *primary pollutant,* and those that are formed *within the atmosphere*, due to chemistry involving the emissions of "precursor" compounds. The latter are referred to as *secondary pollutants*. Among the NAAQS, only ozone is a purely secondary pollutant formed by the reaction of volatile organic compounds (mostly hydrocarbons from man-made sources such as gasoline or naturally emitted compounds from certain plants) with nitrogen oxides [including both nitric oxide (NO) and nitrogen dioxide] in the presence of sunlight (urban smog chemistry). The other compounds are all emitted by man-made (anthropogenic) sources although particulate matter can also be formed by the same chemistry that produces ozone.

Policies aimed at limiting the human health impacts of air pollutants are aimed at reducing human exposure to these compounds. This can be done in two general ways. The first is by limiting emissions of either the compounds themselves (for primary pollutants) or their precursor compounds (for secondary pollutants) from the source in terms of either quantity and/or reactivity. The second approach is to increase the dilution of these emissions before they encounter a sensitive human population. The former approach is mostly related to the types of fuels used and the presence and effectiveness of emissions control devices while the second is mostly related to where sources of emissions relative to these sensitive populations.

Human health and mobile source emissions controls in the United States

For the implementation of the provisions of the Clean Air Act of 1970 related to traffic-related (i.e., from passenger cars and commercial vehicles) emissions, EPA's initial thrust was in the implementation of emissions standards for new vehicles and in the modification of existing vehicle fuels. The first was clear as motor vehicles were a major source of both volatile organic compounds (VOC) and nitrogen oxides, the precursors to ozone formation and carbon monoxide being other criteria pollutants. The second emphasis on motor fuels was also understandable as, at the time the major sources for atmospheric emissions of lead, one of the primary pollutants identified by the act, have been combustion of motor fuels.

Tetraethyl lead

Lead in the form of tetraethyl lead had been introduced into motor fuels in the 1920s as an inexpensive octane enhancer. Gasoline at that time was generally of low quality and mostly came from direct distillation of crude oils. Octane enhancers were necessary to ensure that engines did not "knock" (i.e., to undergo compression ignition or "dieseling") during operations or continue to "run on" after the engine was stopped. Tetraethyl lead as the octane enhancer of choice was controversial from the beginning due to its potential health consequences (in particular Dr. Alice Hamilton objected); its availability relative to other alternatives made it dominant in the marketplace [see, for example, Rosner & Markowitz, 1985].

Enhancements in refining in the years after World War II including catalytic cracking of petroleum and steam reforming had both increased the range of crude oil supplies that could be converted into gasoline and diesel fuel and made octane enhancers unnecessary as other organic compounds could now be used to improve octane ratings. However, most gasoline suppliers continued to use tetraethyl lead for economic reasons.

The use of lead in motor fuels had vastly increased the quantity of lead in the urban environment. Lead had been known from ancient times to produce neurological symptoms and had more recently been shown to have even more adverse impacts on neurological development in children. From a health perspective, removal of lead from motor fuels was a priority for clean air act policies. However, there were practical difficulties with rapid removal of lead from fuels. The existing motor vehicle fleet had been designed to operate with leaded fuels. This was significant in that, in addition to its octane enhancement characteristics, tetraethyl lead had lubricating properties that manufacturers had anticipated in the design of existing vehicles. In addition, although the production of high octane fuels with lead was technically feasible, not all refineries, especially the smaller ones, were equipped to manufacture such fuels in the 1970s.

Alternatively, if both leaded and unleaded fuels were to be offered at the same time, the retail fuels industry would have to have separate infrastructure available to store, deliver, and dispense the fuels. This infrastructure was also not universally available at the time. However, removal of lead from fuels was essential to meet the EPA's other objective in motor vehicle emissions controls. While technologies such as positive crankcase ventilation (PCV valves) and emissions gas recirculation (EGR valves) could reduce some emissions of volatile organic compounds and nitrogen oxides, respectively, significant reduction in motor vehicle emissions would require additional emissions controls. These controls were available in the form of catalytic convertors that use precious-metal-based catalysts to oxidize VOC and carbon monoxide into carbon dioxide. Later (three-way) catalytic convertors would also be able to reduce nitrogen oxides to nitrogen gas. However, these catalysts were contaminated (fouled) by the presence of lead in gasoline thus making unleaded fuels essential to achieving these necessary emissions reductions (EPA, 1973).

This is a recurrent theme associated with policies associated with mobile source emissions. The transportation industry is vast as is its supporting infrastructure. Vehicles need to be designed and lead times extend several years. Likewise, investments in refinery, pipeline, storage, and refueling infrastructure are typically made years in advance of the date in which they will be needed. Similarly, significant changes in vehicle design, for example, the addition of new emissions control systems, will usually require the construction of new facilities to build the components and training of maintenance personnel on their repair or replacement in the field. For this reason, policies often need to have long phase in periods or have implementation dates a significant time into the future.

EPA thus mandated a dual strategy for the removal of lead from gasoline in 1973 (EPA, 1973). Catalytic converters first appeared on US cars during the 1975 model year and were mandatory on all new vehicles beginning with the 1981 model year (catalytic convertor phase in). Similarly, EPA required the retailers to make available at least some unleaded fuels at that time while still allowing the sale of leaded fuels. To avoid misfueling of vehicles, vehicles requiring unleaded fuels were required to be marked "unleaded fuels only" near the refueling port (a requirement that still exists today) that was designed to be smaller than the standard refueling nozzle of the time (i.e., the leaded gasoline nozzle would not fit into the smaller refueling port on a vehicle equipped with a catalytic convertor). Pumps providing unleaded fuels were equipped with a new nozzle that would fit these smaller unleaded refueling ports. At the same time, EPA mandated a gradual reduction in lead content of "leaded" gasoline to allow refiners time to increase their capacity to make unleaded fuels. The mandate, originally overturned by a lawsuit by Ethyl corporation but reinstated on appeal, was for a 91% reduction in lead content by 1986. The remaining lead was allowed in fuel to meet the lubrication needs of pre-1975 model year vehicles that were designed to operate on leaded fuel. Eventually, other additives were found to meet these lubrication requirements and tetraethyl lead was finally banned in 1996. Other countries followed suit on different time frames notably China and India in 2000. The lead removal program can be considered one of the great successes of pollution control with airborne levels of lead reduced by more than 98% since the passage of the clean air act (EPA, 2019).

New vehicle emissions requirements

The 1977 and the 1990 amendments to the Clean Air Act and periodic lowering of the NAAQS for ozone and particulate matter lead to a long-term program of reduction in emissions standards, expressed in grams/mile of activity, throughout the world. When coupled with requirements for improved fuel economy (corporate average fuel economy, or CAFÉ, standards), it became clear that earlier vehicle technologies would be unable to meet the emissions standards of the future and during much of the 1980s manufacturers struggled to meet the new standards.

Eventually, advances in computerized emissions control greatly enhanced our ability to operate vehicles in a low emissions mode. The combination of advanced sensors, multiport fuel injection, improved three-way catalysts, and other technologies enabled progressively more stringent emissions and fuel economy standards to be achieved. The cumulative impact of these changes is substantial. Criteria pollutant emissions from a new passenger car, SUV or light duty truck today are roughly 1% as much as a typical vehicle at the time of the Clean Air Act of 1970, a 99% reduction (EPA, 2019). While not quite as large, the emissions from heavy-duty vehicles have also been drastically reduced. Fig. 14.2 shows the history of EPA's on road emissions standards for heavy-duty diesel vehicles (HEI, 2015).

Not surprisingly, no single policy or program is responsible for these reductions. For example, some technologies for the reduction of emissions and increasing fuel economy in gasoline vehicles available in the late 1990s (e.g., direct injection of fuel into cylinders) could not be effectively implemented due to the relatively high levels of sulfur allowed in gasoline

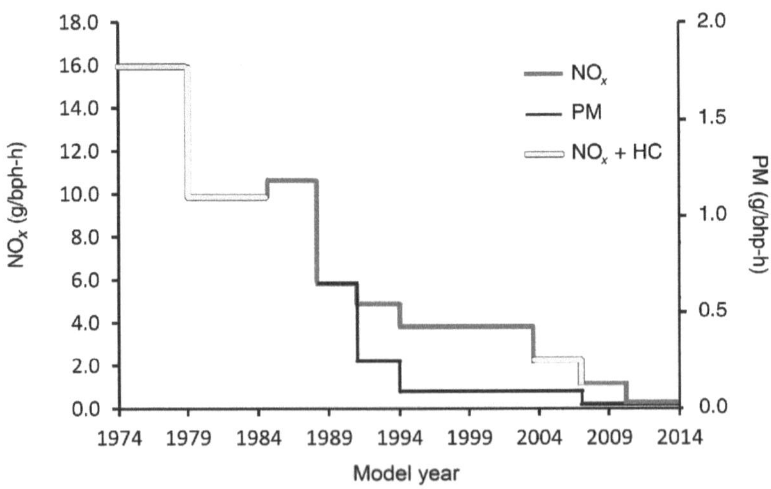

Fig. 14.2 US Emissions standards for heavy-duty vehicles from 1974 to 2015. *(Reproduced from U.S. Environmental Protection Agency. (2013). Heavy-duty highway compression-ignition engines and urban buses—exhaust emission standards. Available: www.epa.gov/oms/standards/heavy-duty/hdci-exhaust.htm.)*

at the time (up to 500 ppm). As was the case for lead more than two decades ago, many smaller refineries lacked the desulfurization capacity to deliver very low sulfur gasoline. Beginning in 2004, EPA mandated a cap for 300 ppm for sulfur in gasoline and required the average fuel to meet a standard of 120 ppm. These standards were reduced to an 80-ppm cap and a 30-ppm average 2 years later although EPA provided waivers for some small refiners until 2007. Beginning in 2017, these standards were further lowered to an average of 10 ppm of sulfur with the same 80 ppm cap for gasoline (EPA, 2020).

A similar situation existed for diesel emissions reductions where high-efficiency selective catalytic reduction (SCR) systems for the reduction of NO_x emissions were inhibited by high sulfur levels in much the same way that lead had fouled early gasoline catalytic convertors. In June 2010, the ultralow sulfur diesel (ULSD) rule became effective that established a limit of 15 ppm in on-road diesel fuels. Collectively, these new vehicle emissions standards have begun to reach practical technological limits for the technology and further reductions in pollutant exposures will require different approaches (EPA, 2020).

Exposure controls

While source-based emissions reductions have been significant, in many cases high vehicle activity can still lead to unhealthy levels of pollutant exposures. Efforts to encourage active transportation modes including bicycling and walking offer positive health benefits to the participants through increased physical activity, but only to the extent that participants can conduct these activities safely and are not exposed to unhealthy levels of traffic-related air pollutants. As stated earlier, a second approach to controlling exposure is by increasing the dispersion of pollutants between sources (vehicles) and receptors (the general public or active transportation participants).

Particularly in densely populated areas, transportation control measures and urban planning can influence the number and configuration and flow of transportation routes and types of transportation available. Each of these options should be considered within the context of both human mobility and the resulting health impacts of these decisions (see, e.g., Wernham, 2011).

Surrounding urban highways with green spaces, particularly with tree species that can absorb fine airborne particulate matter (Chen, Liu, Zhang, Zou, & Zhang, 2017), can effectively reduce local pollutant concentrations.

Similarly, increasing the separation between pedestrian and cycling routes and traffic (see, e.g., Brugge et al., 2015) by both distance and through physical barriers effectively increases safety and can help to buffer the effects of traffic-related air pollution. Encouraging public transport, shared rides, carpooling, and active transportation can reduce the number of vehicles on the road and thus reduce health and safety risks.

Throughout the world, there are a wide variety of approaches to encourage alternative transportation modes. In particular, the Dutch have emphasized cycling as a transportation mode and design much of their transportation infrastructure with an emphasis on cycling safety and convenience. Not surprisingly, cycling has a vastly higher total trips in the Netherlands than in the United States. Similar policies for walkable cities and pedestrian only facilities within otherwise automobile-centric cities have had substantial success. For example, Car-Free Sundays in Mexico City offer pedestrians and bicyclists free rein on some of the city's major thoroughfares. In another vein, the Atlanta Beltline, a repurposed railroad line dedicated to non-automobile traffic, is rapidly becoming one of the city's most popular destinations. Elsewhere, policies to encourage transit-oriented development offer other ways to isolate people-oriented activities from major transportation emissions sources.

While such examples offer a glimpse of what is possible, the reality is that the benefits of many such developments are not equally shared across all portions of society. The elderly, those with limited personal mobility, and children (especially those without access to transport) may have limited opportunities to take advantage of such developments. In this case, it is often the responsibility of urban planning organizations, hopefully with the help of public health professionals, to ensure that sensitive populations are not unduly exposed to traffic-related air pollutants. For example, while it may be desirable from a vehicle access perspective to locate an elementary school adjacent to a major arterial highway, does such a location serve broader societal objectives? Similarly, many major hospitals and rehabilitation centers are located in the vicinity of major transportation facilities, normally interstate highways. Concerns regarding location of major transportation facilities extends beyond the question of institutions to the broader question of environmental equity. In other words, do members of certain socioeconomic or ethnic groups bear a disproportionate share of the health risks associated with transportation? Answer to these questions require the development of policies that effectively consider the broad range of issues that the transportation system must address.

Traffic-related air pollution and human health worldwide

Historically, the development of policies for the control of mobile source air pollution is similar in many ways to that of the "horse manure" problem described at the beginning of the chapter, that is,

1. The problem existed from the initial use of the technology but it is not sufficiently prevalent to strongly impact either society or human health.
2. The use of the technology significantly expands as it proves superior in some important way or ways. At this point, the potential issues with the technology have been identified and the problem is growing but has yet to become a societal priority.
3. The problem has reached a stage where action needs to be taken in terms of immediate remedial actions as well as establishing a path forward to avoid the same issue in the future.

The formation of policy is thus mostly about establishing what sort of remedial actions need to be taken to address the immediate problem and establishing a set of long-term policies to avoid similar issues in the future. Not surprisingly, not every geographic location reaches the same stage of policy development at the same time. While the United States reached its "Stage 3" about 1970 and most European countries and Japan arrived at that point about decade later, many locations throughout the world have yet to reach, or are just reaching the point where public pressures are forcing the development of policies to address these issues.

In fact, much of the world is currently struggling with developing effective policy responses to air pollution induced by mobile and other sources. The World Health Organization (WHO, 2016) rates ambient air pollution as the leading environmental risk to human health worldwide and estimated that three million deaths worldwide in 2012 were due to outdoor, or ambient, air pollution. In a systematic scoping review of 799 studies conducted primarily in North America (21%), Europe (27%), and Asia (35%), Sun and Zhu (2019) found that respiratory diseases, particularly childhood asthma, and mortality based on health records were the most common outcomes associated with ambient air pollution. For their review, outdoor air pollution was conceptualized into three categories: general air pollution exposure (ozone, carbon monoxide, nitrogen dioxide), particulate matter exposure ($PM_{2.5}$ and PM_{10}), and exposure to other hazardous substances including mobile source air toxics (MSAT) and included a range of health outcome categories including respiratory diseases, chronic diseases, cardiovascular diseases, health records (e.g., morbidity, outpatient visits, mortality), and other diseases (e.g., skin, Alzheimer's disease, Parkinson's disease). In evaluating these studies, the researchers found that 95% of the studies reviewed reported at least one statistically significant effect of ambient pollution on negative health outcomes.

For example, recent research revealed ambient air pollution impacts on a variety of health issues including cerebrovascular and cardiovascular diseases (Perez, Hazari, & Farraj, 2015; Williams et al., 2012), respiratory diseases (Goodman et al., 2017; Orellano, Quaranta, Reynoso, Balbi, & Vasquez, 2017), lung and other cancers (Pope III, 2015), neurodevelopmental disorders (Payne-Sturges et al., 2019), dementias (Chen, Kwong, Copes, et al., 2017), Parkinson's disease (Ritz, Lee, Hansen, et al., 2016), and pregnancies and birth outcomes (Centers for Disease Control and Prevention, 2016; Ha et al., 2018; Lamichhane, Leem, Lee, & Kim, 2015; Stieb, Chen, Eshoul, & Judek, 2012). While ambient air pollution affects all geographic areas and socioeconomic groups, residents of some geographic areas experience higher levels of air pollution than those living in other regions. In 2015, 194 member states of the WHO met to address pollution-related health effects and adopted a resolution to monitor and report air pollution; enhance systems, structures, and processes for measuring and reporting sources of ambient pollution and health statistics related to air pollution. While many western nations, including the United States have been dealing with this issue for many decades, the process of controlling the health impacts of transportation is far from complete. For example, Wernham (2011) and collaborators in the *Health Impact Project* emphasized the need for health impact assessment in the decision-making processes for federal, state, and local urban planning, land use, and environmental regulation both within the United States and worldwide.

Conclusions

Like many other societal issues, policies, and decision-making regarding traffic-related air pollution is driven by broader societal issues. When examining historical trends in these decisions, it is often very important to realize what other societal issues need to be addressed at the same time. As history has shown us, it is quite easy for a problem to escalate to have a significant societal impact before effective actions are undertaken. As issues with technology and human behavior become more global in scale, our ability to "wait for problems to arise" becomes increasingly problematic. It is thus essential that we develop a better ability to anticipate potential future issues before they expand into global concerns. Our future may depend on it.

References

American Thoracic Society. (2010). *Breathing in America*. Washington: American Thoracic Society. Available at: https://www.thoracic.org/patients/patient-resources/breathing-in-america/.

Brugge, D., Patton, A. P., Alex, B., Reisner, E., Lowe, L., Bright, O., et al. (2015). Developing community-level policy and practice to reduce traffic-related air pollution exposure. *Environmental Justice, 8*(3), 95–104. https://doi.org/10.1089/env.2015.0007.

Carson, R., Darling, L., & Darling, L. (1962). *Silent spring*. Boston, Cambridge, Mass.: Houghton Mifflin.

CDC. (2020). *1918 Pandemic (H1N1 virus)*. Centers for Disease Control and Prevention. Available at: https://www.cdc.gov/flu/pandemic-resources/1918-pandemic-h1n1.html.

Centers for Disease Control and Prevention. (2016). *Low birth weight and the environment*. Available at: https://ephtracking.cdc.gov/showRbLBWGrowthRetardationEnv.action. Google Scholar.

Chen, H., Kwong, J. C., Copes, R., et al. (2017). Exposure to ambient air pollution and the incidence of dementia: A population-based cohort study. *Environment International, 108*, 271–277. Crossref, Medline, Google Scholar.

Chen, L., Liu, C., Zhang, L., Zou, R., & Zhang, Z. (2017). Variation in tree species ability to capture and retain airborne fine particulate matter (PM2.5). *Scientific Reports*, 7(1), 3206. Crossref, Medline, Google Scholar.

Earthday. (2020). *History of earth day*. Available at: https://www.earthday.org/history/.

EPA. (1973). *EPA requires phase out of lead in all grades of gasoline*. Washington, DC: U.S. Environmental Protection Agency. Available at: https://archive.epa.gov/epa/aboutepa/epa-requires-phase-out-lead-all-grades-gasoline.html.

EPA. (2019). *National air quality: Status and trends of key air pollutants*. Washington, DC: U.S. Environmental Protection Agency. Available at: https://www.epa.gov/air-trends.

EPA. (2020). *Diesel fuel standards and rulemaking*. Washington, DC: U.S. Environmental Protection Agency. Available at: https://www.epa.gov/diesel-fuel-standards/diesel-fuel-standards-and-rulemakings.

Evelyn, J. (1661). *Fumifugium, or, The inconveniencie of the aer and smoak of London dissipated together with some remedies humbly proposed by J.E. esq. to His Sacred Majestie, and to the Parliament now assembled, pamphlet published London.*

Freese, B. (2005). *Coal: A human history* (p. 371). New York: Basic Books. ISBN-13: 978–0465057931.

Goodman, J. E., Zu, K., Loftus, C. T., Tao, G., Liu, X., & Lange, S. (2017). Ambient ozone and asthma hospital admissions in Texas: A time-series analysis. *Asthma Research and Practice, 3*, 6. https://doi.org/10.1186/s40733-017-0034-1. PMID: 28794889.

Ha, S., Sundaram, R., Buck Louis, G. M., Nobles, C., Seeni, I., Sherman, S., et al. (2018). Ambient air pollution and the risk of pregnancy loss: A prospective cohort study. *Fertility and Sterility, 109*, 148–153. https://doi.org/10.1016/j.fertnstert.2017.09.037.

HEI. (2015). *Executive summary. The advanced collaborative emissions study (ACES)*. Boston, MA: Health Effects Institute. Available at: https://www.healtheffects.org/publication/executive-summary-advanced-collaborative-emissions-study-aces.

History Channel, & Waring, G. (2018). Available at: https://www.history.com/topics/landmarks/george-waring.

Johnson, B. (2020). *The great horse manure crisis of 1894, U.K. history magazine*. Historic UK. available at: https://www.historic-uk.com/HistoryUK/HistoryofBritain/Great-Horse-Manure-Crisis-of-1894/.

Kohlstedt, K. (2017). *"The big crappie": NYC transit pollution from horse manure to the horseless carriage, 99% invisible*. Available at: https://99percentinvisible.org/article/cities-paved-dung-urban-design-great-horse-manure-crisis-1894/.

Lamichhane, D. K., Leem, J. H., Lee, J. Y., & Kim, H. C. (2015). A meta-analysis of exposure to particulate matter and adverse birth outcomes. *Environmental Health and Toxicology, 30*, e2015011. Crossref, Medline, Google Scholar.

Mongomerie, R. (2013). *The structural and elemental composition of inhaled particles in ancient Egyptian mummified lung*. Doctoral Dissertation Manchester, UK: University of Manchester.

NHTSA. (2016). *Motor vehicle traffic fatalities and fatality rates (1899–2015)*. Washington: National Highway Traffic Safety Administration. Available at: https://cdan.nhtsa.gov/TSFTables/Fatalities%20and%20Fatality%20Rates%20(1899-2015).pdf.

Orellano, P., Quaranta, N., Reynoso, J., Balbi, B., & Vasquez, J. (2017). Effect of outdoor air pollution on asthma exacerbations in children and adults: Systematic review and multilevel meta-analysis. *PLoS One*, *12*(3). https://doi.org/10.1371/journal.pone.0174050.

Payne-Sturges, D. C., et al. (2019). Healthy air, healthy brains: Advancing air pollution policy to protect Children's health. *American Journal of Public Health*, *109*(4), 550–554.

Perez, C. M., Hazari, M. S., & Farraj, A. K. (2015). Role of autonomic reflex arcs in cardiovascular responses to air pollution exposure. *Cardiovascular Toxicology*, *15*, 69–78. https://doi.org/10.1007/s12012-014-9272-0. PMID: 25123706.

Pope, C. A., III. (2015). Epidemiology of lung cancer associated with outdoor air pollution. *Environmental and Molecular Mutagenesis*, *56*, S41.

Rienow, R., & Rienow, L. (1967). *Moment in the Sun: A report on the deteriorating quality of the American environment* (p. 365). New York: Ballentine Books.

Ritz, B., Lee, P. C., Hansen, J., et al. (2016). Traffic-related air pollution and Parkinson's disease in Denmark: A case-control study. *Environmental Health Perspectives*, *124*(3), 351–356. Crossref, Medline, Google Scholar.

Rosner, F. (1987). Moses maimonides and air pollution. *New York State Journal of Medicine*, *87*(10), 574. PMID:3317154.

Rosner, D., & Markowitz, G. (1985). A 'gift of god'?: The public health controversy over leaded gasoline during the 1920s. *American Journal of Public Health*, *75*(4), 344–352. American Public Health Association.

Smith, R. A. (1872). *Air and rain: The beginning of a chemical climatology*. London: Longman Greens.

Sperling, D., et al. (2018). *The three revolutions*. Washington, DC: Island Press. LOC: 2017951060.

Stieb, D. M., Chen, L., Eshoul, M., & Judek, S. (2012). Ambient air pollution, birth weight and preterm birth: A systematic review and meta-analysis. *Environmental Research*, *117*, 100–111. Crossref, Medline, Google Scholar.

Sun, Z., & Zhu, D. (2019). Exposure to outdoor air pollution and its human health outcomes: A scoping review. *PLoS One*, *14*(5). https://doi.org/10.1371/journal.pone.0216550.

Turvey, R. (2005). Horse traction in Victorian London. *The Journal of Transport History*, *26*(2), 38–59. https://doi.org/10.7227/TJTH.26.2.3.

USDHEW. (1964). *Smoking and health: Report of the Advisory Committee to the Surgeon General of the Public Health Service*. Washington: U.S: Department of Health, Education, and Welfare, Public Health Service, Center for Disease Control. PHS Publication No. 1103.

Wernham, A. (2011). Health impact assessments are needed in decision making about environmental and land-use policy. *Health Affairs*, *30*(5), 947–956. https://doi.org/10.1377/hthaff.2011.0050.

WHO. (2016). *Ambient air pollution: A global assessment of exposure and burden of disease*. Available on WHO website http://www.who.int.

Williams, R., Brook, R., Bard, R., Conner, T., Shin, H., & Burnett, R. (2012). Impact of personal and ambient-level exposures to nitrogen dioxide and particulate matter on cardiovascular function. *International Journal of Environmental Health Research*, *22*, 71–91. https://doi.org/10.1080/09603123.2011.588437. PMID: 21711166.

Further reading

Bourcier, E., Charbonneau, D., Cahill, C., & Dannenberg, A. L. (2015). An evaluation of health impact assessments in the United States, 2011–2014. *Preventing Chronic Disease, 12*, 140376. https://doi.org/10.5888/pcd12.140376.

Krzyzanowski, M. (2008). WHO air quality guidelines for Europe. *Journal of Toxicology & Environmental Health Part A: Current Issues, 71*, 47–50. https://doi.org/10.1080/15287390701557834. PMID: 18080893.

CHAPTER 15

Policy option generation and selection

Anthony D. May
University of Leeds, Leeds, United Kingdom

Abbreviations

ASIF	activity, share, intensity, fuel mix
EC	European Commission
KonSULT	knowledgebase on sustainable urban land use and transport

Introduction

In this chapter we consider the transport policies which might be adopted to address the challenge of traffic-related air pollution, and how policy options might best be identified and preferred policies selected. A policy will typically involve an overall *strategy*, which will include a set of policy *measures*, which will usually form part of a wider *package* of measures. Packages may be designed simply to reinforce the impacts of individual measures, or also to help overcome *constraints* on their implementation. The process of identifying suitable measures and packages is one of *option generation*. Selected measures will typically be implemented in the form of specific *projects*, involving both detailed design and *option appraisal*. These terms may have different meanings in different parts of the world. For the avoidance of doubt, the definitions used throughout this chapter are listed below.

A *strategy* is a direction of change which a government agency, such as a city, wants to achieve in the transport system. Two examples of strategies relevant to reducing traffic-related air pollution are reducing the need to travel and improving the vehicle fleet.

A policy *measure* is a broad type of action which can be taken to contribute to one or more policy objectives in a transport plan, to overcome one or more identified problems, or as part of an overall strategy. In the context of this chapter, one objective would be the enhancement of public health, and one problem that of pollution. But cities will typically have a wider range of policy objectives and face a larger number of problems. Two examples

Traffic-Related Air Pollution
https://doi.org/10.1016/B978-0-12-818122-5.00015-6

within the example strategies above are designing high-density mixed development and providing facilities for electric vehicle charging.

A *package* is a combination of different measures which have been grouped together in a package to contribute more effectively to policy objectives or to the resolution of problems. It could, for example, involve provision both of electric vehicle charging points and of a low-emission zone.

A *constraint* is an obstacle which prevents a given policy measure being implemented or limits the way in which it can be implemented. Examples include limited finance and a lack of public acceptance.

Option generation is the process by which possible measures or packages are identified. The most common sources of suggestions are the existing knowledge and preconceived ideas of policy makers and professionals. However, there are a number of more formalized techniques for stimulating suggestions.

A *project* is a specific application of a type of measure to a given city or area and includes questions such as where and when a measure should be applied, to whom it should be applied, and how intensively it should be applied. Examples include determining the density of provision of charging points and deciding on the area and level of charge for a low-emission zone.

Option appraisal is the process by which a proposed project or package is assessed in advance of its implementation. Effective appraisal involves assessing likely performance against each of the city's objectives (effectiveness), the likelihood of being approved (acceptability), and implications for the city's budget (value for money). It is covered in more detail in Chapter 18, and usually requires predictive modeling, which is covered in Chapter 3.

We consider each of these in more detail in this chapter, but first, consider why option generation is a particular challenge for cities.

The challenge of option generation

The case for an objective approach to policy option generation was set out particularly clearly in a report on transport investment for the UK government (Eddington, 2006). "Unless a wide range of appropriate options is considered, there is a risk that the best options are overlooked and money could be wasted. A good option generation process is crucial to ensure that the transport interventions that offer the highest returns can be found. The full range of options should look across all modes and include making better use of the existing transport system, including better pricing; investing in assets that increase capacity…; investment in fixed infrastructure; and combinations of these options."The UK Department for Transport adopted this advice in its guidance to local authorities that cities "should consider a wide range of options funded through either capital or revenue expenditure [and] should not assume that schemes which have been under consideration for a long period ... are still the most appropriate solution to identified challenges." (Department for Transport, 2009).

Despite this advice, there is ample evidence that, in many cities, measure selection is not a rational process, but is often politically driven and led by the interests of particular actors such as developers and service providers (May, 2013). The European Conference of Ministers of Transport (ECMT, 2002) surveyed 168 cities and found that, while they generally understood which measures they should ideally be using, including them in their strategies was "more easily said than done." A study by Atkins (2007) for the UK Department for Transport of its Local Transport Plan process suggests that local authorities, in England at least, tend not to innovate, but rather to pursue schemes which have been under consideration for a long period, and to focus on infrastructure projects and traffic management to improve the infrastructure, rather than considering enhancements to public transport or ways of managing demand.

The European Commission's manual on Measure Selection (May, 2016) identifies the following challenges to option generation:

1. If cities do not start by identifying the problems to be overcome, they do not have a clear justification for identifying suitable measures. As a result, they may overlook appropriate solutions and may find it harder to justify their proposals to a critical public.
2. Many stakeholders, and some politicians, will have their own preconceived ideas of what should be done. There is a danger that, by focusing on these, cities overlook other potentially more cost-effective solutions.
3. There is a very wide range of types of a policy measure, including providing new infrastructure, managing the transport system, providing new services, improving information, encouraging behavioral change, and charging for use of the transport system. Choosing among these measures is thus difficult.
4. For many of these measures, information on their effectiveness and applicability in different contexts is limited.
5. For most measures, there will be barriers to their implementation, including who is responsible, what funding is available and how acceptable they are. There is limited guidance on how to overcome these barriers.

6. One approach to overcoming these barriers is to use packages of measures, yet there is even less guidance available on how to design packages.
7. It is not sufficient to decide in principle to use a particular measure. Each measure will be implemented as a series of projects, which need to be specified to suit the particular context. Once again, there is relatively little guidance available on this design process.
8. Before deciding to implement a specific project or package, an assessment is needed of its likely impacts and how cost-effective it is likely to be. These processes of prediction and appraisal require specific skills.

The context for option generation

The European Commission's manual on Measure Selection (May, 2016) stresses the importance of specifying the study area, time frame and starting point (typically specified as a "do-nothing" or do-minimum" strategy) as a context for selecting suitable policy measures and packages. In considering a strategy for tackling traffic-related air pollution, these considerations will typically be relatively straightforward. However, the types of appropriate measure will differ depending on whether the focus is a city center, the whole of an urban area, or a nation or region. The time frame will determine whether emerging technologies can be considered, and some preexisting policy measures, such as regulatory controls on car use, encouragement of walking and cycling and the availability of a high-quality public transport service will influence what can be achieved.

The next step in defining the context is to be clear as to the overall policy objectives which the policy measures and packages are designed to meet. In the case of measures to tackle traffic-related air pollution, the principal objective will be one of improving public health. Traditionally, transport policies have focused on objectives such as improving the efficiency of the transport system, reducing carbon emissions environmental damage and accidents, enhancing liveability, improving fairness in the provision and impacts of transport, and supporting the urban economy (DfT, 2009). These can all be linked to the principles of environmental, economic, and social sustainability. To some extent, each of these will also contribute to enhanced public health, but a direct focus on public health has arisen more recently, for example in the analysis of Nieuwenhuijsen and Khreis (2019) and in the practical pursuit of healthy streets in the current Mayor's Transport Strategy for London (TfL, 2018).

While public health is thus the principal objective, it will be important to understand the other objectives being pursued. Different stakeholders will have different objectives and priorities, which may conflict with one another, and compromises may be needed (POLIS and WYCA, 2016). To this end, it may help to adopt a hierarchy of objectives, so that if conflicts arise, decisions can focus on the priority objectives. This may lead to objective statements of the kind: "improving public health so long as doing so does not adversely affect the city's economy." Conversely, measures selected to improve public health may also achieve well the improvements in terms of other objectives, such as efficiency, liveability, and equity. Such co-benefits may help to enhance the case for, and acceptability of, measures focused on public health. This concept of co-benefits is considered further in Chapter 19.

The final step in defining the context is to specify the problems to be overcome. A clearly specified list of problems is the most suitable basis for identifying potential solutions. One of the easiest ways of specifying problems is by reference to the set of policy objectives. This enables the question "how do we know we have got a problem?" to be answered more easily. For example, the efficiency objective relates to problems of congestion and unreliability; the safety objective to accidents and casualties.

Problems may be identified in a number of ways:

Consultation: Transport users and residents can identify the problems that they encounter when travelling and which result from other people travelling. Transport providers can be consulted about the operational problems which they face. This is a key element of the participation process (Rupprecht Consult, 2016). Users and residents will be well placed to identify current problems.

Objective analysis: Objective analysis of problems requires the adoption of an appropriate set of indicators and targets (Gühnemann, 2016). When targets are defined, they can be used, with current data, to identify current problems.

Monitoring: Regular monitoring of conditions, using similar indicators and targets, is another valuable way of identifying problems (Gühnemann, 2016). As well as enabling problems, and their severity, to be specified, a regular monitoring program enables trends to be observed, and those problems which are becoming worse to be singled out for treatment.

The choice of strategy

Cities may find it easier to think about the overall strategy which they wish to pursue than to list the measures which they want to use. At its simplest, a strategy is simply a combination of measures to address a city's objectives.

More specifically, a strategy can be a direction of change which a city wants to achieve in the transport system. Such strategies are not objectives in their own right, but changes which should contribute to the city's chosen objectives.

For example, a city might wish to reduce car use. Presenting this as an objective is likely to attract criticism that the city is "anti-car." But demonstrating that reducing car use should help to improve public health, the environment, liveability, and safety links the strategy directly to objectives and hence helps to justify it.

Typical strategies pursued by cities include:

- reducing the need to travel;
- reducing car use;
- improving public transport;
- improving road network performance;
- improving walking and cycling; and
- improving freight operations.

Table 15.1 shows how these contribute to different policy objectives (May, 2016).

In tackling traffic-related air pollution and its impacts on air quality and health, all six strategies have a part to play, but with the added emphasis, when improving public transport, road network performance and freight operations, of achieving a less polluting vehicle fleet.

Table 15.1 Contribution of strategies to objectives.

	Efficiency	Liveability	Environment	Equity	Safety	Economy
Reducing need to travel	•••	•••••	••••	•••	••••	•
Reducing car use	•••••	••••	•••••	•••	••••	••
Improving public transport	•••	••••	•••	•••••	•••	••••
Improving road use	•••••	•••	•••	•••	•••••	••••
Improving walking + cycling	•••	•••••	••••	•••••	•••	•••
Improving freight	••••	•	•••	•	••	•••••

Source: May, A.D. (2016). Measure selection: Selecting the most effective packages of measures for sustainable urban mobility plans, http://www.sump-challenges.eu/kits.

The available policy measures

There is a growing range of measures available to transport professionals. A total of 64 measures are included in the European Commission's Measure Option Generator (May, 2016). Some of these, such as low-emission zones, bike-sharing, and crowd-sourcing are relatively new. In all, planners have access to around twice as many measures as they did 30 years ago.

There are several ways of categorizing these measures. One is the distinction between "supply-side" and "demand-side." On the supply side are measures which add to the capacity of the transport system to move people and freight. On the demand side are measures which affect how people and freight operators use the transport system. Demand-side measures are often grouped under the title Transport Demand Management (VTPI, 2019).

A second approach, developed initially for reducing greenhouse gas emissions, is the ASIF Framework:

- A: Activity (total travel activity)
- S: Share (share of modes)
- I: Intensity (energy intensity of each mode)
- F: Fuel mix (carbon intensity of the fuel mix) (Eichhorst et al., 2012)

This could potentially be applied to the reduction of traffic-related air pollution, with vehicle technology measures influencing both intensity and fuel mix.

Another categorization considers the type of impact which the measure has. Nakamura, Hayashi, and May (2004) distinguish between technology, regulation, information, and economic measures and consider their impacts on supply, demand, and cost, while Banister (2005) lists urban form, fiscal measures, regulatory measures, and technology. The classification in KonSULT (May, 2016) adds management and behavioral measures.

Unfortunately, evidence on the performance of many of these policy measures is very incomplete. Some measures are novel, and experience is still limited. Collecting the evidence on impacts has often failed. This is particularly true of new roads; the realization, too late, that they generate additional demand is one reason for the abrupt change in policy on them. It is important to take the opportunity to measure and evaluate the impacts of new measures and make that information available to others. In particular, information on policies which have been less successful than planned can help others avoid making the same mistakes.

Even where experience is available it may not be directly relevant in another context. Light rail will work better in larger cities than in smaller ones. Walking and cycling provision is more important in high-density areas than in lower density ones. Parking controls are likely to be more effective in city centers than elsewhere. Regulatory controls will be more acceptable in some cultures than in others (Stead, 2008). For all of these reasons, it can be difficult to judge how transferable experience with successful policy measures will be (Rose, 2005). This is a further reason for encouraging as much experience as possible to be recorded.

Each policy measure will affect the performance of the transport system in one or more of three ways:

- by changing the demand for travel;
- by changing the supply of transport facilities;
- by changing the cost of provision and operation of the transport system.

Initial responses (e.g., changes in mode) may lead to secondary ones (e.g., increases in overcrowding). Each of these types of change will, in turn, affect performance against the objectives and hence reduce (or increase) problems. Tracing all these impacts can be difficult, and causal chain diagrams (Fig. 15.1) can help to understand them. It is through these changes that any policy measure will influence performance against the agreed objectives.

Changes in demand. When faced with a new policy measure such as a low-emission zone, or with a change in an existing one, such as a fare increase, the individual traveler has a number of options, which may include changing mode or destination, reducing the frequency of travel and changing vehicle ownership. The scale of response will depend on the circumstances. Those who are directly exposed to a change will respond more strongly than those for whom the impact is indirect. Those who have fewer alternatives will be more reluctant to change. Longer-term responses may well be stronger, as people have more time to respond. People are more likely to change when they experience life cycle changes, such as having a baby or changing jobs. Elasticities of demand are often used to understand the scale of such responses (e.g., Paulley et al., 2006).

Changes in supply. Changes in the supply of transport can take a number of forms, including changes in capacity and the allocation of that capacity, regulatory, and information changes. Some of which will have a direct influence on travelers, while others will only affect them if they are perceived. For most policy measures, it will be clear how they affect supply, but the scale of this impact may be difficult to assess.

Changes in costs. The principal types of financial cost are capital costs of new infrastructure, operating, administrative and enforcement costs, and costs of maintenance and replacement. Pricing measures will, in addition, generate a revenue stream which will reduce the net cost of the measure and influence demand. Changes in these costs and revenues are crucial in

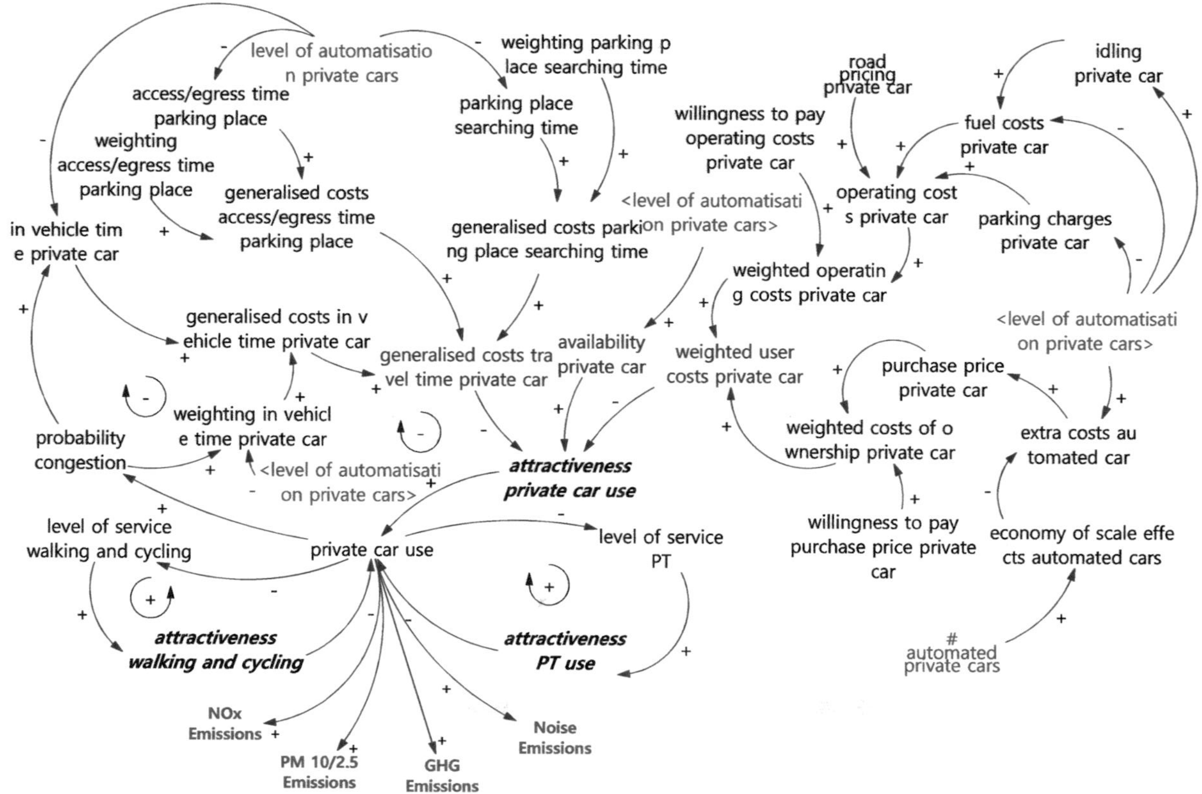

Fig. 15.1 A causal chain diagram for autonomous vehicles. *Source: May, A.D., Shepherd, S.P., Pfaffenbichler, P., Emberger, G. (2019). The potential impacts of automated cars on urban transport: An exploratory analysis. Transport Policy (under review).*

determining whether an individual policy measure, or the overall strategy, provides value for money. Low-cost measures typically offer greater value for money than major infrastructure projects (Goodwin, 2010).

The contribution of different policy measures

There are typically three sources of information on these impacts, which are used in deciding whether a particular measure might be suitable: advocacy, empirical evidence, and predictive computer models. Each of these has its limitations.

Advocacy is the approach adopted by campaign groups, providers of services and technology, and city enthusiasts. A group campaigning for cycling, for example, will promote the advantages of cycling, but may not admit to its limitations. A commercial body which constructs light rail lines will want to promote its benefits, but may not indicate the locations for which it is suitable. A city which has a reputation for car sharing may be interested in promoting the city's image rather than in considering the alternatives. Claims from such organizations should be treated with caution.

Empirical evidence can be collected from before and after studies of the implementation of a particular measure or a package. Ideally, such studies will be carried out whenever a new type of measure is implemented, or a measure is implemented in a different context. However, cities are often reluctant to spend scarce resources on such studies, and national governments rarely invest in them. Even where studies are conducted, they are often less than comprehensive. Ideally, information is needed on resulting changes in demand, supply, and costs of travel, as well as on changes in outcome indicators for each of the objectives of interest to other cities. Moreover, as noted above, performance will be affected by context, and may not be transferable.

Predictive models can in principle overcome these constraints by enabling impacts on demand, and hence on outcome indicators, to be predicted in a number of different contexts. But models themselves have limitations, as discussed further in Chapter 3.

There are several useful sources of evidence from those studies which have been conducted, including CIVITAS (2019), ELTIS (2019), and Evidence (2015). Eltis case studies are added at the rate of perhaps 10 per month and cover both policy measures and planning practice. The case studies on policy measures are a valuable source of additional information on specific measures.

The development of packages

No one measure on its own will be sufficient to achieve a city's objectives or overcome its problems. This is likely to be the case even with a specific objective such as public health, or the specific problem of traffic-related air pollution. The set of measures selected can be thought of as a package, and in designing a package it is necessary to understand how the different measures might interact (Givoni, Macmillen, Banister, & Feitelson, 2013).

The key to developing a package is to identify which policy measures will work well together, or may be needed to make other measures viable. Thus, within a policy package, policy measures can interact in one of two different ways (May, Kelly, & Shepherd, 2006):

- they can achieve more together than either would on its own; this is the principle of synergy;
- they can facilitate other measures in the package by overcoming the constraints on their implementation, which we consider below.

The introduction of congestion charging in London (TfL, 2007) offers a good example of both principles of packaging. Congestion charging had two particular constraints; it was unpopular and was expected to affect lower-income car users. However, its revenue was used to finance an increase in bus services in inner London. These bus services helped overcome the constraints on congestion charging, while congestion charging overcame the financial constraint on increasing bus services. Moreover, the two together achieved a greater switch away from car use than either would have done on its own.

It is difficult to find empirical evidence on the performance of packages given the problems of needing to implement several measures together and of isolating their effects from external changes. An alternative approach is to use predictive models to assess how measures might operate together. This was the approach adopted in determining the synergy scores which are used in the European Commission's Measure Option Generator (May et al., 2016).

Understanding the constraints on implementing specific measures

Most policy measures will face constraints in their implementation. As a result, some measures may be rejected, making the policy less effective. For example, demand management measures in larger cities can control congestion and improve the environment. But cities may be tempted to reject them simply because they will be unpopular. The emphasis should, therefore, be on how to overcome these constraints, rather than simply how to avoid them.

The literature (Atkins, 2007; ECMT, 2006; Gudmundsson, 2007; May, 2009) indicates that the principal constraints are:

- legal and regulatory,
- financial,
- governance and institutional,
- political acceptability,
- public acceptability, and
- technical.

Legal barriers include lack of legal powers to implement a particular measure, legal responsibilities which are split between agencies, and regulations which require the involvement of the private sector. A survey of European cities (May & Matthews, 2007) indicates that land-use, road building, and pricing are the policy areas most commonly subject to legal and institutional constraints.

Finance for transport policy can be obtained from a number of sources, including national and local government, charges on users, and taxes on land owners and developers. Some of these sources of funding will be assigned to certain types of project; for example, the French *Versement Transport* charge on local firms can only be used to improve public transport (Cerema, 2015). Some will be for infrastructure (capital funding) rather than management measures (revenue funding). Both of these constraints are likely to lead to less cost-effective strategies. Infrastructure projects typically have a much lower benefit/cost ratios than management projects (Goodwin, 2010).

In terms of governance, cities are rarely able to make decisions on transport strategies on their own, but the constraints on them differ from city to city. May and Matthews (2007) identified three types of constraint: lack of direct control, particularly for land use and public transport, intervention from other levels of government and involvement of other stakeholders. In Europe it is typically medium-sized cities which suffer most from these governance constraints; smaller cities often have more freedom, while larger ones often have more power. The US DoT (2015) and the European Commission guidance on institutional cooperation (POLIS and WYCA, 2016) provide further guidance on managing shared and split responsibilities.

Political acceptability constraints arise where politicians fear of lack of public acceptance when different political parties hold opposing views, or where pressure groups or the media oppose a given measure. Public acceptability may well differ by socioeconomic group and can be influenced by cultural attributes, such as attitudes to enforcement (IAPP, 2007). It is not uncommon for political and public acceptance of a given policy measure to differ, particularly if politicians have not kept in touch with changes in the public's views. The European Commission guidance on participation (Rupprecht Consult, 2016) provides further guidance on involving the public and stakeholders in policy formulation.

Technical constraints are more obvious. For land use and infrastructure, these may well include land acquisition. For management and pricing, enforcement and administration are key issues. For vehicle fleets, management and information systems, engineering design, and availability of technology may limit progress. Generally, the lack of key skills and expertise can be a significant barrier to progress and is aggravated by rapid changes in the types of policy being considered and the emergence of new technologies.

Of these six types of constraint, experience shows that effective package design can help overcome all but legal and technical ones. Chapter 18 discusses constraints in more detail.

Approaches to option generation

Jones, Kelly, May, and Cinderby (2009) reviewed the literature on option generation methods for both overall strategies and specific schemes. They identified four groups of methods, of which only the first, "inside the box" methods which select from preexisting options, appeared to be appropriate for the generation of overall strategies. This category can be further subdivided into simple library methods which list the options available, and directed library methods which assist the user to identify the items in the library which are of most relevance. The CIVITAS tool inventory (CiViTAS, 2019) is an example of the former. It includes over 100 policy measures (tools), but only provides an eight-level categorization as a means to help the user identify suitable measures.

Four directed library methods are referenced in the literature. The TDM Encyclopedia (VTPI, 2019) lists travel demand management measures under four broad categories and identifies those which are most relevant for different objectives and user groups. The MAXExplorer (EPOMM, 2019) allows users to specify their organization type, objectives and location type, and then lists measures by relevance, with a very brief description for each. The WBCSD mobility planning tool (WBCSD, 2017) includes a solution finder, which proposes potential solutions based on the user's specification of performance indicators and priorities. The European Commission's Measure Option Generator, KonSULT, is similar to these in being objective-based, but goes further in allowing users to specify their preferred strategies and by proposing policy packages (May, Khreis, & Mullen, 2018).

The Measure Option Generator developed for the EC guidance on Measure Selection (May et al., 2018) has been incorporated into the Knowledgebase on Sustainable Urban Land use and Transport (www.konsult.leeds.ac.uk). KonSULT itself was developed, with support from the UK government and the EC, with the aim of assisting policy makers, professionals, and interest groups to understand the challenges of achieving sustainability in urban transport, and to identify appropriate policy measures and packages for their specific contexts. It consists of three elements: a Measure Option Generator, a Policy Guidebook, which contains the information on each of the policy measures in the knowledgebase, and a Decision-Makers' Guidebook.

In the Policy Guidebook, policy measures are grouped into six high-level categories of land use interventions, infrastructure projects, management and service measures, attitudinal and behavioral measures, information provision, and pricing interventions. Each measure is described following a standard structure:

- Summary: a one page summary of the description and findings;
- Taxonomy and description, which describes what the measure is, how it works, what it tries to do, and how it contributes to different strategies;
- First principles assessment, which assesses from first principles how it affects demand, supply, and finance, how, through these impacts, it might contribute to policy objectives and the resolution of policy problems, and what the barriers are to its implementation;
- Evidence on performance, which summarizes a series of case studies which provide empirical evidence on their contribution to policy objectives and problem resolution;
- Policy contribution, which combines the findings of the previous two sections to summarize the measure's contribution to policy objectives and to the resolution of policy problems, and identifies the areas of a city in which it might most usefully operate; and
- References

To ensure consistency of treatment, a standard 11-point scoring method is applied, ranging from +5 (a highly positive contribution) to −5 (a highly negative contribution) throughout the knowledgebase. These scores underpin the operation of the Measure Option Generator. Each of the concepts used, including objectives, problems, strategies, and barriers, is more fully described in the Decision-Makers' Guidebook. The Policy Guidebook currently contains 64 policy measures and over 200 case studies.

The Measure Option Generator allows cities quickly to identify those policy measures which may be of particular value in their context. Users specify their context, including their objectives and strategy, and the measure option generator provides an ordered list of the 64 measures contained in the Policy Guidebook. From the Measure Option Generator screen, the user begins this process by specifying the type of area they are concerned with (corridor, town center, outer suburb, etc.).

The next screen then prompts the user to decide whether to base their search on objectives such as improving the environment, or problems, such as air pollution. An objective-led search and a problem-oriented one should lead to the same overall strategy, provided that the problems identified are consistent with the objectives set. The user is thus required to adopt one of these approaches, to avoid double counting. The user can also assign weights ranging from 0 to 5 to each of the chosen objectives (or problems) to indicate their relative importance in the user's local context. This addresses the concern that objectives may be in conflict, and that it may help to specify a hierarchy of objectives (or problems). This stage is one to which stakeholders might usefully contribute, and the Measure Option Generator is designed to be used interactively.

The third screen prompts the user to select the strategies they envisage adopting. The strategies included in the Measure Option Generator describe broad directions of policy, such as reducing the need to travel, or improving walking and cycling. Users can reflect a mixed approach by assigning weights from 0 to 5 to indicate the relative importance of each selected strategy.

Based on these input values which specify the context of interest to the user, KonSULT's Measure Option Generator produces a list of the 64 available policy measures ranked according to their potential relevance and ability to contribute to the specified context, using a 0–100 scale to indicate relative contribution. This output also provides a broad indication of the cost for each measure and the timescale for implementation. Users can thus limit their search to low cost or rapidly implemented measures.

This output is not intended to be prescriptive, but to prompt the user to investigate measures which might not previously have been considered. Once again, this feature can be used interactively with stakeholders, who may be prompted to debate the relative merits of the more highly ranked measures. At any stage, the user can click on any of the measures listed and transfer immediately to the fuller information on that measure in the KonSULT Policy Guidebook.

As a next step, this ordered list of policy measures can be used to develop packages of measures. The KonSULT Package Option Generator allows the user to consider packaging at one of two levels. The first involves taking a preferred policy measure (such as a low-emission zone) and identifying other measures which might support it. These are referred to as complementary measures. The second involves true packaging, where several measures are chosen which work well together. Computationally, assessing packages of several measures from a long list can rapidly become complex, so the packaging option is limited to packages of up to five measures chosen from a list of up to ten measures.

The Package Option Generator offers two approaches to packaging, as discussed above: to achieve synergy or to helping to overcome constraints. Users can choose either of these approaches in searching for complementary measures or packages. The calculation of synergy is based on detailed research using predictive models to assess the interaction of different pairs, and sets, of policy measures (May et al., 2016). The assessment of the overcoming of constraints is based on the scores for each measure against the constraints of governance, political acceptability, public acceptability, and finance in the Policy Guidebook.

Designing specific projects

Measure selection does not end with the identification of possible measures and packages. Each measure needs to be specified in detail, often by defining one or more projects. In doing this, cities need to consider:

- where the measure should operate
- when it should operate
- who will use it
- how intensively it should be used.

Where it should operate. The obvious answer to this is that a measure should be implemented in the locations where the problems it is designed to overcome are most serious. So the first step is to map these problems and decide how best to apply the measure throughout the affected area, and far enough outside it to avoid simply transferring the problem. Some measures such as fare structures and information campaigns are probably better applied throughout the city or the travel to work area. Most other measures will be more expensive to implement when the area of coverage is larger. For these, it may make sense to consider two or three different areas of operation, and decide which is the most cost-effective.

When it should operate. This is only a relevant question for management and pricing measures, which could in principle be used at certain times of the day. Management measures such as bus priorities, pedestrian areas, and parking controls should be applied when the problems are most serious, which will usually be the peak periods, and extended sufficiently into the shoulders of the peaks to avoid simply transferring problems to these times. But care has to be taken to ensure that users, such as those needing to deliver goods, can carry out their activities at other times. For pricing measures such as bus fares, parking charges and road pricing, charges will again typically be higher in the peaks and in the shoulders of the peaks.

Who will use it. Many management measures are designed to restrict the movement of certain classes of vehicle. Thus low-emission zones might focus solely on buses and commercial vehicles, or could be extended to diesel or petrol cars. Pedestrian areas are for pedestrians but could be designed to permit buses, trams, and bicycles. It will also be necessary in these cases to specify exemptions—e.g., for emergency vehicles and perhaps disabled drivers. Where exemptions are needed, misuse will need to be effectively enforced. Behavioral and information measures can be targeted on those whose behavior the city particularly wants to influence. The challenge here is typically how to influence hard to reach groups who do not use particular communication technology, or find information hard to understand.

How intensively it should be used. This is a rather more complex question, which needs to be answered differently for different types of measure. The design process should consider a number of options, assess their strengths and weaknesses, and produce a shortlist for more detailed assessment. For infrastructure measures intensity will be a question of the capacity provided, the length of the facility and the places stations served. Increases in capacity and length will increase accessibility but also cost more. Closer stations will increase accessibility but may increase travel time and costs. For management measures such as bus priorities and pedestrian areas, intensity is typically related to area of coverage (how many roads have bus lanes; how many streets are pedestrian-only). For measures such as traffic calming it may be a question of how low a speed limit is imposed. For pricing measures, the question is typically how much should be charged, with benefits of different kinds for higher and lower charges. Higher bus fares will increase revenue and make the service more financially sustainable but will reduce patronage. Higher parking charges will help reduce car use further and generate more revenue but may reduce activity in the area.

For many of these questions, the principal input will be professional judgment, including that of stakeholders, and detailed design will often benefit from stakeholder involvement and public consultation. But there is also an increasing number of design tools to help with the process. Jones et al. (2009) summarize a number of them, and give an example of an interactive tool used successfully in the reallocation of street space in a shopping center.

The approach to option appraisal

An effective approach to option generation will identify a long list of possible policy measures and differing suggestions for how they might be packaged. While that long list can be reduced to a shortlist using a judgmental assessment involving professionals, politicians, stakeholders, and the public, it will be necessary to adopt a more analytical approach to choosing between the measures and potential projects in the shortlist. That assessment needs to consider how each project and package might contribute to the policy objectives (or to overcome identified problems), how cost-effective it is likely to be, and how acceptable it will be to stakeholders and the public. It is often the trade-off between effectiveness and acceptability which determines the final policy.

To assess the contribution to policy objectives it is necessary to understand how a given project might influence demand and usage, bearing in mind that some user responses, such as rerouting, will be immediate, while others like changing vehicle may take some time. Because these interactions are complex, it can be difficult to predict the effects of a specific project. Models are often used for this purpose. A model should be a representation of the real world. It could represent how people's travel behavior responds to changes in the transport system provided; how the performance of the system changes as patterns of use change; how these changes, in turn, affect where people choose to live and work and where developers choose to build; and how these land use changes, in turn, affect demand. It will need to represent changes across the city; for example, whether shoppers switch from city center to out of town shops. It will also need to reflect changes over time from the immediate to the long term. Finally, it needs to generate the outcome indicators needed for appraisal. However, at the same time, a model needs to be a simplification of the real system, to keep the costs of building and validating it within bounds and to make it easy and, ideally, quick to use. There is thus a trade-off between complexity and ease of use. Approaches to resolving this trade-off are discussed in more detail in Chapter 3.

Once individual projects have been designed and modeled, it is possible to assess their likely effectiveness. This is the process of appraisal. An appraisal is the *ex ante* process of assessing how well a measure or package will perform. It is thus different from the evaluation which is the *ex post* process of deciding how well a measure or package has performed in practice. While evaluation can use empirical before and after data, appraisal has to use predictive data from models. However, in both cases the question "how well?" should be assessed in terms of an agreed set of performance indicators.

Further guidance on the selection of performance indicators is given in the European Commission manual on monitoring and evaluation (Gühnemann, 2016) but some key pointers (Marsden & Snell, 2009) are:

- outcome indicators, like noise levels and accidents, which reflect resulting changes in chosen objectives, are of most value in the appraisal; but it is important to identify outcome indicators for every chosen objective;
- intermediate outcome indicators, like changes in modal shares, are of less direct use, and should not be used on their own, but can help explain how the transport system is operating;
- output indicators which indicate what has been implemented, and input indicators of the resources used are of less value;
- it may also help to determine the relative importance of different indicators, in case it proves difficult to achieve improvements against all objectives.

Appraisal can be used to:

- reduce a long list of possible measures and projects to a sensible shortlist
- choose the best option for a particular measure
- choose between measures
- choose between packages
- identify weaknesses in any of these which could be overcome by returning to the design stage.

The last of these is a particularly important element in the design process and can help ensure that the final measures and projects are as effective as possible. It is important that any such assessments consider all objectives, and hence all performance indicators. An appraisal framework is, at its simplest, a table in which each option forms a column and each row an indicator. Chapter 18 considers in more detail the methods for developing and using such appraisal frameworks.

Conclusions

In this chapter, we have considered the need for a logical approach to option generation in order to identify the most appropriate transport-related strategies and policy measures for tackling traffic-related air pollution and its impacts in urban settings. The key steps in such a process can be summarized as:

1. Before considering possible measures, make sure that you are clear in your study area, timeframe and your current measures, and committed schemes.
2. Avoid thinking about solutions before you have agreed on your objectives. Bear in mind that you will have other objectives in addition to pollution reduction, which may offer co-benefits for the actions which you take.
3. Identifying these objectives will help you to understand what problems you face. Measures can then be thought of as ways of overcoming those problems.
4. In looking at possible measures, cast your net as widely as possible. Look at the different types of measure and the information on them. Try to understand how each works and can thus contribute to your objectives.
5. Decide whether there are particular strategies that you want to pursue (like reducing the need to travel).
6. Think about the principles of packaging the measures that you are interested in; packaging can help in achieving enhanced performance, but it can also help to overcome barriers to implementation.
7. Be clear as to the constraints that you face. Who is responsible for each of the types of measure that you are considering? What level of funding is available? How acceptable are different measures likely to be? But do not take these constraints as reasons for not pursuing a given measure; you can use packaging and careful design to overcome them.
8. Use a formal tool, such as the Measure Option Generator, to identify possible measures and packages. Involve your stakeholders and the public in selecting the measures and packages which you might adopt.
9. Ensure that each shortlisted measure is designed in sufficient detail to ensure that it can be implemented and that stakeholders and the public know what to expect.
10. Assess the likely impacts (on objectives and problems) of each of these detailed designs. This will require an ability to predict what might happen and can be assisted by predictive models.
11. Use these predictions to appraise each detailed measure and package against your objectives. This will help you to prioritize the measures which you adopt, and may suggest ways in which individual designs can be enhanced.

References

Atkins. (2007). *Long term process and impact evaluation of the local transport plan policy: Final report*. London: Department for Transport.

Banister, D. (2005). *Unsustainable transport*. London: Routledge.

Cerema. (2015). *Le versement transport: une contribution essentielle au financement des transports urbains, Collection Essentiel, "Le point sur"—Mobilité et transports*. Lyon: Cerema.

CIVITAS. (2019). *The urban mobility tool inventory*. www.civitas.eu/tool-inventory. (Accessed February 26, 2019).

Department for Transport. (2009). *Guidance on local transport plans*. London: DfT.

Department of Transportation. (2015). *A guide to transportation decision making*. https://www.planning.dot.gov/documents/GuidetoTransportationDecisionmaking.pdf. (Accessed October 11, 2017).

Eddington, R. (2006). *The Eddington Transport Study. Main report: Transport's role in sustaining the UK's productivity and competitiveness*. London: The Stationery Office.

Eichhorst, U., Sterk, K., Wehnert, T., Bleckmann, J., Henkel, A., Lauko, P., et al. (2012). Navigating transport NAMAs. In *TRANSfer project online handbook*: GIZ, BMU, and Wuppertal Institut. Available at: http://www.transferproject.org/index.php/hb.

ELTIS. (2019). *The European Local Transport Information Service*. www.eltis.org.

European Conference of Ministers of Transport. (2002). *Implementing sustainable urban travel policies*. Paris: OECD.

European Conference of Ministers of Transport. (2006). *Sustainable urban travel: Implementing sustainable urban travel policies: Applying the 2001 key messages*. Paris: ECMT.

European Platform on Mobility Management. (2019). *MAXExplorer*. www.epomm.eu. (Accessed November 26, 2019).

Evidence. (2015). www.evidence-project.eu. (Accessed December 5, 2015).

Givoni, M., Macmillen, J., Banister, D., & Feitelson, E. (2013). From policy measures to policy packages. *Transport Reviews*, *33*(1), 1–20.

Goodwin, P. B. (2010). *Opportunities for improving transport and getting better value for money, by changing the allocation of public expenditures to transport*. UWE: Bristol.

Gudmundsson, H. (2007). Sustainable urban transport in Scandinavia. In *Proceedings of the 86th Transportation Research Board Conference, Washington*.

Gühnemann, A. (2016). *SUMP manual on monitoring and evaluation: Assessing the impact of measures and evaluating mobility planning processes*. www.eltis.org.

International Association for Public Participation. (2007). *IAP2 Spectrum of Public Participation*. www.iap2.org.

Jones, P., Kelly, C., May, A. D., & Cinderby, S. (2009). Innovative approaches to option generation. *European Journal of Transport and Infrastructure Research*, *9*(3), 237–258.

Marsden, G., & Snell, C. (2009). The role of indicators, targets and monitoring in decision-support for transport. *European Journal of Transport and Infrastructure Research*, *9*(3), 219–236.

May, A. D. (2009). Improving decision-making for sustainable urban transport. *European Journal of Transport and Infrastructure Research*, *9*(3), 184–201.

May, A. D. (2013). Encouraging good practice in the development of sustainable urban mobility plans. In *Proceedings of the 13th World Conference on Transport Research, Rio de Janeiro*.

May, A. D. (2016). *Measure selection: Selecting the most effective packages of measures for sustainable urban mobility plans*. http://www.sump-challenges.eu/kits.

May, A. D., Kelly, C., & Shepherd, S. P. (2006). The principles of integration in urban transport strategies. *Transport Policy*, *13*, 319–327.

May, A. D., Khreis, H., & Mullen, C. (2016). Option generation for policy measures and packages: The role of the KonSULT knowledgebase. In *Proceedings of the 14th world conference on transport research, Shanghai*.

May, A. D., Khreis, H., & Mullen, C. (2018). Option generation for policy measures and packages: An assessment of the KonSULT knowledgebase. *Case Studies on Transport Policy*, *6*, 311–318.

May, A. D., & Matthews, B. (2007). Decision making processes. In S. Marshall & D. Banister (Eds.), *Land use and transport planning: European perspectives on integrated policies*. Oxford: Elsevier.

Nakamura, H., Hayashi, Y., & May, A. D. (2004). *Urban transport and the environment—An international perspective*. Oxford: Elsevier.

Nieuwenhuijsen, M., & Khreis, H. (2019). *Integrating human health into urban and transport planning*. Springer.

Paulley, N., et al. (2006). *The demand for public transport*. Crowthorne: TRL.

Promotion of Operational Links with Integrated Services (POLIS), & West Yorkshire Combined Authority (WYCA). (2016). *SUMP Manual on Institutional Cooperation: Working jointly with institutional partners in the context of sustainable urban mobility plans.* www.eltis.org.

Rose, R. (2005). *Learning from comparative public policy: A practical guide*. Oxon: Routledge.

Rupprecht Consult. (2016). *SUMP manual on participation*. www.eltis.org.

Stead, D. (2008). Effectiveness and acceptability of urban transport policies in Europe. *International Journal of Sustainable Transportation, 2*(1), 3–18.

Transport for London. (2007). *Central London congestion charging—Impacts monitoring*. London: TfL.

Transport for London. (2018). *The mayor's transport strategy*. London: TfL.

VTPI. (2019). *On line TDM encyclopaedia*. www.vtpi.org/tdm. (Accessed February 26, 2019).

WBCSD (World Business Council on Sustainable Development). (2017). *Sustainable mobility plan fact-based mobility planning tool.* www.wbcsd.org/Projects/smp2. (Accessed November 12, 2017).

CHAPTER 16

Best practices for air quality and active transportation

Andrew Glazener[a,b,c,d] **and Haneen Khreis**[a,b,e,f,g]
[a]Center for Advancing Research in Transportation Emissions, Energy, and Health (CARTEEH), Texas A&M Transportation Institute (TTI), Bryan, TX, United States
[b]Texas A&M | TTI Headquarters, Bryan, TX, United States
[c]Langford College of Architecture, Texas A&M University, College Station, TX, United States
[d]School of Architecture, University of Texas at Austin, Austin, TX, United States
[e]ISGlobal, Centre for Research in Environmental Epidemiology (CREAL), Barcelona, Spain
[f]Universitat Pompeu Fabra (UPF), Barcelona, Spain
[g]CIBER Epidemiologia y Salud Publica (CIBERESP), Madrid, Spain

Abbreviations

AV	autonomous vehicle
BC	black carbon
BMI	body mass index
CO	carbon monoxide
CO_2	carbon dioxide
EV	electric vehicle
GHG	greenhouse gas
HIA	health impact assessment
NO_2	nitrogen dioxide
NO_x	nitrogen oxide
O_3	ozone
PM	particulate matter
SAV	shared autonomous vehicles
SGIS	Skyrise Greenery Infrastructure Scheme
SO_2	sulfur dioxide
TRAP	traffic-related air pollution
UFP	ultrafine particles
VMT	vehicle miles traveled
WHO	World Health Organization

Introduction

Air pollution is the fifth leading cause of death worldwide, accounting for nearly 4.9 million deaths each year (Health Effects Institute, 2019). Traffic-related air pollution (TRAP) is a contributing factor to this growing mortality burden, causing 184,000 deaths annually, in conservative estimates (Bhalla et al., 2014). Key elements of TRAP include particulate

Traffic-Related Air Pollution
https://doi.org/10.1016/B978-0-12-818122-5.00016-8

matter (PM), nitrogen oxides (NO_x), carbon monoxide (CO), carbon dioxide (CO_2), sulfur dioxide (SO_2), and ozone (O_3) (Crayton & Meier, 2017).All have been identified as causing detrimental health effects including cardiovascular disease (Lu, Xu, Cheng, et al., 2015), stroke (Beelen, Raaschou-Nielsen, Stafoggia, et al., 2014), lung cancer (Raaschou-Nielsen, Andersen, Beelen, et al., 2013), and others (Barcelona Institute for Global Health, 2018b). In the United States, on-road transportation is responsible for 23% of all greenhouse gas (GHG) emissions (Environmental Protection Agency, 2018), while on-road transportation is responsible for just under 20% of all GHG emissions in Europe (EEA, 2018). Up to 30% of PM comes from transportation sources in Europe (WHO, 2018), while this percentage drops to 10% when observing the United States (Environmental Protection Agency, 2019) however, emerging research has shown that negative health outcomes are still experienced at PM concentrations below World Health Organization (WHO) guidelines, necessitating a reevaluation of WHO guidelines (Kelly & Fussell, 2015). In low-income and developing countries, transportation activity typically accounts for a greater share of air pollution (~67% of PM in Sao Paulo, Brazil (Miller, Du, & Façanha, 2017), ~33% of all air pollutants throughout India (International Council on Clean Transportation, 2018), and up to 40% of air pollution in Chinese cities (World Resources Institute, 2018)) due to the use of older and less efficient vehicles (WHO, 2018).

Globally, urban populations are swelling—over 50% of the world's population now lives in cities (United Nations, 2018a)—which is increasing the number of individuals that are exposed to air pollution. The WHO reported that 92% of the world's population lives in cities where air pollution levels exceed the WHO air quality guidelines (Health Effects Institute, 2019). Low-income and developing countries are characterized by poor air quality levels (Health Effects Institute, 2019 #125) and are also the places where the majority of future urban population growth is expected to occur (United Nations, 2018a #26). To reduce human exposure to air pollution, a transition towards nonmotorized transport and more efficient and cleaner vehicles must occur to achieve a reduction in vehicle miles traveled (VMT) and TRAP. The consideration of multiple factors—including motor vehicle use, emerging technologies, urban design, and mobility patterns—is integral to encouraging this transition to mitigate exposure-related health risks. The increasing awareness of TRAP as a major public health concern has provoked policy makers, public health officials, urban and transportation planners, and experts from other fields to research strategies to mitigate its detrimental health effects.

TRAP has become a health concern largely because of the way many cities adapted the built environment to accommodate the automobile. This accommodation fostered a culture of car-dependency that has fueled a consistent increase in the number of vehicles on the road and VMT, leading to polluted urban environments. One action to combat car-dependency, and reduce air pollution, is to invest in active transportation. Active transportation, which includes walking, cycling, and public transportation (due to walking/cycling to and from public transit stops) produce much less, or none, of the harmful pollutants that motor vehicles do.

Further, the sedentary lifestyle that accompanies car-dependency has contributed to an alarming reduction in physical activity. An estimated 31% of adults (those aged 15 and older) worldwide do not achieve the recommended level of physical activity (World Health Organization, 2018) which has been linked to breast cancer, colon cancer, diabetes, heart disease, stroke, and other health problems (Kyu, Bachman, Alexander, et al., 2016). As a result, physical inactivity is the 12th leading cause of mortality worldwide, contributing to over 1 million deaths annually (Health Effects Institute, 2019). The design of many urban built environments inadequately supports active transportation, necessitating investment in active transportation infrastructure to reduce motor vehicle dependence and the associated health burdens of TRAP and physical inactivity in urban areas.

Air pollution exposure during active transportation

A common concern regarding shifts from motorized to active transportation is the heightened exposure to air pollution as a result of increased ventilation rates and the associated health risks (Giles & Koehle, 2014). However, research has shown that the health benefits from increased physical activity due to active transportation outweigh the health detriments of increased air pollution exposure (Mueller, Rojas-Rueda, Cole-Hunter, et al., 2015). Andersen, de Nazelle, Mendez, et al., 2015 concluded that health benefits resulting from physical activity were not significantly reduced by exposure to nitrogen dioxide (NO_2) (Andersen et al., 2015). Rabl and de Nazelle (2012)similarly concluded that health benefits derived from increased physical activity due to mode shifts from passive to active travel outweighed the costs associated with air pollution exposure and crash risk (Rabl & de Nazelle, 2012). Another study concluded that at the global average urban $PM_{2.5}$ concentration (22 $\mu g/m^3$), the benefits of physical activity on reducing all-cause mortality outweigh risks from air pollution, even under the most extreme levels of active transportation. In areas with $PM_{2.5}$ concentrations above 100 $\mu g/m^3$, the health benefits of cycling and walking would be outweighed by exposure to $PM_{2.5}$ only after 1.5 and 10h per day, respectively (Tainio, de Nazelle, Götschi, et al., 2016). A study of bike-sharing in Barcelona estimated that increased physical activity rates due to greater frequency of active travel resulted in ~12 fewer annual deaths, despite the slight increase in air pollution exposure and risk incidence, and eliminated 9 million kg of CO_2 emissions (Rojas-Rueda, de Nazelle, Tainio, et al., 2011). It is worth noting that studies will often focus solely on CO_2 or GHG emissions, but these reductions may be used as indicators to assume that there was a reduction of additional pollutants produced by motor vehicles.

A systematic review of studies examining the difference in air pollution exposure between transportation modes found those who traveled by motor vehicles were more exposed to air pollutants than active commuters in 30 of the 42 comparisons in the included studies, resulting in up to a 1 year reduction in life expectancy compared to cyclists (Cepeda, Schoufour, Freak-Poli, et al., 2017). A quantitative review of 10 studies that compared air pollution exposure—specifically to $PM_{2.5}$, CO, black carbon (BC), and ultrafine particles (UFP)—among active and passive travelers concluded that pedestrians are consistently the least exposed, followed by bus passengers, cyclists, and car passengers (de Nazelle, Bode, & Orjuela, 2017). Time of day and route choice (Ham, Vijayan, Schulte, et al., 2017; Karanasiou, Viana, Querol, et al., 2014; MacNaughton, Melly, Vallarino, et al., 2014) are two factors that can influence exposure to air pollution for active commuters and these may have contributed to instances when active commuters had greater exposure to air pollution than motor vehicle commuters.

Best practices for active transportation and clean air

The best practices to encourage active transportation and improve air quality that are included in this chapter have been separated into six categories: car-free policies, vehicle technologies, urban design interventions, active transportation investments, green spaces, and integrated policy strategies.

Car-free policies

Around the world, actions are being taken to reduce car-dependency and TRAP. Cities have adopted measures to become "car-free" to reduce vehicle trips and emissions (Nieuwenhuijsen & Khreis, 2016), simultaneously allowing active transportation to become the prominent mobility option. In turn, the sedentary lifestyle of car-dependency is replaced by increased physical activity, social interaction, and improved health (Nieuwenhuijsen &

Khreis, 2016). There are several strategies that cities have employed to discourage automobile use and make alternative modes of transportation more attractive: vehicle bans and road closures, congestion pricing, distance-based tolls, low-emission zones, and parking pricing.

Car-free cities

Several cities are adopting measures to become car-free. Oslo, Norway has been augmenting its city center by removing parking spaces and prohibiting passenger vehicles access on certain roads to encourage walking and cycling and reduce air pollution (Peters, 2019). Madrid, Spain is following in the footsteps of Oslo by enacting an ultra-low-emission zone (a geographic region that is accessible only to low-emission vehicles) that restricts motor vehicle access to nonresidents (about 58,000 vehicles daily) in the city center (O'Sullivan, 2018). This ban includes all vehicles manufactured before 2000 and diesel vehicles manufactured before 2006 and will extend in 2020 to all vehicles built before 2006 and diesel vehicles predating 2014 (O'Sullivan, 2018).

Hamburg, Germany intends to ban motor vehicle access on certain streets to create transportation corridors that are amenable for pedestrians and cyclists (Nieuwenhuijsen & Khreis, 2016). These changes, combined with an ambitious place for green infrastructure investment, will propel Hamburg towards achieving its goal of an 80% reduction in GHG emissions by 2050 (Nieuwenhuijsen & Khreis, 2016) and being private car-free by 2034 (Nuwer, 2014). In the German city of Freiburg, the Vauban neighborhood has aimed to be car-free since its development in 1998 (Field, 2014). Although state law required the neighborhood to develop parking spaces for each residential unit, grassroot efforts by residents have maintained a low ratio (<0.5) of parking spaces to residential units (Field, 2014). Limited parking availability, coupled with restricted motor vehicle through-access to surrounding neighborhoods and land use practices that encourage active transportation, has largely influenced transportation trends in Vauban. Of the total, 64% of travel is conducted through nonmotorized means, while another 19% is carried out through public transportation, and there are only 160 cars per 1000 residents (Field, 2014).

Car-free days

In Paris, the scale of car-free policy is different. The first Sunday of every month, cars are not allowed access into the city center, limiting access to those who walk, cycle, or use scooters (Coffey, 2018). NO_2 levels dropped by 40% during a car-free day in Paris (Willsher, 2015) while a quasi-car-free day in Leeds (due to the Tour de France) resulted in a 20% reduction in NO_2 (The City Talking, 2015). It has been noted that car-free days and zones do not always provide city-wide air quality benefits, often because motor vehicle traffic is simply diverted to different parts of the city (Masiol, Agostinelli, Formenton, et al., 2014). Investment in active and alternative modes of transportation can replace motor vehicle trips and assist in attaining broad air quality benefits.

Los Angeles adopted the South American idea of "Ciclovia," which is a car-free day that sponsors an event to promote active transportation instead. During the program, known as "CicLAvia," on-road UFP and $PM_{2.5}$ concentrations were found to be reduced by 21% and 49%, respectively, compared to normal concentrations (Shu, Batteate, Cole, et al., 2016). On a community-wide scale, a 12% reduction in $PM_{2.5}$ was measured (Shu et al., 2016).

Low-emission zones

Low-emission zones are designated regions that permit vehicle access to low-emission vehicles only, sometimes levying a fee for the use of other vehicles in the designated regions. Several cities in Germany have implemented low-emission zones to restrict the use of diesel and older vehicles after nearly 70 German cities exceeded the annual NO_x limits prescribed by the European Union (BBC, 2018). In Stuttgart, the home of Audi, Mercedes-Benz, and Porsche, over 190,000 of the registered 530,000 motor vehicles have been excluded from designated low-emission zones in the city center since the beginning of 2019 (Deutsche Welle, 2018). A study of German low-emission zones reported an average 9% decrease in PM_{10} while air pollution did not increase in areas outside of the low-emission zone (Wolff, 2014). A low-emission zone in Milan, Italy called the Ecopass Zone charges an entry fee to cars with engines predating the Euro 4 emissions standard (Invernizzi, Ruprecht, Mazza, et al., 2011). While reductions in PM have been difficult to measure in the Ecopass, one study measured a 57% decrease in BC concentrations, a pollutant that can easily be traced back to motor vehicle exhaust and contributes to PM_{10} levels (Invernizzi et al., 2011).

Market-based strategies

There are also financial tools that can incentivize reduced motorized travel and produce air quality benefits. Milan and Stockholm have successfully introduced congestion charging zones to alter driver behavior in hopes of improving air quality. Congestion charging requires a fee to enter an area by car, with the aim of reducing traffic congestion. In Milan, CO and PM_{10} concentrations increased by 6% and 17%, respectively, when congestion charging was suspended (Gibson & Carnovale, 2015). NO_x emissions in the cordoned area in Stockholm were reduced by 8.5% while CO_2 emissions across the entire metropolitan area were reduced by 2%–3% (Eliasson, 2014). By 2021, New York city is expected to become the first city in the United States to utilize congestion charging (USA Today, 2019).

Distance-based road pricing is another strategy used to reduce motor vehicle activity and improve air quality by levying a fee on drivers on a per-kilometer/-mile basis. A review of mainly simulated distance-based road pricing schemes in cities within Europe, New Zealand, and the United States produced encouraging estimates of CO_2 reductions ranging from 4.7% to 36.0% (Cavallaro, Giaretta, & Nocera, 2018).

Parking pricing has been used as a deterrent for motor vehicle use. Ten studies that observed traffic behavior in 11 cities around the world between 1927 and 2001 found that, on average, 30% of vehicle traffic was due to people cruising for parking (Shoup, 2007). In Los Angeles, cruising for parking was estimated to account for an additional 950,000 VMT and 730 tons of CO_2 each year (Shoup, 2007). To address this, studies have analyzed the relationship between parking pricing, parking demand, and traffic congestion. Results from these studies show increasing parking price has reduced parking demand (Hensher & King, 2001; Kelly & Clinch, 2009) and congestion (Shoup, 2005). San Francisco is piloting a US Department of Transportation funded adaptive parking pricing program to manage parking demand (Pierce & Shoup, 2013). Adaptive pricing aims to keep on-street parking between 60% and 80% occupied at all times, and the price to park will adjust based on parking occupancy to maintain desired parking availability (Pierce & Shoup, 2013). Analysis has shown that cruising decreased by 50% where adaptive parking pricing is used in comparison to surrounding streets that do not have adaptive parking pricing (Millard-Ball, Weinberger, & Hampshire, 2014). While emissions were not included in this study, it is logical that a reduction in cruising is associated with a reduction in TRAP. Furthermore, a reduction in cruising for parking (and overall vehicle traffic) could make streets more inviting for people utilizing active transportation modes. As a result, local air quality and physical activity could be improved.

Vehicle technologies

In the last century, transportation has evolved from horse-drawn carriages to motorized automobiles, accelerating the pace of trade and travel at an unprecedented rate. Current technologies—Autonomous (AV), connected (CV), and electric vehicles (EV)—have enabled "smart," more fuel and energy efficient, and safer transportation options. These vehicles have the potential to make cities healthier places through improved vehicle efficiency and reduced emissions, but there remains a risk that adverse air quality and physical activity outcomes could be the result of their integration into the automotive market. This is known as the "rebound effect" which refers to increased consumption (i.e., VMT) following improved efficiency (low-emission vehicles, electric vehicles, convenience of autonomous and connected vehicles) and lowered costs (Litman, 2010). Two opposing schools of thought have emerged regarding the potential impacts of AVs: one espouses the benefits of reduced road congestion, improved access for mobility-dependent populations who cannot operate vehicles independently and/or have difficulty utilizing active and public transportation (children, senior citizens, persons with disabilities, etc.) (Alessandrini, Campagna, Site, et al., 2015), and the redevelopment and densification of urban spaces (Papa & Ferreira, 2018), while the other warns against the potential perils of increased comfort and convenience of automobile travel that could lead to increased VMTs and sprawl of urban development (Papa & Ferreira, 2018; Stead & Vaddadi, 2019).

Autonomous and connected vehicles

A total of 22 states in the United States, (Brookings Institute, 2018) as well as the United Kingdom, Germany, South Korea, and Singapore (Bloomberg, 2018), have passed legislation permitting the use of fully-autonomous (Level 4 according to the US Department of Transportation) (NHTSA, 2013) vehicles on their roads, which necessitates continued research to understand how autonomous vehicles will impact environmental and public health. The environmental impacts of AVs, specifically GHG emissions, have been well discussed in the literature (Crayton & Meier, 2017; Faisal, Yigitcanlar, Kamruzzaman, et al., 2019; Wadud, MacKenzie, & Leiby, 2016). Wadud et al. (2016) described 12 potential GHG emission impacts of AVs and examined them in four scenarios. Ultimately, the potential impacts on GHG emissions range from a nearly 50% reduction to more than doubling as a result of AV integration (Wadud et al., 2016). This wide range is dependent on the scale of integration, fuel type, vehicle use patterns, and level of automation (Wadud et al., 2016).

Another function of AVs that could contribute to reduced emissions is car sharing (Faisal et al., 2019). Shared AVs (SAVs) could remove more than twice as many vehicles compared to conventional car-sharing systems, which would reduce congestion, travel time, pollution sources, and emissions (Fagnant & Kockelman, 2015; Paradatheth, 2015). Liu, Kockelman, Boesch, et al. (2017) estimated vehicle emission reductions due to SAVs replacing conventional motor vehicles (Liu et al., 2017). While substantial vehicle emission reductions were estimated (−16.8% GHG, −24.3% PM, −30.7% NO_x, −24.3% SO_2, −42.7% CO), VMT was assumed to increase, offsetting some reductions (Liu et al., 2017). Similarly, a 2017 simulation of autonomous vehicle integration (37.5% vehicle mode share) in Boston estimated that the number of vehicles on the road would decrease by 15%, but VMT would increase by 16% as a result of additional trips to pickup or drop off passengers in mobility-on-demand offerings and from empty miles driven by mobility-on-demand vehicles between passenger rides (World Economic Forum & The Boston Consulting Group, 2018). These findings echo the sentiment of the rebound effect that SAVs may reduce the temporal (Wadud et al., 2016) and monetary costs (Paradatheth, 2015; Wadud et al., 2016) of motor vehicle travel and contribute to an increase in total automobile travel (number of trips and VMT) (Wadud et al., 2016), degrading air quality and reducing active transportation.

Connected vehicles, those that have the ability to communicate with other vehicles, street infrastructure, and the internet, are expected to facilitate more efficient travel (Lu, Cheng, Zhang, et al., 2014). The improved flow of traffic due to constant communication could lead to reduced congestion and air pollution. Similar to the risks of AVs, however, it should not be overlooked that by increasing the convenience and efficiency of motor vehicle travel and lowering the costs, induced travel demand for automobiles could have adverse effects on congestion leading to higher transportation emissions (Anderson et al., 2016) and reduced physical activity.

Electric vehicles

As previously mentioned, about 23% and 20% of all GHG emissions in the United States (Environmental Protection Agency, 2018) and Europe (EEA, 2018), respectively, are produced by on-road transportation activity. There is potential for significant reductions in GHG and exhaust TRAP through the electrification of vehicles. Air quality benefits will not be achieved, however, unless electricity used for charging the vehicles is generated by clean energy methods, such as solar, wind, hydro, etc. Norway is a global leader in the production of clean energy (Global Citizen, 2018) and EV fleet size (World Economic Forum, 2018) per capita. Norway has the fifth lowest average concentration of fine particles in urban areas and the fourth lowest air pollution-related mortality rate when compared to the rest of Europe (Bertrand, 2017). While Norway is a sterling example of how vehicle technologies may contribute to clean air through reduced exhaust pollution, EVs still pose air pollution and health risks through nonexhaust pollution (dust, break, tire, and road surface abrasions, and electricity generation emissions) (Thorpe & Harrison, 2008; Timmers & Achten, 2016). Nonexhaust pollution accounts for 90% and 85% of PM_{10} and $PM_{2.5}$ produced by vehicle traffic, respectively, and the difference in nonexhaust pollution between internal combustion motor vehicles and EVs is negligible (Timmers & Achten, 2016).

Overall, AV, CV, and EV technologies could provide air quality benefits, but the scale of integration would have to be wide (Soret, Guevara, & Baldasano, 2014) in addition to the development of relevant infrastructure to accommodate these new vehicles. Furthermore, auxiliary measures such as active transportation infrastructure investment, improving accessibility of alternative transportation modes, and altering urban land-use patterns will be necessary to ensure that active transportation is not displaced by the allure of emerging vehicle technologies that may increase motor vehicle use and produce excessive air pollution as a result of increased VMT and energy demands.

Urban design interventions

As automobiles became more accessible in the United States, cities sprawled outward as greater distances could be traveled quicker and more conveniently. The urban form then began catering to cars instead of people which resulted in poor urban air quality and a lack of human-scale development to facilitate active travel, leaving almost no alternative to motor vehicle travel in many areas. Urban populations have suffered under these conditions, which is why this section is dedicated to actions that cities have taken, and can take, to successfully reclaim urban space from cars and roads to provide walkable, bikeable, and healthy environments for their citizens.

There is an increasing amount of literature that discusses how the built environment influences both travel behavior and mode choice (Ding, Wang, Liu, et al., 2017). Perceived walkability, parks and open spaces, and active transportation infrastructure are all built environment qualities that have a positive impact on active transportation and physical activity (Smith, Hosking, Woodward, et al., 2017). A systematic review found that land use mix, population (housing) density, and neighborhood design (walkability, active transportation infrastructure, etc.) also positively correlate with adult physical activity rates, especially when these built environment characteristics are intended to promote active transportation (McCormack & Shiell, 2011). Safety, an additional factor that influences the decision to travel actively, can be improved through traffic calming measures (Speck, 2012), installing dedicated cycling lanes and sidewalks (Hull & O'Holleran, 2014; Speck, 2012), and by increasing the number of active commuters (Elvik & Bjørnskau, 2017). An analysis of the built environment in 14 cities concluded that up to 89 min of additional physical activity a week can be attributed to urban design characteristics (Sallis, Cerin, Conway, et al., 2016).

Of the total, 45% of all trips taken by vehicles in the United States were shorter than 3 miles (Federal Highway Administration, n.d.) and 50% of all vehicle trips in Europe in 2005 were shorter than 5 km (WHO, 2006), distances that could feasibly be traversed via active transportation. One health impact assessment (HIA) quantified the health benefits that could be accrued due to modal shifts from motor vehicles to walking and cycling. Rojas-Rueda, de Nazelle, Andersen, et al. (2016) estimated health impacts of a scenario where mode shifts occurred to the extent that walking and cycling accounted for 50% and 35% of all trips, respectively, in six European Cities (Barcelona, Basel, Copenhagen, Paris, Prague, and Warsaw) (Rojas-Rueda et al., 2016). Cumulatively, CO_2 reductions across all six cities would exceed 100,000 metric tons as a result, with the greatest reductions caused by a mode shift to cycling (Rojas-Rueda et al., 2016). The physical activity benefits in both scenarios reduced annual mortality in the six cities by 295 deaths, with the greatest reductions being caused by a mode shift to cycling (Rojas-Rueda et al., 2016).

Another HIA explored the potential health impacts of expanding cycling networks in 167 European cities. The study concluded that a cycling network of 315 km/100,000 persons maximizes the mode share of bicycles, which was estimated to be 24.6% of all trips taken (Mueller, Rojas-Rueda, Salmon, et al., 2018). The health impacts of extending cycling network infrastructure are greatest in cities that have relatively limited cycling networks, while positive health impacts are lower in cities with well-established cycling networks (Mueller et al., 2018). This study also examined the health impacts in seven European cities if a bicycle mode share of 24.6% was achieved. The HIA estimated premature mortality would decrease by over 10,000 deaths annually (Mueller et al., 2018).

Copenhagen is the self-proclaimed "cycling capital of the world." There are 346 km of bicycle paths, in addition to 23 km of bike lanes and 43 km of green bike paths (Nielsen, Skov-Petersen, & Carstensen, 2013). A "superhighway" for cyclists is being built to service those whose daily commutes are between 5 and 20 km to expand bicycle connectivity between the center of Copenhagen to surrounding communities (Nielsen et al., 2013). From 2010 to 2016, the share of all trips to work or education increased from 35% (Gössling, 2013) to 41% (Velo City, 2016) while cyclist deaths fell by 50% from 2008 to 2016 (Denmark CEo, 2010; Statistics Denmak, 2018). Unsurprisingly, car ownership is strikingly low in Denmark compared to other European countries (European Automobile Manufacturers Association, 2017).

In Barcelona, the development of "Superillas" (Superblocks) is transforming mobility patterns. Like many urban areas, Barcelona has struggled with maintaining healthy air quality levels. A report released by the Barcelona Public Health Agency estimated that 95% of the city's population is exposed to PM levels that exceed WHO air quality recommendations (Agencia de Salut Publica, 2016). Expressed concerns from the European Union about the nonattainment of healthy air quality and a lack of air quality improvement action plans—specifically in urban areas—prompted discussion about mitigation plans in Spain (Transport and Environment, 2018). The superblock design was developed from this convention and it has since gained international notoriety. The superblock design is a 3×3 grid composed of nine city blocks whose street pattern prohibits vehicles from traveling through the superblock and calms vehicle traffic within the superblock by reducing the speed limits from 50 to 10 km/h. Safe, clean, and healthy public spaces will replace roadways inside the superblock allowing cyclist and pedestrian activity while limiting vehicle traffic to the roads surrounding the superblock. Other auxiliary policies that are planned to be adopted in conjunction with the superblocks include the addition of segregated bus, cycling, and pedestrian lanes on the basic street network, the addition of bus stops at each superblock intersection, increasing the frequency of buses, and the development of public open and green space. The development of Superblocks and the implementation of these auxiliary policies form Barcelona's New Urban Mobility Plan. The New Urban Mobility Plan aims to decrease the use of private vehicles by 21%, transform 60% of the land that was previously used by cars into public space, and cut back on CO_2 emissions by 40% (Klause, 2018). A HIA of superblocks estimated that premature mortality could be reduced by 667 deaths, most of which were a result of lower air pollution concentrations (Mueller, Rojas-Rueda, Khreis, et al., 2019). One study estimated that vehicle kilometers traveled could be reduced by 32% and traffic-related NO_x emissions would be reduced by 68% due to superblock implementation (Soret et al., 2013).

Complete streets

Complete streets are "designed and operated to enable safe access for all users, including pedestrians, bicyclists, motorists, and transit riders of all ages and abilities." (Smart Growth America, 2019) Over 1300 complete streets policies have been adopted in the United States since 2005, when there were only 35 (Smart Growth America, 2018). This approach to the streetscape and urban design prioritizes accessibility, safety, and convenience around the human, not the car.

Complete Streets are oriented towards reducing motor vehicle travel through the provision of active transportation opportunities, and studies have proven complete streets to be successful in this venture. A study conducted in Dane County, Wisconsin, which has a population of 450,000, estimated a VMT reduction of 0.6 miles/person per each mile of constructed sidewalk infrastructure every day, which nets a significant reduction in VMT (182 million VMT/year) and air pollution when aggregated at the population level in a scenario where sidewalks have been constructed along all roadways (Guo & Gandavarapu, 2010). Another study of six complete streets in Los Angeles found that vehicle traffic decreased by 16% on one street and 26% on another, while the concentrations of PM_{10} and UFP were reduced by 2% and 7%, respectively, across all six streets (Zhu, Wang, Shu, et al., 2016).

Parking standards

Parking minimums, stipulations of local policy that require a certain amount of parking spots for new developments, have had a significant impact on strengthening car-dependency in cities. There are cities, however, that are reversing the trend by replacing parking minimums with maximum parking standards. By limiting the amount of parking spaces, active transportation infrastructure can be developed to promote mode shifts away from motor vehicles.

Several European cities have been progressive in attempts to reduce parking availability. In Zurich, the SihlCity development demonstrated that the parking demand induced by parking maximums can be quelled by providing adequate pedestrian, cycling, and public transportation infrastructure. As a result, 65% of trips to the SihlCity mall are made by walking, cycling, and public transportation (Sihlcity, 2018). In Paris, the provision of parking is prohibited for new developments that are within 500 m of a public transportation stop, which happens to encompass nearly the entire city center (Kodransky & Hermann, 2011). By cleansing streetscapes of parking spaces and encouraging people to travel actively or via public transportation, streets can become more lively, safer, and health-promoting.

It is important to note that restrictive parking measures do not reduce economic activity. Studies have shown that those who travel by transit or active modes save more money ($9000–$9500/year in Dallas and Cleveland, respectively, for examples) than those who travel predominantly by motor vehicle (Smart Growth America, 2012a). Additionally, areas that experience an influx in pedestrian activity also experience increased commercial spending. A study on the economic impact of replacing parking spots with bike-share docking stations in New York City estimated a 52% increase in commercial spending (Peters, Davidson, & Santiago, 2013), while the addition of a bike lane on a street in San Francisco boosted local business sales by 60% (Smart Growth America, 2012b). Similarly, Lawlor (2014) reviewed the effect of the reduction of cars on existing business performance (footfall and retail); urban regeneration (new business, rental income, employment, social exclusion, etc.); improved consumer and business perceptions, and business diversity (Lawlor, 2014). They found evidence that well-planned improvements to these public spaces can boost footfall and trading by up to 40% and that investing in better streets and spaces for walking can provide a competitive return compared to other transport projects; walking and cycling projects can increase retail sales by 30% (Lawlor, 2014).

Green spaces

Green spaces (parks, gardens, urban forests, trees, vegetation) provide a place of relaxation, respite, and recreation within busy and congested cities. The health effects of green spaces have been well documented and are largely positive (Gascon, Triguero-Mas, Martínez, et al., 2016; Ulrich, 1984). Specifically, green spaces have been associated with improved air quality (Nowak, Crane, & Stevens, 2006) and physical activity (Lee, Jordan, & Horsley, 2015), with some added benefit to active transportation (Wahlgren & Schantz, 2014). There is a body of literature regarding potential negative impacts that vegetation can have on air quality if certain vegetation types (mainly trees) are placed in such a way that it traps air pollutants underneath the tree canopy (also known as the "green paradox") (Vos, Maiheu, Vankerkom, et al., 2013).

Air pollution uptake is positively correlated with leaf surface area making trees and tall plants the most efficient vegetation types for pollutant removal (Derkzen, van Teeffelen, & Verburg, 2015), although the composition, structure, and maintenance of green spaces (Vieira, Matos, Mexia, et al., 2018) and the built environment determine the effectiveness of air pollution uptake in green spaces. Perhaps the most well-known study of green space effects on air quality is that by Nowak et al. (2006). The authors modeled the removal of air pollution due to trees in 55 US cities. In total, an estimated 711,000 tons of air pollution are removed by trees in 55 US cities each year, with the greatest uptake being in Jacksonville, Florida (11,100 tons) (Nowak et al., 2006). A similar study of CO_2 uptake by vegetation in Leicester, England estimated 230,000 tons of CO_2 were removed, 97% of which were absorbed by trees (Davies, Edmondson, Heinemeyer, et al., 2011).

In Santiago, Chile, air pollution uptake by trees and vegetation were assessed in different socioeconomic subregions. Unsurprisingly, these green elements were most present in high-income regions and sparsest in the poorest region monitored. Overall, the greatest impact on air pollution during the study was through the uptake of PM_{10} ($-35\,g\,m^{-2}$/year) followed by O_3, NO_2, SO_2, and CO. (Escobedo & Nowak, 2009) A more recent study in Strasbourg, France discovered that trees in public green spaces remove about 88 tons of air pollution each year, mostly O_3 and PM_{10} (7% of all PM_{10} in the city) (Selmi, Weber, Rivière, et al., 2016).

Active transportation has a more variable relationship with green space, but the health benefits are still prevalent. A survey of residents in Barcelona found that proximity to green space was a significant indicator of bicycle commuting (Cole-Hunter, Donaire-Gonzalez, Curto, et al., 2015). Similarly, a study of adolescent pedestrian travel activity in San Diego and Minneapolis found that greenness greatly determined pedestrian activity (Rodríguez, Merlin, Prato, et al., 2015).

Singapore is the self-proclaimed "garden city" as 47% of the landcover is public green space (Gan, 2015). This was accomplished partially through an incentives program that compensated for the removal of street-level green space due to urban development through the installation of rooftops gardens and high-rise terraces. The Skyrise Greenery Incentive Scheme (SGIS) will fund 50% of the installation costs of rooftop or vertical greenery to provide environmental benefits (improved air quality) (National Parks Board, n.d.). Since 2009, 110 buildings have utilized SGIS to install green roofs, rooftop gardens, and green walls (National Parks Board, n.d.).

Singapore is also creating a park network that will provide greater access and connectivity to natural environments and hopefully encourage active transportation. A qualitative study of residents in Singapore from 2017 found that while cycling is perceived as a more enjoyable activity than walking, cycling infrastructure (cycle lanes and bike parking) is a considerable limitation to cycling as a significant travel mode (Rojas López & Wong, 2017). This is supported by a 2011 report by the Land Transport Authority in Singapore: 23% of all trips are completed via active modes (and an additional 44% through public transportation), but cycling only accounted for 1% of all trips (Land Transport Authority, 2011). Singapore adopted a National Cycling Plan in 2010 that aims to develop 700 km of cycling infrastructure by 2030 to improve connectivity, largely through the park network connector (Land Transport Authority & Urban Redevelopment Authority, 2018). The introduction of bike-share programs has also increased the use of bicycles, with 50% of users reporting that they replace at least one to three car trips each week with cycling, and 30% saying they replace at least five car trips per week by cycling (Sutton, 2018).

Other cities are investing in public green spaces to increase the convenience of active transportation and removal of air pollution. Hamburg intends to make walking and cycling the dominant modes of transportation by converting 40% of its land into a "green network" by 2035 (Garfield, 2018). Another city that has plans to undergo a green revolution is Hangzhou, a Chinese city with a population of over 6 million. Hangzhou is replacing abandoned buildings and derelict properties with green spaces and adorning streets, rail lines, and public areas with vegetation. In 2012, 40% of the city was covered by green space, providing "easy access" to green space for 90% of the city's residents (Lijie, Yonggang, Weiping, et al., 2013; Wolch, Byrne, & Newell, 2014), with an additional 13 million m^2 of green space expected to be developed (Wolch et al., 2014).

Public transportation investments

Public transportation is key to creating healthy cities due to its potential to reduce motor vehicle traffic and increase physical activity. Traffic congestion, a compounding factor in the degradation of urban air quality, is the leading concern that Americans express about the future of their cities (Morrison & Lin Lawell, 2016). Traffic congestion costs in the United States in 2011 were estimated to be $120 billion, while congestion-induced travel delay quintupled from 1982 to 2011 (1.1–5.5 billion hours) (Schrank, Eisele, & Lomax, 2012). Congestion delay relief provided by public transportation services only doubled during the same time period, alleviating 865 million hours of congestion delay in 2011 (Schrank et al., 2012). Still, public transportation ridership accounts for only 5% of all trips taken in the United States, (Bureau of Transportation Statistics, 2015) despite 70% of all public transportation ballot measures being passed from 2000 to 2009 allocating over $100 billion to be invested in public transportation (Leinberger, 2009). New York city has the greatest public transportation mode share of all US cities at 9%, but this pales to other world cities like Vancouver (14%), Barcelona (35%), Tokyo (60%), and Hong Kong (73%) (Newman, Beatley, & Boyer, 2017). While public transportation ridership does not solely determine air quality (New York and Vancouver have lower average annual $PM_{2.5}$ concentrations than Barcelona (Barcelona Institute for Global Health, 2018a)), it can have dramatic effects on transportation-related emissions. For example, a new bus system in Mexico City reduced CO_2 emissions by an estimated 100,000 tons each year after implementation (Harvey & Segafredo, 2011). Similarly, a bus rapid transit system in Johannesburg was estimated to reduce CO_2 emissions by 40,000 tons in 1 year (Joburg Innovation and Knowledge Exchange, 2012).

A common method for measuring the congestion and emissions impacts of public transportation is observing changes in traffic flow and air pollution that occur during public transportation shutdowns. In 2003, Los Angeles public transportation employees went on strike for 35 days. Anderson (2014) analyzed hourly traffic speed data of major freeways during those 35 days and found that the lack of public transportation resulted in delays lasting 47% longer than average (Anderson, 2014). Bauernschuster, Hener, and Rainer (2017) observed 71 one-day strikes in five German cities (Berlin, Frankfurt, Cologne, Munich, and Hamburg). Travel times increased by 8.4% during morning peak travel hours and almost 4% during the evening, resulting in a 4% increase in NO_2 emissions and a 14% spike in PM_{10} (Bauernschuster e al., 2017). A study of two subway system shutdowns in Sao Paulo, Brazil in 2003 and 2006 saw increases in PM_{10} of 60.34 and 34.03 $\mu g/m^3$, respectively, which translated to an estimated 8 and 6 deaths attributable to each day of excessive air pollution in each year, respectively (Silva, Saldiva, Amato-Lourenço, et al., 2012). Lastly, there were 208 days between 2009 and 2016 that the public transportation system in Barcelona shut down. There was an average city-wide increase in air pollution (NO_2, NO, PM_{10}, BC) of 4%–8% during these strikes, but the air pollution effects of subway system closures were most pronounced. NO_x spiked nearly 50% while BC and PM saw significant increases ranging from 8% to 30% (Basagaña, Triguero-Mas, Agis, et al., 2018).

Not only does public transportation reduce congestion and TRAP, it also offers the opportunity to be physically active (Besser & Dannenberg, 2005).

The average public transportation commuter in Seattle walk 14 min each day explicitly as first- and last-mile connections (Saelens, Vernez Moudon, Kang, et al., 2014). A systematic review of new public transportation systems in the United States, United Kingdom, Mexico, and Canada concluded that new public transportation provides an additional ~30 min of low- to moderate-intensity physical activity per week (Xiao, Goryakin, & Cecchini, 2019). In Charlotte, North Carolina, light rail commuters were found to have a 1.18 kg/m^2 lower body mass index (BMI)—a metric that can measure body fat percentage and may indicate health problems such as diabetes, heart disease, and cancer—and they were 81% less likely to become obese when compared to nonlight rail users (MacDonald, Stokes, Cohen, et al., 2010). It is worth noting that the positive perception of neighborhood and the built environment correlated with lower BMI and obesity risk too (MacDonald et al., 2010). Similarly, BMI was found to be lower (1 kg/m^2 for men and 0.67 kg/m^2 for women) for those who utilized public transportation and other active transportation modes compared to those who commuted solely via motor vehicles in a cross-sectional study in the United Kingdom that included over 150,000 adult participants (Flint & Cummins, 2016).

Concerns about high air pollution concentrations measured in the subway and underground metro stations have become the focus of some air quality research. PM has been traced back to the friction between subway wheels, brake pads, and tracks (Querol, Moreno, Karanasiou, et al., 2012). In Barcelona, for example, concentrations of certain air pollutants have been recorded that are 300× higher than street level concentrations (Moreno et al., 2017). Similar to the "green paradox" mentioned previously, planning interventions that are intended to be health-promoting (such as investment in active transportation infrastructure) must be thoroughly examined before and after implementation to ensure unintended consequences are not counteracting the intended benefits. Studies have found that regular cleaning of subway platforms (Johansson & Johansson, 2003) and tunnels (Salma, Weidinger, & Maenhaut, 2007), updating tunnel ventilation systems (Querol et al., 2012), and installing platform doors (Kim, Ho, Jeon, et al., 2012) can reduce air pollution concentrations and exposure in these isolated spaces.

Integrated policy packages

While each of the previous sections has highlighted a singular policy or practice that produce air quality and/or physical activity benefits, multifaceted policy approaches offer the greatest improvement. Policy packages that integrate interventions can address the health burdens associated with air quality and physical activity holistically and comprehensively.

A report presented several scenarios of reduced transportation GHG emissions in the United States based on the adoption of policies that included improved vehicle efficiency, technology innovations, a variety of pricing schemes to deter driving, changes in the built environment and land use, and behavioral travel changes. Simultaneous implementation of these policies would result in an estimated 48% reduction in transportation-related GHG emissions by 2030 (Greene & Schafer, 2003). The greatest reductions were a result of improved vehicle efficiency (18% reduction in transportation GHG emissions by 2030), pricing policies (6%–9%), and the use of alternative fuels (7%) (Greene & Schafer, 2003).

In 2008, the United Kingdom adopted the Climate Change Act which pledged an 80% reduction in GHG emissions by 2050. Whitelegg, Haq, Cambridge, et al. (2010) identified four policy pathways that would enable the United Kingdom to achieve a zero-carbon transportation future by 2050. These four pathways (spatial planning, fiscal, behavioral change, and technology) incur changes such as pedestrian-oriented development, road user charges, car sharing, vehicle electrification, and improved vehicle efficiency, which were estimated to reduce transportation CO_2 emissions by 100% by 2030 (92% of which are attributed to road transportation) (Whitelegg et al., 2010). Ignoring the adoption of new vehicle technologies (which included EVs which effectively removed all CO_2 from on-road emissions), road CO_2 emissions were still projected to decline by nearly 75% as a result of measures related to the other three pathways (Whitelegg et al., 2010).

In Vancouver, an integrated action plan was adopted that aimed to improve air quality and reduce GHG emissions to help meet city-wide sustainability goals adopted in 2008 (Metro Vancouver, 2011). In total, 18 policies and actions were adopted in 2011 and a progress report was produced in 2014. When compared to vehicle emissions in 2005, NO_2 was expected to decrease by 51% by 2015, SO_2 by 60%, volatile organic compounds by 40%, and $PM_{2.5}$ by 58% (Metro Vancouver, 2014).

Lastly, Woodcock, Edwards, Tonne, et al. (2009) published a seminal paper on health outcomes related to integrated policy solutions in London and Delhi (Woodcock et al., 2009). The study compared GHG emissions in scenarios which included wide use of low-carbon-emission vehicles and increased active travel with a business-as-usual projection. Increased active travel coupled with the use of low-carbon-emission vehicles was estimated to reduce CO_2 emissions by 60% from 1990 levels and reduce premature mortality by 541 deaths/million persons in London (Woodcock et al., 2009). In Delhi, the cumulative effect of increased active transport and the use of low-carbon-emission vehicles on CO_2 emissions is less straightforward.

CO_2 emissions were estimated to increase by 199% compared to 1990 levels, but per person CO_2 emissions were lower than 2010 levels (Woodcock et al., 2009). Premature mortality reductions were estimated at 532 deaths/million persons in Delhi (Woodcock et al., 2009). Disability adjusted life years were reduced by 12,995/million persons and 7439/million persons in Delhi and London, respectively, as well (Woodcock et al., 2009).

Conclusions

Urban air quality has long been considered unhealthy due in part to traffic-related air pollution. Attempts to reduce traffic-related air pollution have targeted several key areas. Policies and designs that restrict motor vehicle use in favor of active transportation modes have made cities safer and healthier places. Improvements in vehicle technologies have increased the efficiency of motor vehicles and reduced their emissions footprint. Efforts to prioritize the placement of green space and vegetation in cities have improved air quality and provided safe and pleasant places for people to actively travel off the road. Cities have invested in public transportation systems to reduce the burden of traffic congestion and provide alternative modes of transportation. In many cases, policies are adopted that integrate strategies from each of these areas to comprehensively mitigate the effects of traffic-related air pollution.

Still, there is a need for further research to bridge the knowledge gaps regarding various practices and their overall impact on urban air quality and physical activity. The WHO air quality guidelines are being updated to account for health effects that occur at air pollution concentrations below the current air quality guidelines. Until the air quality guidelines are updated, air pollution mitigation measures will fail to attain air pollution reductions that appropriately limit the health risks posed by air pollution. There are still uncertainties about the net impact of green spaces on air pollution, specifically the "green paradox." Furthermore, the integration of vehicle technologies will first occur in developed nations, however, 90% of the population growth in the next 30 years will occur in developing nations (United Nations, 2018b). The adoption of modern technologies will be delayed in these developing nations, creating a period where increasing populations will put a stress on transportation systems, potentially increasing emissions and exposure hotspots. There is the potential that the proven best practices displayed in developed urban areas can be replicated in developing countries but determining these best practices will rely on the development of studies that can trace air pollution back to transportation sources. Table 16.1 outlines the impacts of the policies and practices discussed in this chapter.

Table 16.1 Overview of policies and practices.

Policy/practice	Targeted outcome	Benefits air quality?	Benefits active transportation?	Overall impact
Car-free policies				
Car-free cities/days (Nieuwenhuijsen & Khreis, 2016)	Reduce motor vehicle use and emissions	Yes	Yes	Reducing motor vehicle use will improve air quality and result in more physically active populations through shifts to active transportation
Low-emission zones (O'Sullivan, 2018)	Restrict motor vehicle access to high polluting cars	Yes	No	Restricting access to certain parts of the city to high polluting vehicles can improve air quality
CicLAvia (Shu et al., 2016)	Encourage cycling	Yes	Yes	Restricting car access in certain areas or specific streets in Los Angeles to promote cycling and also reduce TRAP
Congestion charging (Eliasson, 2014)	Disincentivize motor vehicle use through increased cost	Yes	Potentially	Reduce motor vehicle use by requiring a fee to drive on designated roads and potentially induce a mode shift
Distance-based road pricing (Cavallaro et al., 2018)	Disincentivize motor vehicle use through increased cost	Yes	Potentially	Charging a fee based on distance driven can discourage total VMT and related emissions
Parking pricing (Pierce & Shoup, 2013)	Disincentivize motor vehicle use through increased cost	Yes	Potentially	San Francisco has introduced a fluctuating pay-to-park systems that limits the amount of time spent cruising for parking
Vehicle technologies				
Autonomous vehicles	Improve vehicle operation efficiency	Potentially	No	Improved efficiency of driving patterns can increase fuel efficiency and reduce the total number of vehicles on the road
Electric vehicles	Reduce vehicle exhaust pollution	Potentially	No	By removing tailpipe emissions from TRAP air quality can be improved

Continued

Table 16.1 Overview of policies and practices.—cont'd

Policy/practice	Targeted outcome	Benefits air quality?	Benefits active transportation?	Overall impact
Urban design interventions				
Active transportation infrastructure (Smith et al., 2017)	Increase active transportation	Yes	Yes	Constructing sidewalks and cycling lanes to increase safety and accessibility for traveling by active modes
Built environment design (Sallis et al., 2016)	Create a built environment more conducive for walking and cycling	Potentially	Yes	Walkability, density, land use mix, and active transportation infrastructure can reduce vehicle use
Traffic calming measures (Speck, 2012)	Slow motor vehicle traffic	Potentially	Potentially	Shrinking lane width, adding medians, and lowering speed limits can reduce vehicle traffic speed and make active transportation safer
Segregated bike lanes (Hull & O'Holleran, 2014)	Increase safety for cyclist	Potentially	Yes	Constructing bike lanes that are separated from the street can improve safety and increase cycling rates and potentially induce a modal shift
Mode shift (Rojas-Rueda et al., 2016)	Encourage the use of alternative transportation modes to motor vehicles	Yes	Yes	Increased active transportation can derive health benefits and reduce motor vehicle use and TRAP
Superblocks (Klause, 2018)	Restrict motor vehicle access and provide space for active transportation	Yes	Yes	Superblocks aim to restrict motor vehicle access and provide more space for active transportation
Extended cycle networks (Mueller et al., 2018)	Increase cycling connectivity	Potentially	Yes	Further developing cycling networks can increase cycling mode share in cities

Complete streets (Zhu et al., 2016)	Improve safety and connectivity for pedestrians and cyclists	Yes	Yes	Adding infrastructure on streetscapes to slow vehicle traffic and allow active transportation can produce healthier outcomes
Parking standards (Kodransky & Hermann, 2011)	Limit the amount of parking provided	Potentially	Potentially	Enforcing parking maximums, or banning the development of parking spaces, forces people to travel by active and alternative transportation
Green spaces				
Tree planting (Nowak et al., 2006)	Increase pollution uptake by trees	Yes	No	Trees can uptake air pollution and have a positive impact on air quality
Green space provision (Wolch et al., 2014)	Provide green spaces in urban areas	Yes	Yes	Green spaces can improve local air quality and invite people to be more active if green spaces are perceived to be safe and aesthetically pleasing
Skyrise greenery incentive scheme (SGIS) (National Parks Board, n.d.)	Vegetation that can uptake air pollution	Yes	No	Substituting the loss of green ground cover by installing green roofs, green walls, and gardens
Public transportation investment				
Developing public transportation systems (Anderson, 2014; Saelens et al., 2014)	Improving/Introducing public transportation connectivity and access	Yes	Yes	Public transportation users generally exhibit higher physical activity rates than motor vehicle commuters and public transportation can replace motor vehicle trips and congestion, reducing emissions and TRAP

References

Agencia de Salut Publica. (2016). *Evaluation of air quality in the city of Barcelona.* https://www.aspb.cat/wp-content/uploads/2016/07/Avaluacio-de-la-qualitat-aire-a-la-ciutat-de-barcelona-2016-PRV.pdf.

Alessandrini, A., Campagna, A., Site, P. D., et al. (2015). Automated vehicles and the rethinking of mobility and cities. *Transportation Research Procedia, 5*, 145–160. https://doi.org/10.1016/j.trpro.2015.01.002.

Andersen, Z. J., de Nazelle, A., Mendez, M. A., et al. (2015). A study of the combined effects of physical activity and air pollution on mortality in elderly urban residents: The Danish Diet, Cancer, and Health Cohort. *Environmental Health Perspectives, 123*, 557–563. https://doi.org/10.1289/ehp.1408698.

Anderson, M. L. (2014). Subways, strikes, and slowdowns: The impacts of public transit on traffic congestion. *American Economic Review, 104*, 2763–2796. https://doi.org/10.1257/aer.104.9.2763.

Anderson, J. M., Kalra, N., Stanley, K. D., Sorensen, P., Samaras, C., & Oluwatola, T. A. (2016). *Autonomous vehicle technology: A guide for policymakers*. RAND Corporation.

Barcelona Institute for Global Health. (2018a). *5 keys to healthier cities*. https://www.isglobal.org/en/ciudadesquequeremos.

Barcelona Institute for Global Health. (2018b). *Urban and transport planning related exposures and health outcomes*. https://www.isglobal.org/evidencia-cientifica-exposiciones-ciudad.

Basagaña, X., Triguero-Mas, M., Agis, D., et al. (2018). Effect of public transport strikes on air pollution levels in Barcelona (Spain). *Science of the Total Environment, 610–611*, 1076–1082. https://doi.org/10.1016/j.scitotenv.2017.07.263.

Bauernschuster, S., Hener, T., & Rainer, H. (2017). When labor disputes bring cities to a standstill: The impact of public transit strikes on traffic, accidents, air pollution, and health. *American Economic Journal: Economic Policy, 9*, 1–37. https://doi.org/10.1257/pol.20150414.

BBC. (2018). *Diesel ban approved for German cities to cut pollution*. https://www.bbc.com/news/business-43211946.

Beelen, R., Raaschou-Nielsen, O., Stafoggia, M., et al. (2014). Effects of long-term exposure to air pollution on natural-cause mortality: An analysis of 22 European cohorts within the multicentre ESCAPE project. *The Lancet, 383*, 785–795. https://doi.org/10.1016/S0140-6736(13)62158-3.

Bertrand, P. (2017). *Which European countries are the most polluted?*. https://www.euronews.com/2017/05/18/which-european-countries-are-the-most-polluted-who-statistics.

Besser, L. M., & Dannenberg, A. L. (2005). Walking to public transit: Steps to help meet physical activity recommendations. *American Journal of Preventive Medicine, 29*, 273–280. https://doi.org/10.1016/j.amepre.2005.06.010.

Bhalla, K., Shotten, M., Cohen, A., Brauer, M., Shahraz, S., Burnett, R., et al. (2014). *Transport for health: The global burden of disease from motorized road transport*. Washington, DC: World Bank Group.

Bloomberg. (2018). *Self-driving cars find clearer paths in Europe*. https://www.bloomberg.com/news/articles/2018-05-14/california-arizona-may-get-lapped-by-asia-in-self-driving-race.

Brookings Institute. (2018). *The state of self-driving car laws across the U.S.* https://www.brookings.edu/blog/techtank/2018/05/01/the-state-of-self-driving-car-laws-across-the-u-s/.

Bureau of Transportation Statistics. (2015). *Commute mode share: 2015*. https://www.bts.dot.gov/content/commute-mode-share-2015.

Cavallaro, F., Giaretta, F., & Nocera, S. (2018). The potential of road pricing schemes to reduce carbon emissions. *Transport Policy, 67*, 85–92. https://doi.org/10.1016/j.tranpol.2017.03.006.

Cepeda, M., Schoufour, J., Freak-Poli, R., et al. (2017). Levels of ambient air pollution according to mode of transport: A systematic review. *The Lancet Public Health, 2*, e23–e34. https://doi.org/10.1016/S2468-2667(16)30021-4.

Coffey, H. (2018). *Paris to ban cars in city center one sunday a month*. https://www.independent.co.uk/travel/news-and-advice/paris-car-free-sundays-city-centre-france-pedestrian-a8566991.html.

Cole-Hunter, T., Donaire-Gonzalez, D., Curto, A., et al. (2015). Objective correlates and determinants of bicycle commuting propensity in an urban environment. *Transportation Research Part D: Transport and Environment, 40*, 132–143. https://doi.org/10.1016/j.trd.2015.07.004.

Crayton, T. J., & Meier, B. (2017). Autonomous vehicles: Developing a public health research agenda to frame the future of transportation policy. *Journal of Transport & Health*. https://doi.org/10.1016/j.jth.2017.04.004.

Davies, Z. G., Edmondson, J. L., Heinemeyer, A., et al. (2011). Mapping an urban ecosystem service: Quantifying above-ground carbon storage at a city-wide scale. *Journal of Applied Ecology, 48*, 1125–1134. https://doi.org/10.1111/j.1365-2664.2011.02021.x.

de Nazelle, A., Bode, O., & Orjuela, J. P. (2017). Comparison of air pollution exposures in active vs. passive travel modes in European cities: A quantitative review. *Environment International, 99*, 151–160. https://doi.org/10.1016/j.envint.2016.12.023.

Denmark CEo. (2010). *Bicycle statistics from Denmark*.

Derkzen, M. L., van Teeffelen, A. J. A., & Verburg, P. H. (2015). REVIEW: Quantifying urban ecosystem services based on high-resolution data of urban green space: An assessment for Rotterdam, the Netherlands. *Journal of Applied Ecology, 52*, 1020–1032. https://doi.org/10.1111/1365-2664.12469.

Deutsche Welle. (2018). *Stuttgart to introduce diesel driving ban in 2019*. https://www.dw.com/en/stuttgart-to-introduce-diesel-driving-ban-in-2019/a-44634246.

Ding, C., Wang, D., Liu, C., et al. (2017). Exploring the influence of built environment on travel mode choice considering the mediating effects of car ownership and travel distance. *Transportation Research Part A: Policy and Practice, 100*, 65–80. https://doi.org/10.1016/j.tra.2017.04.008.

EEA. (2018). *Greenhouse gas emissions from transport*. https://www.eea.europa.eu/data-and-maps/indicators/transport-emissions-of-greenhouse-gases/transport-emissions-of-greenhouse-gases-11.

Eliasson, J. (2014). *The Stockholm congestion charges: An overview*. http://www.transportportal.se/swopec/cts2014-7.pdf.

Elvik, R., & Bjørnskau, T. (2017). Safety-in-numbers: A systematic review and meta-analysis of evidence. *Safety Science, 92*, 274–282. https://doi.org/10.1016/j.ssci.2015.07.017.

Environmental Protection Agency. (2018). *Fast facts on transportation greenhouse gas emissions*. https://www.epa.gov/greenvehicles/fast-facts-transportation-greenhouse-gas-emissions.

Environmental Protection Agency. (2019). *Smog, soot, and other air pollution from transportation*. https://www.epa.gov/transportation-air-pollution-and-climate-change/smog-soot-and-local-air-pollution.

Escobedo, F. J., & Nowak, D. J. (2009). Spatial heterogeneity and air pollution removal by an urban forest. *Landscape and Urban Planning, 90*, 102–110. https://doi.org/10.1016/j.landurbplan.2008.10.021.

European Automobile Manufacturers Association. (2017). *Vehicles per capita, by country*. https://www.acea.be/statistics/tag/category/vehicles-per-capita-by-country.

Fagnant, D. J., & Kockelman, K. (2015). Preparing a nation for autonomous vehicles: Opportunities, barriers and policy recommendations. *Transportation Research Part A: Policy and Practice*, 77, 167–181. https://doi.org/10.1016/j.tra.2015.04.003.

Faisal, A., Yigitcanlar, T., Kamruzzaman, M., et al. (2019). Understanding autonomous vehicles: A systematic literature review on capability, impact, planning and policy. *Journal of Transport and Land Use, 2019*, 12. https://doi.org/10.5198/jtlu.2019.1405.

Federal Highway Administration (n.d.), https://nhts.ornl.gov/vehicle-trips.

Field, S. (2014). *Vauban—Freiburg, Germany*. ITDP Europe.

Flint, E., & Cummins, S. (2016). Active commuting and obesity in mid-life: Cross-sectional, observational evidence from UK biobank. *The Lancet Diabetes & Endocrinology*, *4*, 420–435. https://doi.org/10.1016/S2213-8587(16)00053-X.

Gan, V. (2015). *The link between green space and well-being isn't as simple as we thought*. https://www.citylab.com/life/2015/08/the-link-between-green-space-and-well-being-isnt-as-simple-as-we-thought/401324/.

Garfield, L. (2018). *13 cities that are starting to ban cars*. https://www.weforum.org/agenda/2018/03/13-cities-that-are-starting-to-ban-cars.

Gascon, M., Triguero-Mas, M., Martínez, D., et al. (2016). Residential green spaces and mortality: A systematic review. *Environment International*, *86*, 60–67. https://doi.org/10.1016/j.envint.2015.10.013.

Gibson, M., & Carnovale, M. (2015). The effects of road pricing on driver behavior and air pollution. *Journal of Urban Economics*, *89*, 62–73. https://doi.org/10.1016/j.jue.2015.06.005.

Giles, L.V., & Koehle, M. S. (2014). The health effects of exercising in air pollution. *Journal of Sports Medicine*, *44*, 223–249. https://doi.org/10.1007/s40279-013-0108-z.

Global Citizen. (2018). *Norway just proved that a future with 100% renewable energy cars is possible*. https://www.globalcitizen.org/en/content/norway-sustainable-energy-green-cars/.

Gössling, S. (2013). Urban transport transitions: Copenhagen, city of cyclists. *Journal of Transport Geography*, *33*, 196–206. https://doi.org/10.1016/j.jtrangeo.2013.10.013.

Greene, D. L., & Schafer, A. (2003). *Reducing greenhouse gas emissions from transportation*. https://www.c2es.org/site/assets/uploads/2003/05/reducing-greenhouse-gas-emissions-us-transportation.pdf.

Guo, J. Y., & Gandavarapu, S. (2010). An economic evaluation of health-promotive built environment changes. *Preventive Medicine*, *50*, S44–S49. https://doi.org/10.1016/j.ypmed.2009.08.019.

Ham, W., Vijayan, A., Schulte, N., et al. (2017). Commuter exposure to PM2.5, BC, and UFP in six common transport microenvironments in Sacramento, California. *Atmospheric Environment*, *167*, 335–345. https://doi.org/10.1016/j.atmosenv.2017.08.024.

Harvey, H., & Segafredo, L. (2011). *Policies that work: How to build a low-emissions economy*. https://energyinnovation.org/wp-content/uploads/2014/11/Policies-That-Work_Overview-Report.pdf.

Health Effects Institute. (2019). *State of global air*. https://www.stateofglobalair.org/report.

Hensher, D. A., & King, J. (2001). Parking demand and responsiveness to supply, pricing and location in the Sydney central business district. *Transportation Research Part A: Policy Practice*, *35*, 177–196.

Hull, A., & O'Holleran, C. (2014). Bicycle infrastructure: Can good design encourage cycling? *Urban Planning and Transport Research*, *2*, 369–406. https://doi.org/10.1080/21650020.2014.955210.

International Council on Clean Transportation. (2018). *India*. https://theicct.org/india.

Invernizzi, G., Ruprecht, A., Mazza, R., et al. (2011). Measurement of black carbon concentration as an indicator of air quality benefits of traffic restriction policies within the ecopass zone in Milan, Italy. *Atmospheric Environment*, *45*, 3522–3527. https://doi.org/10.1016/j.atmosenv.2011.04.008.

Joburg Innovation and Knowledge Exchange. (2012). *Innovating knowledge in the City*.

Johansson, C., & Johansson, P.-Å. (2003). Particulate matter in the underground of Stockholm. *Atmospheric Environment*, *37*, 3–9. https://doi.org/10.1016/S1352-2310(02)00833-6.

Karanasiou, A., Viana, M., Querol, X., et al. (2014). Assessment of personal exposure to particulate air pollution during commuting in European cities—Recommendations and policy implications. *Science of the Total Environment*, *490*, 785–797.

Kelly, J. A., & Clinch, J. P. (2009). Temporal variance of revealed preference on-street parking price elasticity. *Transport Policy*, *16*, 193–199. https://doi.org/10.1016/j.tranpol.2009.06.001.

Kelly, F. J., & Fussell, J. C. (2015). Air pollution and public health: Emerging hazards and improved understanding of risk. *Environmental Geochemistry and Health, 37*, 631–649. https://doi.org/10.1007/s10653-015-9720-1.

Kim, K.-H., Ho, D. X., Jeon, J.-S., et al. (2012). A noticeable shift in particulate matter levels after platform screen door installation in a Korean subway station. *Atmospheric Environment, 49*, 219–223.

Klause, K. (2018). *Barcelona superblocks: How power and politics shape transformational adaptation.* http://www.bcnuej.org/2018/04/06/barcelona-superblocks-how-socio-political-power-struggles-shape-transformational-adaption/.

Kodransky, M., & Hermann, G. (2011). *Europe's parking U-turn: From accommodation to regulation.* https://www.itdp.org/wp-content/uploads/2014/07/Europes_Parking_U-Turn_ITDP.pdf.

Kyu, H. H., Bachman, V. F., Alexander, L. T., et al. (2016). Physical activity and risk of breast cancer, colon cancer, diabetes, ischemic heart disease, and ischemic stroke events: Systematic review and dose-response meta-analysis for the Global Burden of Disease Study 2013. *British Medical Journal, 354*, i3857. https://doi.org/10.1136/bmj.i3857.

Land Transport Authority. (2011). *Passenger transport mode share in world cties.* https://studylib.net/doc/7972629/passenger-transport-mode-shares-in-world-cities.

Land Transport Authority, Urban Redevelopment Authority. (2018). *Walking and cycling design guide.*

Lawlor, E. (2014). *The pedestrian pound. The business case for better streets and places.*

Lee, A. C. K., Jordan, H. C., & Horsley, J. (2015). Value of urban green spaces in promoting healthy living and wellbeing: Prospects for planning. *Risk Management and Healthcare Policy, 8*, 131–137. https://doi.org/10.2147/RMHP.S61654.

Leinberger, C. B. (2009). *The option of urbanism: Investing in a new American dream* (2nd ed.). Island Press.166.

Lijie, S., Yonggang, S., Weiping, Z., et al. (2013). Accessibility of urban green spaces in Hangzhou City. *Progress in Geography, 32*, 950–957.

Litman, T. (2010). *Rebound effects: Implications for transport planning.* http://www.vtpi.org/tdm/tdm64.htm.

Liu, J., Kockelman, K., Boesch, P. M., et al. (2017). Tracking a system of shared autonomous vehicles across the Austin, Texas network using agent-based simulation. *Transportation.*

Lu, N., Cheng, N., Zhang, N., et al. (2014). Connected vehicles: Solutions and challenges. *IEEE Internet of Things Journal, 1*, 289–299. https://doi.org/10.1109/JIOT.2014.2327587.

Lu, F., Xu, D., Cheng, Y., et al. (2015). Systematic review and meta-analysis of the adverse health effects of ambient PM2.5 and PM10 pollution in the Chinese population. *Environmental Research, 136*, 196–204. https://doi.org/10.1016/j.envres.2014.06.029.

MacDonald, J. M., Stokes, R. J., Cohen, D. A., et al. (2010). The effect of light rail transit on body mass index and physical activity. *American Journal of Preventive Medicine, 39*, 105–112. https://doi.org/10.1016/j.amepre.2010.03.016.

MacNaughton, P., Melly, S., Vallarino, J., et al. (2014). Impact of bicycle route type on exposure to traffic-related air pollution. *Science of the Total Environment, 490*, 37–43. https://doi.org/10.1016/j.scitotenv.2014.04.111.

Masiol, M., Agostinelli, C., Formenton, G., et al. (2014). Thirteen years of air pollution hourly monitoring in a large city: Potential sources, trends, cycles and effects of car-free days. *Science of the Total Environment, 494–495*, 84–96. https://doi.org/10.1016/j.scitotenv.2014.06.122.

McCormack, G. R., & Shiell, A. (2011). In search of causality: A systematic review of the relationship between the built environment and physical activity among adults. *International Journal of Behavioral Nutrition Physical Activity, 8*, 125.

Metro Vancouver. (2011). *Integrated air quality and greenhouse gas management plan.* http://www.metrovancouver.org/services/air-quality/AirQualityPublications/IntegratedAirQualityGreenhouseGasManagementPlan-October2011.pdf.

Metro Vancouver. (2014). *Integrated air quality and greenhouse gas management plan progress report.* http://www.metrovancouver.org/services/air-quality/AirQualityPublications/2014IAQGGMPProgressReport.pdf.

Millard-Ball, A., Weinberger, R. R., & Hampshire, R. C. (2014). Is the curb 80% full or 20% empty? Assessing the impacts of San Francisco's parking pricing experiment. *Transportation Research Part A: Policy and Practice, 63*, 76–92. https://doi.org/10.1016/j.tra.2014.02.016.

Miller, J., Du, L., & Façanha, C. (2017). *Vehicle and fuel standards can dramatically reduce air pollution in Brazil, São Paulo's air quality data demonstrate.* https://theicct.org/blogs/staff/vehicle-and-fuel-stds-can-dramatically-reduce-air-pollution-brazil.

Moreno, T., Reche, C., Minguillón, M. C., Querol, X., Martins, V., Amato, F., et al. (2017). *Improving air quality in the subway environment: Technical guide.*

Morrison, G. M., & Lin Lawell, C. Y. C. (2016). Does employment growth increase travel time to work? An empirical analysis using military troop movements. *Regional Science and Urban Economics, 60*, 180–197. https://doi.org/10.1016/j.regsciurbeco.2016.07.007.

Mueller, N., Rojas-Rueda, D., Cole-Hunter, T., et al. (2015). Health impact assessment of active transportation: A systematic review. *Preventive Medicine, 76*, 103–114. https://doi.org/10.1016/j.ypmed.2015.04.010.

Mueller, N., Rojas-Rueda, D., Khreis, H., et al. (2019). Changing the urban design of cities for health: The superblock model. *Environment International, 105132*. https://doi.org/10.1016/j.envint.2019.105132.

Mueller, N., Rojas-Rueda, D., Salmon, M., et al. (2018). Health impact assessment of cycling network expansions in European cities. *Preventive Medicine, 109*, 62–70. https://doi.org/10.1016/j.ypmed.2017.12.011.

National Parks Board, n.d. Skyrise greenery incentive scheme 2.0. https://www.nparks.gov.sg/skyrisegreenery/incentive-scheme.

Newman, P., Beatley, T., & Boyer, H. (2017). *Resilient cities: Overcoming fossil fuel dependence* (2nd ed.). Island Press.

NHTSA. (2013). *U.S. department of transportation releases policy on automated vehicle development.* https://www.transportation.gov/briefing-room/us-department-transportation-releases-policy-automated-vehicle-development.

Nielsen, T. A. S., Skov-Petersen, H., & Carstensen, T. A. (2013). Urban planning practices for bikeable cities—The case of Copenhagen. *Urban Research & Practice, 6*, 110–115. https://doi.org/10.1080/17535069.2013.765108.

Nieuwenhuijsen, M. J., & Khreis, H. (2016). Car free cities: Pathway to healthy urban living. *Environment International, 94*, 251–262. https://doi.org/10.1016/j.envint.2016.05.032.

Nowak, D. J., Crane, D. E., & Stevens, J. C. (2006). Air pollution removal by urban trees and shrubs in the United States. *Urban Forestry & Urban Greening, 4*, 115–123. https://doi.org/10.1016/j.ufug.2006.01.007.

Nuwer, R. (2014). *Hamburg plans to become car free by 2034.* https://www.smithsonianmag.com/smart-news/hamburg-plans-become-car-free-2034-180949780/.

O'Sullivan, F. (2018). *Spain wants to ban cars in dozens of cities, and the public's on board.* https://www.citylab.com/transportation/2018/11/spain-nationwide-car-free-city-center-car-ban/576976/.

Papa, E., & Ferreira, A. (2018). Sustainable accessibility and the implementation of automated vehicles: Identifying critical decisions. *Urban Science, 2*, 5.

Paradatheth, S. (2015). *Car-sharing and public parking in Boston.* https://www.hks.harvard.edu/sites/default/files/centers/mrcbg/files/Paradatheth_final.pdf.

Peters, A. (2019). *What happened when Oslo decided to make its downtown basically car-free?.* https://www.fastcompany.com/90294948/what-happened-when-oslo-decided-to-make-its-downtown-basically-car-free.

Peters, J. R., Davidson, A. L., & Santiago, N.T. (2013). *The economic impacts of transferring curb space from car parking to bike share docks.* https://nacto.org/wp-content/uploads/2015/12/2013_Peters-Davidson-and-Santiago_Economic-Impact-of-Transferring-Curba-Space-form-Car-Parking-to-Bike-Share-Docks.pdf.

Pierce, G., & Shoup, D. (2013). Getting the prices right. *Journal of the American Planning Association, 79*, 67–81. https://doi.org/10.1080/01944363.2013.787307.

Querol, X., Moreno, T., Karanasiou, A., et al. (2012). Variability of levels and composition of PM_{10} and $PM_{2.5}$ in the Barcelona metro system. *Atmospheric Chemistry and Physics, 12*, 5055–5076. https://doi.org/10.5194/acp-12-5055-2012.

Raaschou-Nielsen, O., Andersen, Z. J., Beelen, R., et al. (2013). Air pollution and lung cancer incidence in 17 European cohorts: Prospective analyses from the European study of cohorts for air pollution effects (ESCAPE). *The Lancet Oncology, 14*, 813–822. https://doi.org/10.1016/S1470-2045(13)70279-1.

Rabl, A., & de Nazelle, A. (2012). Benefits of shift from car to active transport. *Transport Policy, 19*, 121–131. https://doi.org/10.1016/j.tranpol.2011.09.008.

Rodríguez, D. A., Merlin, L., Prato, C. G., et al. (2015). Influence of the built environment on pedestrian route choices of adolescent girls. *Environment and Behaviour, 47*, 359–394. https://doi.org/10.1177/0013916513520004.

Rojas López, M. C., & Wong, Y. D. (2017). Attitudes towards active mobility in Singapore: A qualitative study. *Case Studies on Transport Policy, 5*, 662–670. https://doi.org/10.1016/j.cstp.2017.07.002.

Rojas-Rueda, D., de Nazelle, A., Andersen, Z. J., et al. (2016). Health impacts of active transportation in Europe. *PLoS ONE, 11*, e0149990. https://doi.org/10.1371/journal.pone.0149990.

Rojas-Rueda, D., de Nazelle, A., Tainio, M., et al. (2011). The health risks and benefits of cycling in urban environments compared with car use: Health impact assessment study. *BMJ, 343*, d4521. https://doi.org/10.1136/bmj.d4521.

Saelens, B. E., Vernez Moudon, A., Kang, B., et al. (2014). Relation between higher physical activity and public transit use. *American Journal of Public Health, 104*, 854–859. https://doi.org/10.2105/AJPH.2013.301696.

Sallis, J. F., Cerin, E., Conway, T. L., et al. (2016). Physical activity in relation to urban environments in 14 cities worldwide: A cross-sectional study. *The Lancet, 387*, 2207–2217. https://doi.org/10.1016/S0140-6736(15)01284-2.

Salma, I., Weidinger, T., & Maenhaut, W. (2007). Time-resolved mass concentration, composition and sources of aerosol particles in a metropolitan underground railway station. *Atmospheric Environment, 41*, 8391–8405. https://doi.org/10.1016/j.atmosenv.2007.06.017.

Schrank, D., Eisele, B., & Lomax, T. (2012). *TTI's 2012 urban mobility report*. https://static.tti.tamu.edu/tti.tamu.edu/documents/ums/archive/mobility-report-2012.pdf.

Selmi, W., Weber, C., Rivière, E., et al. (2016). Air pollution removal by trees in public green spaces in Strasbourg city, France. *Urban Forestry & Urban Greening, 17*, 192–201. https://doi.org/10.1016/j.ufug.2016.04.010.

Shoup, D. (2005). *Parking cashing out*. http://shoup.bol.ucla.edu/Parking%20Cash%20Out%20Report.pdf.

Shoup, D. (2007). Cruising for parking. *ACCESS*, 16–22.

Shu, S., Batteate, C., Cole, B., et al. (2016). Air quality impacts of a CicLAvia event in Downtown Los Angeles, CA. *Environmental Pollution, 208*, 170–176. https://doi.org/10.1016/j.envpol.2015.09.010.

Sihlcity. (2018). *Media factsheet*. https://sihlcity.ch/storage/information/Medien/Informationen-zu-Sihlcity/Medien_Factsheet_EN.pdf.

Silva, C. B., Saldiva, P. H., Amato-Lourenço, L. F., et al. (2012). Evaluation of the air quality benefits of the subway system in São Paulo, Brazil. *Journal of Environmental Management, 101*, 191–196. https://doi.org/10.1016/j.jenvman.2012.02.009.

Smart Growth America. (2012a). *Benefits of complete streets: Complete streets stimulate the local economy*. https://www.smartgrowthamerica.org/app/legacy/documents/cs/factsheets/cs-economic.pdf.

Smart Growth America. (2012b). *It's a safe decision: Complete streets in california*. https://www.smartgrowthamerica.org/app/legacy/documents/cs/resources/cs-in-california.pdf.

Smart Growth America. (2018). *Best Complete Street Initiative of 2017*. https://smartgrowthamerica.org/app/uploads/2018/03/Best-Complete-Streets-Initiatives-of-2017.pdf.
Smart Growth America. (2019). *What are complete streets?*. https://smartgrowthamerica.org/program/national-complete-streets-coalition/publications/what-are-complete-streets/.
Smith, M., Hosking, J., Woodward, A., et al. (2017). Systematic literature review of built environment effects on physical activity and active transport—An update and new findings on health equity. *International Journal of Behavioral Nutrition Physical Activity*, *14*, 158. https://doi.org/10.1186/s12966-017-0613-9.
Soret, A., Guevara, M., & Baldasano, J. M. (2014). The potential impacts of electric vehicles on air quality in the urban areas of Barcelona and Madrid (Spain). *Atmospheric Environment*, *99*, 51–63. https://doi.org/10.1016/j.atmosenv.2014.09.048.
Soret, A., Jimenez–Guerrero P, Andres D, et al. (2013). Estimation of future emission scenarios for analysing the impact of traffic mobility on a large Mediterranean conurbation in the Barcelona metropolitan area (Spain). *Atmospheric Pollution Research*, *4*, 22–32. https://doi.org/10.5094/APR.2013.003.
Speck, J. (2012). *Walkable city: How downtown can save America, one step at a time*. New York: Farrar, Straus and Giroux.163–209.
Statistics Denmak. (2018). *Traffic Accidents*. https://www.dst.dk/en/Statistik/emner/levevilkaar/trafikulykker.
Stead, D., & Vaddadi, B. (2019). Automated vehicles and how they may affect urban form: A review of recent scenario studies. *Cities*, *92*, 125–133. https://doi.org/10.1016/j.cities.2019.03.020.
Sutton, M. (2018). *Study of Singapore's Mobike users shows shift away from private cars and doubled modal share*. https://cyclingindustry.news/study-of-singapores-mobike-users-shows-shift-away-from-private-cars-and-doubled-modal-share/.
Tainio, M., de Nazelle, A. J., Götschi, T., et al. (2016). Can air pollution negate the health benefits of cycling and walking? *Preventive Medicine*, *87*, 233–236. https://doi.org/10.1016/j.ypmed.2016.02.002.
The City Talking. (2015). *Air quality leeds*. http://airquality.thecitytalking.com/.
Thorpe, A., & Harrison, R. M. (2008). Sources and properties of non-exhaust particulate matter from road traffic: A review. *Science of the Total Environment*, *400*, 270–282. https://doi.org/10.1016/j.scitotenv.2008.06.007.
Timmers, V. R. J. H., & Achten, P. A. J. (2016). Non-exhaust PM emissions from electric vehicles. *Atmospheric Environment*, *134*, 10–17. https://doi.org/10.1016/j.atmosenv.2016.03.017.
Transport and Environment. (2018). *Six EU governments face legal action over air pollution*. https://www.transportenvironment.org/news/six-eu-governments-finally-face-legal-action-over-air-pollution.
Ulrich, R. (1984). View through a window may influence recovery from surgery. *Science*, *224*, 420–421. https://doi.org/10.1126/science.6143402.
United Nations. (2018a). *68% of the world population projected to live in urban areas by 2050, says UN*. https://www.un.org/development/desa/en/news/population/2018-revision-of-world-urbanization-prospects.html.
United Nations. (2018b). *Revision of world urbanization prospects*. https://www.un.org/development/desa/publications/2018-revision-of-world-urbanization-prospects.html.
USA Today. (2019). *New York City drivers will soon have to pay for the privilege of sitting in traffic*. https://www.usatoday.com/story/news/nation/2019/04/03/congestion-pricing-new-york-city-drivers-soon-face-new-toll/3350401002/.
Velo City. (2016). http://www.cycling-embassy.dk/wp-content/uploads/2017/07/Velo-city_handout.pdf.
Vieira, J., Matos, P., Mexia, T., et al. (2018). Green spaces are not all the same for the provision of air purification and climate regulation services: The case of urban parks. *Environmental Research*, *160*, 306–313. https://doi.org/10.1016/j.envres.2017.10.006.

Vos, P. E. J., Maiheu, B., Vankerkom, J., et al. (2013). Improving local air quality in cities: To tree or not to tree? *Environmental Pollution, 183*, 113–122. https://doi.org/10.1016/j.envpol.2012.10.021.

Wadud, Z., MacKenzie, D., & Leiby, P. (2016). Help or hindrance? The travel, energy and carbon impacts of highly automated vehicles. *Transportation Research Part A: Policy and Practice, 86*, 1–18. https://doi.org/10.1016/j.tra.2015.12.001.

Wahlgren, L., & Schantz, P. (2014). Exploring bikeability in a suburban metropolitan area using the active commuting route environment scale (ACRES). *International Journal of Environmental Research and Public Health, 11*, 8276–8300. https://doi.org/10.3390/ijerph110808276.

Whitelegg, J., Haq, G., Cambridge, H., et al. (2010). *Towards a zero carbon vision for UK transport*. https://www.jstor.org/stable/resrep00495.

WHO. (2006). *Promoting physical activity and active living in urban environments*. http://www.euro.who.int/__data/assets/pdf_file/0009/98424/E89498.pdf.

WHO. (2018). *Health and sustainable development. Air pollution*. http://www.who.int/sustainable-development/transport/health-risks/air-pollution/en/ (Accessed October 12, 2018).

Willsher, K. (2015). *Paris's first attempt at car-free day brings big dro in air and noise pollution*. https://www.theguardian.com/world/2015/oct/03/pariss-first-attempt-at-car-free-day-brings-big-drop-in-air-and-noise-pollution.

Wolch, J. R., Byrne, J., & Newell, J. P. (2014). Urban green space, public health, and environmental justice: The challenge of making cities 'just green enough'. *Landscape and Urban Planning, 125*, 234–244. https://doi.org/10.1016/j.landurbplan.2014.01.017.

Wolff, H. (2014). Keep your clunker in the suburb: Low-emission zones and adoption of green vehicles. *The Economic Journal, 124*, F481–F512. https://doi.org/10.1111/ecoj.12091.

Woodcock, J., Edwards, P., Tonne, C., et al. (2009). Public health benefits of strategies to reduce greenhouse-gas emissions: Urban land transport. *The Lancet, 374*, 1930–1943. https://doi.org/10.1016/S0140-6736(09)61714-1.

World Economic Forum. (2018). *The Oslo model: How to prepare your city for the electric-vehicle surge*. https://www.weforum.org/agenda/2018/08/the-oslo-model-how-to-prepare-your-city-for-electric-vehicles/.

World Economic Forum, The Boston Consulting Group. (2018). *Reshaping urban mobility with autonomous vehicles: Lessons from the city of Boston*. http://www3.weforum.org/docs/WEF_Reshaping_Urban_Mobility_with_Autonomous_Vehicles_2018.pdf.

World Health Organization. (2018). *Physical inactivity: A global public health problem*. http://www.who.int/dietphysicalactivity/factsheet_inactivity/en/.

World Resources Institute. (2018). *Reducing China's urban air pollution with transport solutions at Beijing workshop*. https://wrirosscities.org/news/reducing-china%E2%80%99s-urban-air-pollution-transport-solutions-beijing-workshop.

Xiao, C., Goryakin, Y., & Cecchini, M. (2019). Physical activity levels and new public transit: A systematic review and meta-analysis. *American Journal of Preventive Medicine, 56*, 464–473. https://doi.org/10.1016/j.amepre.2018.10.022.

Zhu, Y., Wang, R., Shu, S., et al. (2016). *Effects of complete streets on travel behavior and exposure to vehicular emissions*.

CHAPTER 17

Air pollution mitigation through vegetation barriers and green space

Richard Baldauf
U.S. Environmental Protection Agency, Research Triangle Park, NC, United States

Abbreviations

BC	black carbon
CO	carbon monoxide
NO_2	nitrogen dioxide
O_3	ozone
PM	particulate matter
SO_2	sulfur dioxide
UFP	ultrafine particulate matter
US	United States

Introduction

Public health concerns related to urban residents' exposures to poor air quality near busy roads and other large transportation facilities have become an important environmental issue. A growing body of research has linked adverse human health effects with people living, working, and going to school near these sources (e.g., HEI, 2010). Many of the adverse human health effects have been attributed to the increased exposures experienced by urban populations to particulate matter, gaseous criteria pollutants, and air toxic contaminants emitted by traffic activity on the road or within a facility such as a rail yard, airport, or marine port.

The significant impact of transportation emissions on urban populations across the world has motivated the need to identify and implement methods to reduce people's exposures to high levels of air pollution. While motor vehicle emission control standards and regulatory programs that directly reduce pollutants emitted to the air from transportation sources remain a vital component of air quality management strategies, these programs often take a long time and significant capital investment to fully implement. Thus, additional mitigation options are necessary to complement these programs.

Traffic-Related Air Pollution
https://doi.org/10.1016/B978-0-12-818122-5.00017-X 2020 Published by Elsevier Inc.

The preservation and planting of vegetation along the road, as well as the construction of roadside structures such as sound walls, have recently been investigated as a local mitigation strategy. These methods provide some of the only, relatively short-term mitigation options available to urban planners, developers, and community residents to reduce air pollution exposures at existing locations or planned developments experiencing high pollution levels near high-traffic roads. These air pollution mitigation methods, if successful, can not only provide additional benefits to existing pollution control programs and regulations but also provide a mechanism to reduce impacts from air pollution sources that have limited or no other pollution control standards such as brake and tire wear and re-entrained road dust.

This chapter provides a summary of how urban vegetation and green space can affect air quality and public health. While some background is provided on general health and air quality benefits of urban green infrastructure, the chapter emphasizes the influence roadside vegetation has on local, near-road air quality since this subject is less documented and most relevant to traffic-related air pollution emissions and exposures.

Benefits of urban green space

Urban green space includes a variety of locations with trees, bushes, grasses, and other vegetation within a city. Green spaces can include such diverse land uses such as woods, meadows, wetlands, and other preserved natural ecosystems, as well as parks, sports fields, and other human-designed and planted elements. In general, urban vegetation and green space have been shown to contribute to human health benefits. Many positive health outcomes have been associated with exposure to urban green space and infrastructure including the promotion and increase in physical activity, lowering of obesity, improvements in mental health, improvements in birth outcomes, lowering of adverse cardiovascular illness, and a decreasing of mortality rates (Coutts & Hahn, 2015; James, Banay, Hart, & Laden, 2015; James, Hart, Banay, & Laden, 2016; Nieuwenhuijsen, Khreis, Triguero-Mas, Gascon, & Dadvand, 2017; Tzoulas et al., 2007; Wolch, Byrne, & Newell, 2014). Research studies have also shown improvements in school children's cognitive development associated with an increase in surrounding green space, particularly on the school grounds (Dadvand et al., 2015). These health benefits have been attributed to a number of the benefits attributed to green space and vegetation assets, including reductions in nearby air pollution levels.

Among the many potential benefits green space and roadside vegetation can provide are improved aesthetics, increased property values, reduced urban heat island effects, increased carbon sequestration, and control of surface water runoff and flooding. In addition, roadside vegetation may reduce noise pollution with proper density, thickness, and height although vegetation alone is typically not used as a primary noise reduction strategy. Roadside vegetation can also create challenges that need to be considered in the design, planting, and maintenance process when being planted or preserved within a community. Such challenges can include obstructing driver sight lines, trees and plants protruding into clear zones along the highway right-of-way, leaves and debris falling onto the road surface, vegetation creating local fire hazards, and serving as pathways for pests and invasive species into surrounding urban neighborhoods. In addition, increase in property values and housing costs may lead to the displacement of existing residents and neighborhood gentrification. Thus, the benefits and potential unintended consequences of urban green space, including roadside vegetation, need to be considered for any application.

Vegetation effects on air quality

Trees and other types of vegetation can contribute to improvements in regional air quality, primarily through the removal of airborne particles onto leaf, needle or branch surfaces, and with the uptake of gaseous air contaminants through leaf stomata on the plant surface (Gallagher et al., 2015; Janhäll, 2015). Air pollution removal by urban trees has been calculated for the continental United States (US) for the air contaminants: particulate matter (PM), ozone (O_3), nitrogen dioxide (NO_2), sulfur dioxide (SO_2), and carbon monoxide (CO) using the US Forest Service's i-Tree model (Nowak, Hirabayashi, Bodine, & Greenfield, 2014). Vegetation was estimated to reduce air pollution concentrations across the US by approximately 1%, equal to over 10 million tons of air contaminants removed. Removal of gaseous pollutants by trees can be permanent, while trees may provide only a temporary retention site for airborne particles. After PM has been removed from the atmosphere onto the plant surface, particles can re-enter the environment through resuspension to the air during high, turbulent wind conditions, precipitation washing off the particles to the ground, or the particles dropping to the ground when the leaf, needle, or twig fall

(Nowak et al., 2000). These removal mechanisms can impact local air, water, and soil quality; thus, careful evaluation of the land uses and human activity that surround the roadside vegetation will be an important consideration when choosing plant species for a roadside application.

At the local level, the impact of trees and bushes on air quality can be complex and variable. Trees can act as barriers between sources and people, creating a barrier that disrupts airflow between the pollutant emissions and people. If preserving or planting vegetation results in development occurring further from the pollution source, then these additional trees and bushes will extend the distance between the emission of the pollutants by traffic and the exposure to these pollutants. Since traffic-related air pollution generally decreases exponentially with distance, the extra space created by roadside vegetation can lead to a reduction in ambient air pollutant concentrations alone (Karner, Eisinger, & Niemeier, 2010). The presence of the vegetation barrier will also alter airflow from the road and provide an opportunity for the plants to filter PM (Deshmukh et al., 2019). The effectiveness of vegetative barriers at reducing ultrafine particles (UFP; PM less than 100 nm in aerodynamic diameter) concentrations has been shown to be variable (Hagler et al., 2012; Janhäll, 2015; Pataki et al., 2011; Tong, Whitlow, MacRae, Landers, & Harada, 2015). This variability can be attributed to multiple confounding factors related to the physical characteristics of a roadside vegetation barrier. The complex and porous structure of trees and bushes can modify the near-road pollutant concentrations via pollutant capture; however, this structure also alters airflow, which can result in either reduced dispersion through the reduction of wind speed and boundary layer heights (Nowak et al., 2000; Vos, Maiheu, Vankerkom, & Janssen, 2013; Wania, Brse, Blond, & Weber, 2012) or in enhanced dispersion due to increased air turbulence and mixing as the pollutant plume is lofted up, over, and around the vegetation (Bowker, Baldauf, Isakov, Khlystov, & Petersen, 2007). Recirculation zones have also been observed directly downwind of vegetation with a flow structure consistent with an intermittent recirculation pattern that can allow for air pollutant buildup (Detto, Katul, Siqueira, Juang, & Stoy, 2008; Frank and Ruck, 2008). Janhäll (2015) summarized that low vegetation like hedges can filter out PM when located close to an emission source such as a road, while high vegetation like trees can reduce mixing and turbulence and result in increased concentration levels. Deshmukh et al. (2019) highlighted the importance of the vegetation characteristics on near-road air quality, with downwind pollutant reductions occurring when the vegetation barrier was dense and tall, while pollutant increases occurred when the vegetation barrier was highly porous. Thus, type, height, and thickness of roadside vegetation all influence the extent of mixing and pollutant deposition experienced along and behind the barrier (Deshmukh et al., 2019; Tiwari et al., 2019).

The built environment can also significantly impact the effects of urban vegetation on local air quality. Airflow, and the resulting air pollution impacts around trees, can be substantially different for a street canyon environment compared with an open highway environment due to the influence of the surrounding buildings on microscale airflow in addition to the presence of the urban trees (Abhijith et al., 2017; Buccolieri et al., 2011; Buccolieri, Gromke, Di Sabatinoa, & Ruck, 2009; Gromke, Buccolieria, Di Sabatinoa, & Ruck, 2008; Gromke, Jamarkattel, & Ruck, 2016; Li et al., 2016; Pugh, MacKenzie, Whyatt, & Hewitt, 2012).

Several studies have shown significant reductions in air pollution concentrations behind roadside vegetation barriers in both street canyon and open road configurations (Al-Dabbous & Kumar, 2014; Brantley, Hagler, Deshmukh, & Baldauf, 2014; Deshmukh et al., 2019; Gromke et al., 2016; Steffens, Wang, & Zhang, 2012; Tong, Baldauf, Isakov, Deshmukh, & Zhang, 2016). Each of these studies compared air quality concentrations in the vicinity of a large roadway with and without the presence of roadside vegetation. Reductions in PM and PM components like black carbon (BC) and UFP ranged from approximately 10% to over 50%. Gaseous pollutants like NO_2 and CO were generally reduced by 10%–30% (Abhijith et al., 2017; Baldauf, 2017). In all of these studies, the roadside vegetation was dense and usually consisted of a mixture of trees and bushes with full coverage from the ground to the top of the canopy (although the heights, thickness, and species varied). For street canyons, in particular, modeling demonstrated that hedgerows can improve air quality since these types of bushes also provide full coverage from the ground to the top with no openings or gaps (Gromke et al., 2016; Li et al., 2016). Alternatively, other studies have shown air pollution near a highway increased compared to measurements in similar areas with no vegetation (Tong et al., 2015; Yli-Pelkonen, Setälä, & Viippola, 2017). For these studies, the vegetation consisted of scattered, ornamental trees with open space underneath the canopy, or the vegetation was highly porous.

Vegetation characteristics impacts on local air quality

The following sections provide a summary of roadside vegetation characteristics that have been shown to mitigate near-road air pollution concentrations. A primary characteristic is that the planted or preserved vegetation

(A)
(B)
(C)

Fig. 17.1 Vegetation characteristics that can result in effective, downwind air pollution mitigation. Example (A) shows trees and small bushes grown together to provide complete coverage from the ground to the top of the canopy similar to the conditions described by Brantley et al. (2014). Example (B) shows primarily bushes with some scattered trees that provide dense vegetation with complete coverage as highlighted by Deshmukh et al. (2019) as an effective barrier condition. Example (C) shows large

(Continued)

must not have gaps or open spaces between the ground and the top of the canopy. Polluted air must not be able to easily pass through or underneath the plantsto achieve pollution reductions due to increased dilution and/or PM filtering. For vegetation stands with complete coverage, a higher and thicker vegetation barrier between the traffic source and the locations where people spend time will generally achieve greater reductions in downwind air pollutant concentrations. While studies evaluating varying heights of vegetation barriers have been limited, several studies have investigated the effect of height on pollutant reductions behind solid sound walls. Computational fluid dynamic (CFD) modeling of solid sound walls of varying heights indicated that higher barriers require additional plume transport and dispersion to go above the structure, resulting in greater downwind pollutant reductions (Hagler et al., 2012). While the porosity of vegetation allows some air movement through the barrier, the height of the structure still forces airflow up and over the vegetation, increasing the plume transport distance and dispersion. The porosity and thickness of the vegetation will also affect the amount of airflow allowed through the plants compared with flow forced up and over. Generally, the lower the porosity and thicker the vegetation barrier, the more airflow forced over the structure. At extremely low porosities, the vegetation will similarly affect pollutant transport and dispersion as a solid sound wall. However, vegetation barrier design should allow some airflow through the vegetation to enhance particulate removal. Previous studies suggest porosities between 0.5% and 0.9% coverage to be most effective (Deshmukh et al., 2019; Tong et al., 2016).

The integrity of the vegetation barrier must be maintained to allow for pollutant reductions downwind. Gaps in vegetation barriers can lead to increased pollutant concentrations downwind, sometimes higher than concentrations would be if no barrier were present (Deshmukh et al., 2019; Hagler et al., 2012). These increases can occur because pollutant emissions from the road concentrate and funnel through the gaps. In addition, the highly porous vegetation can cause winds to slow and stagnate, also leading to higher downwind air pollutant concentrations. Fig. 17.1 provides

Fig. 17.1, Cont'd hedges that can be planted along roadways as a safety and air pollution barrier similar to the analyses by Gromke et al. (2016) and Al-Dabbous and Kumar (2014). *(From Baldauf, R. (2017). Roadside vegetation design characteristics that can improve local, near-road air quality. Transportation Research Part D: Transport and Environment, 52, 354–361; U.S. Environmental Protection Agency (2016). Recommendations for constructing roadside vegetation barriers for improving near-road air quality, EPA/600/R-16/072. Office of Research and Development, Washington, DC.)*

examples of effective vegetation barriers that have full coverage from the ground to the top of the canopy. This type of barrier can be formed by combinations of trees and bushes, often situated in multiple rows. Hedges can also provide a complete, solid structure of vegetation that can either be planted alone or in combination with other vegetation.

On the other hand, Fig. 17.2 provides examples of ineffective vegetation barriers due to gaps that may result in higher downwind pollutant concentrations. These gaps can be due to the types of trees chosen, such as ornamental trees with only trunks at the ground, the spacing of trees planted or already existing along the roadside, or the result of dead or poorly maintained vegetation.

Multiple rows and types of vegetation may be needed to achieve the physical characteristics of a vegetation barrier required to achieve air pollution reductions. For example, a vegetation barrier could consist of rows of bushy plants and shrubs followed by rows of trees to create a barrier with full coverage from the ground to top of the canopy at the initial planting, yet achieve higher canopy heights than feasible by bushy plants alone. In addition, rows of multiple vegetation types may maintain a robust and healthy barrier that allows for sufficient downwind pollutant removal while the vegetation grows over time after first planting. This approach will ensure sufficient density for pollutant removal at the initial planting while allowing for increased pollutant removal as the vegetation matures. This design will also limit concerns of promoting plant monocultures such as soil degradation, pest resistance, and lack of biological diversity.

In addition to passing through gaps or highly porous plants, air pollutants can meander around the edges of a roadside vegetative barrier. Thus, if a vegetative barrier will be constructed for a specific facility (e.g., school, daycare, and elderly care housing) or neighborhood, the ends of the barrier should extend sufficiently beyond the area being protected. Research on solid sound walls suggests that the barrier should extend at least 50 m (approximately 150 ft) laterally beyond the area of concern to maximize reductions in downwind concentrations (Baldauf et al., 2016, 2008). If extending the barrier laterally is not feasible or desirable, extending the barrier perpendicular to the road to cover the area of interest on two or three sides, has been shown to be effective as well (Brantley et al., 2014).

Certain types and species of vegetation will provide more air quality benefits compared to others. Since the availability and robustness of plant species will vary by region, this section focuses on key physical characteristics needed for the vegetation to provide an effective barrier.

(A)

(B)

(C)

Fig. 17.2 Vegetation characteristics that can result in no air quality improvements or even higher downwind air pollution concentrations compared to locations with no vegetation. Example (A) shows scattered, ornamental trees with open space under the canopy similar to the conditions of Yli-Pelkonen et al. (2017). Examples (B) and (C) show highly porous vegetation with large gaps due to dying or poorly maintained vegetation (B) or sparsely planted vegetation (C) similar to conditions described by Hagler et al. (2012). *(From Baldauf, R. (2017). Roadside vegetation design characteristics that can improve local, near-road air quality. Transportation Research Part D: Transport and Environment, 52, 354–361; U.S. Environmental Protection Agency (2016). Recommendations for constructing roadside vegetation barriers for improving near-road air quality, EPA/600/R-16/072. Office of Research and Development, Washington, DC.)*

When considering the design and construction of a vegetation barrier, optimal physical characteristics should be favored to the extent feasible. Tools for choosing appropriate vegetation species for a roadside barrier include the US Forest Service's i-Tree model (https://www.itreetools.org) which can provide a list of potential species that best meet the factors listed below, with users determining whether particular vegetation types can survive and prosper in the specific area of interest.

- *Seasonal effects*: The vegetation chosen for a roadside barrier should not experience significant changes in characteristics and integrity during different climatic seasons. Deciduous trees that lose leaves during cold temperatures should not be considered for a vegetative barrier to mitigate air pollution impacts. Instead, trees that are not subject to significant seasonal changes, such as coniferous plants, should be considered. Other shrubs and bushes that are not subject to seasonal changes can also be effective components of a roadside barrier.
- *Leaf surfaces*: The type and structure of the leaf surface can also enhance particulate and gaseous air pollutant removal through diffusion and interception. Trees and bushes with waxy and/or hairy surfaces have been shown to preferentially remove and retain particulates compared to smooth leaf surfaces. In addition, vegetation with leaf and branch structures that provide increased surface area for particle diffusion can increase removal (Tong et al., 2016).
- *Vegetation air emissions*: When selecting vegetation for a roadside barrier, especially at locations where sensitive populations may be spending significant amounts of time, care must be taken to choose species that do not emit compounds which can increase allergic responses or air pollution. Compounds that can be emitted by vegetation include high-allergy pollens and volatile organic compounds (VOCs), which can enhance the formation of ozone. Both of these constituents can exacerbate respiratory effects and should be avoided for roadside barriers.
- *Resistant to air pollution and other environmental stressors*: Vegetation used in a roadside barrier will likely experience higher levels and more intense environmental stressors. Thus, the chosen plants must be resistant to higher concentrations of air pollution and other traffic stressors. If the vegetation cannot resist these stressors and maintain its integrity, gaps will form in the barrier, potentially leading to increased pollutant concentrations downwind as discussed previously. Air pollutants emitted by traffic can include the typical tailpipe emissions like CO, NOx, and PM; materials from brake and tire wear; re-entrained road dust; and salt and sand used for road surface treatment during winter weather conditions.

Additional considerations

Besides air quality, other potentially beneficial and adverse aspects of roadside vegetation need to be considered in the development and use of a vegetative barrier. These considerations include general physical and species-specific factors. While location-specific issues will need to be addressed on an individual basis, some general considerations include

- *Vegetation maintenance*: The roadside vegetation will need to be maintained to provide a protective barrier from air pollution exposures yet not lead to safety concerns as described previously, such as reduced visibility or falling debris. Maintenance requirements will depend on vegetation type and species, so a plan should be in place when either preserving existing vegetation or selecting and constructing a new vegetative barrier to ensure long-term effectiveness. Maintenance requirements include watering and fertilization needs, trimming and other pruning requirements, and overall plant care. These maintenance practices should also include vegetation replacement due to die-off, disease, or damage from accidents.
- *Water runoff control*: An additional benefit of a roadside vegetation barrier can be the control and containment of surface water runoff from the impervious road and surrounding built infrastructure. Roadside vegetative barriers constructed to provide water runoff control can prevent localized flooding as well as improve water quality in the area. For certain regions of the country, drought-resistant vegetation that can also resist high-water events may be required.
- *Native species*: Whenever feasible, native plant species should be considered for the roadside barrier. Native species will most likely be more robust and have evolved to local climatic conditions, as well as require less maintenance.
- *Noninvasive species*: Vegetation barriers should not be constructed using invasive species that may not be contained within the project area and may create problems at other locations or the roadside.
- *Nonpoisonous species*: For roadside vegetation barriers, nonpoisonous species should be used, especially when near sensitive populations or located in areas that have the potential to cause harm in other ways. If the barrier can be isolated to avoid the potential for any physical contact or harm, this factor may not be a concern; however, nonhazardous alternatives should be strongly considered.
- *Roadway safety*: Planting on or near a highway right-of-way (ROW) requires consideration of potential safety issues. In most cases, the applicable highway department will require approvals for planting near roads

due to these issues. Concerns may include creating undesirable wildlife habitat near roadways (e.g., deer and other animals that can exacerbate auto accidents), preserving safe lines of sight and views for drivers on the road, maintaining compatibility of the chosen vegetation species with existing species and not obstructing outdoor advertising.

Vegetation collocated with sound walls

Research studies evaluating vegetation collocated with a solid sound wall suggest that this combination provides even further downwind pollutant reductions than either vegetation or a solid sound wall alone (Baldauf et al., 2008; Lee et al., 2018). For vegetation planted with a solid sound wall, the overall considerations should be the same as for vegetation alone. However, for the vegetation to have an additive benefit for pollutant reductions, the vegetation height should exceed the top of the sound wall by a sufficient length to allow airflow through and over the plants to enhance pollutant removal and air dispersion.

Solid barriers can vary in height; research on near-road air pollution reductions from these structures has been conducted for heights between 4.5 and 6 m (approximately 14–18 ft.; see review by Baldauf et al., 2016). A vegetation barrier should extend at least 1 m (3 ft) above the barrier although the higher and thicker the plants, the greater the downwind reduction. For shorter solid barriers, vegetation should extend above the barrier, generally to a height of at least 4 m (12 ft) to maximize the potential for downwind pollutant reductions. Fig. 17.3 provides an example of combining vegetation with a solid sound wall that could lead to increased reductions in downwind air pollutant concentrations.

Previous research studies have focused on vegetation planted behind the sound wall (opposite side from the road). Although not studied, plants in front of solid barriers would also likely provide an added air pollution reduction if the vegetation protruded sufficiently away from the solid wall to allow air to flow through. Some modeling studies suggest that "green walls" such as ivy or other climbing vegetation on sound walls may improve local air quality; however, no air quality measurement studies have been conducted to confirm or negate these modeling results (Pugh et al., 2012).

When combining vegetation with solid sound walls, research suggests that gaps or spaces in the vegetation still have an additive air quality benefit and will not lead to increased downwind pollutant concentrations as can occur with vegetation alone (Ranasinghe et al., 2019). Since solid sound

Fig. 17.3 Examples of an effective combination of vegetation with solid sound walls as studied in Baldauf et al. (2008).

walls alone can reduce downwind pollutant concentrations, gaps in accompanying vegetation can still provide PM filtering benefits, while the solid wall causes increased dilution of the air pollutants in the traffic plume.

Applications on urban commercial and residential streets

The designs and recommendations for roadside vegetation planting or preservation can be used to improve air quality and enhance community livability along urban streets. Fig. 17.4 depicts a conceptual drawing of the use of urban green infrastructure to provide multiple benefits to a residential and commercial area in a city. The vegetation along the road is thick, tall, and provides full coverage from the ground to the top in the form of hedges to achieve an air quality benefit. Trees and other vegetation can be interspersed to provide shade and aesthetic value as well as increase the vegetation height above the hedges for further air pollution reductions. If designed properly, the vegetation will also reduce stormwater runoff and flooding compared to a location with only impervious paved streets and sidewalks. In addition, the vegetation separates pedestrians from car traffic, which is more aesthetically pleasing, may promote physical activity, can provide some noise reduction and has the potential to increase safety from cars unintentionally or intentionally going off the road.

If the vegetation cannot provide full coverage from the ground to the top of the hedge/trees, a solid wall or partition could be constructed at the

Fig. 17.4 Conceptual drawing of the implementation of urban green infrastructure along a city street designed to improve local air quality while achieving other environmental and community benefits.

edge or in between the vegetation. Adding a wall can provide the benefits of a solid barrier and vegetation combination if some airflow occurs through the bushes and trees above and/or on either side of the wall. This wall can be integrated into the surrounding building designs and construction, incorporated into the vegetation aesthetics, or even be constructed of clear materials if views of the road are desired. Overall, implementing this type of conceptual design on an urban street can provide physical and mental health benefits, as well as improve the general usability of the street.

Summary and conclusions

Research shows that roadside vegetation affects nearby air quality. If properly designed, vegetation barriers can be preserved or constructed to reduce air pollution concentrations near roads and other transportation sources, either alone or in combination with solid sound walls which will likely amplify the benefits. To achieve air quality benefits, important physical and species-specific characteristics are needed relative to height, thickness, and the porosity of the vegetation that can be maintained under harsh roadside conditions and changing weather and seasons. These characteristics can be incorporated into comprehensive urban and transportation plans to provide multiple health benefits to a city's population.

While strong evidence exists on the ability of roadside vegetation to provide local and regional air quality benefits, research remains on developing methods to quantify these benefits, particularly at the local scale. The ability to model air pollution impacts from roadside vegetation, both potential decreases and increases in downwind concentrations, may allow for this technique to be used for regulatory purposes in addition to community enhancements. For modeling vegetation effects on local air quality, research is also needed on characterizing the key factors and characteristics of vegetation that contribute to air pollution reductions for both increased pollutant transport and dispersion as well as deposition of particles onto the vegetation surfaces.

Disclaimer

This chapter has been subjected to the US Environmental Protection Agency's review process and has been approved for publication. These are the views of the author and do not necessarily reflect the official policy of the Agency. Mention of trade names or commercial products does not constitute endorsement or recommendation for use.

References

Abhijith, K.V., Kumar, P., Gallagher, J., McNabola, A., Baldauf, R., Pilla, F., et al. (2017). Air pollution abatement performances of green infrastructure in open road and built-up street canyon environments—A review. *Atmospheric Environment, 162*, 71–86.

Al-Dabbous, A. N., & Kumar, P. (2014). The influence of roadside vegetation barriers on airborne nanoparticles and pedestrians exposure under varying wind conditions. *Atmospheric Environment, 90*, 113–124.

Baldauf, R. (2017). Roadside vegetation design characteristics that can improve local, near-road air quality. *Transportation Research Part D: Transport and Environment, 52*, 354–361.

Baldauf, R. W., Isakov, V., Deshmukh, P. J., Venkatram, A., Yang, B., & Zhang, K. M. (2016). Influence of solid noise barriers on near-road and on-road air quality. *Atmospheric Environment, 129*, 265–276.

Baldauf, R. W., Thoma, E., Khlystov, A., Isakov, V., Bowker, G., & Long, T. (2008). Impacts of noise barriers on near-road air quality. *Atmospheric Environment, 42*, 7502–7507.

Bowker, G. E., Baldauf, R. W., Isakov, V., Khlystov, A., & Petersen, W. (2007). The effects of roadside structures on the transport and dispersion of ultrafine particles from highways. *Atmospheric Environment, 41*, 8128–8139.

Brantley, H. L., Hagler, G. S. W., Deshmukh, P. J., & Baldauf, R. W. (2014). Field assessment of the effects of roadside vegetation on near-road black carbon and particulate matter. *Science of the Total Environment, 468–469*, 120–129.

Buccolieri, R., Gromke, C., Di Sabatinoa, S., & Ruck, B. (2009). Aerodynamic effects of trees on pollutant concentration in street canyons. *Science of the Total Environment, 407*, 5247–5256.

Buccolieri, R., Salimb, S. M., Leoa, L. S., Di Sabatinoa, S., Chanb, A., Ielpoc, P., et al. (2011). Analysis of local scale tree–atmosphere interaction on pollutant concentration in idealized street canyons and application to a real urban junction. *Atmospheric Environment, 45*, 1702–1713.

Coutts, C., & Hahn, M. (2015). Green infrastructure, ecosystem services, and human health. *International Journal of Environmental Research and Public Health, 12*(8), 9768–9798.

Dadvand, P., Nieuwenhuijsen, M. J., Esnaola, M., Forns, J., Basagaña, X., Alvarez-Pedrerol, M., et al. (2015). Green spaces and cognitive development in primary schoolchildren. *Proceedings of the National Academy of Sciences, 112*(26), 7937–7942.

Deshmukh, P., Isakov, V., Venkatram, A., Yang, B., Zhang, K. M., Logan, R., et al. (2019). The effects of roadside vegetation characteristics on local, near-road air quality. *Air Quality, Atmosphere & Health*, 1–12.

Detto, M., Katul, G. G., Siqueira, M., Juang, J.-Y., & Stoy, P. (2008). The structure of turbulence near a tall forest edge: The backward-facing step flow analogy revisited. *Ecological Applications, 18*, 1420–1435.

Frank, C., & Ruck, B. (2008). Numerical study of the airflow over forest clearings. *Forestry, 81*, 259–277.

Gallagher, J., Baldauf, R., Fuller, C. H., Kumar, P., Gill, L.W., & McNabola, A. (2015). Passive methods for improving air quality in the built environment: A review of porous and solid barriers. *Atmospheric Environment, 120*, 61–70.

Gromke, C., Buccolieria, R., Di Sabatinoa, S., & Ruck, B. (2008). Dispersion study in a street canyon with tree planting by means of wind tunnel and numerical investigations—Evaluation of CFD data with experimental data. *Atmospheric Environment, 42*, 8640–8650.

Gromke, C., Jamarkattel, N., & Ruck, B. (2016). Influence of roadside hedgerows on air quality in urban street canyons. *Atmospheric Environment, 139*, 75–86.

Hagler, G. S. W., Lin, M.-Y., Khlystov, A., Baldauf, R. W., Isakov, V., & Faircloth, J. (2012). Field investigation of roadside vegetative and structural barrier impact on near-road ultrafine particle concentrations under a variety of wind conditions. *Science of the Total Environment, 419*, 7–15.

Health Effects Institute (HEI). (2010). Traffic-related air pollution: A critical review of the literature on emissions, exposure, and health effects. In *HEI Special Report 17*. Boston, MA: Health Effects Institute.

James, P., Banay, R. F., Hart, J. E., & Laden, F. (2015). A review of the health benefits of greenness. *Current Epidemiology Reports, 2*, 131–142.

James, P., Hart, J. E., Banay, R. F., & Laden, F. (2016). Exposure to greenness and mortality in a nationwide prospective cohort study of women. *Environ Health Perspect*, https://doi.org/10.1289/ehp.1510363.

Janhäll, S. (2015). Review on urban vegetation and particle air pollution—Deposition and dispersion. *Atmospheric Environment, 105*, 130–137.

Karner, A. A., Eisinger, D. S., & Niemeier, D. A. (2010). Near-roadway air quality: Synthesizing the findings from real-world data. *Environmental Science & Technology, 44*(14), 5334–5344.

Lee, E. S., Ranasinghe, D. R., Ahangar, F. E., Amini, S., Mara, S., Choi, W., et al. (2018). Field evaluation of vegetation and noise barriers for mitigation of near-freeway air pollution under variable wind conditions. *Atmospheric Environment, 175*, 92–99.

Li, X. B., Lu, Q. C., Lu, S. J., He, H. D., Peng, Z. R., Gao, Y., et al. (2016). The impacts of roadside vegetation barriers on the dispersion of gaseous traffic pollution in urban street canyons. *Urban Forestry & Urban Greening, 17*, 80–91.

Nieuwenhuijsen, M. J., Khreis, H., Triguero-Mas, M., Gascon, M., & Dadvand, P. (2017). Fifty shades of green: Pathway to healthy urban living. *Epidemiology, 28*, 63–71.

Nowak, D. J., Civerolo, K. L., Trivikrama Rao, S., Gopal, S., Luley, C. J., & E. Crane D. (2000). A modeling study of the impact of urban trees on ozone. *Atmospheric Environment, 34*, 1601–1613.

Nowak, D. J., Hirabayashi, S., Bodine, A., & Greenfield, E. (2014). Tree and forest effects on air quality and human health in the United States. *Environmental Pollution, 193*, 119–129.

Pataki, D. E., Carreiro, M. M., Cherrier, J., Grulke, N. E., Jennings, V., Pincetl, S., et al. (2011). Coupling biogeochemical cycles in urban environments: Ecosystem services, green solutions, and misconceptions. *Frontiers in Ecology and the Environment, 9*(1), 27–36.

Pugh, T. A., MacKenzie, A. R., Whyatt, J. D., & Hewitt, C. N. (2012). Effectiveness of green infrastructure for improvement of air quality in urban street canyons. *Environmental Science & Technology, 46*(14), 7692–7699.

Ranasinghe, D., Lee, E. S., Zhu, Y., Frausto-Vicencio, I., Choi, W., Sun, W., et al. (2019). Effectiveness of vegetation and sound wall-vegetation combination barriers on pollution dispersion from freeways under early morning conditions. *Science of the Total Environment, 658*, 1549–1558.

Steffens, J. T., Wang, Y. J., & Zhang, K. M. (2012). Exploration of effects of a vegetation barrier on particle size distributions in a near-road environment. *Atmospheric Environment, 50*, 120–128.

Tiwari, A., Kumar, P., Baldauf, R., Zhang, K. M., Pilla, F., Di Sabatino, S., et al. (2019). Considerations for evaluating green infrastructure impacts in microscale and macroscale air pollution dispersion models. *Science of the Total Environment, 672*, 410–426.

Tong, Z., Baldauf, R. W., Isakov, V., Deshmukh, P., & Zhang, K. M. (2016). Roadside vegetation barrier designs to mitigate near-road air pollution impacts. *Science of the Total Environment, 541*, 920–927.

Tong, Z., Whitlow, T. H., MacRae, P. F., Landers, A. J., & Harada, Y. (2015). Quantifying the effect of vegetation on near-road air quality using brief campaigns. *Environmental Pollution, 201*, 141–149.

Tzoulas, K., Korpela, K., Venn, S., Yli-Pelkonen, V., Kaźmierczak, A., Niemela, J., et al. (2007). Promoting ecosystem and human health in urban areas using green infrastructure: A literature review. *Landscape and Urban Planning, 81*(3), 167–178.

Vos, P. E., Maiheu, B., Vankerkom, J., & Janssen, S. (2013). Improving local air quality in cities: To tree or not to tree? *Environmental Pollution, 183*, 113–122.

Wania, A., Brse, M., Blond, N., & Weber, C. (2012). Analysing the influence of different street vegetation on traffic-induced particle dispersion using microscale simulations. *Journal of Environmental Management, 94*, 91–101.

Wolch, J. R., Byrne, J., & Newell, J. P. (2014). Urban green space, public health, and environmental justice: The challenge of making cities 'just green enough'. *Landscape and Urban Planning, 125*, 234–244.

Yli-Pelkonen, V., Setälä, H., & Viippola, V. (2017). Urban forests near roads do not reduce gaseous air pollutant concentrations but have an impact on particles levels. *Landscape and Urban Planning, 158*, 39–47.

Further reading

Brode, R., Wesson, K., & Thurman, J. (2008). AERMOD sensitivity to the choice of surface characteristics. In *101st annual conference of the air & waste management association, Portland, OR*.

Nowak, D. J. (2005). Strategic tree planting as an EPA encouraged pollutant reduction strategy: How urban trees can obtain credit in state implementation plans. *Sylvan Communities*, 23–27.

Petroff, A., Zhang, L., Pryor, S. C., & Belot, Y. (2009). An extended dry deposition model for aerosols onto broadleaf canopies. *Journal of Aerosol Science, 40*, 218–240.

Stone, B., & Norman, J. M. (2006). Land use planning and surface heat island formation: A parcel-based radiation flux approach. *Atmospheric Environment, 40*, 3561–3573.

U.S. Environmental Protection Agency. (2016). *Recommendations for constructing roadside vegetation barriers for improving near-road air quality, EPA/600/R-16/072*. Washington, DC: Office of Research and Development.

CHAPTER 18

Cost-effectiveness of projects and policies

Mark Burris
Texas A&M University, College Station, TX, United States

Abbreviations

AASHTO	American Association of State Highway Transportation Officials
BCA	Benefit-Cost Analysis
CARB	California Air Resources Board
CRD	congestion reduction demonstration
EMFAC	emission factors
EPA	Environmental Protection Agency
FHWA	Federal Highway Administration
HEAT	Health Economic Assessment Tool
MOVES	MOtor Vehicle Emission Simulator
NOx	nitrogen oxides
SART	Sickness Absence Reduction Tool
UPA	urban partnership agreement
VMT	vehicle miles traveled

Introduction

Many of the policies and ideas discussed in this book have a positive impact on both air quality and health. A key question to ask in the context of policy making is if those benefits outweigh the costs of the policy? This can, and often is, done by placing monetary values on clean air and our health. This chapter explores why it is important to set a monetary value on clean air, techniques used to determine the value of clean air, suggested values for benefit-cost analyses, and concludes with an example of estimating the benefits from a reduction in emissions due to a transportation project.

The need to quantify the value of clean air

Why do we need to quantify the value of clean air, our health, even our life? The World Health Organization estimates 4.2 million deaths per year due to ambient air pollution (https://www.who.int/airpollution/en/). How can we place a value on that and thus how much we should reasonably

Traffic-Related Air Pollution
https://doi.org/10.1016/B978-0-12-818122-5.00018-1

pay to improve air quality? Is it enough to say that all of these items are extremely valuable, even priceless? That may make a nice sound bite, but imagine policies or funding decisions using an unreasonably high value of life. Every benefit-cost analysis would favor the option that reduces pollution since reducing deaths and illness from pollution would almost always have a net positive impact regardless of the project's cost. An infrastructure project would become overly expensive due to the extreme lengths that the designers would go to make the infrastructure both as environmentally friendly and as safe as possible. Even those extreme lengths may not lower the number of deaths to an acceptable (and economically logical) level. The next step would be to reduce the speed limit to 5 or 10 miles per hour to minimize (and maybe eliminate) deaths. This would make sense from a benefit-cost standpoint due to the extremely high benefits calculated from the reduced crashes and emissions. However, these decisions would have significant negative impacts—to the point where it may cripple the economy of an entire country. This is one extreme end of the value of life/health spectrum.

The other end of the spectrum is when emissions, air pollution, or health effects are ignored or given little value. This was the case in the United States as recently as the mid-1900s, where air and water pollution went nearly unchecked and led to dense smog in many cities and rivers so polluted that they could catch on fire (https://clevelandhistorical.org/items/show/63). Some excellent pictures from that era can be found at https://www.archives.gov/research/environment/documerica-topics, like Fig. 18.1. Some key legislation, such as the initial Air Pollution Control Act of 1955 to the Clean Air Act of 1970 (https://www.epa.gov/clean-air-act-overview/clean-air-act-requirements-and-history), as well as the Motor Vehicle Air Pollution Control Act of 1965, and the Clean Water Act (https://www.epa.gov/laws-regulations/summary-clean-water-act), were all developed to combat these problems, and resulted in setting a value on clean air.

Policy decisions based on either of these two extremes (no value or an overly high value placed on health) are clearly not optimal. Therefore, the challenge is to price emissions, air pollution, health, life itself, at reasonable levels such that decisions and policies can be developed for the greatest societal benefit. This concept is represented graphically in Fig. 18.2. The left side of Fig. 18.2 depicts the less controlled situation, such as the case of low safety and emissions standards for vehicles back in the 1950s and 1960s. Policy changes implemented at the left side of the graph have a large positive impact at a relatively low cost. Examples include the addition of catalytic converters, seat belts on automobiles, or the removal of

Fig. 18.1 Smog over Salt Lake City (by Bruce McAllister, 1936), photographer (NARA record: 3823134)—US National Archives and Records Administration, Public Domain, https://commons.wikimedia.org/w/index.php?curid=16913367).

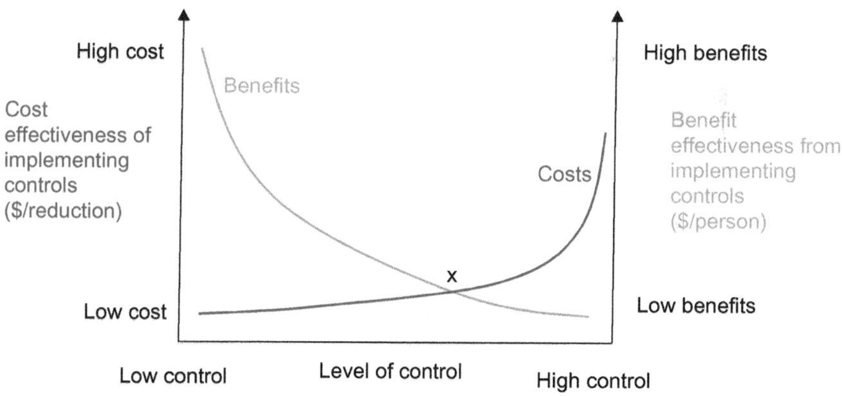

Fig. 18.2 Theoretical costs and benefits of levels of controls over pollution.

lead from gasoline. One analysis of removing lead from gasoline found that the benefits were more than 10 times the costs (Schwartz, Pitcher, Levin, Ostro, & Nichols, 1985). The benefits were primarily health related. As more and more high-impact, low-cost, solutions are adopted, we move to the right side of the graph where higher-cost, lower-impact solutions remain. One recent paper looked at the cost of a range of greenhouse gas reducing policies as compared to the amount of greenhouse gas reduced (Gillingham & Stock, 2018). Some had very low cost to reduce emissions and, not surprisingly, were already done—such as blending up to 10% ethanol with gasoline. Other policies were expensive per ton reduced and thus should be considered only after the more cost-efficient options—if at all. At some point, the marginal cost of that solution/policy exceeds the marginal benefit (point x on the graph). To the right of this point, it does not make *economic* sense to implement that solution. Note that x is not a fixed point as technological advances may significantly reduce the cost or increase the impact of a solution, potentially shifting that solution from having a net societal loss to a net societal gain. It is from this perspective that we attempt to place a quantitative value on emissions so that we can identify both (1) uneconomical solutions/policies and (2) which of several solutions/policies have the best benefit to cost ratios.

Cost-effectiveness methodologies

The focus of this chapter is on the cost-effectiveness of policies or projects that impact emissions, and thus impact air quality and health. One of the key issues is how to monetize (place a monetary value on) changes in emissions. The following section details some of the challenges in determining this monetary value. Before delving into those details, this section provides an overview of how those monetary values are used in project or policy evaluation.

Transportation projects (and to a lessor extent policies) are often evaluated against one another for many reasons, including identifying the most beneficial projects to fund. The evaluation techniques can involve some aspect of cost-effectiveness, or may be based on the projects ability to meet specific goals, or a combination of the two.

A common technique used to evaluate cost-effectiveness is benefit-cost analysis (BCA). To perform a BCA, the cost of the project or policy plus the monetary value of the impacts (benefits) of that project or policy must be determined to the greatest extent possible. Generally, the costs are relatively straight forward as they are the actual costs of implementing the policy or project, plus operation, maintenance, replacement, and other costs over the life of the project, minus any salvage value at the end of the project's life. Many of these costs occur at the start of the project, but other costs occur well into the future. To account for this, all costs (and benefits) are adjusted to a common year using a real discount rate. Current guidance in the United States is to use a discount rate of 7%, but with flexibility to use 3% as well (https://www.transportation.gov/regulations/omb-circular-94, https://www.transportation.gov/sites/dot.gov/files/docs/Tiger_Benefit-Cost_Analysis_%28BCA%29_Resource_Guide_1.pdf). This is similar to the rates used in many industrialized countries (Mackie & Worsley, 2013).

The calculation of benefits is generally not as straight forward since benefits include many reductions in nonmonetary costs associated with travel. These include such items as reductions in travel time, reductions in crashes and associated injuries and deaths, and of particular interest in this book—reductions in emissions leading to improved air quality and health. Although not easy to monetize, monetary values for each of these have been developed so that these benefits can be compared to the cost of the project or policy. The Tiger BCA resource guide noted above is a good source for these values used in the United States. Transport for London provides similar guidance through its transport analysis guidance (https://www.gov.uk/guidance/transport-analysis-guidance-webtag), and a comparison of transport BCA in several countries can be found in the report: International Comparisons of Transport Appraisal Practice: Overview Report (Mackie & Worsley, 2013).

The focus of this chapter is on the health impacts of reduced emissions, but more recently there has been efforts to also establish the health impacts of more active transportation options such as walking and biking. The World Health Organization has developed a tool to estimate benefits from reduced deaths due to increases in physical activity due to transportation projects or policies called the Health Economic Assessment Tool (HEAT) (http://www.euro.who.int/en/health-topics/environment-and-health/Transport-and-health/activities/guidance-and-tools/health-economic-assessment-tool-heat-for-cycling-and-walking). Similarly, the Sickness Absence Reduction Tool (SART) (http://www.hse.gov.uk/sicknessabsence/sart/index.htm) estimates the monetary benefits from reduced sick days due to increased physical activity brought about by the transport policy or project.

Once all of the benefits and costs have been determined and those values brought to a common year using a discount rate then the ratio of benefits to costs can be calculated. Generally, the higher the ratio the better. However, when comparing very different projects—particularly very large and small projects—the ratio will not tell the whole story. Another common analysis is to simply subtract the costs from the benefits to obtain the net benefit. An excellent source for information on benefit-cost analysis can be found at: http://bca.transportationeconomics.org/.

Another common method used to rank transportation projects is multiple criteria decision-making. This method does not necessarily require monetizing the impacts of a project. Rather it looks at how well the project achieves specific goals and scores each project on its ability to meet those goals. Those scores are often weighted with more important goals getting a higher weight (greater emphasis). Then, the weighted scores are totaled with the best projects obtaining the highest scores. Quite often some measure of cost-effectiveness (like BCA) will be included as one of the goals that are measured and scored. This method has the advantage of not needing to convert everything to a monetary value, but adds subjectivity in the weighting process. The techniques for scoring and evaluating these decision-making methods can be extremely complex. However, the simple method outlined above is the most common method for transportation project selection since it is easy for nontechnical decision makers and the general public to follow.

How to quantify costs (impacts) and benefits

Placing a monetary value on a nonmonetary item like an emission [e.g., a gram of nitrogen oxides (NOx)], a statistical life, or an adverse health outcome has a range of challenges. This chapter focuses on the benefits and costs of emissions, which are particularly difficult to value. Two of the largest challenges stem from the fact:

(1) Emissions interact with the environment and disperse differently based on topography and weather patterns. For example, a valley may trap emissions where an open area near the ocean may help disperse emissions. Also, warm, sunny areas have more issues with specific air pollutants like ozone as the heat and sunlight help to speedup the chemical reactions between NOx and volatile organic compounds (VOC) to produce ground-level ozone. Thus, a given quantity of emissions may lead to dramatically different impacts on air quality based on location.

(2) Ambient air pollution impacts people differently. Children, elderly people, and those with respiratory and cardiovascular problems are particularly susceptible and will be the most impacted by air pollution.

Therefore, the bottom line impact of emissions on health varies greatly by location. Topography, weather, and socio-demographic characteristics of the population greatly influence the monetary impact of a given amount of emissions.

Quantifying impacts—The method commonly used in transportation benefit-cost analyses

Therefore, the values discussed in this chapter are generic values developed for an average population living in a typical area of the United States. Understanding that this is far from perfect, we look at ways to estimate a monetary value for emissions. Delucchi (2000) has done seminal work in the area of developing reasonable monetary values for air pollution. His often-cited methodology includes the following five steps:

1. Assume some change in vehicle use (an alternative would be a change in the fleet of vehicles)
2. Estimate the change in emissions resulting from that change in vehicle use
3. Estimate the change in air quality due to the change in emissions
4. Estimate the impacts (with health being the key impact) of the change in air quality
5. Value the impacts in dollars

Each of these steps involves considerable uncertainty and requires the values to be estimated. The results are also highly dependent on location (ambient air quality and weather being major factors) and population (health of the individuals). Despite the uncertainty, researchers (such as Delucchi) have developed reasonable monetary values for emissions that are used to conduct the benefit-cost analysis (BCA).

An all-encompassing BCA would include costs associated with any impact, including items such as:

- surface water runoff; hazardous chemicals build up on roadways and wash into the surrounding environment
- leaks of transportation-related chemicals and fuels into the surrounding environment
- noise
- barrier effects and severance. When the transportation system acts as a hindrance or barrier to travel for residents near the facility.

However, these items are often not included due to the difficulty of measuring and valuing these items, plus the health impacts from poor air quality dwarf the impacts from these other items. In addition, construction and design techniques are used to minimize the impacts of these items. They are, however, required as part of an environmental impact statement (AASHTO, 2010) for a project.

Next the travel impacts of the project(s) or policy change(s) must be estimated. Transportation planning models (such as TransCAD) are often used to estimate the change in travel for large-scale projects with far-reaching impacts. Smaller-scale projects with more limited impacts may be modeled with a microsimulation tool such as VISSIM. In rare cases, the BCA is performed ex post, allowing the analyst to gather data on actual impacts of the project. Even ex post, the data available are unlikely to be perfect as one of two things is likely to occur: (1) the project was only recently completed thus data on future year impacts must be estimated or (2) the project occurred long enough ago that all traffic impacts are known—but exogenous factors, such as the price of gas or a dramatic change in employment, have also significantly impacted traffic volumes.

Despite these limitations, the best estimate of how the project or policy will impact traffic is used to estimate the change in emissions. In the United States, this is often done using EPA's MOtor Vehicle Emission Simulator (MOVES—https://www.epa.gov/moves) or similar software that converts the combination of total vehicle mile traveled (VMT), speeds, and road type into quantities of emissions produced. The focus is usually the difference between the mass of emissions produced in a base or existing case (no change in policy or no new projects) and an alternative or future case, where the policy is implemented or the project is completed. The mass of emissions is usually calculated for every year over the life of the project considering the future changes in travel demand.

The next two steps (3—estimating the change in air quality and 4—estimating the impacts of the change in air quality) are generally not performed in a BCA. This would be extremely difficult and time consuming since those impacts, and thus the benefits per kilogram of pollutant reduced, would be different for every location based on topography, weather, and population. Instead, the analyst uses standard impact values (e.g., dollars per kilogram of emissions) for several criteria pollutants based on their expected impact on health. These standard impact values were derived using a process that follows steps 3 and 4, with an excellent example being the Delucchi (2000) effort. Finally, the difference in the mass of each of the various pollutants produced between the base scenario and the alternative scenario is multiplied by it's impact value to complete step 5.

Quantifying impacts—Recommended values

Current Federal Highway Administration (FHWA) guidance on the impact value of pollutants is provided in the document "Benefit-Cost Analysis Guidance for Discretionary Grant Programs" at https://www.transportation.gov/office-policy/transportation-policy/benefit-cost-analysis-guidance-2017. The recommended values for pollutants, from this document, are shown in Table 18.1.

Table 18.1 Damage costs for pollutant emissions (BCA Guidance Document).

Emission type	Cost ($/kg) (in 2017 dollars)
Volatile organic compounds (VOCs)	2.205
Nitrogen oxides (NOx)	9.149
Particulate matter ($PM_{2.5}$)	416.453
Sulfur dioxide (SO_2)	53.903

The values in Table 18.1 cite The Safer Affordable Fuel-Efficient (SAFE) Vehicles Rule for Model Year 2021–2026 Passenger Cars and Light Trucks (https://www.nhtsa.gov/sites/nhtsa.dot.gov/files/documents/ld_cafe_co2_nhtsa_2127-al76_epa_pria_181016.pdf) as the source. That document, in turn, refers readers to the Environmental Protection Agency's (EPA's) Regulatory Impact Analysis: Final Rulemaking for 2017–2025 Light-Duty Vehicle Greenhouse Gas Emission Standards and Corporate Average Fuel Economy Standards, EPA-420-R-12-016, August, 2012, Section 6.3, pp. 6–72 to 6–105 (https://19january2017snapshot.epa.gov/regulations-emissions-vehicles-and-engines/final-rule-model-year-2017-and-later-light-duty-vehicle_.html). In Table 6.3-14 of the Regulatory Impact Analysis document, the monetized values are presented. In this table, it is noted that the values were derived from Pope et al. (2002) for $PM_{2.5}$, SO_2, and NOx impacts and Bell, McDermott, Zeger, Samet, and Dominici (2004) for ozone-related impacts. All of these references discuss the methodology for determining monetized values. The focus is on the health impacts of each of these pollutants. The Safer Affordable Fuel-Efficient (SAFE) Vehicles Rule for Model Year 2021–2026 Passenger Cars and Light Trucks discusses these impacts in depth in Chapter 10.3. The Regulatory Impact Analysis: Final Rulemaking for 2017–2025 Light-Duty Vehicle Greenhouse Gas Emission Standards and Corporate Average Fuel Economy Standards also discusses these plus provides Table 6.3-8, which lists the impacts accounted for in the costs. These include items such as premature mortality, chronic bronchitis, nonfatal heart attacks, respiratory hospital visits, missed work days, missed school days, and minor restricted activity days. Since these values are based on research from the early 2000s they do not include more recent findings on how air pollution impacts health.

The domestic social cost of CO_2 values was obtained from the document "Benefit-Cost Analysis Guidance for Discretionary Grant Programs" at https://www.transportation.gov/office-policy/transportation-policy/benefit-cost-analysis-guidance-2017. Those values were obtained from

Table 8-24 in "The Safer Affordable Fuel-Efficient Vehicles Rule for MY2021-MY2026 Passenger Cars and Light Trucks Preliminary Regulatory Impact Analysis (October 2018)" (https://www.nhtsa.gov/sites/nhtsa.dot.gov/files/documents/ld_cafe_co2_nhtsa_2127-al76_epa_pria_181016.pdf) and were developed based on the report "Valuing Climate Damages: Updating Estimation of the Social Cost of Carbon Dioxide (2017)" (https://www.nap.edu/download/24651). They are shown in Table 18.2. This cost includes expected damages to domestic physical and economic systems due to changes in global climate as detailed in the appendix to Chapter 8 in the Valuing Climate Damages (2017) document.

Quantifying impacts—Other potential methods

The above paragraphs detail the standard methodology and values for major air pollutants in the United States. The following paragraphs discuss three other potential methods of valuing the impact of an air pollutant including its impact in health. Since air pollution is an externality, the economic solutions to externalities, and their associated costs, may help identify the value people place on clean air. Internalizing the externality through pricing is the optimal solution to deal with the externality. This could involve ownership of air molecules at the microlevel or, more realistically, collective bargaining of the right to pollute air molecules at the macrolevel (Delucchi, 2000; Howitt & Altshuler, 1999).

At the microlevel, in theory, individuals would own a set amount of air molecules and they can buy and sell those molecules. Factory owners might purchase a considerable amount of clean air molecules from many individuals so that they can pollute them. Conversely, someone could

Table 18.2 Domestic social cost of CO_2, 2015–2050 (in 2016 dollars per metric ton).

Year	Cost ($/ton) (in 2016 dollars)	
	Discount Rate	
	3%	7%
2015	6	1
2020	7	1
2025	7	1
2030	8	1
2035	9	2
2040	9	2
2045	10	2
2050	10	2

buy polluted air at a reduced price, clean it at their expense, and sell the clean air at a profit. The price of these air molecules would quickly establish the value people place on clean air and, in turn, the cost of polluting that air.

More realistically, macrolevel property rights would involve collective bargaining over the right to pollute the air. This could be similar to the EPA's cap and trade programs (https://www.epa.gov/emissions-trading-resources/what-emissions-trading), where companies buy the right to pollute a given number of tons of pollution. This operates in many states and, at one point, provided an estimate of the market value of a ton of SO_2 and NO_x. However, changes in clear air act regulations and lawsuits have rendered the program nonbinding and caused prices to drop to a few pennies per ton (Schmalensee & Stavins, 2013) (https://www.epa.gov/airmarkets/so2-allowance-auctions).

The final alternative method to estimate the value of emissions discussed in this chapter is called hedonic pricing. This method uses the value of a good or service that is bought and sold in an open and competitive market to infer the value of components of that good or service.

The most common example is the housing market. Typically, the sale price of homes in a community, or similar communities, would be the dependent variable in a regression equation. The independent variables would be key variables that influence a home's price—such as square footage, number of bedrooms, if there is a pool, and, for this purpose, the air quality. In theory, the coefficients estimated for each independent variable indicate the impact on sales price of that feature of the home.

For air quality, there may be some thresholds such as the measured amount of pollutants in the air, or similarly, the number of ozone alert days per year. The air quality variable could have a value of one for homes in areas that exceed the threshold and a zero for homes in areas that do not exceed the threshold. Then, whatever coefficient is estimated for this air quality variable is the impact, in cost, of that poor air quality on homes in the area.

Using hedonic pricing in this way assumes home buyers are knowledgeable about air quality issues and, consciously or subconsciously, those air quality issues impact their willingness to pay for the home. There are multiple studies using this method (e.g., Harrison & Rubinfeld, 1978; Kim, Cho, Lambert, & Roberts, 2010; Smith & Huang, 1995) but the standard values used in BCA for FHWA are obtained by the five-step method mentioned at the beginning of this chapter.

SFpark—An example of quantifying impacts

SFpark was one of six urban partnership agreement (UPA)/congestion reduction demonstration (CRD) projects funded by the Federal Highway Administration in 2007 to showcase how innovative technology, transit, telecommuting, and tolling could be used to reduce congestion (https://ops.fhwa.dot.gov/congestionpricing/docs/fhwajpo11040/index.htm). SFpark combined innovative parking pricing with real-time parking information dissemination to enhance the parking experience for users and reduce congestion caused by travelers cruising for a parking spot in San Francisco. As part of this effort, a team of researchers evaluated each of the six UPA/CRD projects. An important part of that evaluation was the benefit-cost analysis conducted by the author of this chapter. In this example, the environmental impact portion of the BCA for SFpark is explored.

In theory, changes in the price of parking and the ease in which a person can find parking impacts travel behavior and traffic congestion. The changes in travel behavior would impact the entire transportation system. Therefore, to capture the true, wide-scale, impacts of SFpark would require the changes to be modeled and measured in the local transportation planning model and then verified using field data. However, the innovative nature of this project, and many of the UPA/CRD projects, meant it was difficult for the traditional travel demand models to incorporate and model changes in travel due to the changes brought about by the projects. Therefore, changes in travel times and speeds were based on empirical data measured both before and after the implementation of SFpark. These data included travel speeds measured by roadway sensors, travel times based on bus data, and manual observations of the time it took drivers to find parking.

SFpark pricing was in effect from Monday to Saturday. Due to the different traffic conditions during the weekdays and weekends, the Monday through Friday traffic changes were measured separately from the Saturday traffic changes and Sunday was not included in the analysis. Changes in speeds and vehicle miles traveled (VMT) due to improved ability to find a parking spot lead to a decrease in emissions as calculated by the EMFAC2011 model. EMFAC is the California Air Resources Board's (CARB) tool for estimating emissions from on-road vehicles and is the EPA-approved method for doing so in California (the MOVES model and variations of MOVES, discussed above, are used in other states).

The EMFAC2011 model takes into account the fleet of motor vehicles in San Francisco County to estimate the emissions from the vehicles. The emissions for the vehicle fleet, based on the speed of the vehicles, are shown in Table 18.3. To estimate the change in grams of emissions, the change in VMT

Table 18.3 Emission Factors (grams per mile traveled) from EMFAC2011 for San Francisco County, 2012 passenger vehicles.

Pollutant	Grams of pollution per mile traveled at speed	
	5 mph	**10 mph**
VOC	0.34	0.22
NOx	0.34	0.29
CO_2	1194.70	886.39
$PM_{2.5}$	0.048	0.011

Source: San Francisco Urban Partnership Agreement, National Evaluation Report (https://rosap.ntl.bts.gov/view/dot/3535).

at each speed was multiplied by the emission factor for each pollutant in Table 18.3. In this example, the changes in VMT were due to reduced searching for parking which occurred at low speeds and thus only the 5 and 10 mph emissions factors were needed. The amount of emissions reduced (in pounds per day and annually) due to the SFpark project is shown in Table 18.4. Note that SO_2 emissions were not included as the environmental team indicated they would be minimal compared to other emissions, as SO_2 from newer vehicles is minimal (AASHTO, 2010; U.S. Department of Transportation, 2014).

The amount of emissions reduction shown in Table 18.4 was based on field observations of changes in parking searches not long after SFpark was fully operational in 2013. Ideally, those would be used to calibrate and validate the local travel demand model. Then, the travel demand model would be used to predict future travel changes due to SFpark, and future emissions changes would be obtained from model runs incorporating future traffic flows. As noted previously, the urban planning model was not able to model the impacts of SFpark, so only observed data were available. Therefore, it was assumed that future years emission changes would be the same as the observed data. This is a conservative approach to estimating future impacts

Table 18.4 Amount of emissions reduction from SFpark pricing.

Pollutant	Reduction in emissions (pounds per weekday)	Reduction in emissions (pounds per Saturday)	Reduction in emissions (pounds per year)
VOC	2.5	2.3	734
NO_X	2.5	2.3	746
$PM_{2.5}$	31.7	29.4	9456
CO_2	8714	8076.7	2,598,488

Source: San Francisco Urban Partnership Agreement, National Evaluation Report (https://rosap.ntl.bts.gov/view/dot/3535).

as it is likely that traffic congestion will worsen and the impact (benefits) of SFpark will increase over time. The number of years to include in the BCA is primarily dependent on the useful life of the project/infrastructure. In this case, the project was technology based and thus subject to a fairly short life span. A 10-year of operational life was chosen, to compute the benefits from reduced emissions.

Next, the pounds of emissions for each of the 10 years (shown in Table 18.4 and the same for every year) were multiplied by the monetary value of each emission. (Note that in the UPA National Evaluation those values are different than what is shown here as the national evaluation was done in 2013—but this example uses the latest guidance as shown in Tables 18.1 and 18.2 for consistency within this chapter.) The dollar value per kg shown in Table 18.1 was in 2017 dollars, and all emissions benefits were calculated to year 2017 dollars. All emissions benefits not accrued in 2017 are adjusted to that year using a 7% discount rate based on the federal government guidance (Office of Management and Budget Guidance [http://www.whitehouse.gov/omb/assets/a94/a094.pdf (p. 9)] and FHWA guidance [Federal Register, Vol. 75, No. 104, p. 30476, n.d.)]. For example, the VOC cost per kg in 2017 dollars was $2.205. This value is multiplied by the reduction in VOC produced (734 pounds or 333 kg) resulting in a benefit of reduced VOC of $734 in 2017. These values are adjusted by 7% per year to obtain values over the 10-year time frame required (see Table 18.5).

The final step is to adjust the total benefit ($243,128) to the base year for the project. This would be the one common year for all costs

Table 18.5 Health benefits from reduced emissions in the SFpark project.

	Benefit of reduced pollutant in 2017 dollars				
Year	**VOC**	**NO_x**	**$PM_{2.5}$**	**CO_2**	**Total**
2013	962	4058	25,786	1545	32,351
2014	899	3793	24,099	1444	30,235
2015	841	3544	22,523	1349	28,257
2016	786	3313	21,049	1261	26,408
2017	734	3096	19,672	1179	24,681
2018	686	2893	18,385	1102	23,066
2019	641	2704	17,182	1029	21,557
2020	599	2527	16,058	962	20,147
2021	560	2362	15,008	899	18,829
2022	523	2207	14,026	840	17,597
Total	7232	$30,497	$193,788	$11,611	$243,128

and benefits. The base year is often the start of the project—but could just as easily be a different year like 2017. For this example, assume all costs and benefit values are to be in 2013 dollars. Then, the health benefits from reduced emissions would be \$318,691 ($243{,}128 \times 1.07^4$). When compared to the benefits derived from travel time savings (over \$28 million—https://rosap.ntl.bts.gov/view/dot/3535), the benefits from reduced emissions were relatively small. The travel time savings were derived from travelers having to spend less time looking for parking spots. This is not unusual for transportation projects—where environmental benefits often represent a small, but an important, portion of total benefits from the project.

Summary and conclusions

This chapter explored how and why it is important to place a value on emissions and air pollution. There are many challenges in estimating these values, including considerable uncertainty in estimating the values and the knowledge the values used will undoubtedly change as our ability to model emission dispersion and health impacts improves. Nevertheless, the values are needed to quantify the benefits of projects or policies that impact air quality. Using reasonable values that reflect health benefits from reduced emissions allow for transportation planners and decision makers to better understand the impact of the project or policy and are better able to weigh those impacts versus the cost of the project or policy.

This chapter focused on the use of BCA to evaluate the health impacts of changes in emissions. A key strength, and key weakness, of a BCA is that all items must be given a monetary value. The strength lies in the ability to readily compare projects and options since all benefits and costs are converted to a single unit—money. This can also be considered a weakness if the monetary values placed on emissions are difficult to estimate correctly or are incomplete. In the case of emissions impact on health, estimating monetary values of these emissions is extremely difficult and our rapidly increasing knowledge regarding the impact of air pollution on health means the values rapidly become out of date.

Despite these challenges, it is important to continue to attempt to measure the benefits and costs of potential projects. The key will be to get the value of the impact of emissions as accurate as possible so that we can focus transportation funding efforts on the most beneficial policies or projects.

References

American Association of State Highway and Transportation Officials (AASHTO). (2010). *User and non-user benefit analysis for highways, United States.*

Bell, M., McDermott, A., Zeger, S., Samet, J., & Dominici, F. (2004). Ozone and short-term mortality in 95 US urban communities, 1987–2000. *Journal of the American Medical Association, 292*(19), 2372–2378.

Delucchi, M. (2000). Environmental externalities of motor-vehicle use in the US. *Journal of Transport Economics and Policy, 34*(2), 135–168.

Federal Register, Office of the Federal Register, National Archives and Records Administration. Vol. 77 (199), 62623–63200, https://www.federalregister.gov/documents/2012/10/15/2012-21972/2017-and-later-model-year-light-duty-vehicle-greenhouse-gas-emissions-and-corporate-average-fuel.

Gillingham, K., & Stock, J. (2018). The cost of reducing greenhouse gas emissions. *Journal of Economic Perspectives, 32*(4), 53–72.

Harrison, D., & Rubinfeld, D. (1978). Hedonic housing prices and the demand for clean air. *Journal of Environmental Economics and Management, 5*(1), 81–102. https://doi.org/10.1016/0095-0696(78)90006-2.

Howitt, A. M., & Altshuler, A. (1999). The politics of controlling auto air pollution. In *Chapter 7 of essays in transportation economics and policy*. Washington, DC: Brookings Institution Press.

Kim, S. G., Cho, S. H., Lambert, D. M., & Roberts, R. K. (2010). Measuring the value of air quality: Application of the spatial hedonic model. *Air Quality, Atmosphere, & Health, 3*(1), 41–51. https://doi.org/10.1007/s11869-009-0049-8.

Mackie, P., & Worsley, T. (2013). *International comparisons of transport appraisal practice: Overview report.* Institute for Transport Studies, University of Leeds: UK. https://assets.publishing.service.gov.uk/government/uploads/system/uploads/attachment_data/file/209530/final-overview-report.pdf.

Pope, C. A., III, Burnett, R. T., Thun, M. J., Calle, E. E., Krewski, D., Ito, K., et al. (2002). Lung cancer, cardiopulmonary mortality, and long-term exposure to fine particulate air pollution. *Journal of the American Medical Association, 287*, 1132–1141.

Schmalensee, R., & Stavins, R. (2013). The SO_2 allowance trading system: The ironic history of a grand policy experiment. *Journal of Economic Perspectives, 27*(1 (Winter)), 103–122.

Schwartz, J., Pitcher, H., Levin, R., Ostro, B., & Nichols, A. (1985). *Costs and benefits of reducing lead in gasoline: Final regulatory impact analysis.* Washington, DC: Office of Policy Analysis, U.S. Environmental Protection Agency.

Smith, V., & Huang, J. (1995). Can markets value air quality? A meta-analysis of hedonic property value models. *Journal of Political Economy, 103*, 209–227. https://doi.org/10.1086/261981.

U.S. Department of Transportation. (2014). *San Francisco Urban partnership agreement, national evaluation report.* https://rosap.ntl.bts.gov/view/dot/3535. FHWA-JPO-14-128.

U.S. Environmental Protection Agency. (2012). *Regulatory impact analysis: Final Rulemaking for 2017–2025 light-duty vehicle greenhouse gas emission standards and corporate average fuel economy standards, EPA-420-R-12-016.* Office of Transportation and Air Quality. https://19january2017snapshot.epa.gov/regulations-emissions-vehicles-and-engines/final-rule-model-year-2017-and-later-light-duty-vehicle_html.

Valuing Climate Damages: Updating Estimation of the Social Cost of Carbon Dioxide. (2017). *Committee on assessing approaches to updating the social cost of carbon, board on environmental change and society, and division of behavioral and social sciences and education.* National Academies of Science, Engineering and Medicine. https://www.nap.edu/download/24651.

CHAPTER 19

The social, environmental, health, and economic impacts of low carbon transport policy: A review of the evidence

Andrew Sudmant[a,*], Ersilia Verlinghieri[b,*], Haneen Khreis[c,d,e,f,g], and Andrew Gouldson[a]

[a]School of Earth and Environment, University of Leeds, Leeds, United Kingdom
[b]Transport Studies Unit, School of Geography and the Environment, University of Oxford, Oxford, United Kingdom
[c]Center for Advancing Research in Transportation Emissions, Energy, and Health (CARTEEH), Texas A&M Transportation Institute (TTI), Bryan, TX, United States
[d]Texas A&M | TTI Headquarters, Bryan, TX, United States
[e]ISGlobal, Centre for Research in Environmental Epidemiology (CREAL), Barcelona, Spain
[f]Universitat Pompeu Fabra (UPF), Barcelona, Spain
[g]CIBER Epidemiologia y Salud Publica (CIBERESP), Madrid, Spain

Abbreviations

BNEF	Bloomberg New Energy Finance
BRT	Bus Rapid Transit
CI	confidence interval
DALYs	Disability-Adjusted Life Year
EVs	electric vehicles
GDP	gross domestic product
GHG	greenhouse gas
ICE	internal combustion engine
LCUTP	low carbon urban transport policies
NO_2	nitrogen dioxide
NO_x	nitrogen oxides
PM	particulate matter
UK	United Kingdom
UN DESA	United Nations Department of Economic and Social Affairs
USA	Unites States of America

Introduction

While the largest sources of greenhouse gas (GHG) emissions—from the electricity sector, buildings, and industry—are peaking or in decline in many of the world's highest emitting nations, emissions from the transport sector

* These authors contributed equally.

Traffic-Related Air Pollution
https://doi.org/10.1016/B978-0-12-818122-5.00019-3

have stubbornly continued to grow as demand for motorized mobility outpaces increased vehicle efficiency (Rogelj et al., 2018). At the same time, continuing urbanization, which will see two-thirds of all people living in urban areas by 2050 (UN DeSA, 2018) and commitments from a growing number of urban areas to reduce GHG emissions (Grafakos et al., 2020) are increasing the focus on cities as places for low carbon action.

In this context, *low carbon urban transport policies* (LCUTP), defined as urban interventions designed to reduce GHG emissions from the transport sector while simultaneously addressing urban mobility challenges, is a topic of critical importance. Focusing narrowly on their impact on GHG emissions, a growing literature demonstrates the opportunity for LCUTP (Colenbrander et al., 2016; Hidalgo & Graftieaux, 2008; Sudmant, Colenbrander, Gouldson, & Chilundika, 2017). Complementing this literature there is a growing awareness that LCUTP are not only less energy and emission-intensive, but can improve safety, promote economically efficient urban development, increase social inclusivity, reduce congestion, and contribute to cleaner air, among other wider benefits (Gouldson, Sudmant, Khreis, & Papargyropoulou, 2018; Litman, 2018; Rode, 2018). Synergies between these benefits suggest that the overall cost of LCUPT policies may be greatly less than the wider social, economic, health, and environmental benefits they can realize (Gouldson et al., 2018).

Developed poorly, however, or ignored amidst competing challenges on the time and resources of policymakers and wider urban actors, and transport networks can lock in transport patterns that smother a region's economy, encourage sprawl, and exacerbate existing social, economic, health, and environmental inequalities. Understanding the contexts and conditions that support effective low carbon transport interventions is therefore critical. But on a more fundamental level, research also reveals that the benefits of LCUTP need to be celebrated carefully, placed within the context of wider urban processes, and understood with an awareness of impacts that can bring advancement to one set of priorities while ignoring or creating setbacks for others. Cycling infrastructure, for example, may only benefit able-bodied members of society, land use changes can displace members of the informal economy and public transport networks can prioritize travel to and from the urban center over travel in other parts of an urban area. What impacts are associated with LCUTP and for whom may be negative, when are the advancement of one set of social, health, environmental, or economic criteria at the cost of other worthy goals, and how are different criteria and metrics are compared and made comparable are among the critical questions policymakers need to have in the front of their minds.

After reviewing evidence on the wider impacts of three key LCUTP, this chapter considers the barriers and facilitators to catalyzing best practice in urban low carbon transport interventions. Specifically, the first part of the chapter provides an analysis of the impacts associated with three core categories of LCUTP: *Land use changes, modal shift and public transport improvements, and fleet improvement and transport electrification.* These LCUTP were identified in Gouldson et al. (2018) as three modes of policy intervention that cover the primary means of public and private intervention in urban transport. Drawing on the systematic review of core papers in the field of transport studies, this chapter highlights their potential benefits and negative impacts in terms of health, congestion, employment, and poverty. Subsequently, the second part of the chapter analyses the barriers and the facilitators to policy implementation for LCUTP, in the context of mitigating traffic-related emissions, air pollution, exposures, and adverse health effects and social impacts.

This chapter draws and extends on an existing systematic review of the literature and evidence on the co-benefits of different forms of low-carbon urban development (Gouldson et al., 2018). This systematic review was conducted across 11 mitigation measures in three sectors (transport, buildings, and waste), which, in turn, were matched with 16 pathways through which measures were seen to have a potential impact on the higher-level objectives of health, congestion and time, employment and the green economy, and poverty alleviation and inequality. For example, nonmotorized transport options were linked with public health through "pathways" that include physical activity and improved air quality. These pathways and measures were identified through expert consultation and following an initial review of the literature on co-benefits. Of the 11 mitigation measures, three measures specific to the transport sector are considered in this analysis: *Land use changes, modal shift and public transport improvements, and fleet improvement and transport electrification.* Collectively these are referred to as LCUTP.

For each combination of a LCUTP and pathway, a keyword combination was developed and used in database searches of the literature. The keyword combinations included (1) search terms related to the specific policy; (2) search terms related to the pathway; and (3) search terms related to the wider objectives relating to public health, congestion and time, employment, and the green economy and poverty alleviation. The specific keywords used in each search can be found in Gouldson et al. (2018).

In this chapter, the literature relating to urban transport is reevaluated with the intention of (1) highlighting the whole range of associated social, environmental, health, and economic impacts, and (2) showing the key barriers and enabling conditions for measures to be implemented. To support the first aspect, additional literature known to the authors was also included.

Land use measures

Land-use decisions, made by public authorities and private actors, but also through the actions and choices of urban citizens and households (sometimes in the absence of any public or private "authority"), are part of the basic context against which transport interventions need to be assessed. Relatively higher levels of density, for example, are often considered a prerequisite for large-scale public transport investments. At the same time, choices made around land use in urban areas heavily influence private transport choices, and can "lock" a region to a particular travel pattern. Low-density "exurbs" around the periphery of many cities in North America are laced with roads and highways that are expensive to remove (physical/infrastructural lock-in), that make nonmotorized transport virtually impossible for many trips (behavioral lock-in) and that lead to high levels of dependency on private vehicles that policymakers may be cautious to challenge (institutional lock-in).

Actions to affect land use consequentially present a tremendous challenge. However, where policies can be effectively implemented the benefits are commensurately large: reductions in GHG emissions, improved air quality, reduced travel times, increased economic productivity, and job creation are among the benefits.

Authors focusing on urban areas in the Global North have suggested that density alone can be a proxy for the effective use of urban space. Reviewing more than 300 academic papers Ahlfedlt and Pietrostefani (n.d.) find that more than 70% of literature reviewed found positive effects from increased economic density (the number of people living or working in an area) and 58% of literature found overall positive impacts from a higher density of the built environment, with papers highlighting net positive impacts across a range of social, health, economic, and environmental indicators.

Urban density, through its impact on the available pool of workers and by increasing the opportunity for knowledge sharing and interactions, is strongly connected with higher levels of economic productivity (Fujita, Krugman, & Venables, 2001; Glaeser et al., 1991). For example, Carlino and Hunt (2007) find that a 2× increase in employment density is associated with a 20% increase in patent intensity and Haughwout (2000) finds that a

200% increase in density is associated with a 6% increase in labor productivity in wealthy countries. Policies in the US limiting density have been found to cost the economy US$1 trillion a year (Hsieh & Moretti, 2014). Across literature reviewed in Gouldson et al. (2018), elasticities were found to range from −0.007 to 0.084 with a mean of 0.03, implying that a doubling of urban agglomeration (by population) would be expected to lead to a 3% increase in economic productivity.

Health impacts from density can arise from improved air quality and increased use of nonmotorized transport. A systematic review of 46 studies showed that increasing the density of retail and service destinations can directly and effectively promote the adoption of active travel over motorized travel (Sugiyama, Neuhaus, Cole, Giles-Corti, & Owen, 2012)[a] and wider literature finds strong correlations between levels of density, physical activity, and health indicators (Ewing, Meakins, Hamidi, & Nelson, 2014; Frank, Schmid, Sallis, Chapman, & Saelens, 2005). The scale of these health benefits is substantial. Reviewing compact development measures in six cities (Melbourne, Australia; Boston, United States; Copenhagen, Denmark; São Paulo, Brazil; and Delhi, India), Stevenson et al. (2016) find overall health gains of 420–826 Disability-Adjusted Life Year (DALYs) per 100,000 population from measures. Changes in urban density and land use have also important impacts on broader aspects of health and well-being, having strong effects on patterns of social inclusion, social capital, and livability (Fleuret & Atkinson, 2007).

Finally, density is also associated with reductions in transport GHG emissions. Ewing, Bartholomew, Winkelman, Walters, and Anderson (2008) found that increasing density in the USA by shifting to a type of urban development with a 50-unit increase per hectare could reduce vehicle kilometers per capita by up to 40%. Comparisons of urban centers have found that dense, highly connected urban centers like Hong Kong produce only one-third of the carbon emissions per capita of European cities, while European cities produce only one-fifth of the carbon emissions of sprawling and poorly connected cities like Houston (Rode et al., 2014). Conversely, the rapid expansion of metropolitan areas, or urban sprawl, and the resulting homogenous land use and low-density development patterns in the USA reinforce the need and convenience for extensive road networks and private car travel, leading to higher emissions (Frumkin, 2002).

a Half of the studies came from North America, 11 from Australia, 8 from Europe, 3 from South America, and 1 from Japan.

A related set of literature has emphasized building residential, commercial, and light industrial in close proximity ("mixed-use design") and building higher areas of density around transit hubs ("transit-oriented development") in land-use decisions. By facilitating access to jobs, services and amenities, mixed-use design, and transit-oriented development increase the "effective density" of urban areas, by making it easier for people to travel in cities. In so doing these actions can create the same benefits as increased density. Unique from policies focused narrowly on density, however, these policies do not necessarily increase the number of people or infrastructure in a given area, therefore leaving more space for parks and other amenities. In addition, these policies place a specific emphasis on public transit options, leading to some unique impact on wider social, health, economic, and environmental variables (Conlan, Fraser, Vedrenne, Tate, & Whittles, 2016; Giles-Corti et al., 2016; Guttikunda & Mohan, 2014; Stevenson et al., 2016).

For example, the development of public transport can significantly reduce traffic accidents, benefiting both passengers and members of the public (Litman, 2011). Cities with the highest share of mass transport users have the lowest share of traffic fatalities. Indeed, public transit is shown to have one-tenth the per km traffic casualty rate compared with motorized travel over the same distance. As a result, public transit-oriented communities have one-fifth of the accidents of car-oriented communities (American Public Transportation Association, 2016). In addition, research suggests that investments in multimodal transit create relatively more employment than investment in conventional transport. For example, a review of the literature found that for each US$1 million invested, 11.4 jobs were generated by cycling projects, 9.9 jobs by pedestrian projects, and 7.8 jobs by road projects (Bertaud & Richardson, 2004; Graham, 2007; Litman, 2009; Rode et al., 2014).

Land-use decisions can be costly and challenging to change in already built-up areas. Clever planning measures, however, can lead to dramatic changes in the way residents live and work in urban areas. In Barcelona, "superblocks," neighborhoods entirely closed to traffic or with harsh restrictions on motorized vehicles that are surrounded by traffic corridors were estimated to have a significant impact on the health and well-being of residents. According to Mueller et al. (2019), Barcelona's 503 existing and proposed "superblocks" could prevent 667 premature deaths annually, primarily through reductions in nitrogen dioxide (NO_2) exposures, noise, and heat.

While the literature discussed above is compelling, the limitations of the evidence base, the documented negative impacts of increased density (and other land-use policies) in some contexts, and the potential for negative unintended consequences, are important to note.

It is important to note that the majority of research is based on case studies from the United States and European countries and results might apply differently to other countries where the culture around walking, cycling, and the use of public transportation options is significantly different (see, e.g., Almahmood, Scharnhorst, Carstensen, Jørgensen, & Schulze, 2017; Uteng & Cresswell, 2016). In this regard, specific attention should be made to how new land-use policies interact with preexisting territorial conditions, as they might intensify socioeconomic inequalities. As shown by Blanco and Apaolaza (2018) this is, for example, the case when infrastructure development is unequally spatially distributed and facilitating access for high-income groups that already have access to private mobility. Guzman et al. (2017, p. 238) stress how in Bogotá "centralization of economic activities in the expanded city center, land regulations and housing market dynamics have made housing close to major centers of employment unaffordable for the poorest population." Similarly, the concentration of employment and education opportunities in specific locations of the city has caused major problems of inequality in terms of access to these activities, particularly affecting low-income groups. This has led to the growth of informal housing, a pattern of development common in expanding cities in the South (Bocarejo & Oviedo, 2012).

The negative impacts of some policies in some contexts are also important to note. Land use mix and proximity to services can facilitate walking (Rothman, Buliung, Macarthur, To, & Howard, 2013), but have also been found to increase the incidence and severity of injuries, especially for cyclists and pedestrians (Rothman et al., 2013). Similarly, where higher density is designed without transport options, or investments in transport do not sufficiently meet residents' needs, compact development can lead to congestion, levels of air pollution and noise, and reduced green space and ability of the landscape to absorb water (and therefore prevent flooding) (de Hartog, Boogaard, Nijland, & Hoek, 2010; Hankey & Brauer, 2012; Hedley et al., 2002; Takeshita, 2012).

Finally, unintended negative consequences need further research and attention from policymakers. This is particularly significant around questions of equity and justice. Evidence suggests that the benefits of new infrastructure are rarely shared equally. For example, new green areas or biking facilities might not have direct positive effects for those people that are unable to access them, due to limited time or economic resources or, more broadly, capabilities.[b] Therefore, a specific land use arrangement might be beneficial to a group but not to another (Blanco & Apaolaza, 2018;

b On the possibility and importance of centering policy evaluation and assessment strategies on people's capabilities, see the work of Amartya Sen and Marta Nussbaum (Nussbaum, 2011; Robeyns, 2017; Sen, 2010).

Kaufmann, Bergman, & Joye, 2004; Schwanen et al., 2015). This means that the fairness of certain land use and transport arrangement is evaluated appropriately only when the assessment is not limited to actual demand, but accounts also for those people that do not access or participate in it (Martens, 2016).[c] As it will be discussed for public transport policies, factors beyond "location" affect the accessibility and "usefulness" of transport and land use infrastructures for different populations, emphasizing the need for context-specific analysis, especially in highly socioeconomically unequal contexts such as cities in the Global South (Blanco, Lucas, Schafran, Verlinghieri, & Apaolaza, 2018).

The nature and extent of the wider social, health, economic, and environmental impacts of land-use policies, therefore, depend critically on policymakers considering their local context, considering trade-offs and complementarities between different approaches and on the development of integrated policies that can approach problems from different angles. A discussion around the barriers and facilitators of effective policy can be found in the concluding section.

Mode shift and public transport investments

The challenge of improving urban mobility while reducing GHG emissions is paradoxically both highly complex and painfully simple. It is complex because it involves coordinating the action of almost every individual inhabitant of an urban area (and many who are not inhabitants) several times a day, every day of the year. It is simple because a single action underpins almost every intervention: A shift from private cars to other modes of transport.

Public transport investments form a core approach for urban policymakers in this context. For cities that are established (large and growing at a relatively slow pace), for example, London (UK), New York (USA), Tokyo (Japan), Shanghai (China), Delhi (India), or Johannesburg (South Africa), changing the structure of the city can be costly and time-consuming, making public transport investments a compelling alternative, or complement, to longer-term changes in land use. In cities that are rapidly growing, such as Lagos (Nigeria), Tianjin (China), and Hyderabad (India), public transport investments can sometimes be made in advance of demand for travel and prevent the development of car-dependent commuting patterns.

c To know more on available methods to assess accessibility to different transport options and how these vary in terms of mode, location, social groups, and activity see for example https://www.sciencedirect.com/science/article/pii/S0966692303000607.

Some of the most substantial benefits of public transport investments and mode shift away from private vehicles come from reducing vehicle traffic in urban areas. Congestion has been linked with significant reductions in employment growth in urban areas. For example, for cities already facing high levels of congestion (such as Los Angeles), Hymel (2009) found that a 1% increase in congestion reduced employment growth by 0.4%. Reduced vehicle traffic is also associated with reduced road deaths and injuries, motor vehicle accidents, and higher levels of walking and bicycling (Ewing & Hamidi, 2015; Nakahara, Ichikawa, & Kimura, 2011; Wei & Lovegrove, 2012; Yiannakoulias & Scott, 2013). Further, eliminating private vehicle trips can have a substantial effect on air quality. Among a very substantial literature in this area, Grabow et al. (2012) finds that the elimination of automobile round trips in 11 metropolitan areas of the US would reduce $PM_{2.5}$ by 0.1 $\mu g/m^3$ resulting in net health benefits of US$4.94 billion per year.[d] Finally, reducing private vehicle travel can have very large impacts on energy expenditure, leaving more money for households to spend in the local economy and reducing rates of transport poverty and forced car ownership (Lucas, Mattioli, Verlinghieri, & Guzman, 2016; Mattioli, Nicolas, & Gertz, 2018). Research suggests this can range from a few hundred dollars for the average household in developing contexts where private vehicle ownership and usage is relatedly lower (Colenbrander, Sudmant, Chilundika, & Gouldson, 2019; Sudmant et al., 2017), to significantly larger amounts in middle- and higher-income cities (Gouldson et al., 2015).

Bus Rapid Transit (BRT) systems have been popularized as a relatively low-cost rapid transit investment, with successful schemes in Mexico City (Mexico), Ahmadabad (Iran), Bogotá (Colombia), Guangzhou (China), Lagos (Nigeria), and Johannesburg (South Africa), and demonstrating the potential for reducing air pollution and urban congestion (Ang & Marchal, 2013). For example, in Mexico City, the Metrobus BRT reduced 110,000 tons of greenhouse gas (GHG) emissions after 3 years of service. Along specific routes PM concentrations were reduced by up to one-half of the previous level, travel times fell by 40% and traffic accidents by 84% (Francke, Macías, & Schmid, 2012). Similar studies have modeled the health benefits of new public transport routes in Adelaide (Xia et al., 2015) and Thessaloniki (Sabel et al., 2016).

Public transport systems can also have positive impacts on the urban economy. Sanchez, Shen, and Peng (2004) and Yi (2006) find that public transport networks are associated with higher labor force participation

d 95% confidence interval (CI): US$0.2 billion, US$13.5 billion rates and higher rates of employment growth. In addition, public transport networks raise employment by shifting consumption patterns and generating induced employment. In US cities with high-quality public transport, Litman (2014) found that residents typically spent approximately US$3000 (12% of income) on transport, while residents of cities with poor networks spent approximately US$3300 (14.9% of income). Similarly, McCann (2000) found auto-dependent communities in the US spent US$8500 per house household on travel, while those in cities with effective transport networks spend less than US$5500. These savings can generate additional employment if they are spent in sectors with a relatively higher labor intensity: US$1 million spent on petrol or other vehicle expenses generate 12.8–13.7 jobs across the US economy, while the same expenditure across a typical bundle of goods purchased by a household generated 17.0–17.3 jobs Chmelynski (2008).

Similar benefits come from actions that lead to higher levels of non-motorized transport. These actions can include investments in sidewalks and bike lanes, but increases in walking and cycling also frequently result from the development of public transport or from land-use decisions that promote density, diversity, and connectivity. Bicycle infrastructure, including bike lanes, bike stands, and storage, can be a low-cost alternative to major investments in transport and generate substantial health benefits. Assessing the impact of increasing cycling in 167 of the largest cities in Europe, Mueller et al. (2018) find that if cycling rates were at least 24.7% of trips, more than 10,000 lives would be saved annually. Of critical importance to the wider transport network, cycling infrastructure can provide "last-mile mobility," between people's homes or workplaces and public transport.

While the potential benefits are substantial, realizing them depends on careful consideration of the local context and effective implementation. Although in the long-term LCUTP might result in higher rates of local employment, frictional unemployment, that is, unemployment created when workers move between jobs, may result in the short-term. More generally, in the context of the rapid advancement of some transport technologies, tremendous long-term uncertainty exists around the future of transport networks in cities, adding a note of caution to the most optimistic projections of the impact of public transport and mode shift measures. This is something that needs to be specifically considered in the case of developing countries where informal transport constitutes an important source of employment (Rodríguez, Khattak, & Evenson, 2006).

Specific to BRT development, recent research in developing contexts emphasizes that possible benefits, including improved mobility, health, access to employment, and other benefits, are rarely equitably distributed across society (Rizzo, 2015, 2017) and frequently do not reach the poorest groups (Venter, Jennings, Hidalgo, & Pineda, 2018). In wealthier regions research also highlights the link between the development of new public transport infrastructures, gentrification, and displacement of the poor (Enright, 2013; Luckey, 2018; Suzuki, Murakami, Hong, & Tamayose, 2015). And in both wealthy and developing contexts there are numerous cases of BRT projects failing to be financially self-sufficient and creating a strain on local transport agencies (Rizzo, 2015, 2017; Rumé, 2018).

The potential for public transport investment to have positive impacts on local social, health, environmental, and economic factors is therefore substantial, but there is both scope for negative impacts on specific populations and net negative impacts from projects overall if actions are poorly planned and implemented. Some of the factors that need to be considered by policymakers are explored in the concluding section.

Fleet efficiency improvement and electrification

Improving fleet efficiency and moving toward an electrified (or hydrogen) fleet are increasingly prominent strategies in urban low carbon plans. In the UK, for example, sales of vehicles with internal combustion engines (ICE) will be banned from 2040, India is targeting an all-electric fleet by 2030, and in China sales of ICE vehicles have been discouraged with taxes and incentives to both the public and vehicle manufacturers (Bloomberg New Energy Finance, 2019). At the same time, policies to improve vehicle efficiency are becoming increasingly stringent.

Underlying these actions is a growing concern around climate change, but also a belief that such actions can contribute to improving urban air quality, a pressing concern in both developing and wealthy cities (Ji, Cherry, Bechle, Wu, & Marshall, 2012). Reviewing the literature finds support for each of efficiency improvements and electrification to improve air quality and reduce GHG emissions. However, the appropriate balance of policies will depend on local factors, including how policymakers prioritize each of air quality and GHG mitigation, and wider economic and social benefits from these policies will depend on the extent that efficiency and electrification are supported and integrated into a wider program of interventions. Further, results raise concerns that incomplete or out of date evidence may be underpinning some urban policies. Three key factors contribute to this conclusion.

First, a shift to electric vehicles may lead to higher levels of particulate pollution from electricity generation and from vehicle braking (as a result of electric vehicles being heavier than vehicles with internal combustion engines). Timmers and Achten (2016), for example, find that total PM_{10} emissions from electric vehicles are not significantly less than their non-electric counterparts and reduction in $PM_{2.5}$ emissions from electric vehicles (EVs) is estimated to be negligible (1%–3%). Similarly, Ji et al. (2012) find that the particulate emissions and associated health impacts in China would be higher from electric than gasoline cars and that the health burden would be shifted from urban dwellers to relatively socioeconomically disadvantaged rural populations. Evidence from the US, conversely, shows that lifecycle air pollutant emissions may be dramatically lower where electric vehicles (EVs) are powered by renewable electricity or natural gas (Tessum, Hill, & Marshall, 2014). These results underline the importance of lifecycle and region-wide assessments of the impacts from EV policies, particularly as EVs batteries are highly energy-intensive to produce and production is currently concentrated in China where the electricity grid is both carbon and air pollution intensive. It should be noted that the specific health effects of non-exhaust particulate pollution are beyond the scope of this chapter; however, research suggests distinct adverse health effects from non-exhaust particulate emissions (Gasser et al., 2009).

Second, levels of EVs uptake would need to be relatively high before significant impacts are seen on GHG emissions or air pollution. Soret, Jimenez-Guerrero, and Baldasano (2011), for example, finds that NO_x emissions would decline by 11% in Barcelona and 17% in Spain only after more than 40% of the fleet is electrified while Timmers and Achten (2016) note that the impact on particulate emissions is likely negligible even at high levels of uptake. Given that the most optimistic projections do not see these high levels of EVs uptake until 2030s (Bloomberg New Energy Finance, 2019), anticipated benefits maybe a decade or further away.

Thirdly, stricter ICE vehicle emission standards may allow consumers to continue buying and driving ICE vehicles further into the future than some modeling optimistically predicts. Focusing on the lifecycle emissions from EV adoption in the UK, research has found that emission impacts are concentrated in the period after 2030 despite the UK's low carbon electricity grid, in part due to rising vehicle efficiencies (Hill, Heidrich, Creutzig, & Blythe, 2019). Across the European market, diesel sales are falling rapidly following the Volkswagen scandal and as a result of rising emissions standards, but the rate of EVs uptake is much slower due to the limited number of vehicles and infrastructure limitations (Bloomberg New Energy Finance, 2019). In the most recent year (2018), EVs sales were found to have declined in the UK while sales of large and more emissions-intensive SUV vehicles rose, underlining the challenge of convincing a large proportion of drivers to shift away from internal combustion technology (Bloomberg New Energy Finance, 2019).

The longer-term necessity of shifting to EVs (or other alternative fuel vehicles) to achieve GHG emission reductions need to be considered alongside policies that can achieve nearer-term air quality benefits, including reducing the number of diesel vehicles, restricting vehicles during specific times of day, and congestion and parking charges (among other actions). Xue et al. (2015), for example, finds that while new energy vehicles (electric and hydrogen-powered vehicles) improve air quality in Xiamen, China, much larger and more immediate impacts come from wider policy options, including reducing the number of diesel trucks in the city center. In Dublin, Ireland, a ban on older diesel vehicles was found to have the potential to reduce the health burden (in terms of the number of hospital admissions from poor air quality) by more than half by 2030 (Dey, Caulfield, & Ghosh, 2018). Policies to ban or tax diesel vehicles are now being implemented or recommended in a growing number of cities.

Continued advancement in EVs technology, and grid decarbonization in many regions of the world, will improve the case for fleet electrification—both from GHG and air pollution perspectives. The results from this research may therefore not hold further than the early 2020s, at which point EVs may be superior due to efficiency improvement in each of these areas (Bloomberg New Energy Finance, 2019). Wider benefits of these actions, however, depend both today and in the future on the way these policies are integrated into wider transport planning. This is specifically important when considering issues of affordability for vehicles utilizing new technologies.

From an economic standpoint, efficiency and electrification have tremendous potential to generate economic returns. Assessing global urban areas, Sudmant, Millward-Hopkins, Colenbrander, and Gouldson (2016) found that electrification and efficiency measures could generate $210 billion in energy savings globally by 2030 and more than $700 billion by 2050. Case study literature similarly shows that at the urban level, the economic benefits of efficiency and electrification in the transport sector can be 1%–2% of urban Gross Domestic Product (GDP) (Colenbrander et al., 2016, 2019; Sudmant et al., 2017). Whether these benefits materialize, however, depends on strong assumptions about the extent of rebound effects—to what extent drivers use fuel savings to drive further or engage in other forms of consumption that generate emissions. Furthermore, these savings will only exist for those who own vehicles in the first place and will be largest for those who can afford the upfront cost of electric vehicles, limiting their benefit for economically disadvantaged populations.

In addition, the economic benefits of higher efficiency and EVs may encourage car ownership and private transport options, preventing the development of multimodal and nonmotorized transport options. Beyond the vehicles themselves, investments in charging networks and the wider infrastructure for a new generation of private vehicles may generate GHG and air quality benefits over the coming decade only to prevent the deeper transformations needed to meet the targets of the Paris Agreement (for climate change), to eliminate urban transport air pollution, or to achieve a number of the Sustainable Development Goals related to transport and mobility. Efficiency and electrification measures should, therefore, be seen as a complementary to wider programs of action but potentially limited as self-standing policies.

Barriers and facilitators to LCUTP implementation

A growing evidence base details the potential for low carbon transport options to have substantial, and in many cases broadly felt, impacts on social, health, economic, and (non-GHG) environmental objectives. These are critical for policymaking as climate policy exists among a wider set of priorities, particularly so in rapidly developing cities of the Global South where more than 1 million new urban residents each week need access to shelter, basic services, and employment (UN DeSA, 2018). Importantly, the key interventions described above rely on established "technologies." Interventions, in other words, that have been tested and shown to be successful in specific contexts. Importantly, however, as also discussed, changes in this context can lead to a wide range of negative impacts. The extent to which learning from successful interventions can inform action in other places, therefore, depends on a range of factors, including, existing policies and infrastructure, local politics, geography, and cultural norms. This means that it is necessary to carefully assess contextual factors before supporting the implementation of policies.

Implementation of actions and interventions will, therefore, require multilevel coordination to be successful. Only by establishing an enabling environment, can policymakers maximize the opportunity for effective action. Policies around taxes, efficiency standards, and revenue sharing are nearly always set at the national level, but the interests of urban and national governments may not be aligned. For national governments, vehicle sales and imports can be a lucrative source of revenue and are often seen as important to the growth of the economy, while at the urban level rising vehicle numbers increase congestion and encourage sprawl. Urban governments can encourage nonmotorized transport options like cycling and walking, and conduct planning for larger interventions like BRT networks, but rarely can conduct the detailed engineering studies that are needed for implementation, nor do they typically have access to the financing for these investments. In these cases, national governments can directly support local governments, but can also establish the conditions that can allow local governments to lead action. This can involve the devolution of powers, for example by allowing urban areas to raise bonds, collaboration with researchers and academia to deliver the necessary studies, or establishing rules for public-private partnerships. This last option is only recommended when institutions can carefully monitor and compensate for the risks that this model poses in terms of equity and effective local and national development (Miraftab, 2004).

Policymakers will also need to consider cross-sectoral coordination. Urban planning decisions play an important role in determining the locations of where people live and work and the viability of different transport options. Similarly, the location of schools, hospitals, park, and other public institutions and amenities significantly affect transport decision-making, even though they are rarely determined by transport departments. Importantly, many decisions that affect urban landscapes have long-reaching influence. While private vehicles are replaced every 10–20 years, infrastructure lifetimes are typically 30 years at a minimum, and land-use decisions can last for centuries. Cross-sectoral decision-making processes to coordinate action are for this reason especially important in rapidly growing cities of the Global South, where transit systems can sometimes be designed in advance of population growth and where initial levels of inequality and deprivation are highly different from the Global North (Blanco et al., 2018; Uteng & Lucas, 2017).

Finally, policymakers will also need to consider the barriers facing non-governmental actors, both private businesses and members of the public, as they consider low carbon transport options. Relative to other sectors of the economy, such the energy sector, where public firms are frequently key actors, or parts of the industrial sector, where there may only be a small number of private actors, the transport sector is characterized by a large number of individual private actors, most employing several different modes of transport on a daily basis. Strict policy regimes can achieve coordinated action, but less invasive actions to enable individuals to take up lower carbon transport modes are generally more politically feasible. Policymakers need to consider both financial and nonfinancial barriers.

Financial barriers preventing individuals from switching to a low carbon transport option or an engineering firm from participating in a tender for a new metro system are in cases perceived to be higher risks and lower returns from low carbon investments. The cost of low carbon actions has been falling: EVs, for example, are anticipated to cost the same as conventional vehicles by 2022 (Bloomberg New Energy Finance, 2019). In many cases, however, upfront costs remain higher for low carbon technologies, reducing financial returns. Investment risks include financial risks, such as exchange rate uncertainty, while nonfinancial risks including policy and legal risks. The potential for a new government to change regulations around subsidies for electric vehicles or requirements for new building standards are examples of policy risk. Such risks, however, can also relate more generally to the way government policies and approaches affect firms' access to global supply chains and the ability to operate their businesses. Whether a new technology can be patented and how and by whom it is regulated are examples of legal risks. Combined, these factors manifest in constraints over the amount of finance available, the rate it is available at and the timescales over which investments are to be paid back.

A range of policy options, including many detailed above, can help to address financial barriers to low carbon action. Tax incentives for certain kinds of investments, charges on certain types of vehicles and modes of transport, subsidies for public transport, and loan guarantees or exclusive operating rights are just a few of the options policymakers have at their disposal to overcome these barriers. These measures can be employed by different levels of government (depending on the context), but frequently require the support of national or regional governments who have financial leverage.

There are also critical considerations beyond the narrow confines of the "financial case" for low carbon transport. Uncertainties around new technologies, government policies and legal codes make up one aspect of these barriers. More widely, cultural and social norms embedded in transport practices can also serve as barriers to effective policymaking. Cycling needs to be seen as a legitimate transport option, and not just a means of exercise or transport for those without cars before it can be expected to be taken up by the masses. Public transit needs to be understood as safe and reliable before people will use it widely for their daily commute. Similarly, pedestrians' right to walk and safety has to be guaranteed widely and ensured by all actors on the roads before moving on foot could become a core part of transport systems. Addressing these barriers requires a more nuanced and place-based approach relative to the challenge of overcoming financial barriers. At the same time, it requires an ongoing process of stakeholders and citizens' participation, which can allow policymakers to have constantly updated evidence of the impacts that policies are having and to effectively design shared strategies for mitigation (Vanclay, Esteves, Aucamp, Research, & Franks, 2015).

The successful future of LCUTP depends on the ability of a variety of stakeholders to coordinate and embrace an approach to transport governance that is flexible, adaptable, and that combines localized interventions and integrated planning while preparing for negative consequences (Guzman et al., 2017). Increasing evidence shows that successful approaches are ones that integrate different measures and carefully account for the specific contextual, cultural, and societal aspects. Going beyond transport and thinking about mobilities as a series of socially and territorially contextualized practices might allow policymakers to uncover more subtle effects that policies and technologies have on overall societal well-being. We have stressed how shifting to more sustainable transport systems requires the provision of infrastructures to be accompanied by a number of legislative interventions and societal changes that take into account the different abilities and needs of different groups. The variety of available LCUTP can be a promising step in this direction.

Summary and conclusions

LCUTP holds tremendous promise, both for reducing GHG emissions and for improving the liveability of urban areas and the health of their citizens. Land-use policies can discourage the uptake of private vehicles and enable a shift toward nonmotorized mobility and mass public transport. These actions hold particularly large promise in rapidly growing cities where the expense and disruption of knocking down old infrastructure can sometimes be avoided. Public transport investments can increase the "effective density" of neighborhoods and provide a viable alternative to motorized transport for residents unable to walk or cycle to work. Electrifying transport, both private cars and public vehicles can reduce both GHG emissions and air pollutants as a complement to land use and public transport investments, or where these options have not yet been developed.

While the evidence base for these actions continues to grow, policymakers need to balance the urgency of addressing climate change, as well as other urban issues, with a need to take the time to develop actions that draw on the latest evidence, are contextually appropriate, and that employs a comprehensive approach to the challenge, rather than relying on any one action or technology. In this chapter, we have highlighted the importance of considering the whole range of social, health, environmental, and economic impacts associated with LCUTP and have also stressed how negative impacts and the potential for negative unintended consequences, need to be carefully considered.

Where assessment that includes these elements provides a case for action, this chapter has also discussed the barriers and facilitators to LCUTP implementation. Municipal actors rarely have the opportunity or capacity to unilaterally implement low carbon transport interventions at a meaningful scale. To the extent they can, however, it would be wise to develop and implement actions with the engagement and support of a wider array of actions. Collaboration and partnership between transport policymakers vertically (to regional, national, and international policymakers, as well as local and community organizations), and horizontally (between cities and departments within cities), can facilitate learning, provide much needed capacity, and help coordinate the transport planning and policy process. Urban policymakers can also work to address the barriers affecting private and community actors as they look to support LCUTP. This can include helping to address financial barriers to action, but it can also relate to the socially and territorially contextualized practices that are much a part of any transport system as cars, buses, and trains and other infrastructures. In this respect, policymakers need to think about the diverse needs of citizens and the ways that they can engage with the nuanced and sometimes place-specific barriers and enablers of behavioral and social change.

References

Ahlfedlt, G., and Pietrostefani, E. The effects of compact urban form: A qualitative and quantitative evidence review. Coalition for Urban Transitions, London/Washington, DC. Available at: http://newclimateeconomy.net/content/cities-working-papers.

Almahmood, M., Scharnhorst, E., Carstensen, T. A., Jørgensen, G., & Schulze, O. (2017). Mapping the gendered city: Investigating the socio-cultural influence on the practice of walking and the meaning of walkscapes among young Saudi adults in Riyadh. *Journal of Urban Design*, *22*, 229–248. https://doi.org/10.1080/13574809.2016.1273742.

American Public Transportation Association. (2016). *The hidden traffic safety solution: Public transportation*. Available at: https://www.apta.com/resources/reportsandpublications/Documents/APTA-Hidden-Traffic-Safety-Solution-Public-Transportation.pdf.

Ang, G., & Marchal, V. (2013). *Mobilising private investment in sustainable transport*. Paris: Organisation for Economic Co-operation and Development. Available at: http://www.oecd-ilibrary.org/content/workingpaper/5k46hjm8jpmv-en.

Bertaud, A., & Richardson, A. W. (2004). *Transit and density: Atlanta, the United States and Western Europe*. Available at: http://courses.washington.edu/gmforum/Readings/Bertaud_Transit_US_Europe.pdf.

Blanco, J., & Apaolaza, R. (2018). Socio-territorial inequality and differential mobility. Three key issues in the Buenos Aires Metropolitan Region. *Journal of Transport Geography*, *67*, 76–84. https://doi.org/10.1016/j.jtrangeo.2017.07.008.

Blanco, J., Lucas, K., Schafran, A., Verlinghieri, E., & Apaolaza, R. (2018). Contested mobilities in the Latin American context. *Journal of Transport Geography*, *67*, 73–75. https://doi.org/10.1016/j.jtrangeo.2018.01.006.

BNEF. (2019). *Electric vehicle outlook 2019 | Bloomberg New Energy Finance. [online]*. Available at: https://about.bnef.com/electric-vehicle-outlook/#toc-viewreport. (Accessed16 August 2019).

Bocarejo, S. J. P., & Oviedo, H. D. R. (2012). Transport accessibility and social inequities: A tool for identification of mobility needs and evaluation of transport investments. *Journal of Transport Geography*, *24*, 142–154.

Carlino, G., & Hunt, R. (2007). *Innovation across US industries: The effects of local economic characteristics. Working Paper 07-28*. Federal Reserve Bank of Philadelphia. Available at: www.philadelphiafed.org/research-and-data/publications/working-papers/2007/wp07-28.pdf.

Chmelynski, H. (2008). *National economic impacts per $1 million household expenditures (2006). Spreadsheet based on IMPLAN Input-Output Model*. Jack Faucett Associates.

Colenbrander, S., Gouldson, A., Sudmant, A. H., Papargyropoulou, E., Chau, L. W., & Ho, C. S. (2016). Exploring the economic case for early investment in climate change mitigation in middle-income countries: A case study of Johor Bahru. *Climate and Development*, *8*, 351–364. https://doi.org/10.1080/17565529.2015.1040367.

Colenbrander, S., Sudmant, A., Chilundika, N., & Gouldson, A. (2019). The scope for low-carbon development in Kigali, Rwanda: An economic appraisal. *Sustainable Development*, *27*(3), 349–365.

Conlan, B., Fraser, A., Vedrenne, M., Tate, J., & Whittles, A. (2016). *Evidence review on effectiveness of transport measures in reducing nitrogen dioxide*. Department for Environment, Food and Rural Affairs (DEFRA).

de Hartog, J. J., Boogaard, H., Nijland, H., & Hoek, G. (2010). Do the health benefits of cycling outweigh the risks? *Environmental Health Perspectives*, *118*, 1109–1116. https://doi.org/10.1289/ehp.0901747.

Dey, S., Caulfield, B., & Ghosh, B. (2018). Potential health and economic benefits of banning diesel traffic in Dublin, Ireland. *Journal of Transport & Health*, *10*, 156–166.

Enright, T. E. (2013). Mass transportation in the Neoliberal City: The mobilizing myths of the grand Paris express. *Environment and Planning A: Economy and Space*, *45*(4), 797–813.

Ewing, R., Bartholomew, K., Winkelman, S., Walters, J., & Anderson, G. (2008). Urban development and climate change. *Journal of Urbanism: International Research on Placemaking and Urban Sustainability*, *1*(3), 201–216.

Ewing, R., & Hamidi, S. (2015). Compactness versus sprawl. A review of recent evidence from the United States. *Journal of Planning Literature*, *30*(4), 413–432.

Ewing, R., Meakins, G., Hamidi, S., & Nelson, A. C. (2014). Relationship between urban sprawl and physical activity, obesity, and morbidity—Update and refinement. *Health and Place, 26*, 118–126.
Fleuret, S., & Atkinson, S. (2007). Wellbeing, health and geography: A critical review and research agenda. *New Zealand Geographer, 63*, 106–118. https://doi.org/10.1111/j.1745-7939.2007.00093.x.
Francke, E., Macías, J., & Schmid, G. (2012). *Mobilising private Investment for bus Rapid Transit systems: The case of Metrobus*. Mexico City: CTS EMBARQ. https://www.oecd.org/env/cc/Case%20study%20Mexico.pdf.
Frank, L. D., Schmid, T. L., Sallis, J. F., Chapman, J., & Saelens, B. E. (2005). Linking objectively measured physical activity with objectively measured urban form: Findings from SMARTRAQ. *American Journal of Preventive Medicine, 28*(2), 117–125.
Frumkin, H. (2002). Urban sprawl and public health. *Public Health Reports, 117*(3), 201.
Fujita, M., Krugman, P. R., & Venables, A. J. (2001). *The spatial economy: Cities, regions, and international trade*. MIT Press.
Gasser, M., Riediker, M., Mueller, L., Perrenoud, A., Blank, F., Gehr, P., et al. (2009). Toxic effects of brake wear particles on epithelial lung cells in vitro. *Particle and Fibre Toxicology, 6*(1).
Giles-Corti, B., Vernez-Moudon, A., Reis, R., Turrell, G., Dannenberg, A. L., et al. (2016). City planning and population health: A global challenge. *The Lancet, 388*(10062), 2912–2924.
Glaeser, E. L., et al. (1991). *Growth in cities. No. w3787*. National Bureau of Economic Research.
Gouldson, A., Colenbrander, S., Sudmant, A., McAnulla, F., Kerr, N., Sakai, P., et al. (2015). Exploring the economic case for climate action in cities. *Global Environmental Change, 35*, 93–105.
Gouldson, A., Sudmant, A., Khreis, H., & Papargyropoulou, E. (2018). *The economic and social benefits of low-carbon cities: A systematic review of the evidence*. London/Washington, DC: Coalition for Urban Transitions. http://newclimateeconomy.net/content/cities-working-papers.
Grabow, M. L., Spak, S. N., Holloway, T., Stone, B., Jr., Mednick, A. C., & Patz, J. A. (2012). Air quality and exercise-related health benefits from reduced car travel in the midwestern United States. *Environmental Health Perspectives, 120*(1), 68.
Grafakos, S., Viero, G., Reckien, D., Trigg, K., Viguie, V., Sudmant, A., et al. (2020). Integration of mitigation and adaptation in urban climate change action plans in Europe: A systematic assessment. *Renewable and Sustainable Energy Reviews, 121*. https://doi.org/10.1016/j.rser.2019.109623. 109623.
Graham, D. J. (2007). Agglomeration, productivity and transport investment. *Journal of Transport Economics and Policy, 41*(3), 317–343.
Guttikunda, S. K., & Mohan, D. (2014). Re-fueling road transport for better air quality in India. *Energy Policy, 68*, 556–561.
Guzman, L. A., Oviedo, D., & Rivera, C. (2017). Assessing equity in transport accessibility to work and study: The Bogotá region. *Journal of Transport Geography, 58*, 236–246. https://doi.org/10.1016/j.jtrangeo.2016.12.016.
Hankey, J. M. J., & Brauer, M. (2012). Health impacts of the built environment: Within-urban variability in physical inactivity, air pollution, and ischemic heart disease mortality. *Environmental Health Perspectives, 120*(2), 247–252.
Haughwout, A. (2000). The paradox of infrastructure investment: Can a productive good reduce productivity? *Brookings Review, Summer*, 40–43. Available at: www.brookings.edu/articles/2000/summer_productivity.aspx.
Hedley, A. J., Wong, C. M., Thach, T. Q., Ma, S., Lam, T. H., & Anderson, H. R. (2002). Cardiorespiratory and all-cause mortality after restrictions on sulphur content of fuel in Hong Kong: An intervention study. *The Lancet, 360*(9346), 1646–1652. https://doi.org/10.1016/S0140-6736(02)11612-6.

Hidalgo, D., & Graftieaux, P. (2008). Bus rapid transit systems in Latin America and Asia: Results and difficulties in 11 cities. *Transportation Research Record Journal of the Transportation Research Board, 2072*, 77–88. https://doi.org/10.3141/2072-09.

Hill, G., Heidrich, O., Creutzig, F., & Blythe, P. (2019). The role of electric vehicles in near-term mitigation pathways and achieving the UK's carbon budget. *Applied Energy, 251*, 113111.

Hsieh, C., & Moretti, E. (2014). *Growth in cities and countries*. Cambridge, MA: National Bureau of Economic Research.

Hymel, K. (2009). Does traffic congestion reduce employment growth? *Journal of Urban Economics, 65*(2), 127–135.

Ji, S., Cherry, C. R., Bechle, M. J., Wu, Y., & Marshall, J. D. (2012). Electric vehicles in China: Emissions and health impacts. *Environmental Science & Technology, 46*(4), 2018–2024.

Kaufmann, V., Bergman, M. M., & Joye, D. (2004). Motility: Mobility as capital. *International Journal of Urban and Regional Research, 28*(4), 745–756.

Litman, T. (2009–2010). *Transportation cost and benefit analysis* (p. 31). Victoria Transport Policy Institute.

Litman, T. (2011). *Pricing for traffic safety: How efficient transport pricing can reduce roadway crash risk*. Victoria, Canada: Victoria Transport Policy Institute.

Litman, T. (2014). A new transit safety narrative. *Journal of Public Transportation, 17*(4), 8.

Litman, T. (2018). Toward more comprehensive evaluation of traffic risks and safety strategies. *Research in Transportation Business & Management, 29*, 127–135. https://doi.org/10.1016/j.rtbm.2019.01.003. Special Issue title: [Re]evaluating how we value transportation.

Lucas, K., Mattioli, G., Verlinghieri, E., & Guzman, A. (2016). Transport poverty and its adverse social consequences. *Proceedings of the Institution of Civil Engineers—Transport, 169*, 353–365. https://doi.org/10.1680/jtran.15.00073.

Luckey, K. S. (2018). Affordable for whom? Introducing an improved measure for assessing impacts of transportation decisions on housing affordability for households with limited means. *Research in Transportation Business & Management*. https://doi.org/10.1016/j.rtbm.2018.04.003.

Martens, K. (2016). *Transport justice: Designing fair transportation systems*. Routledge.

Mattioli, G., Nicolas, J.-P., & Gertz, C. (2018). Household transport costs, economic stress and energy vulnerability. *Transport Policy, 65*, 1–4. https://doi.org/10.1016/j.tranpol.2017.11.002.

McCann, B. (2000). *Driven to spend: The impact of sprawl on household transportation expenses*. Surface Transportation Policy Partnership.

Miraftab, F. (2004). Public-private partnerships: The Trojan horse of neoliberal development? *Journal of Planning Education and Research, 24*, 89–101. https://doi.org/10.1177/0739456X04267173.

Mueller, N., Rojas-Rueda, D., Khreis, H., Cirach, M., Andrés, D., Ballester, J., et al. (2019). Changing the urban design of cities for health: The superblock model. *Environment International*, 105132.

Mueller, N., Rojas-Rueda, D., Salmon, M., Martinez, D., Ambros, A., Brand, C., et al. (2018). Health impact assessment of cycling network expansions in European cities. *Preventive Medicine, 109*, 62–70.

Nakahara, S., Ichikawa, M., & Kimura, A. (2011). Population strategies and high-risk-individual strategies for road safety in Japan. *Health Policy, 100*(2), 247–255.

Nussbaum, M. C. (2011). *Creating capabilities*. Harvard University Press.

Rizzo, M. (2015). The political economy of an urban megaproject: The bus rapid transit project in Tanzania. *African Affairs (London), 114*, 249–270. https://doi.org/10.1093/afraf/adu084.

Rizzo, M. (2017). *Taken for a ride: Grounding neoliberalism, precarious labour, and public transport in an African Metropolis, critical frontiers of theory, research, and policy in international development studies*. Oxford, New York: Oxford University Press.

Robeyns, I. (2017). *Wellbeing, freedom and social justice: The capability approach re-examined*. Open Book Publishers. https://doi.org/10.11647/OBP.0130.
Rode, P. (2018). *Governing compact cities: How to connect planning*. Design and Transport: Edward Elgar Publishing.
Rode, P., Floater, G., Thomopoulos, N., Docherty, J., Schwinger, P., Mahendra, A., et al. (2014). *Accessibility in cities: Transport and urban form. NCE cities paper 03*. LSE Cities. London School of Economics and Political Science.
Rodríguez, D. A., Khattak, A. J., & Evenson, K. R. (2006). Can new urbanism encourage physical activity?: Comparing a new urbanist neighborhood with conventional suburbs. *Journal of the American Planning Association, 72*(1), 43–54.
Rogelj, J., Shindell, D., Jiang, K., Fifita, S., Forster, P., Ginzburg, V., et al. (2018). Mitigation pathways compatible with 1.5°C in the context of sustainable development. In V. Masson-Delmotte, P. Zhai, H.-O. Pörtner, D. Roberts, J. Skea, P. R. Shukla, & T. Waterfield (Eds.), *Global Warming of 1.5°C. An IPCC Special Report on the impacts of global warming of 1.5°C above pre-industrial levels and related global greenhouse gas emission pathways, in the context of strengthening the global response to the threat of climate change, sustainable development, and efforts to eradicate poverty*, IPCC.
Rothman, L., Buliung, R., Macarthur, C., To, T., & Howard, A. (2013). Walking and child pedestrian injury: A systematic review of built environment correlates of safe walking. *Injury Prevention, 20*(1), 41–49.
Rumé, S. (2018). Reflexiones antropológicas sobre la difícil ejecución del proyecto tranvía en Cuenca. *CIVITIC Revista Interuniversitaria de Estudios Urbanos de Ecuador, 4*, 25–34.
Sabel, C. E., Hiscock, R., Asikainen, A., Bi, J., Depledge, M., et al. (2016). Public health impacts of city policies to reduce climate change: Findings from the URGENCHE EU-China project. *Environmental Health, 15*(1), 5.
Sanchez, T., Shen, Q., & Peng, Z. (2004). Transit mobility, jobs access and low-income labour participation in US Metropolitan areas. *Urban Studies, 41*(7), 1313–1331.
Schwanen, T., Lucas, K., Akyelken, N., Cisternas Solsona, D., Carrasco, J.-A., & Neutens, T. (2015). Rethinking the links between social exclusion and transport disadvantage through the lens of social capital. *Transportation Research Part A: Policy and Practice, 74*, 123–135. https://doi.org/10.1016/j.tra.2015.02.012.
Sen, A. (2010). *The idea of justice*. London: Penguin UK.
Soret, A., Jimenez-Guerrero, P., & Baldasano, J. M. (2011). Comprehensive air quality planning for the Barcelona Metropolitan Area through traffic management. *Atmospheric Pollution Research, 2*(3), 255–266.
Stevenson, M., Thompson, J., de Sá, T. H., Ewing, R., Mohan, D., et al. (2016). Land use, transport, and population health: Estimating the health benefits of compact cities. *The Lancet, 388*(10062), 2925–2935.
Sudmant, A., Colenbrander, S., Gouldson, A., & Chilundika, N. (2017). Private opportunities, public benefits? The scope for private finance to deliver low-carbon transport systems in Kigali, Rwanda. *Urban Climate, 20*, 59–74.
Sudmant, A., Millward-Hopkins, J., Colenbrander, S., & Gouldson, A. (2016). Low carbon cities: Is ambitious action affordable? *Climatic Change, 138*(3–4), 681–688.
Sugiyama, T., Neuhaus, M., Cole, R., Giles-Corti, B., & Owen, N. (2012). Destination and route attributes associated with adults' walking: A review. *Medicine and Science in Sports and Exercise, 44*(7), 1275–1286.
Suzuki, H., Murakami, J., Hong, Y. H., & Tamayose, B. (2015). *Financing transit-oriented development with land values: Adapting land value capture in developing countries*. The World Bank.
Takeshita, T. (2012). Assessing the co-benefits of CO_2 mitigation on air pollutants emissions from road vehicles. *Applied Energy, 97*, 225–237.
Tessum, C. W., Hill, J. D., & Marshall, J. D. (2014). Life cycle air quality impacts of conventional and alternative light-duty transportation in the United States. *Proceedings of the National Academy of Sciences, 111*(52), 18490–18495.

Timmers, V. R., & Achten, P. A. (2016). Non-exhaust PM emissions from electric vehicles. *Atmospheric Environment, 134*, 10–17.

UN DeSA. (2018). *World population prospects: The 2018 revision. Population division of the department of economic and social affairs of the United Nations secretariat* (p. 18). New York: UN DeSA.

Uteng, T. P., & Cresswell, T. (2016). *Gendered mobilities*. Aldershot, England: Routledge.

Uteng, T. P., & Lucas, K. (Eds.). (2017). *Urban mobilities in the global south* Routledge.

Vanclay, F., Esteves, A. M., Aucamp, I., Research, E., & Franks, D. M. (2015). *Social impact assessment: Guidance for assessing and managing the social impacts of projects*. Fargo, ND: International Association for Impact Assessment.

Venter, C., Jennings, G., Hidalgo, D., & Pineda, A. F.V. (2018). The equity impacts of bus rapid transit: A review of the evidence and implications for sustainable transport. *International Journal of Sustainable Transportation, 12*, 140–152. https://doi.org/10.1080/15568318.2017.1340528.

Wei, V. F., & Lovegrove, G. (2012). Sustainable road safety: A new (?) neighbourhood road pattern that saves VRU lives. *Accident Analysis and Prevention, 44*(1), 140–148.

Xia, T., Nitschke, M., Zhang, Y., Shah, P., Crabb, S., & Hansen, A. (2015). Traffic-related air pollution and health co- benefits of alternative transport in Adelaide, South Australia. *Environment International, 74*, 281–290.

Xue, X., Ren, Y., Cui, S., Lin, J., Huang, W., & Zhou, J. (2015). Integrated analysis of GHGs and public health damage mitigation for developing urban road transportation strategies. *Transportation Research Part D: Transport and Environment, 35*, 84–103.

Yi, C. (2006). The impact of public transit on employment status: Disaggregate analysis of Houston. *Transportation Research, 1986*(1), 137–144.

Yiannakoulias, N., & Scott, D. M. (2013). The effects of local and non-local traffic on child pedestrian safety. *Social Science & Medicine, 80*, 96–104.

CHAPTER 20

Environmental justice

Disproportionate impacts of transportation on vulnerable communities

Christina H. Fuller[a] and Doug Brugge[b]
[a]Department of Population Health Sciences, Georgia State University School of Public Health, Atlanta, GA, United States
[b]Department of Public Health Sciences, University of Connecticut, Farmington, CT, United States

Abbreviations

DOT	Department of Transportation
E.O.	Executive Order
EJ	environmental justice
EPA	Environmental Protection Agency
GAO	General Accounting Office
HIA	Health Impact Assessment
NO	nitric oxide
NO_2	nitrogen dioxide
NO_x	nitrogen oxides
PCB	polychlorinated biphenyls
PM	particulate matter
PM_{10}	particulate matter $< 10\,\mu m$ in aerodynamic diameter
$PM_{2.5}$	particulate matter $< 2.5\,\mu m$ in aerodynamic diameter
SES	socioeconomic status
TRAP	traffic-related air pollution
UCC	United Church of Christ
UFP	ultrafine particles
UK	United Kingdom
US	United States

Definition and historical context of environmental justice

The environmental justice (EJ) movement is broad and diverse, involving people from a variety of backgrounds across the globe. EJ has no single figurehead, but instead, its principles are moved forward by a network of grassroots activists and supporters. The movement emerged from multiple struggles including US Civil Rights actions, antitoxics campaigns, labor movements, Native American land rights, and traditional environmentalism (Cole & Foster, 2001). The Civil Rights movement, in

Traffic-Related Air Pollution
https://doi.org/10.1016/B978-0-12-818122-5.00020-X

particular, played a significant role because inequalities in areas of civil life such as basic rights, health care, discrimination, segregation also extend into the environmental realm. Communities of color, in particular, as well as poor and immigrant neighborhoods, continue to be targeted for the placement of industrial facilities such as chemical plants, oil refineries, and hazardous waste landfills. Many thriving communities have been razed or divided to make way for highways and transportation hubs. The focus of EJ groups has been to address chemical exposures and disease risks as well as maintain and restore natural ecosystems; as compared to mainstream, largely white, environmental organizations that have traditionally focused more narrowly on the care and conservation of natural ecosystems.

A watershed moment in the formation of the EJ movement was the siting of a hazardous waste landfill in Warren County, North Carolina in 1982. The purpose of the site was to hold 32,000 cubic yards of polychlorinated biphenyl (PCB) contaminated soil that had been illegally spread across a large swath of roadways in the state (Bullard, 1994). Warren County was rural, predominately African American and poor, located in the far northern portion of the state. Community members, led by Dollie Burwell and Benjamin Chavis, organized against the landfill, staging multiple acts of nonviolent protest. Approximately 500 people were arrested during several weeks of direct action, including 94 children, which brought nationwide attention to the injustice. Despite efforts from the county's residents, the landfill was opened in 1982; however, residents persisted in the face of adversity and continued to organize. Years later, the landfill suffered a breach contaminating nearby land and groundwater. The continued actions of the residents and supporters led to the eventual closure and cleanup of the site and surrounding contaminated areas. After many years of remediation, the site was declared clean in 2003, over 20 years after its initial construction.

Grassroots organizing related to the Warren County landfill prompted the General Accounting Office (GAO), as instructed by Walter Fauntroy (Delegate from the District of Columbia), to study the linkages between race and class with locations of hazardous waste sites. The 1983 GAO study documented in its examination that although African Americans comprised 20% of the population in the southeastern states studied, predominantly African American communities were home to three out of four hazardous waste landfills (General Accounting Office, 1983). In 1987, The United Church of Christ (UCC) Commission for Racial

Justice expanded this examination to the entire United States and included other races/ethnicities. The report, "Toxic Wastes and Race in the United States," documented that African American and Hispanic communities were overburdened by toxic sites. The UCC study revealed that the most important factor in the siting of hazardous waste facilities was race (United Church of Christ, 1987). In all, 60% of African Americans lived in a community that had one or more hazardous waste sites. Socioeconomic status was the second most important factor related to the placement of hazardous waste facilities. The strength of these findings was surprising for some and validating to the lived experience of many people that had been subjected to environmental injustice.

In October 1991, the First National People of Color Environmental Leadership Summit brought together communities from the United States and across the globe to Washington, DC to discuss issues and draft a pathway to achieve justice. Through the work of the attendees, the EJ movement achieved identity and developed a list of 17 guiding principles that united them in their pursuit of a healthy, just, and equitable world. The Principles of Environmental Justice provide a framework, until this day, for activists engaged in the struggle. The principles not only seek to improve human health and maintain natural systems, but also demand justice, equity, and self-determination, as evidenced by its preamble:

> *WE, THE PEOPLE OF COLOR, gathered together at this multinational People of Color Environmental Leadership Summit, to begin to build a national and international movement of all peoples of color to fight the destruction and taking of our lands and communities, do hereby re-establish our spiritual interdependence to the sacredness of our Mother Earth; to respect and celebrate each of our cultures, languages and beliefs about the natural world and our roles in healing ourselves; to ensure environmental justice; to promote economic alternatives which would contribute to the development of environmentally safe livelihoods; and, to secure our political, economic and cultural liberation that has been denied for over 500 years of colonization and oppression, resulting in the poisoning of our communities and land and the genocide of our peoples, do affirm and adopt these Principles of Environmental Justice.*

The success of the summit spurred a response from the US government. In 1994, President Bill Clinton issued Executive Order 12898, "Federal Actions to Address Environmental Justice in Minority Populations and Low-Income Populations" (E.O. 12898). This executive order mandated that each federal agency, "identifies and addresses disproportionately high and adverse human health or environmental effects of its programs, policies, and activities on minority populations and low-income populations."

This order resulted in the creation of the US Environmental Protection Agency's (EPA) Office of Environmental Justice. In compliance with E.O. 12898, the US Department of Transportation (DOT) published its own Environmental Justice Order on April 15, 1997 and updated the order in 2012. With the E.O. 12898 mandate in hand, communities, researchers and government agencies began to rigorously assess the distribution of environmental risks beyond hazardous waste sites to include power plants, industrial facilities, and transportation corridors. The results revealed systematic biases that over time and have created the inequalities in exposures and health impacts that we see today. Full implementation of the order has not been realized, however, advances have been made. Subsequent years have seen the increased engagement of community groups, researchers, and agencies in the evaluation of exposure disparities related to transportation as well as collaboration to identify solutions to current and potential issues. In particular, representatives from agencies such as EPA and DOT have collaborated with communities to develop Health Impact Assessments (HIAs) for proposed development projects. This allows stakeholders to have the ability to maximize benefits and reduce hazards proposed by new projects.

The US EJ movement spurred activism in many other countries across the globe, however, only a small number have been chronicled in the academic literature. There has been significant engagement around EJ in the United Kingdom (UK), focused mostly on injustice based on income and class designations. Issues concerning exposure to traffic-related air pollution (TRAP) have been at the forefront there. In South Africa, EJ took root in struggles against the inequitable placement of industry such as oil refineries in mostly poor and black townships. Similar to the link between EJ and Civil Rights in the United States, the South African EJ movement was heavily influenced by the anti-apartheid movement (Walker, 2012). Context has molded the EJ framework to fit local issues and challenges, which has been welcome within the movement. Tailoring strategies as well as acknowledging the overlap and branching of concerns are a hallmark to EJ being a grassroots movement.

The national legacy of the transportation system in urban areas

Transportation has a long and negative history with low-income, immigrant, and people of color communities that have been adversely impacted in the name of "urban renewal." The leader and the most prominent purveyor of urban transportation development was Robert Moses who over 44 years redrew the map of New York City disrupting the lives of thousands of people (Bullard, Johnson, & Torres, 2004; Caro, 1974). Moses was never elected to any public office but held many government positions that he manipulated in order to refashion the city as he saw fit. Beginning in 1920, he built miles of highways throughout the boroughs of New York displacing approximately 250,000 people from their homes. Moses created a city for vehicles and not people, putting high-speed traffic corridors straight through formerly bustling areas of commerce, recreation, and residences. Neither the physical health nor the economic health of the minority, working-class, and immigrant neighborhoods in the Bronx, Manhattan, and Brooklyn was considered in his sweeping plans for regional transportation. Highways provided limited access for the communities through which they passed, leaving behind pollution, noise, and crime. The available public transit, usually diesel-fueled buses, was unreliable and often serviced and housed in low-income, minority areas.

The strategy pioneered by Robert Moses became the de facto plan for other modernizing cities across the country. Cities such as Los Angeles, Milwaukee, and Atlanta built highways leading from downtown jobs into affluent and predominantly white suburbs. Many times, well-utilized light rail systems were dismantled to make space for highways such as in Los Angeles. The pollution, noise, and unsightliness of highways caused neighboring businesses and residents to flee leaving openings for crime and neglect. Those without sufficient resources to leave remained in the declining neighborhoods surviving as best they could. Many of these areas have remained in disrepair to this day and due to relatively low-cost living has drawn additional people from marginalized segments of the population. Older European cities—like London, Paris, and Barcelona—have also struggled with an increased burden in air pollution due to vehicles. These cities were built for travel by foot and horse, before the development of the car and are now choked with traffic. In these cities as well, new highways were built through old neighborhoods in order to connect urban areas by car.

Near-roadway air pollution gradients

Air pollutants are often distinguished as either local (primary) or regional (secondary) species. It is well established that some pollutants are elevated more than others next to highways and major roadways. Regional pollutants usually form from chemical reactions of primary emissions in the atmosphere. Fine particulate matter, $PM_{2.5}$ (aerodynamic diameter $<2.5\,\mu m$) is a prominent example of a secondary pollutant that tends to be distributed over tens or hundreds of kilometers. While sometimes elevated near roadways, the gradients of $PM_{2.5}$ in these locations are usually slight (Karner, Eisinger, & Niemeier, 2010). Hence, while there are vast disparities in $PM_{2.5}$ exposure globally, variation is more subtle within cities. Because there is substantial evidence in the literature that $PM_{2.5}$ is a leading cause of poor health outcomes globally, regional pollution has received much more attention from policy makers and regulators than locally elevated pollutants (EPA, 2018).

Pollutants that are regularly elevated within 100–200 m of roadways with high traffic volume include ultrafine particles (UFP; $<100\,nm$), oxides of nitrogen (including NO_x, NO_2, and NO), and PM_{10} (aerodynamic diameter $<10\,\mu m$). Particularly during cool weather before sunrise, when air can be stable, elevated concentrations can extend further away than 200 m (Hu et al., 2009). UFP, in particular, has the potential to cause more damage than $PM_{2.5}$ due to its small size and chemical composition. In contrast, the regional pollutant ozone has an inverse association with concentrations of traffic since NO reacts with ozone to destroy it. Thus, ozone tends to be lower in urban locations and higher in the surrounding areas.

In addition to local spatial variation, primary traffic emissions also vary rapidly in time (Durant et al., 2010). Traffic, and therefore TRAP, vary based on the number of cars and trucks on the road. Thus, very high peaks are often seen during or directly after morning and evening rush hours. Meteorological patterns also vary by day including temperature; wind speed and direction; and mixing height. In some locations, the highest concentrations are measured in the early morning when the boundary layer is low causing pollutants from the exhaust to build up. The extremely fine grain spatial and temporal patterns of elevation of near roadway pollutants present problems for assigning exposure. In order to monitor and assign exposure to these pollutants, one must either use many monitors, mobile monitoring or both to capture the variation in levels.

Because UFP levels fluctuate rapidly in space and time, it is important to capture as much variation as possible. In research with the Community Assessment of Freeway Exposure and Health study (CAFEH), we set out to do as much as we could to reduce factors that would contribute to uncertainty in assigning exposure for use in an epidemiology study (Fuller et al., 2013). We started by using gold standard methods for geocoding addresses in order to reduce errors in a residential position which can be up to 100 m by traditional address matching methods (Lane et al., 2013). We also built fine-grain models of UFP (measured as particle number count) at hourly temporal resolution and 20-m geographic resolution (Patton et al., 2015). These were combined with time-activity data of study participants to estimate their exposure for every hour over the course of a year (Lane et al., 2015).

One clear lesson that emerged from this work was that for locally elevated pollutants like UFP that rise and fall rapidly in space and time, there can be considerable error in assigning exposure. The problem of assigning exposure to local traffic pollutants such as those mentioned above is further complicated by the movement of people into and out of fields of higher and lower exposure. If one were to assume, as many epidemiology studies do, that exposure is represented by the level of pollution at a person's residential address, there would likely be an error in the assigned exposure. This is because the resident may very well spend large amounts of time at work, school, or otherwise away from home (Lane et al., 2015). Reducing that error requires a tremendous effort and success at doing so is difficult to estimate. We think the lesson for those interested in EJ related to traffic is that there is a need for well-designed and extremely careful data collection and analysis. After which it will be possible to assess exposures accurately across socioeconomic and racial/ethnic lines.

Pollutants that are elevated near traffic, such as UFP, lag behind regional pollutants like $PM_{2.5}$ and ozone in terms of assessment of risk and creation and enforcement of policy and public health protection. Assessing the environmental impacts of these exposures is not always straight forward.

Exposures and health disparities

There is a substantial and, in both our opinion and that of others, convincing literature that shows that people living adjacent to highways and major roadways experience a greater burden of adverse health outcomes than people living farther away. The increased exposure faced by some communities results in disparities in health end points. Health issues are elevated next to busy roadways include cardiovascular, respiratory, neurological, and birth outcomes. Along with total mortality, specific illnesses linked to road proximity include some that might surprise the naïve reader, such as autism in children and heart disease in adults (Brugge, Durant, & Rioux, 2007; HEI Panel on the Health Effects of Traffic-Related Air Pollution, 2010). Prevalence of asthma, diabetes, and hypertension are higher in low-income and minority communities when compared to overall averages. Other analyses have shown an increased risk for cancer in EJ communities based on air pollution exposures, especially those from local TRAP (Morello-Frosch, Pastor, & Sadd, 2001).

SES is a particularly relevant factor with regard to EJ concerns. Land next to highways, for example, is often inexpensive and used to develop affordable or low-income housing. Racial discrimination can result in minority populations being relegated to these locations as well. Since SES and racial/ethnic identity are strongly associated with ill health independent of traffic exposure, attributing effects of traffic must be made with caution. Epidemiology studies of proximity to highways/major roadways or that assign exposure to pollution that is elevated near roadways nearly universally control for SES. However, it is a potential problem that there is not a perfect measure of SES, leaving possible residual confounding in these studies.

A study of cardiopulmonary hospitalizations in New York City examined TRAP along with race/ethnicity and SES. They found nonsignificant increases in risk for Hispanics compared to whites. They also found that medical insurance was the strongest predictor as those with little or no coverage experienced greater risks across racial/ethnic groups. Another study identified greater effects of traffic density on poor and minority groups separately. However, the greatest impacts were found for those that were poor and minority. A similar finding was identified for TRAP and birth weights. The greatest difference in birth weight existed for those women who were lower income and minority (Schweitzer & Valenzuela, 2004).

While EJ emerged as an issue in the United States, it has since been taken up internationally. Most, but not all, of the international EJ studies have focused on disparities based on SES rather than race or ethnicity. This may be because few countries have a similar racial history as the United States, may not be as diverse, or do not record demographic information on race/ethnicity. A review of EJ studies and SES globally found that European studies were more mixed in terms of identifying associations, while North American, Asian, and African studies mostly reported increased air pollution exposure for lower SES populations (Hajat, Hsia, & O'Neill, 2015).

Mitchell and Dorling (Mitchell & Dorling, 2003) conducted a study of ward-level SES and NO_2 across the United Kingdom. They identified a roughly linear inverse association between car ownership and NO_2, showing that the highest burden of pollution was shouldered by areas with the lowest car ownership. The association between poverty and NO_2 was J-shaped, with those in highest poverty having the greatest pollution, but also a less dramatic variation among the least disadvantaged. This can be explained by high-income areas in the city center with high volume traffic and increased pollution compared to outlying areas. Effects on mortality were in line with this finding as identified by Pearce et al. (Pearce, Richardson, Mitchell, & Shortt, 2010). They used ward-level income and novel multiple variable deprivation indexes as predictors of mortality in the United Kingdom. The deprivation index was designed to represent the physical built environment and was built with four measures of air pollution, climate, industrial facilities, UV radiation, and greenspace. They found that the highest mortality rates were associated with the wards of the highest deprivation.

Two interesting studies, one from Belgium, the other from New Zealand, did report pollution exposure relative to immigrant population and race/ethnicity, respectively. The Belgian study used rigorous methods to address spatial autocorrelation that can affect the validity of ordinary regression models. Doing so, they found that factors associated with housing, rather than numbers of immigrants better predicted higher pollution exposures (Verbeek, 2019). The New Zealand study, which is older and did not use sophisticated statistical methods, documented that there was higher pollution in areas with more people identifying as Maori, Pacific Islander, or Asian (Kingham, Pearce, & Zawar-Reza, 2007).

While we contend that proximity to high volume traffic is well established as a risk factor for numerous adverse health outcomes, the aspect of the near highway environment that is causally responsible is not yet firmly established. The leading suspect is locally elevated TRAP from motor vehicles; however, associated vehicle noise and areas of lower SES near highways and major roadways may also play a part (Babisch, 2014; Morello-Frosch & Jesdale, 2006). Teasing apart the relative effects is complicated by correlations between these factors, although, separating them is possible since their geographic and temporal patterns do not overlap completely.

From an EJ perspective, the correlation between measures of SES and/or minority status and exposure to environmental hazards is of interest and concern. However, even if these populations are differentially proximal to heavily trafficked corridors, and that would in and of itself constitute an injustice, this does not necessarily mean that their health problems are due solely to the traffic pollution exposure. This is because SES influences health in other ways such as the likelihood of smoking, poor diet, and health care access, which could confound the effect of local air pollution exposure. Being a member of a minority group is also associated with daily chronic psychosocial stress, which is linked to mortality and chronic conditions such as hypertension (Williams, Lawrence, & Davis, 2019). SES and minority status can also modify exposure to TRAP. Many low-income residents, for example in public housing, might live in drafty buildings and open windows for ventilation potentially increasing exposure to outdoor pollution (Brugge et al., 2017). Even if ambient pollution levels are similar for low and high-income residents, better ventilation systems in market-rate and luxury housing likely result in disparate exposures along socioeconomic lines.

Contradictory examples

While it is often the case that low-income or affordable housing is disproportionately developed on land deemed undesirable for other purposes, which includes properties close to highways and major city streets, this is not uniformly true. Since the beginning of the new millennium, there has been a movement of higher-income people, who are predominantly white, from outer ring suburbs into central cities. An example would be large parts of Brooklyn, Queens, and the Bronx. Years of disinvestment based on the past transportation decisions of Robert Moses has made these areas cheap and available for new development. Housing prices have skyrocketed due to the demand to be close to Manhattan. However, those living in high-end real estate developments are also impacted by the substantial burden of traffic, much of which is concentrated in street canyons that trap air pollution, increasing exposure (Clougherty et al., 2013). An analysis of elevated exposures in transitioning cities may find that those most affected can be both high and lower income as Mitchell and Dorling noted a J-shaped association between poverty and NO_2 across the United Kingdom (Mitchell & Dorling, 2003). This does not mean that EJ should shift its focus, but it illustrates the necessity to understand the local context. In an analysis of the city of Boston, we also identified higher exposures among higher-income census tracts (Fuller, Thayer, Simon, & Brugge, 2019).

Concurrently, with the flux of suburbanites into the inner city, urban planning and development have pursued intentional strategies that might inadvertently increase traffic pollution exposures. Among the development strategies that we think could do this are "smart growth" strategies such as high-density development close to transportation. These design strategies seek to reduce driving time for commuters by concentrating housing near transportation corridors, which for many cities means roadways and fossil fuel-powered public transportation. While such a strategy could reduce regional air pollution, they may place more people, including some high-income populations, close to TRAP sources. Again, understanding local conditions and resources will be key to determine what type of development can provide benefits to local health. There is a need for greater attention of advocates and researchers to estimate the tradeoffs in exposure and health that result from such urban planning decisions.

Strategies to mitigate disproportionate exposures and health outcomes

Strategies to eliminate disparities in exposure to TRAP can be divided into primary and secondary groups. Primary mitigation would be strategies that would reduce TRAP exposure by reducing emissions and contact with sources. First and foremost, emissions could be reduced through improving gas mileage on diesel and gasoline-powered vehicles; encouraging the use of hybrid and electric vehicles; and promoting travel on cleaner fuel burning transit. It is important to ensure that additional burden is not displaced to new communities. A novel aspect of transportation inequalities internationally is the case in China. In China, there is an effort to increase the use of electric vehicles. One of the benefits of replacing combustion engines with electric vehicles is reduced emissions from vehicles, including eliminating tailpipe emissions. However, the use of electric vehicles could increase pollution at distant locations if the power source is, for example, coal-fired power plants. A study modeled the impact of increased electric vehicle use on pollution in cities where the vehicles would be mostly driven and the rural areas that host power plants. The results were that lower-income rural populations would likely experience increases in pollution exposure, while higher-income urban populations would experience reductions. The authors point to the obvious, although potentially slow to develop, solution of replacing fossil fuel combustion for electricity with renewables that present far less of an air pollution problem (Ji et al., 2015). Another strategy to proactively reduce exposure, although not always possible, would be choosing to site buildings distant from highways and busy roadways. Current guidelines for California restrict the construction of new schools within 500 ft (~150 m) of a major road or highway (Section 17213 of the California Education Code and section 21151.8 of the California Public Resources Code, 2003). By definition, primary mitigation is usually limited to new development, although cases of substantial redevelopment may allow for the upgrade or replacement of ventilation systems in ways that render them more protective (Brugge et al., 2015).

Secondary efforts seek to mitigate pollution for existing communities and are particularly important with respect to EJ considerations because most low-income housing built in the past has had few, if any, design features that protect residents from ambient pollution. To our knowledge, there has been little attention to the role that mechanical air handling systems might play in mitigating exposures in public housing. However, higher-end housing in multifamily buildings, especially larger developments, usually have forced air ventilation systems. Combined with a tighter building envelope and filters that can remove small particles from air, these buildings can reduce ambient exposures considerably (Stephens & Siegel, 2013). Inclusionary housing (low-income units in market-rate multifamily developments) can benefit from this approach. Unfortunately, resources for extending these approaches to single-family homes of lower-income residents are lacking at this time. Options for secondary mitigation of low-income housing are often limited by the financial resources required. At the more individual level, there is widespread interest in, and growing use of, freestanding filtration units as a means to reduce indoor particle exposures. Common uses for these filters are to reduce allergens or indoor particles and there is a substantial literature that suggests they have benefits with respect to acute respiratory reactions (Fisk, 2013). While some evidence is emerging that these filters might also be beneficial for combustion products, both those generated indoors and outdoors, the findings have been mixed and less than fully convincing (Brugge et al., 2017). There is a need for more, and better, studies of room-based filters in homes near ambient traffic exposures. Public funds might be needed for low-income residents to install this technology, including covering electrical costs.

Other approaches to reducing exposure at homes that are already close to highways or major roadways would be to add physical barriers between the housing and traffic sources. Noise barriers are increasingly common in the United States, although they are built to reduce sound propagation from the roadway, they can also reduce pollution at nearby residences (Gallagher et al., 2015). Orientation relative to prevailing wind direction, the height of houses around the barrier and other factors must be considered when assessing potential efficacy for pollution exposure reduction. Greening roads through tree barriers can also reduce exposures. There is evidence that barriers of trees and shrubs can reduce the concentrations of particulate matter and some gases in areas adjacent to major roadways. However, to be effective barriers, vegetation must extend to the ground, be sufficiently dense and evergreen. The EPA has established guidelines to facilitate the inclusion of vegetative barriers in planning and revitalizing roadways, as discussed in Chapter 17 (Baldauf, 2016).

Summary and conclusions

Over the space of several decades, EJ has grown from a nascent US-based movement to one that is recognized and promoted globally. While its origins lay in hazardous waste siting, today an appreciable emphasis is the disproportionate impact of transportation and transportation policy on communities of color and the poor. As we have described above, while there is some variation in findings, occasional contrary examples, and appreciable gaps in the literature, the bulk of evidence points toward traffic on highways and major roadways disproportionately affecting racial/ethnic minority and lower SES communities. Accordingly, we would contend that there is a pressing need to both mitigate adverse impacts of TRAP with an emphasis on disadvantaged neighborhoods and to take into consideration social justice and equity in transportation and housing policy and planning.

Resources

Book

Bullard, R. D. (1990). *Dumping in dixie: Race, class, and environmental quality*. Westview, Boulder. Available for purchase at: http://amazon.com/Dumping-Dixie-Class-Environmental-Quality/dp/0813367921.

Bullard, R., Johnson, G., & Torres, A. (2004). *Highway robbery: Transportation racism and new routes to equity*. Boston: Sound End Press. Available for purchase at: http://amazon.com/Highway-Robbery-Transportation-Racism-Routes/dp/0896087042.

Caro, R. A. (1974). *The power broker: Robert Moses and the fall of New York*. New York: Knopf. Available for purchase at: http://amazon.com/Power-Broker-Robert-Moses-Fall/dp/0394720245.

Walker, G. (2012). *Environmental justice: Concepts, evidence and politics* (1st ed.). Routledge. Available for purchase at: https://www.routledge.com/Environmental-Justice-Concepts-Evidence-and-Politics-1st-Edition/Walker/p/book/9780415589741.

Documents

Exec. Order No. 12898, 59 C.F.R. 7629 (February 16, 1994). Accessible at https://www.archives.gov/files/federal-register/executive-orders/pdf/12898.pdf.

General Accounting Office (GAO) (1983). *Siting of hazardous waste landfills and their correlation with racial and economic status of surrounding communities.* Accessible at http://archive.gao.gov/d48t13/121648.pdf.

United Church of Christ Commission for Racial Justice (1987). *Toxic wastes and race in the United States.* Accessible at https://www.nrc.gov/docs/ML1310/ML13109A339.pdf.

United Church of Christ Justice and Witness Ministries (2007). *Toxic wastes and race at twenty.* Accessible at https://www.ucc.org/environmental-ministries_toxic-waste-20.

Environmental Protection Agency (EPA) (2016). *Recommendations for constructing roadside vegetation barriers to improve near-road air quality.* Accessible at https://cfpub.epa.gov/si/si_public_record_report.cfm?Lab=NRMRL&dirEntryId=321772&simpleSearch=1&searchAll=Recommendations+for+constructing+roadside+vegetation+barriers+to+improve+near+road+air+quality.

References

Babisch, W. (2014). Updated exposure-response relationship between road traffic noise and coronary heart diseases: A meta-analysis. *Noise & Health, 16*(68), 1–9.

Baldauf, R. (2016). *Recommendations for constructing roadside vegetation barriers to improve near-road air quality* (p. 13). Washington, DC: Office of Research and Development NRML, Air Pollution Prevention and Control Division.

Brugge, D., Durant, J. L., & Rioux, C. (2007). Near-highway pollutants in motor vehicle exhaust: A review of epidemiologic evidence of cardiac and pulmonary health risks. *Environmental Health, 6*, 23.

Brugge, D., Patton, A. P., Bob, A., Reisner, E., Lowe, L., Bright, O. J. M., et al. (2015). Developing community-level policy and practice to reduce traffic-related air pollution exposure. *Environmental Justice, 8*(3), 95–104.

Brugge, D., Simon, M. C., Hudda, N., Zellmer, M., Corlin, L., Cleland, S., et al. (2017). Lessons from in-home air filtration intervention trials to reduce urban ultrafine particle number concentrations. *Build and Environment, 126*, 266–275.

Bullard, R. D. (1994). *Dumping in Dixie: Race, class, and environmental quality* (2nd ed.). Boulder: Westview Press.

Bullard, R. D., Johnson, G. S., & Torres, A. O. (2004). *Highway robbery: Transportation racism & new routes to equity* (p. ix). Cambridge, MA: South End Press [245 pp.].

Caro, R. A. (1974). *The power broker: Robert Moses and the fall of New York* (1st ed.). New York: Knopf [ix, 1246, xxxiv pp.].

Clougherty, J. E., Kheirbek, I., Eisl, H. M., Ross, Z., Pezeshki, G., Gorczynski, J. E., et al. (2013). Intra-urban spatial variability in wintertime street-level concentrations of multiple combustion-related air pollutants: The new York City Community air survey (NYCCAS). *Journal of Exposure Science & Environmental Epidemiology, 23*(3), 232–240.

Cole, L. W., & Foster, S. R. (2001). *From the ground up: Environmental racism and the rise of the environmental justice movement* (p. xii). New York: New York University Press [244 pp.].

Durant, J. L., Ash, C. A., Wood, E. C., Herndon, S. C., Jayne, J. T., Knighton, W. B., et al. (2010). Short-term variation in near-highway air pollutant gradients on a winter morning. *Atmospheric Chemistry and Physics, 10*(2), 5599–5626.

EPA, U. S. (2018). Integrated science assessment (ISA) for particulate matter (external review draft). In *Agency*. Washington, DC: USEP.

Fisk, W. J. (2013). Health benefits of particle filtration. *Indoor Air, 23*(5), 357–368.

Fuller, C. H., Patton, A. P., Lane, K., Laws, M. B., Marden, A., Carrasco, E., et al. (2013). A community participatory study of cardiovascular health and exposure to near-highway air pollution: Study design and methods. *Reviews on Environmental Health, 28*(1), 21–35.

Fuller, C. H., Thayer, K., Simon, M. C., & Brugge, D. (2019). *An environmental justice analysis of ultrafine particle exposure in Boston, MA, USA*. Utrecht, The Netherlands: International Society of Environmental Epidemiologists.

Gallagher, J., Baldauf, R., Fuller, C. H., Kumar, P., Gill, L. W., & McNabola, A. (2015). Passive methods for improving air quality in the built environment: A review of porous and solid barriers. *Atmospheric Environment, 120*, 61–70.

General Accounting Office. (1983). *Siting of hazardous waste landfills and their correlation with racial and economic status of surrounding communities* (p. 25).

Hajat, A., Hsia, C., & O'Neill, M. S. (2015). Socioeconomic disparities and air pollution exposure: A global review. *Current Environmental Health Reports, 2*(4), 440–450.

HEI Panel on the Health Effects of Traffic-Related Air Pollution. (2010). *Traffic-related air pollution: A critical review of the literature on emissions, exposure, and health effects*. Boston, MA: Health Effects Institute.

Hu, S., Fruin, S., Kozawa, K., Mara, S., Paulson, S., & Winer, A. M. (2009). A wide area of pollutant impact downwind of a freeway during pre-sunrise hours. *Atmospheric Environment, 43*, 2541–2549.

Ji, S. G., Cherry, C. R., Zhou, W. J., Sawhney, R., Wu, Y., Cai, S.Y., et al. (2015). Environmental justice aspects of exposure to PM2.5 emissions from electric vehicle use in China. *Environmental Science & Technology, 49*(24), 13912–13920.

Karner, A. A., Eisinger, D. S., & Niemeier, D. A. (2010). Near-roadway air quality: Synthesizing the findings from real-world data. *Environmental Science & Technology, 44*(14), 5334–5344.

Kingham, S., Pearce, J., & Zawar-Reza, P. (2007). Driven to injustice? Environmental justice and vehicle pollution in Christchurch, New Zealand. *Transportation Research Part D: Transport and Environment, 12*(4), 254–263.

Lane, K. J., Kangsen Scammell, M., Levy, J. I., Fuller, C. H., Parambi, R., Zamore, W., et al. (2013). Positional error and time-activity patterns in near-highway proximity studies: An exposure misclassification analysis. *Environmental Health: A Global Access Science Source, 12*(1), 75.

Lane, K. J., Levy, J. I., Scammell, M. K., Patton, A. P., Durant, J. L., Mwamburi, M., et al. (2015). Effect of time-activity adjustment on exposure assessment for traffic-related ultrafine particles. *Journal of Exposure Science & Environmental Epidemiology, 25*(5), 506–516.

Mitchell, G., & Dorling, D. (2003). An environmental justice analysis of British air quality. *Environment and Planning A, 35*(5), 909–929.

Morello-Frosch, R., & Jesdale, B. M. (2006). Separate and unequal: Residential segregation and estimated cancer risks associated with ambient air toxics in U.S. metropolitan areas. *Environmental Health Perspectives, 114*(3), 386–393.

Morello-Frosch, R., Pastor, M., & Sadd, J. (2001). Environmental justice and Southern California's "riskscape": The distribution of air toxics exposures and health risks among diverse communities. *Urban Affairs Review, 36*(4), 551–578.

Patton, A. P., Zamore, W., Naumova, E. N., Levy, J. I., Brugge, D., & Durant, J. L. (2015). Transferability and generalizability of regression models of ultrafine particles in urban neighborhoods in the Boston area. *Environmental Science & Technology, 49*(10), 6051–6060.

Pearce, J. R., Richardson, E. A., Mitchell, R. J., & Shortt, N. K. (2010). Environmental justice and health: The implications of the socio-spatial distribution of multiple environmental deprivation for health inequalities in the United Kingdom. *Transactions of the Institute of British Geographers, 35*(4), 522–539.

Schweitzer, L., & Valenzuela, A. (2004). Environmental injustice and transportation: The claims and the evidence. *Journal of Planning Literature, 18*(4), 383–398.
Section 17213 of the California Education Code and section 21151.8 of the California Public Resources Code, (2003).
Stephens, B., & Siegel, J. A. (2013). Ultrafine particle removal by residential heating, ventilating, and air-conditioning filters. *Indoor Air, 23*(6), 488–497.
United Church of Christ. (1987). *Commission for racial justice. Toxic wastes and race in the United States: A national report on the racial and socio-economic characteristics of communities with hazardous waste sites* (p. xvi). New York, NY: Public Data Access: Inquiries to the Commission [69 pp.].
Verbeek, T. (2019). Unequal residential exposure to air pollution and noise: A geospatial environmental justice analysis for Ghent, Belgium. *SSM—Population Health*, 7, 100340.
Walker, G. (2012). *Environmental justice: Concepts, evidence and politics*. Routledge.
Williams, D. R., Lawrence, J. A., & Davis, B. A. (2019). Racism and health: Evidence and needed research. *Annual Review of Public Health, 40*, 14.1–14.21.

CHAPTER 21

Emerging transportation technologies and implications for traffic-related emissions, air pollution exposure, and health

Shams Tanvir, Peng Hao, and Kanok Boriboonsomsin
College of Engineering—Center for Environmental Research and Technology, University of California at Riverside, Riverside, CA, United States

Abbreviations

AERIS	applications for the environment: real-time information synthesis
BEV	battery electric vehicle
CACC	cooperative adaptive cruise control
CAV	connected and automated vehicle
CO	carbon monoxide
CO_2	carbon dioxide
CV	connected vehicle
C-V2X	cellular vehicle-to-everything
DSRC	dedicated short range communications
EAD	eco-approach and departure
EF	emissions factor
EV	electric vehicle
FCEV	fuel cell electric vehicle
GHG	greenhouse gases
HC	hydrocarbons
IEA	International Energy Agency
NHTS	national household transportation survey
NOx	nitrogen oxides
ODM	on-demand mobility services
PHEV	plug-in hybrid electric vehicle
SAE	society of automotive engineers
SPaT	signal phase and timing
TNC	transport network companies
V2I	vehicle-to-infrastructure
V2V	vehicle-to-vehicle
VMT	vehicle miles traveled
ZEV	zero emission vehicle

Traffic-Related Air Pollution
https://doi.org/10.1016/B978-0-12-818122-5.00021-1

Introduction

The transportation sector is going through a major revolution with the emergence of three potentially transformative technologies—connectivity and automation, electrification, and shared mobility (Fulton, Mason, & Meroux, 2017). Connectivity and automation together are the driving force for many applications that promote safety and efficiency of transportation systems (McGuckin, Lambert, Newton, Pearmine, & Hubbard, 2017). Electrification of transportation systems involves no direct combustion of fossil fuel onboard, and instead relies on the electrical grid for fuel. Shared mobility is achieved through sharing a single mode of travel among multiple travelers. In addition to the safety and mobility benefits, energy efficiency improvement, and emissions reduction are often touted as co-benefits from connected and autonomous vehicles (Rios-Torres & Malikopoulos, 2017). However, proper estimation of emissions reduction benefits is difficult because of uncertainties regarding the final form of technology as well as end-user adoption and adaptation rates. Similar uncertainties exist in the estimation of emissions reduction benefits of electrification and shared mobility.

Traffic-related emissions are a function of total vehicle miles traveled (VMT) and emission factors for the particular pollutant per unit distance. Once released into the air, these emissions are dispersed and exposed by the nearby population, potentially affecting their health. The level of exposure to traffic-related emissions and any consequent health impacts depends on several factors such as proximity to major roadways, meteorological conditions, and demographics. Any technological change which increases VMT and/or induces a condition that results in a higher emission factor will lead to an increase in traffic-related emissions. Emission factors depend not only on the specific vehicle under consideration but also on the traffic and environmental conditions under which the vehicle operates. For example, emission factors are often represented as a function of average traffic speed on the roadway link under consideration. Typically, the function has a "U" shape with higher emission factors at very low and very high average speeds.

Several previous studies have attempted to estimate energy savings potential of the three emerging technologies (Bauer, Greenblatt, & Gerke, 2018; Brown, Gonder, & Repac, 2014; Wadud, MacKenzie, & Leiby, 2016; Wenzel, Rames, Kontou, & Henao, 2019). However, energy savings does not always translate to emissions reduction. Fig. 21.1 illustrates how a technology that increases average traffic speed from 25 to 35 mph would reduce total energy consumption but increase nitrogen oxides (NO_x) emissions.

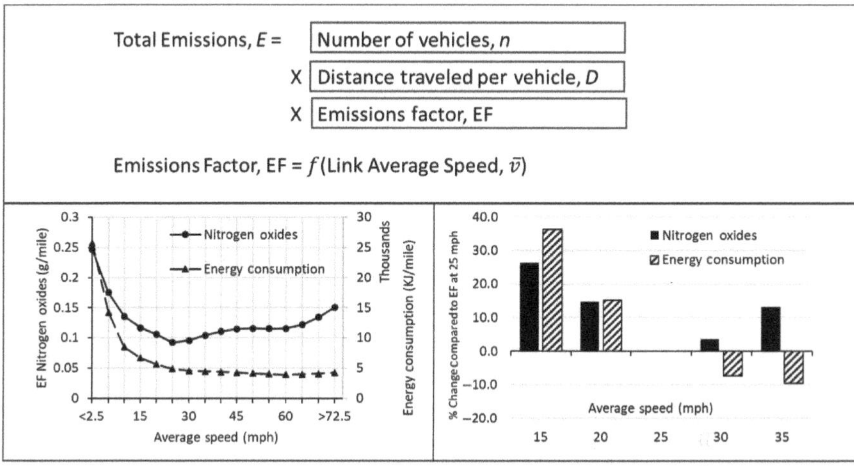

Fig. 21.1 Estimation of traffic-related emissions. Emissions factors are taken from MOVES 2014b model for a typical weekday in summer 2019 for Riverside County, CA.

In this chapter, we will:

- describe the three emerging transportation technologies,
- review adoption patterns of these technologies where available,
- discuss implications of these technologies for traffic-related emissions, air pollution, exposures, and health,
- analyze potential or documented unintended consequences of these technologies including the spatial and temporal shifts of emissions, air pollution, exposures, and adverse health impacts, and
- examine implementation considerations and contingencies for these technologies.

Emerging transportation technologies

We have categorized the emerging transportation technologies into three groups-connected and automated vehicles, on-demand mobility services, and electric and low emission vehicles. The following subsections briefly review the three technologies.

Connected and automated vehicles

The US personal mobility market is valued at around 1 trillion dollars (Walker & Johnson, 2016). In recent years, the promises of sensing, communication, and artificial intelligence technologies have attracted billions of dollars in investments to develop connected and automated vehicles. These investments were predominantly made by 'Silicon Valley' companies, and almost all the traditional automotive manufacturers have since followed suit.

The Society of Automotive Engineers (SAE) defines five levels of driving automation in its J3016 taxonomy (SAE international, 2016). The levels increase from 0 to 5 with a higher level representing lesser intervention and attention from the human driver. Level 3 to level 5 automation, according to SAE, represents automated driving features such as traffic jam chauffeur, local driverless taxi, and steering wheel-free driving.

Connectivity allows for the exchange of information between vehicle and infrastructure. Information exchange among vehicles termed as vehicle-to-vehicle (V2V) communication has enabled several applications in the safety, mobility, and environmental domains. Examples of such applications are forward-collision warning (safety), advanced traveler information (mobility), and eco-driving (environment). Vehicle to infrastructure (V2I) communication has capacitated additional applications such as red-light violation warning (safety), transit signal priority (mobility), and eco-approach and departure (environment). The US National Highway Traffic Safety Administration proposed the inclusion of V2V connectivity in all new-light duty vehicles and trucks in 2016 (U.S. National Highway Traffic Safety Administration, 2016).

On-demand mobility services

On-demand mobility services involve arranging for travel on-demand. Usually, the travelers request their "rides" on automobile or paratransit using a smartphone app. The service provider typically manages a fleet of vehicles and optimizes the vehicles' locations and route assignments based on the dynamic demand of subscribers. Greenblatt and Shaheen (2015) classified on-demand mobility services in four major categories—transport network companies (TNCs), carsharing, ridesharing, and e-hailing.

TNCs such as Uber and Lyft match drivers who subscribe to their platform with trip requests from individuals. Most of the trip requests are from single travelers; therefore, the majority of the TNC trips are not shared rides. For example, according to a recent study in Boston, the average occupancy rate for TNC trips is 1.5, including the driver (MAPC, 2018). To promote ridesharing, major TNCs offer shared ride service options such as UberPOOL and Lyft Line in major cities. However, only a small fraction of users who have access to these shared-ride options request them, and even a smaller number of those trip requests are matched as shared rides.

Carsharing is subscription-based mobility where subscribers pay a short-term fee. Zipcar is an example of typical carsharing services. On the other hand, the traditional carpooling and vanpooling fall under the ridesharing category. Ridesharing through microtransit services such as Via, Chariot, and Flex is considered a potential mobility solution in highly dispersed urban areas. *E*-hailing services match traditional for-hire taxis to the nearest passengers using apps like TNCs. As in the case of TNCs, these on-demand mobility services only reduce the total number of trips when shared rides happen.

Although on-demand mobility services include other modes such as bicycle and scooter, we have purposefully limited the discussion in this chapter to exclude those modes. We assume that mobility decisions taken by the users of those modes will not significantly impact vehicle emissions (e.g., through mode shift). In addition, TNCs will be a focal point of our discussion regarding the impacts of on-demand mobility services because the current market share of all the other categories combined is inappreciable compared to that of TNCs.

Zero-emission vehicles

In recent years, the impetus for zero-emission vehicles came from the climate targets adopted in the 2015 Paris Agreement (Rogelj et al., 2016). According to the pact, participating countries have agreed to limit the rise of the global temperature well below 2°C above the preindustrial level. To achieve this goal, the International Energy Agency (IEA) recommends at least 20% of all on-road vehicles be driven electrically by 2030 in addition to many other measures. Several countries in the European Union and Asia have started to act on the IEA recommendation. In the United States, California is leading the push to achieve low emission mobility through a state regulation called the zero-emission vehicle (ZEV) program. The ZEV program, currently followed by nine other states in the United States, operates by issuing ZEV credits to automakers for selling plug-in hybrid electric, battery electric, and fuel cell electric vehicles. The mandate limits the sale of non-ZEV vehicles based on total ZEV credits accumulated (either by ZEV sale or trade). Complementary to the increasing affordability and range of battery technologies, these regulations and policies are nudging the adoption of electric vehicles.

Vehicles classified as ZEV are technologically diverse. Plug-in hybrid electric vehicles (PHEVs) use electric motor(s) in addition to the conventional internal combustion engine to propel the vehicle. PHEVs can be plugged into the electrical grid so that their batteries can be recharged. In contrast, battery electric vehicles (BEVs) run entirely on the electric motor(s), which rely on electricity from batteries. Fuel cell electric vehicles (FCEVs) use electricity generated from compressed hydrogen gas to run electric motor(s). Similar to gasoline-powered vehicles, FCEVs can be replenished with hydrogen gas at a refueling station and have a better range than BEVs.

Interactions and synergies among technologies

Although the motivations for the aforementioned technologies are different, the interactions and synergies among them provide us with unique opportunities and challenges. For example, the development of shared autonomous vehicles is heavily pursued by TNCs as more than 50% of the overall expense in the current human driver-dependent operation go toward paying driver wages (Anair, 2017). In addition, increased flexibility of shared autonomous vehicles makes it possible to devise more efficient routing, which in turn may increase the potential for shared rides.

As another example, connected and automated vehicles (CAVs) depend on a range of onboard sensing and computational equipment requiring robust electricity supplies, which are already available in electric vehicles. Automation, in turn, can help electric vehicles achieve a more responsive and optimized operation while connectivity can help in the management of charging activities.

On-demand mobility services typically operate longer hours; therefore, shared electric vehicles are still technologically challenged under current electric vehicle range limits. TNC drivers would also be subject to higher opportunity costs for the time spent charging electric vehicles. Considering 5-year total cost of ownership with no electric vehicle incentives, it was estimated that hybrid vehicles would result in the lowest cost of operation for TNC drivers (Slowik, Pavlenko, & Lutsey, 2019). However, the ultimate synergy can be achieved in a shared autonomous electric vehicle. Autonomy can solve several practical issues in shared and electric fleet operations simultaneously such as empty haul period, occupancy, charging times, and locations.

Adoption and adaptation

The emerging technologies mentioned in Section "Emerging transportation technologies" are in different stages of development. Adoption of and adaptation to these technologies are evolving. In this section, we discuss the current adoption rates of these technologies as a means to understand the extent of their impacts.

There has been rapid advancement in automated vehicle technology in recent years. Google's driverless car company, Waymo, released a limited trial of self-driving taxi service called "Waymo One" in Phoenix, Arizona on December 5, 2018 (Krafcik, 2018). Uber was testing its autonomous taxi fleet in Pittsburgh, Pennsylvania (Shepardson, 2018). Ford recently announced that it would start production of steering-wheel-free autonomous vehicles by 2021 (Ford Motor Company, 2018). Several other companies (e.g., Apple, GM, Tesla, and Volkswagen) all have SAE level 4 or level 5 technologies close to being ready for market deployment in the near future.

Connected vehicle technology, on the other hand, is already available. In October 1999, the US Federal Communication Commission allocated 75 MHz spectrum in the 5.9 GHz band to be used for traffic-related applications (Kenney, 2011). Dedicated Short Range Communications (DSRC) operate in this band and have several advantages such as low latency, limited interference, and robust performance in adverse weather conditions. The vehicle-to-vehicle (V2V) communication with DSRC was commercially deployed in Cadillac CTS 2017 interim model year sedans. While several vehicle manufacturers are actively working on integrating DSRC into their vehicles at the production level, a newer communication technology namely, cellular vehicle-to-everything (C-V2X) is currently under development using 5G protocol. Volkswagen has announced to include C-V2X in all of their future autonomous vehicles. The US Department of Transportation has three connected vehicle pilot deployment programs in Wyoming, Florida, and New York to test the operational capabilities of connected vehicle technology and promote partnerships among stakeholders.

The National Household Travel Survey (NHTS) is the only nationwide data available on the usage of different travel modes including on-demand mobility services in the United States. Among the 92,400 reported trips in the 2016–2017 NHTS survey, 3463 were classified as "Taxi/Limo." The 2017 NHTS survey also indicated that people used their private vehicles less compared to previous years, and instead used ride-hailing options more. According to Lyft submitted a prospectus to the US Securities and Exchange Commission (Lyft Inc., 2019), their number of annual rides increased from 162 million in 2016 to 619 million in 2018. A recent article (Gessner, 2019) reported the February 2019 TNC market share of Uber, Lyft, and others at 67.3%, 30.3%, and 2.4%, respectively. Thus, the total estimated TNC rides for 2018 in the United States is about 2.04 billion.

According to the IEA published report (Bunsen et al., 2018), in 2017 the global electric car stock exceeded 3 million vehicles, and the total number of electric buses globally increased to 370,000. The United States had 25% of the total number of electric vehicles (EVs) globally. However, the EV market share of new vehicle sales in the United States was only 1.2%. China is adopting EVs at a rate that surpasses other countries by a large margin; more than 50% of the new electric vehicle sales were in China in 2017. BEVs makeup about 65% of the global EV stock in 2017, while the total number of EV charging outlets reached 3.5 million with more than 85% of those being privately owned.

Early adopters of EVs, at least in Sweden and Germany, tend to be predominantly male, young, wealthy, and college-educated (Langbroek, Franklin, & Susilo, 2017; Plötz, Schneider, Globisch, & Dütschke, 2014). Spurlock et al. (2019) surveyed 1094 people in San Francisco and found that people with household income more than $200,000 are 7% and 13% more likely to adopt BEV and PHEV, respectively, compared to people with household income less than $75,000. The same survey also found ZEV adoption rate to be negatively correlated with the population density of an area.

Implications

Uncertainties regarding the final form of the technologies and their adoption make it a speculative process to project the implications. On one hand, there are champions for these technologies promoting different benefits including a reduction in emissions. On the other hand, some critics are apprehensive of the "rebound effect" of these technologies, and caution that the net benefits may diminish or flip toward negative. In this chapter, we identify potential impacts resulting from each of the three categories of emerging technologies. We limit the identification of these impacts in the realm of their potential to change the level of traffic-related emissions and the equitable distribution of the emissions.

Methodology for classifying impacts

The traffic-related emissions and the associated air quality, exposure, and health impacts ensue from the underlying need for the movement of people and goods. Some of the potential changes in emissions due to the adoption of emerging technologies happen indirectly, at the very root of how people decide to make trips. Other changes affect the level of emissions and their distribution more directly. In this chapter, we categorize the potential changes as follows:

1. *Change in Exposure Time*: Exposure time, in the context of this chapter, refers to the amount of time people spend in the microenvironments impacted by traffic-related emissions. Any changes which increase exposure are likely to increase health-related impacts as well. Increased exposure may or may not be a direct result of adopting the technologies.
2. *Change in Vehicle and Fuel Technologies*: Any increase or decrease in emissions resulting from a change in the vehicle and/or fuel technologies is likely to impact public health in the same direction, given that there are no changes in exposure time. Changes in these technologies may be different geographically and across time.
3. *Change in Short-Term Travel Activity*: Changes in short-term travel-related activity resulting from the adoption of emerging technologies may include trip-specific decisions such as driving more or less, sharing a ride, choosing a different departure time, and adopting a more efficient driving style. They do not include long-term travel-related decisions such as mode choice for the daily commute and residential location choice.
4. *Change in Long-Term Travel Behavior*: In addition to the short-term activity changes, the emerging technologies may also induce lifestyle changes that impact travel behavior such as foregoing vehicle ownership, moving to a new location, and making new types of trips. Such long-term changes may have a lesser immediate effect as compared to the other three categories, but the magnitude of the effect, in the long run, may be significant depending on the technology and the type of behavior influenced.

In the following subsections, we review potential benefits and negative consequences of the three emerging technologies from various studies, with a focus on their effect on emissions and the distribution of emissions.

Connected and automated vehicles

Connected vehicle (CV) technology, which enables V2V and V2I communications, has demonstrated potential in enhancing safety, mobility, and energy efficiency as well as in reducing vehicular emissions. With "Cleaner Air through Smarter Transportation" as its vision, the applications for the environment: real-time information synthesis (AERIS) program was initialized in 2009 to investigate eco-friendly transportation operations in CV environment (U.S. Department of Transportation, 2016). Five operational scenarios, including Eco-Signal Operations, Eco-Lanes, Low Emissions Zones,

Eco-Traveler Information, and Eco-Integrated Corridor Management, were defined in AERIS. Eighteen CV applications were proposed to reduce traffic-related fuel consumption and emissions. Among those applications, Eco-Approach and Departure (EAD) at Signalized Intersections and Cooperative Adaptive Cruise Control (CACC) have been furthest in research and development, as they can show significant efficiency improvements under relatively low technology penetration rates.

The EAD application aims to guide the vehicle to travel through signalized intersections in an eco-friendly manner using signal phase and timing (SPaT) and map information broadcasted by the traffic signal controllers. The hardware system of EAD consists of a DSRC modem to receive SPaT information, an onboard computer to calculate recommended driving speed, and a computer screen to display the recommended driving speed to the driver. In the AERIS program, field tests of the EAD application were conducted in both Riverside, California, and McLean, Virginia, to demonstrate its benefits in terms of energy savings and emissions reduction (Xia, 2014). For the test in Riverside with a 2008 Nissan Altima, the test vehicle achieved 11%–28% carbon dioxide (CO_2) reduction with the help of the EAD application. For the test in McLean with a 2008 Jeep Grand Cherokee, the fuel savings was between 2.5% and 18%. Taking it a step further, in the "GlidePath" project (Altan, Wu, Barth, Boriboonsomsin, & Stark, 2017), the EAD application was integrated with vehicle automation so that the test vehicle automatically followed the recommended driving speed as calculated by EAD. The experiment results showed an average CO_2 reduction of 17% for the test vehicle.

In 2015, another EAD field test was conducted in real-world traffic along the El Camino Real corridor in California (Hao, Wu, Boriboonsomsin, & Barth, 2018). A graphical user interface of the system was designed to provide traffic signal countdown, recommended driving speed, and the presence of preceding vehicle (if any) information to the driver in real-time. For comparison, an uninformed driver in a similar vehicle traveled through the same set of intersections in the adjacent lane simultaneously. The results showed that the EAD application, when activated, reduced CO_2, carbon monoxide (CO), hydrocarbons (HC), and NO_x emissions by 6%, 32%, 30%, and 24%, respectively.

For CACC, most of the research and development has been focused on the mobility improvement resulting from vehicles traveling in a tightly spaced platoon, but there is still some research that investigates its environmental impact. In the partially automated truck CACC platooning test

(Shladover et al., 2018), the results showed 6%–11% fuel savings for the following trucks in the platoon, which also indicates that CO_2 reduction would also be within this range. The test also revealed the synergy between the short following distance and the aerodynamic improvement strategies, as the combined strategy outperformed the sum of the reductions from each individual strategy. Another research showed additional energy savings and emissions reduction from an eco-friendly CACC algorithm over the conventional CACC algorithm. Specifically, CO_2, CO, HC, and NO_x, emissions could be further reduced by 2.2%, 17.0%, 6.7%, and 3.0%, respectively, if eco-friendly CACC was applied (Wang, Wu, Hao, Boriboonsomsin, & Barth, 2017).

While the real-world testing and demonstration of CAVs to date has shown potential for emissions reduction in the near term, primarily as a result of improved vehicle and traffic operational efficiency, the long-term impacts of these technologies on traffic emissions are unclear as there is no real-world evidence at this time. One speculative long-term impact is that the expanded and inexpensive mobility provided by CAVs may cause increased trip frequency and, consequently, higher VMT. Wadud et al. (2016) estimated VMT increase of 4%–60% due to the reduced cost of drivers' time. Meyer, Becker, Bösch, and Axhausen (2017) suggested that autonomous vehicles could increase accessibility in suburban areas by 10%–14%. Such an increase in accessibility may translate to urban sprawl and result in increased trip length. These uncertain yet potentially significant impacts of CAVs may result in an overall increase in emissions in the long run.

In addition, CAVs are expected to provide a new form of mobility to certain population groups that have largely been travel-restricted. These include, for example, children, the elderly, and people with a medical condition. Harper, Hendrickson, Mangones, and Samaras (2016) estimated that those population groups could increase the total VMT in the United States by 14%. Such an increase in travel activity could also mean an increase in the amount of time that these vulnerable population groups are exposed to traffic-related air pollution.

On-demand mobility services

On-demand mobility services, especially those offered by TNCs, operate on a premise that people will not own vehicles and rather use these services to meet their travel needs. In the case of TNCs, users often spend time by curbside waiting for the vehicles to pick them up. This could translate to an increase in the amount of time spent in the microenvironment with a higher level of traffic-related air pollution as compared to using personal vehicles. The extent of the increase in exposure to traffic emissions will depend on a number of factors such as the frequency of service use, the pick-up locations, and the level of traffic emissions at those locations.

Current TNC operation is primarily based on existing vehicle and fuel technologies and is expected to remain the same in the near future. However, TNCs are likely to be early adopters of fuel-efficient and alternative fuel vehicles as fuel cost accounts for a significant portion of their total operating costs. Hu, Dong, Lin, and Yang (2018) studied the potential of replacing New York City taxi trips with TNC services. They found that 94% of the TNC trips required would be shorter than 200 miles. Considering the range of current electric vehicles in the market, TNC drivers can practically switch to electric vehicles.

On-demand mobility services that do not convert nonmotorized and transit trips to vehicle trips and that promote ridesharing are expected to reduce the total VMT on the roadways, which would result in a reduction in emissions. Alonso-Mora, Samaranayake, Wallar, Frazzoli, and Rus (2017) compared 14,000 New York City taxi trips with a hypothetical ride-sharing operation. They found that 98% of the total taxi trips could be replaced by 3000 ride-share vehicles with a capacity of four passengers per vehicle or 2000 vehicles with 10 passengers per vehicle. Bischoff and Maciejewski (2016) simulated trips from 1.1 million personal vehicles in the city of Berlin, Germany, and found that a fleet of 100,000 shared autonomous vehicles would be able to serve those trips sufficiently. In these simulation studies, no barriers to the adoption of ride-sharing were assumed and, therefore, they represent an ideal scenario.

Evidence of the real-world impacts of TNCs, especially on VMT and congestion, has begun to emerge recently. In the short term, TNCs can incur a significant amount of "deadhead" miles, which are the miles where the driver is driving without any passenger, such as from the drop-off location of one trip to the pick-up location of the next trip (Wenzel et al., 2019). A recent paper (Komanduri, Wafa, Proussaloglou, & Jacobs, 2018) using RideAustin data estimated that 37% of all the miles are due to deadheading. It was estimated to be around 20% in San Francisco (Castiglione et al., 2018) and about 50% in New York City (Schaller, 2018). Castillo, Knoepfle, and Weyl (2017) presented a condition of matching failure by TNCs where there is a fewer number of available vehicles nearby than the trip requests. Such failure may contribute to a substantially increased amount of deadhead miles. The number of these empty passenger trips can be reduced significantly by imposing adaptive matching radii within which the driver supply and trip demand matching must occur (Xu,Yin, & Ye, 2019).

In longer-term, TNCs may complement public transit by providing first and last mile services, or decrease transit ridership by competing with it. Hall, Palsson, and Price (2018) found that transit ridership decreased by 5.9% in smaller cities but increased by 0.8% in larger cities due to the entry of Uber. Clewlow and Mishra (2017) surveyed more than 4000 residents in seven large cities in the United States and found that 15% of public transit, 17% walk, and 7% bike trips were replaced by TNC trips, and that induced trips—trips which would not have happened if TNC service had not been there—consisted of 22% of the total TNC trips. A similar study by Henao and Marshall (2018) conducted in Denver, Colorado, found that public transit trips replaced by TNCs and induced TNC trips were 22% and 12%, respectively.

Evidence to date has pointed in the direction that TNCs, in the way they are operated and used currently, increase VMT, which has also resulted in increased traffic congestion in some areas. As a result, traffic-related emissions will likely also increase. However, it should be noted that any increase in VMT caused by TNCs will not increase traffic emissions if those miles are zero-emission miles. And if TNCs' zero-emission miles can displace some of the personal vehicles' gasoline miles, there can be a net reduction in total traffic emissions. For example, Bauer et al. (2018) showed that a centrally managed shared autonomous electric taxi fleet in Manhattan Island of New York City would yield 58% reduction in energy consumption and 73% reduction in GHG emissions as compared to an autonomous fleet of conventional gasoline vehicles.

In terms of equity point of view, there is a large variation in the usage of TNCs among the population in different income groups. According to the 2017 National Household Travel Survey (NHTS) in nine major metropolitan areas, people with annual household income more than $200,000 made 44 TNC trips per year, whereas people with annual household income less than $15,000 made only six TNC trips. Thus, the increase in traffic emissions generated by TNCs is contributed more by high-income population. It is unclear, however, how the air quality and health impacts of those increased emissions are distributed in the environmental justice context.

Zero-emission vehicles

ZEVs represent changes in both vehicle and fuel technologies from the conventional vehicle with an internal combustion engine fueled by gasoline. Replacement of the internal combustion engine with electric motor(s) results in a more energy-efficient vehicle powertrain. For instance, BEVs convert about 59%–62% of energy from the grid to power the wheels whereas gasoline vehicles only convert 17%–21% of energy from the fossil fuel (Chae et al., 2011; Miller, Holmes, Conlon, & Savagian, 2011). In terms of emissions, BEVs produce lower brake wear emissions due to the regenerative braking technology and do not produce any tailpipe emissions. However, emissions at power plants may increase as a result of increased demand for electricity to serve BEVs (Nopmongcol et al., 2017). This means emissions are simply shifted from where BEVs are driven to the power plants that generate electricity for these BEVs. However, that will not be the case if that electricity is generated from renewable sources such as wind and solar. As of 2015, driving an average BEV would result in lower GHG emissions than driving a gasoline vehicle in regions of the United States that cover two-thirds of the population combined (Nealer, Reichmuth, & Anair, 2015).

In the life cycle of a BEV that includes its manufacturing and disposal, other important sources of emissions are the production of the battery and the vehicle itself (Tessum, Hill, & Marshall, 2014). Weis, Jaramillo, and Michalek (2016) conducted a study comparing BEVs and conventional gasoline vehicles and found that BEVs have higher sulfur dioxide but lower CO and volatile organic compound life-cycle emissions. Based on a similar comparative analysis, Nealer et al. (2015) found that, on average, BEVs produce less than half the GHG emissions of comparable gasoline-powered vehicles.

Summary of impacts

The different impacts of the three emerging technologies on emissions as discussed in the previous subsections are summarized in Table 21.1. Although there are many uncertainties in the different types of change, we have used the evidence that has been reviewed along with our best judgment to determine the likelihood, directionality, and magnitude of emissions impact. Three categories of likelihood are used—unlikely, somewhat likely, and extremely likely. The directionality is positive if emissions are expected to increase, and negative if the technology is expected to reduce emissions. Three magnitudes of emissions impact are postulated—low, moderate, and high.

Implementation considerations

The emerging transportation technologies discussed in this chapter have the potential to greatly reduce traffic-related emissions, exposure to these emissions, and the associated health burdens. However, there could be unintended

Table 21.1 Anticipated impacts of emerging technologies on emissions and exposure (Likelihood[a], Directionality[b], and Magnitude[c] of Impact).

Changes in	Emerging technologies								
	CAV[d]			ODM[e]			ZEV[f]		
	L	D	M	L	D	M	L	D	M
Exposure time	◑	⇧/⇩	Low	◑	⇧	Low	○	N/A	N/A
Vehicle/fuel technology	●	⇩	Moderate	◑	⇩	Low	●	⇩	High
Short term activity	●	⇩	High	◑	⇧	Low	○	N/A	N/A
Long term behavior	◑	⇧	High	◑	⇧	Moderate	◑	⇧	Low

[a]Likelihood is noted by L: Unlikely - ○, Somewhat likely - ◑, Extremely likely - ●

[b]Directionality is noted by D: Increase – ⇧, Decrease – ⇩, N/A – Not applicable

[c]Magnitude is noted by M: Low - (Low), Moderate- (Moderate), and High - (High), N/A – Not applicable

[d]CAV – Connected and autonomous vehicles
[e]ODM – On-demand mobility services
[f]ZEV – Zero emission vehicles

consequences that lead to the opposite outcome as well. Transportation policies should be developed to maximize the benefits of these technologies and minimize the effects of unintended consequences.

Emissions-related benefits of CAVs come mostly as a by-product of more efficient vehicle and traffic operations. However, measures and policies need to be put in place to prevent potential negative consequences of CAVs such as increasing VMT. These include, for instance, measures to incentivize shared rides of CAVs and to restrict empty passenger trips. Transportation agencies may provide incentives to companies that use CAVs providing first-mile and last-mile connections with their mass transit systems. Although a city with fully automated vehicles could mean no need for parking spaces, the total removal of public parking facilities would create excess empty passenger trips and tendency to run errands using personal CAVs. To restrict these conditions, city planners may allocate spaces where CAVs could be parked when not in use and allow for system-optimal reorganization. The connectivity in CAVs can be used to enable a variety of pricing mechanisms such as mileage-based charges, cordon pricing, and even emission pricing based on onboard emissions sensing and reporting. Such data transactions will also improve the transparency of CAV operations, allowing any spatiotemporal disparity in emission impacts from CAVs to be analyzed. Regulators can then use that information to equitably distribute those emission impacts to ensure environmental justice for all communities.

Promotion of ridesharing and public transportation are the two most effective measures to reduce the negative consequences of on-demand mobility services. Imposing restrictions on single passenger TNC trips can help reduce VMT generated by TNCs. Using on-demand mobility services to provide first-mile and last-mile connections with public transit systems can help promote the use of public transit. On the other hand, multimodal facilities should be modified or upgraded to better accommodate TNC pick-ups and drop-offs. Similarly, better curb space management is needed in city centers and major trip destinations to reduce traffic congestion and idle emissions in those areas, as well as to reduce the wait time where travelers are exposed to traffic-related air pollution.

Affordability of ZEVs is vital in promoting their adoption. Tax credits and rebates have been effective tools for encouraging the adoption of light-duty ZEVs in California, Norway, and China (Stephens, Zhou, Burnham, & Wang, 2018). Similar policy-level pushes in medium-duty and heavy-duty sectors can help promote the adoption of ZEVs also in these sectors, and further reduce emissions from on-road transportation. Other forms of benefit such as access to high-occupancy vehicle lanes and public refueling stations can also help make ZEVs more attractive. Policies requiring the increasing share of renewable energy sources in electrical grids will reduce emissions at power plants, and help ensure that the life-cycle emissions of ZEVs are lower than their conventional counterparts.

It is important that, in designing measures and policies in response to these emerging transportation technologies, careful considerations are given to the synergies and trade-offs between these technologies. A set of measures and policies that jointly promote ridesharing in ZEVs that are operated highly efficiently as a result of their connected and automated capabilities will lead to a future with less concern about traffic-related emissions and their associated health impacts.

Conclusions

CAVs, OMDs, and ZEVs are emerging technologies that can significantly affect the amount of traffic-related emissions, and consequently, the level of exposure to these emissions and the associated health effects. In this chapter, we have reviewed published work and evidence to date to qualitatively assess the likelihood, directionality, and magnitude of the anticipated impacts of these technologies.

From the vehicle and fuel perspective, all three technologies would shift the vehicle population toward more electrification, which will reduce the overall tailpipe emissions. While the real-world testing and demonstration of CAVs to date has shown emissions reduction in the short term, primarily as a result of improved vehicle and traffic operational efficiency, some studies have pointed to the potential for an increase in traffic-related emissions in the long-term as a result of increased VMT from induced travel demand and certain population groups that have largely been travel-restricted. For OMDs, evidence has recently begun to emerge that TNCs result in increased VMT and traffic congestion in some cities due to deadheading as well as replacing walk and bike trips. On the other hand, it is inconclusive on whether TNCs complement public transit by providing first and last mile services or decrease transit ridership by competing with it, although more data points support the latter. Lastly, while ZEVs have no tailpipe emissions, there may still be emissions from the generation of electricity to fuel these vehicles. Thus, for these emerging technologies to have positive impacts on traffic-related emissions, exposure, and health, it is important to promote strategies that curb an unnecessary increase in VMT from CAVs, promote ridesharing in OMDs, and generate electricity for ZEVs from renewable sources.

Compared to the changes in vehicle and fuel technology, short-term activity, and long-term behavior, the impacts of the three emerging technologies on the exposure time and environmental justice have been given less attention. More research should be directed at studying changes in the exposure to and distribution of traffic-related air pollution as a result of adopting these technologies. In addition, targeted policies or countermeasures should be developed and implemented to prevent undesirable long-term behavioral changes induced by these technologies.

References

Alonso-Mora, J., Samaranayake, S., Wallar, A., Frazzoli, E., & Rus, D. (2017). On-demand high-capacity ride-sharing via dynamic trip-vehicle assignment. *Proceedings of the National Academy of Sciences*, *114*(3), 462–467.

Altan, O. D., Wu, G., Barth, M. J., Boriboonsomsin, K., & Stark, J. A. (2017). GlidePath: Eco-friendly automated approach and departure at signalized intersections. *IEEE Transactions on Intelligent Vehicles*, *2*(4), 266–277.

Anair, D. (2017). *Capturing the climate benefits of autonomous vehicles.* UC Davis Institute of Transportation Studies.

Bauer, G. S., Greenblatt, J. B., & Gerke, B. F. (2018). Cost, energy, and environmental impact of automated electric taxi fleets in Manhattan. *Environmental Science & Technology, 52*(8), 4920–4928.

Bischoff, J., & Maciejewski, M. (2016). Simulation of city-wide replacement of private cars with autonomous taxis in Berlin. *Procedia Computer Science, 83*, 237–244.

Brown, A., Gonder, J., & Repac, B. (2014). An analysis of possible energy impacts of automated vehicles. In *Road vehicle automation.* Cham: Springer. (pp. 137–153).

Bunsen, T., Cazzola, P., Gorner, M., Paoli, L., Scheffer, S., Schuitmaker, R., et al. (2018). *Global EV Outlook 2018: Towards cross-modal electrification.* International Energy Agency.

Castiglione, J., Cooper, D., Sana, B., Tischler, D., Chang, T., Erhardt, G., et al. (2018). *TNCs & congestion.* Draft report San Francisco County Transportation Authority.

Castillo, J. C., Knoepfle, D., & Weyl, G. (2017, June). Surge pricing solves the wild goose chase. In *Proceedings of the 2017 ACM conference on economics and computation.* (pp. 241–242): ACM.

Chae, H. J., Kim, W. Y., Yun, S. Y., Jeong, Y. S., Lee, J. Y., & Moon, H. T. (2011). *3.3 kW on board charger for electric vehicle.* In *8th international conference on power electronics-ECCE Asia.* (pp. 2717–2719): IEEE.

Clewlow, R. R., & Mishra, G. S. (2017). *Disruptive transportation: The adoption, utilization, and impacts of ride-hailing in the United States.* Davis, CA: University of California, Davis, Institute of Transportation Studies. Research Report UCD-ITS-RR-17-07.

Ford Motor Company. (2018). *Looking further.* Accessed on 04.10.2019 from https://corporate.ford.com/articles/autonomous-technology/autonomous-2021.html.

Fulton, L., Mason, J., & Meroux, D. (2017). *Three revolutions in urban transportation.* Davis, CA: Institute for Transportation & Development Policy.

Gessner, K. (2019). *Rideshare: Lyft announces IPO after gaining ground against market leader Uber.* Accessed on 4.4.2019 from https://secondmeasure.com/datapoints/rideshare-industry-overview/.

Greenblatt, J. B., & Shaheen, S. (2015). Automated vehicles, on-demand mobility, and environmental impacts. *Current Sustainable/Renewable Energy Reports, 2*(3), 74–81.

Hall, J. D., Palsson, C., & Price, J. (2018). Is Uber a substitute or complement for public transit? *Journal of Urban Economics, 108*, 36–50.

Hao, P., Wu, G., Boriboonsomsin, K., & Barth, M. (2018). Eco-approach and departure (EAD) application for actuated signals in real-world traffic. *IEEE Transactions on Intelligent Transportation Systems, 20*(1), 30–40.

Harper, C. D., Hendrickson, C. T., Mangones, S., & Samaras, C. (2016). Estimating potential increases in travel with autonomous vehicles for the non-driving, elderly and people with travel-restrictive medical conditions. *Transportation Research Part C: Emerging Technologies, 72*, 1–9.

Henao, A., & Marshall, W. E. (2018). The impact of ride-hailing on vehicle miles traveled. *Transportation, 46*, 1–22.

Hu, L., Dong, J., Lin, Z., & Yang, J. (2018). Analyzing battery electric vehicle feasibility from taxi travel patterns: The case study of New York City, USA. *Transportation Research Part C: Emerging Technologies, 87*, 91–104.

Kenney, J. B. (2011). Dedicated short-range communications (DSRC) standards in the United States. *Proceedings of the IEEE, 99*(7), 1162–1182.

Komanduri, A., Wafa, Z., Proussaloglou, K., & Jacobs, S. (2018). Assessing the impact of app-based ride share systems in an urban context: Findings from Austin. *Transportation Research Record, 2672*(7), 34–46.

Krafcik, J. (2018). *Waymo one: The next step on our self-driving journey.* Accessed on 04.10.2019 from https://medium.com/waymo/waymo-one-the-next-step-on-our-self-driving-journey-6d0c075b0e9b.

Langbroek, J. H., Franklin, J. P., & Susilo, Y. O. (2017). Electric vehicle users and their travel patterns in Greater Stockholm. *Transportation Research Part D: Transport and Environment, 52*, 98–111.
Lyft, Inc. (2019). *Form S-1 registration statement.* United States Securities and Exchange Commission. https://www.sec.gov/Archives/edgar/data/1759509/000119312519059849/d633517ds1.htm. (Accessed 4 March, 2019).
MAPC. (2018). *Fare choice: A survey of ride-hailing passengers in metro Boston. Metropolitan area planning council report* Accessed from:http://www.mapc.org/wp-content/uploads/2018/02/Fare-Choices-MAPC.pdf. (Accessed 29 September, 2019).
McGuckin, T., Lambert, J., Newton, D., Pearmine, A., & Hubbard, E. (2017). *Leveraging the promise of connected and autonomous vehicles to improve integrated corridor management and operations: A primer (no. FHWA-HOP-17-001).* FHWA.
Meyer, J., Becker, H., Bösch, P. M., & Axhausen, K. W. (2017). Autonomous vehicles: The next jump in accessibilities? *Research in Transportation Economics, 62*, 80–91.
Miller, M. A., Holmes, A. G., Conlon, B. M., & Savagian, P. J. (2011). The GM "Voltec" 4ET50 multi-mode electric transaxle. *SAE International Journal of Engines, 4*(1), 1102–1114.
Nealer, R., Reichmuth, D., & Anair, D. (2015). *Cleaner cars from cradle to grave: How electric cars beat gasoline cars on lifetime global warming emissions.* Union of Concerned Scientist.
Nopmongcol, U., Grant, J., Knipping, E., Alexander, M., Schurhoff, R., Young, D., et al. (2017). Air quality impacts of electrifying vehicles and equipment across the United States. *Environmental Science & Technology, 51*(5), 2830–2837.
Plötz, P., Schneider, U., Globisch, J., & Dütschke, E. (2014). Who will buy electric vehicles? Identifying early adopters in Germany. *Transportation Research Part A: Policy and Practice, 67*, 96–109.
Rios-Torres, J., & Malikopoulos, A. A. (2017). A survey on the coordination of connected and automated vehicles at intersections and merging at highway on-ramps. *IEEE Transactions on Intelligent Transportation Systems, 18*(5), 1066–1077.
Rogelj, J., Den Elzen, M., Höhne, N., Fransen, T., Fekete, H., Winkler, H., et al. (2016). Paris agreement climate proposals need a boost to keep warming well below 2 C. *Nature, 534*(7609), 631.
SAE international. (2016). *Taxonomy and definitions for terms related to driving automation systems for on-road motor vehicles.* SAE International [J3016].
Schaller, B. (2018). *Making congestion pricing work for traffic and transit in New York City.* Schaller Consulting.
Shepardson, D. (2018). *Uber resuming self-driving car testing in Pennsylvania.* https://www.reuters.com/article/us-uber-selfdriving/uber-resuming-self-driving-car-testing-in-pennsylvania-idUSKCN1OJ1OD. (Accessed 4 October, 2019).
Shladover, S., Lu, X., Yang, S., Ramezani, H., Spring, J., Nowakowski, C., et al. (2018). *Cooperative adaptive cruise control (CACC) for partially automated truck platooning: Final report.* Federal Highway Administration—Exploratory Advanced Research Program.
Slowik, P., Pavlenko, N., & Lutsey, N. (2019). Emerging policy approaches to electrify ride-hailing in the United States. In *International Council on Clean Transportation (ICCT) policy briefing.* Accessed from https://theicct.org/sites/default/files/publications/EV_ridehailing_policy_approaches_20190108.pdf. (Accessed 29 September, 2019).
Spurlock, C. A., Sears, J., Wong-Parodi, G., Walker, V., Jin, L., Taylor, M., et al. (2019). Describing the users: Understanding adoption of and interest in shared, electrified, and automated transportation in the San Francisco Bay Area. *Transportation Research Part D: Transport and Environment, 71*, 283–301.
Stephens, T., Zhou, Y., Burnham, A., & Wang, M. (2018). *Incentivizing adoption of plug-in electric vehicles: A review of global policies and markets (no. ANL/ESD-18/7).* Argonne, IL (United States): Argonne National Lab.
Tessum, C. W., Hill, J. D., & Marshall, J. D. (2014). Life cycle air quality impacts of conventional and alternative light-duty transportation in the United States. *Proceedings of the National Academy of Sciences of the United States of America, 111*(52), 18490–18495.

U.S. Department of Transportation. (2016). *Applications for the environment: Real-time information synthesis (AERIS)*. [Accessed 1 March 2019] http://www.its.dot.gov/aeris/.

U.S. National Highway Traffic Safety Administration. (2016). *DOT advances deployment of connected vehicle technology to prevent hundreds of thousands of crashes*. https://www.nhtsa.gov/press-releases/us-dot-advances-deployment-connected-vehicle-technology-prevent-hundreds-thousands.

Wadud, Z., MacKenzie, D., & Leiby, P. (2016). Help or hindrance? The travel, energy and carbon impacts of highly automated vehicles. *Transportation Research Part A: Policy and Practice, 86*, 1–18.

Walker, J., & Johnson, C. (2016). *Peak car ownership: The market opportunity of electric automated mobility services*. Rocky Mountain Institute.

Wang, Z., Wu, G., Hao, P., Boriboonsomsin, K., & Barth, M. (2017, June). *Developing a platoon-wide eco-cooperative adaptive cruise control (CACC) system*. In *2017 IEEE intelligent vehicles symposium (IV)*. (pp. 1256–1261): IEEE.

Weis, A., Jaramillo, P., & Michalek, J. (2016). Consequential life cycle air emissions externalities for plug-in electric vehicles in the PJM interconnection. *Environmental Research Letters, 11*(2), 024009.

Wenzel, T., Rames, C., Kontou, E., & Henao, A. (2019). Travel and energy implications of ridesourcing service in Austin, Texas. *Transportation Research Part D: Transport and Environment, 70*, 18–34.

Xia, H. (2014). *Eco-approach and departure techniques for connected vehicles at signalized traffic intersections*. Doctoral DissertationRiverside: University of California.

Xu, Z., Yin, Y., & Ye, J. (2019). On the supply curve of ride-hailing systems. *Transportation Research Part B: Methodological, 132*, 29–43.

CHAPTER 22

Traffic-related air pollution, human exposure, and commercially available market solutions: Perspectives from the developing nation context

S.M. Shiva Nagendra[a], Uwe Schlink[b], V. Dheeraj Alshetty[a], M. Diya[a], and Jyothi S. Menon[c]
[a]Department of Civil Engineering, Indian Institute of Technology Madras, Chennai, India
[b]UFZ Helmholtz Centre for Environmental Research, Leipzig, Germany
[c]Public Health Foundation of India, New Delhi, India

Abbreviations

VEEs	vehicular exhaust emissions
NO_x	oxides of nitrogen
CO	carbon monoxide
HC	hydrocarbon
SOx	oxides of sulfur
PAHs	polycyclic aromatic hydrocarbons

Introduction

Urban air pollution poses a significant threat to human health, environment, and the quality of life of people throughout the world. Deterioration of urban air quality due to vehicular emissions and associated health implications has been a topic of debate for a long time (Srimuruganandam and Nagendra, 2015).Vehicular emissions significantly contribute to climate change on one hand and on the other hand climate conditions can modify the adverse health effects of air pollution (Leitte et al., 2009). The general air pollution sources in urban areas include (i) transport-motor vehicles and railways, (ii) commercial establishments (hotels, convention centers, and large corporations), (iii) industries, (iv) domestic cooking and heating, (v) biomass burning, (vi) fugitive dust, and (vii) natural sources.

Traffic-Related Air Pollution
https://doi.org/10.1016/B978-0-12-818122-5.00022-3

In recent years, in most countries, the air pollution from industrial and domestic sources has markedly decreased due to various control techniques initiated by different governments. However, there has been a substantial increase of air pollution caused by the vehicular exhaust emissions (VEEs) across the globe due to addition of more and more vehicles on transportation networks to meet increase in transportation demand. Therefore, motor vehicle exhaust emissions have been demonstrated as one of the significant source groups of ground-level air pollution in almost all metro cities of the developing countries. Oxides of nitrogen (NO_x), carbon monoxide (CO), hydrocarbons (HCs), oxides of Sulfur (SO_x), polycyclic aromatic hydrocarbons (PAHs), lead (Pb), particulate matter (PM), etc. are some of the important pollutants emitted by vehicles (Jaikumar, Shiva Nagendra, & Sivanandan, 2017). In general, vehicular emission depends on vehicle type (two-wheeler, three-wheeler, cars, buses, trucks, etc.), traffic, and road characteristics such as vehicle weight, mileage, fuel type, age of the vehicle, engine characteristics, emission controlling equipment, road gradient, traffic density and intersections, vehicle kilometer traveled (VKT), traffic congestion, increase in construction activity, poor road infrastructure, poor fuel quality, large proportion of old vehicles, high traffic density, and inadequate inspection and maintenance (I/M) programs (Badami, 2005; Srimuruganandam and Shiva Nagendra, 2011). Further, poor maintenance and age of the vehicle's engine also adds up to the exhaust emissions (Beydouna, 2004). In case of nonexhaust emissions, contribution of resuspendable road dust is abundant (50%–60%) and it is significant in tropical countries like India, because of the dry climate and existence of unpaved roads in the urban areas.

Urban air quality and health impacts

Vehicular exhaust emissions in urban areas and from highways are major concern across the globe since last three decades due to rapid urbanization and increasing vehicular population. They are the largest source of gaseous pollutants in the urban environment and major reason for urban hotspots (Chen & Yuan, 2020). Harmful substances emitted by exhausts and wear from vehicles are accumulated on road pavement (Amato et al., 2009). The chemical constituents of road dust include inorganic ions (e.g., sulfates, nitrates, ammonium, sodium, potassium, calcium, magnesium, and chloride), organic and elemental carbon, particle-bound metals, and PAHs (Tolis et al., 2015).

Air pollution can affect public health in many ways with both short- and long-term effects. Examples of short-term effects include irritation to the eyes, nose and throat, and upper respiratory infections such as bronchitis and pneumonia. Other symptoms can include headaches, nausea, and allergic reactions. Short-term air pollution can aggravate the medical conditions of individuals with asthma and emphysema. Long-term health effects can include chronic respiratory disease, lung cancer, heart disease, and even damage to the brain, nerves, liver, or kidneys. Continual exposure to air pollution affects the lungs of growing children and may aggravate or complicate medical conditions in the elderly. It is estimated that half a million people die prematurely every year in the United States as a result of smoking cigarettes. People with health problems such as asthma, heart, and lung disease may also suffer more when the air is polluted. The extent to which an individual is harmed by air pollution usually depends on the total exposure to the toxic air pollutants.

Personal exposure and health risk

A study by the Health Effects Institute (HEI), indicated an annual mortality of 3000 in Delhi due to personal exposure to vehicular emissions along the major and arterial roads in the city (HEI, 2013). The extent to which urban air pollution leads to personal exposure is determined by a range of factors like proximity to the source, microenvironmental characteristics, time spent in each microenvironment and level of physical activity, and ultimately, the dose reaching the target organs especially lungs. The exposure profile of each individual varies depending on the various microenvironments that a person encounters in their day-to-day life and the time spent in each of these microenvironments as depicted in Fig. 22.1. The pollutant concentration distribution in each of these microenvironments will also vary depending on the source profile. The final exposure will be an integrated exposure of all these individual microenvironment concentrations that the individual gets exposed to. The exposure to air pollution in the developing countries is complex because of the presence of unique sources (open cooking, open solid waste burning, use of biomass as fuel, and use of diesel generator sets), which is not usually found in the developed countries (Pant, Guttikunda, & Peltier, 2016). The socioeconomic status also plays an important role in determining exposure where the certain section of the population with low income are more vulnerable to air quality-related health effects. Recent advancements in mobile and sensor technologies as well as development of light-weight personal samplers are expected to allow greater feasibilities for the use of personal exposure measurements in health studies related to

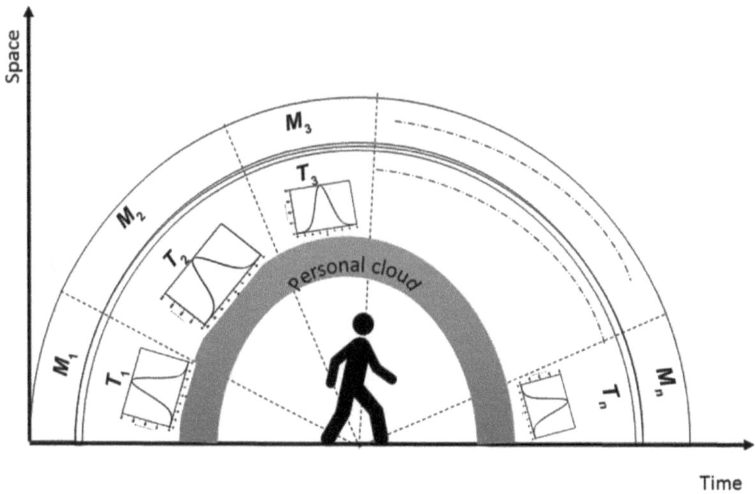

Fig. 22.1 Personal exposure based on the microenvironments and time spent in each microenvironment (M—microenvironment, T—time spent).

transport emissions in the near future (Balakrishnan et al., 2018; Nyarku et al., 2018).

Particle characteristics such as number and size distribution along with the physiological state (breathing parameters such as frequency and rate) of the individual determines the internal dose (uptake of pollutants in the lungs), hence the health effects (Broderick et al., 2015; McNabola, Broderick, & Gill, 2008; Menon & Nagendra, 2018). Estimation of particle dosimetry is an upcoming research area, where the internal dose can be used to determine microscale health effects associated with different modes of transport (Broderick et al., 2015). This has been employed in some of the recent studies to understand the traffic-related air pollution exposure in urban areas (Manigrasso, Natale, Vitali, Protano, & Avino, 2017; Menon & Nagendra, 2018). The effects of various air pollutants on humans are given in Table 22.1.

Control solutions to address traffic-related pollution

In general, there are three different types of measures for the improvement of urban air quality by mitigating traffic-related air pollution. Preventive actions aim to avoid the emission of air pollutants. Primary measures (such as emission filters) focus on a reduction of the emissions at source. Secondary measures are techniques that treat the air for cleaning purposes. In India,

Table 22.1 Effect of air pollutants on humans.

Pollutants	Health effects
CO	Reduces delivery of oxygen to the body, which is particularly serious for those with cardiovascular disease, causes impairment of function in healthy people
NO_x	In high concentrations, irritates lungs and lowers resistance to respiratory infection; an important precursor to ozone and acid precipitation which can damage sensitive ecosystems
VOC	Results in ozone which can damage lung tissue, reduces lung function, and causes irritation (these effects occur even at low levels in healthy people who engage in moderate exercise); also causes ecosystem degradation, mainly through damaging foliage
PM_{10}	In high concentrations aggravates existing respiratory and cardiovascular disease, alters the immune system, can be carcinogenic, and causes lung damage
SO_2	Major contributor to acid rain; degrades lung function and lower lung defenses while aggravating existing respiratory disease
Lead	Accumulates in the body and affects kidney, liver, and nervous system; causes neurological impairments

to address the traffic-related pollution problems, government has initiated several control measures (preventive) at source, such as implementation of the Bharat Stage (BS) Emissions Standards, standards for fuel, encouraging electric vehicles, etc. whereas, significant efforts have also been made to treat the air emissions at the engines (burner modification, over firing system) and at tile pipes (catalytic convertors, exhaust filters). Catalytic converters, diesel particulate filters, exhaust gas recirculation, and selective catalytic redactors are some of the recent advanced technologies being used in vehicles to reduce the ground-level emissions from tail pipes. Recent research studies have also shown that there is a need to monitor the real-time exhaust emissions to predict the impact of pollutants on personal health (Rohit, Shiva Nagendra, & Sivanandan, 2017). Many secondary techniques are also adopted to control the vehicular emissions. This includes development of green belts or emission barriers, urban air purifiers along the major roads, and use of face masks by individuals.

Brake and tire wear contributes a significant amount of emissions. Hence, the road surface must be paved to also reduce emissions. In order to reduce tire wear emission, it is also important to check overloading of vehicles. In order to arrest the PM emission at the source, proper dust controllable barriers have to be designed and constructed on either side of the urban roads to contain PM emission. Further, a thick green belt development all along the roads will contain the emissions.

Currently, very few efforts have been made to combat ambient air pollution at the urban scale. Since, the currently implemented technologies are either large in size and scale, involving high capital and operational cost with high maintenance, or they are so small that they do not have sufficiently large impact. Few of the technologies are energy intensive and require frequent monitoring, which becomes difficult once it is installed at roadside. Over a period of time, these setup becomes obsolete due to poor maintenance. Further, their overall benefit in mitigating air pollution is unclear.

Roadside air purifiers

WAYU (Wind Augmentation and Air Purifying) is ambient air purification system developed by the National Environmental Engineering Research Institute (NEERI) in 2017 to address vehicular air pollution. This system works by converting pollutants such as $PM_{2.5}$ and PM_{10} particles, CO and VOCs into CO_2. It works on the principle of dilution and derives the necessary energy using solar panels. The height at which it will be installed will depend on the traffic volume and the level of pollution at the area.

The City Air Purification System is developed in Hong Kong by Sino Green and ARUP is designed like a small bus stop shelter. In this system, air is drawn from the bottom, passed through the fabric filter to remove the PM, and is given out at the top. The system also includes smart controllers to manage operating hours more efficiently, solar panels for energy, and a mist cooling system for summer months.

Cyclofine

This system has been developed by the Department of Civil Engineering, IIT-Madras and EnviTran, a start-up company at the IIT-M Research Park. This system has three major components. The first one is the air quality monitoring system using wireless network of sensors, which provides real-time air quality information at the hotspots as well as triggers the operation of air pollution control system. The second component consists of the purification system. The third component is the outlet tower designed to remove finer dust particles, gases, and heavy metals and deliver the clean air on the downwind direction (Fig. 22.2).

There are several challenges to implement air purification at urban roadside. This includes (i) location-specific characteristics of airborne particles

Fig. 22.2 CYCLOFINE hotspot ambient air purification system.

and gaseous pollutants, (ii) operation and maintenance of the air purification system, (iii) noise disturbance to the public, and (iv) management of collected filter and filter materials.

Face masks to reduce personal exposure

Due to their proximity to vehicular exhaust emissions, pedestrians and cyclists are being exposed to high level of particulate and gaseous air pollutants. As a protective measure, commercial face mask respirators are being widely used in both developing and developed countries to avoid personal exposure to vehicular air pollution. The performance of face mask (i.e., reduction of mass concentration of black carbon, $PM_{2.5}$, particle number concentrations, and lung-deposited surface area) is strongly dependent on the quality of the mask (Pacitto et al., 2019).

Need for real-world emissions testing and measurement

Emission factors were often widely used by regulatory agencies such as Environmental Protection Agency (EPA) in the United States, Automotive Research Association of India (ARAI) in India for vehicular testing and certification. A recent EPA report on emission testing of diesel cars in the United States revealed that the manufacturers were inserting defeat devices, which can give reduced emissions only during test conditions, while the real-world emissions were found to be several times (35) higher. Further these factors were generally used as inputs for various air quality prediction models, which results in the development of poor air quality management strategies (Nagendra and Khare, 2002). In the past, exhaust emission models were widely used by various regulatory agencies to estimate emissions and to prepare control strategies for urban air quality management. In most of these models, the emissions were measured in laboratory conditions using dynamometer for standard driving cycles. These driving cycles were not usually reflecting the driving behavior of the real-world conditions thereby adding considerable error into the emission estimation. Most of these model results were further integrated into dispersion models for the prediction of air quality in urban areas. The dispersion model predictions were also used in framing various policy decisions (Gulia, Nagendra, Barnes, & Khare, 2018). In general, emissions reported by these models were several times lower than the real-world emissions. Hence, the policy developed based on model predictions showed only marginal improvement in urban air quality. Therefore, the knowledge on real-world emission characteristics is essential for the design of effective exhaust emission control strategies and better traffic management at intersections and busy corridors. Recently, low-cost sensors are gaining importance in air quality monitoring as they are cost effective, compact, and well suited for personal exposure and mobile monitoring.

The vehicular pollution in urban areas can be controlled by taking the following steps (Rao, 1991):

- reduction in amount of pollutants formed during combustion by suitable modifications of the internal combustion engine;
- development of exhaust system reactors that will complete the combustion process and change potential pollutants into more acceptable materials,
- development of alternative fuels that will produce low levels of pollutants upon combustion;
- replacement of internal combustion engine with low pollution engines;
- introduction of an effective inspection and maintenance program; and
- phasing out of old vehicles.

Summary

Motor vehicles are widely known as the prime contributors to deterioration of urban air quality in cities around the world, and air pollution a critical human health problem in the urban areas of both developed and developing counties of the world. In this chapter, the health risk due to personal exposure of urban air pollution is explained, with an emphasis on the context of urban India. It causes a spectrum of health effects ranging from eye irritation to mortality (WHO, 2009). The extent to which urban air pollution leads to personal exposure is determined by a range of factors like proximity to the source, microenvironmental characteristics, time spent in each microenvironment, and level of physical activities. The pollutant concentration distribution in each of these microenvironments will also vary depending on the source profile.

In general, there are three different types of measures namely preventive actions (aim to avoid the emission of air pollutants), primary measures (such as emission filters) focus on a reduction of the emitted pollutants, and secondary measures (techniques that treat the air for cleaning purposes) are useful to control the vehicular emission. In India, to address traffic-related air pollution issues, the government has initiated several control measures (preventive) at source, such as implementation of BS norms, standards for fuel, encouraging electric vehicles, etc. Significant efforts have also been made to treat the air emissions at the engines (burner modification, over firing system) and at the tailpipe pipes (catalytic convertors, exhaust filters). Many secondary techniques and commercially available solutions (development of green belts or emission barriers, urban air purifiers, the use of face masks by individuals) are also adopted to control the vehicular emissions along the major roads in Indian cities.

References

Amato, F., Pandolfi, M., Escrig, A., Querol, X., Alastuey, A., Pey, J., et al. (2009). Quantifying road dust resuspension in urban environment by multilinear engine: A comparison with PMF2. *Atmospheric Environment, 43*(17), 2770–2780.

Badami, M. G. (2005). Transport and urban air pollution in India. *Environmental Management, 36*, 195–204.

Balakrishnan, K., Ghosh, S., Thangavel, G., Sambandam, S., Mukhopadhyay, K., Puttaswamy, N., et al. (2018). Exposures to fine particulate matter (PM2.5) and birthweight in a rural-urban, mother-child cohort in Tamil Nadu, India. *Environmental Research, 161*, 524–531.

Beydouna, M. (2004). Vehicular characteristics and urban air pollution: Socioeconomic and environmental policy issues. *Doctor of Philosophy, Ohio State University*. City and Regional Planning. http://rave.ohiolink.edu/etdc/view?acc_num=osu1085681028.

Broderick, B., Byrne, M., McNabola, A., Gill, L.W., Pilla, F., McGrath, J., et al. (2015). *PALM: A personal activity-location model of exposure to air pollution*. Wexford, Ireland: Environmental Protection Agency.

Chen, X., & Yuan, Z. (2020). Environmentally friendly traffic control strategy—A case study in Xi'an city. *Journal of Cleaner Production, 249*(10), 1–8.

Gulia, S., Nagendra, S. S., Barnes, J., & Khare, M. (2018). Urban local air quality management framework for non-attainment areas in Indian cities. *Science of the Total Environment, 619*, 1308–1318.

HEI. (2013). *Review panel on ultrafine particles. Understanding the health effects of ambient ultrafine particles* (p. 122). Boston, MA: HEI Perspectives 3Health Effects Institute.

Jaikumar, R., Shiva Nagendra, S. M., & Sivanandan, R. (2017). Modal analysis of real-time, real world vehicular exhaust emissions under heterogeneous traffic conditions. *Transportation Research Part D: Transport and Environment, 54*, 397–409.

Leitte, A. M., Petrescu, C., Franck, U., Richter, M., Suciu, O., Ionovicic, R., et al. (2009). Respiratory health, effects of ambient air pollution and its modification by air humidity in Drobeta-Turnu Severin, Romania. *Science of the Total Environment, 407*(13), 4004–4011.

Manigrasso, M., Natale, C., Vitali, M., Protano, C., & Avino, P. (2017). Pedestrians in traffic environments: Ultrafine particle respiratory doses. *International Journal of Environmental Research and Public Health, 14*(3), 288.

McNabola, A., Broderick, B. M., & Gill, L. W. (2008). Relative exposure to fine particulate matter and VOCs between transport microenvironments in Dublin: Personal exposure and uptake. *Atmospheric Environment, 42*, 6496–6512.

Menon, J. S., & Nagendra, S. M. (2018). Statistical distribution and particle dosimetry models to estimate personal exposure at urban sidewalks of tropical climate. *Sustainable Cities and Society, 40*, 254–265. https://doi.org/10.1016/j.scs.2017.09.005.

Nagendra, S. M. S., & Khare, M. (2002). Line source emission modelling: Review. *Atmospheric Environment, 36*(13), 2083–2098.

Nyarku, M., Mazaheri, M., Jayaratne, R., Dunbabin, M., Rahman, M. M., Uhde, E., et al. (2018). Mobile phones as monitors of personal exposure to air pollution: Is this the future? *PLoS One, 13*(2), e0193150.

Pacitto, A., Amato, F., Salmatonidis, A., Moreno, T., Alastuey, A., Reche, C., et al. (2019). Effectiveness of commercial face masks to reduce personal PM exposure. *Science of The Total Environment, 650*, 1582–1590.

Pant, P., Guttikunda, S. K., & Peltier, R. E. (2016). Exposure to particulate matter in India: A synthesis of findings and future directions. *Environmental Research, 147*, 480–496.

Rao, C. S. (1991). *Environmental pollution control engineering*. Delhi. New: New Age International (P) Ltd.

Rohit, J., Shiva Nagendra, S. M., & Sivanandan, R. (2017). Modal analysis of real-time, real world vehicular exhaust emissions under heterogeneous traffic conditions. *Transportation Research Part D: Transport and Environment, 54*, 397–409.

Srimuruganandam, B., & Shiva Nagendra, S. M. (2011). Chemical characterization of PM10 and PM2.5 mass concentrations emitted by heterogeneous traffic. *Science of the Total Environment, 409*(17), 3144–3157.

Tolis, E. I., Saraga, D. E., Lytra, M. K., Papathanasiou, A. Ch., Bougaidis, P. N., Prekas-PatronakisaIoannis, O. E., et al. (2015). Concentration and chemical composition of PM2.5 for a one-year period at Thessaloniki, Greece: A comparison between city and port area. *Atmospheric Environment, 113*, 197–207.

WHO. (2009). *Quantification of the disease burden attributable to environmental risk factors. Department of Public Health and Environment*. Geneva: World Health Organization. http://www.who.int/quantifying_ehimpacts/summaryEBD_updated.pdf.

CHAPTER 23

The state of the literature on traffic-related emissions, air pollution, human exposures, and health

Kristen A. Sanchez[a,b], Tara Ramani[a], Josias Zietsman[a], Mark J. Nieuwenhuijsen[c,d,e], and Haneen Khreis[a,c,d,e]
[a]Center for Advancing Research in Transportation Emissions, Energy, and Health (CARTEEH), Texas A&M Transportation Institute (TTI), College Station, TX, United States
[b]Texas A&M School of Public Health, College Station, TX, United States
[c]ISGlobal, Centre for Research in Environmental Epidemiology (CREAL), Barcelona, Spain
[d]Universitat Pompeu Fabra (UPF), Barcelona, Spain
[e]CIBER Epidemiologia y Salud Publica (CIBERESP), Madrid, Spain

Abbreviations

BC	black carbon
CARTEEH	Center for Advancing Research in Transportation Emissions, Energy, and Health
CO	carbon monoxide
CO_2	carbon dioxide
COPD	chronic obstructive pulmonary disease
HC	hydrocarbons
NO_2	nitrogen dioxide
NO_x	nitrogen oxides
PM_{10}	particulate matter with diameter $< 10\,\mu m$
$PM_{2.5}$	particulate matter with diameter $< 2.5\,\mu m$
Ref ID	reference ID
TRAP	traffic-related air pollution
TRB	Transportation Research Board
TRR	Transportation Research Record

Introduction

Traffic-related air pollution (TRAP) is a term that refers to the contribution of traffic activity to ambient air pollution. Traffic activity includes the use of motorized vehicles such as passenger cars, motorcycles, buses, coaches, light-duty, and heavy-duty vehicles. These vehicles emit air pollutants including

Traffic-Related Air Pollution
https://doi.org/10.1016/B978-0-12-818122-5.00023-5

carbon monoxide (CO), nitrogen oxides (NO_x), particulate matter with a diameter less than 2.5 μm ($PM_{2.5}$), particulate matter with a diameter less than 10 μm (PM_{10}) and hydrocarbons (HC), to name a few (Health Effects Institute Panel on the Health Effects of Traffic-Related Air Pollution, 2010). The aforementioned air pollutants contribute to the degradation of ambient air quality, resulting in numerous adverse health outcomes for exposed populations.

The global population has risen substantially over the past century and is estimated to reach 9.8 billion by the year 2050 from just 7.6 billion in 2017 (United Nations Department of Economic and Social Affairs Population Division, 2017). With population growth and economic growth comes an increase in the number of vehicles being manufactured, purchased, driven, and emitting more air pollutants. Increased transportation activity continues to overpower emission regulation and advancements in vehicle and fuel technology, and therefore TRAP is expected to be on the rise (Barker et al., 2007).

TRAP is associated with a variety of adverse health outcomes including, but not limited to: arrhythmia (Link & Dockery, 2013), autism and child behavior problems (Raz et al., 2015), childhood asthma (Khreis et al., 2017), wheeze (Gasana, Dillikar, Mendy, Forno, & Vieira, 2012), chronic obstructive pulmonary disease (COPD) (Lindgren et al., 2009), coagulation effects (Brook et al., 2010), congenital anomalies (Vrijheid et al., 2011), coronary events (Cesaroni et al., 2014), dementia (Power, Adar, Yanosky, & Weuve, 2016), low birth weight (Fleischer et al., 2014), lung cancer (Raaschou-Nielsen et al., 2013), myocardial infarction (Mustafić et al., 2012), obesity (Jerrett et al., 2014), pneumonia (MacIntyre et al., 2014), preeclampsia (Pedersen et al., 2014), premature mortality (Beelen et al., 2014), preterm birth (Sapkota, Chelikowsky, Nachman, Cohen, & Ritz, 2012), reduced sperm quality (Lafuente, García-Blàquez, Jacquemin, & Checa, 2016), reduced birth weight (Pedersen et al., 2013), stroke (Stafoggia et al., 2014), systemic inflammation (Brook et al., 2010), and type II diabetes (Eze et al., 2015). These health outcomes and their respective references are outlined in Table 23.1.

The impacts of TRAP on human health can be quantified and assessed using numerous approaches, one of which includes a full characterization of the events connecting the air pollution sources to the final health impacts and is referred to as full-chain modeling (Khreis, de Hoogh, & Nieuwenhuijsen, 2018). Full-chain modeling is a process that models the full chain of events between traffic activity and the final health outcomes

Table 23.1 Diseases associated with TRAP: A noncomprehensive list.

Health outcome associated with TRAP	Ref ID
Arrhythmia	Link and Dockery (2013)
Autism and child behavior problems	Raz et al. (2015)
Childhood asthma	Khreis et al. (2017)
Childhood asthma and wheeze	Gasana et al. (2012)
Chronic obstructive pulmonary disease (COPD)	Lindgren et al. (2009)
Coagulation effects	Brook et al. (2010)
Congenital anomalies	Vrijheid et al. (2011)
Coronary events	Cesaroni et al. (2014)
Dementia	Power et al. (2016)
Low birth weight	Fleischer et al. (2014)
Lung cancer	Raaschou-Nielsen et al. (2013)
Myocardial infarction (heart attack)	Mustafić et al. (2012)
Obesity	Jerrett et al. (2014)
Pneumonia	MacIntyre et al. (2014)
Preeclampsia	Pedersen et al. (2014)
Premature mortality	Beelen et al. (2014)
Preterm birth	Sapkota et al. (2012)
Reduced sperm quality	Lafuente et al. (2016)
Reduced birth weight	Pedersen et al. (2013)
Stroke	Stafoggia et al. (2014)
Systemic inflammation	Brook et al. (2010)
Type II diabetes	Eze et al. (2015)

that can be attributed back to that activity's TRAP (Fig. 23.1). This start-to-finish approach highlights the different components of the full chain and links them together to pinpoint the impacts of particular air pollution sources (traffic in this case). The full chain includes traffic activity, traffic emissions, dispersion and resulting air quality, exposure, and human health impacts (Khreis et al., 2018). It is important to note that each element of the full chain can be measured and/or modeled, but for large-scale analyses, modeling is typically more feasible. The strength of the full chain is in how it ties multiple components together rather than simply focusing on one piece of the puzzle. Understanding the relationship between the aforementioned elements of the full-chain paves way for discussion about policy and practice implications and for making more concrete and specific policy recommendations. The full-chain approach also offers a useful framework to understand and explore the literature under each element and the

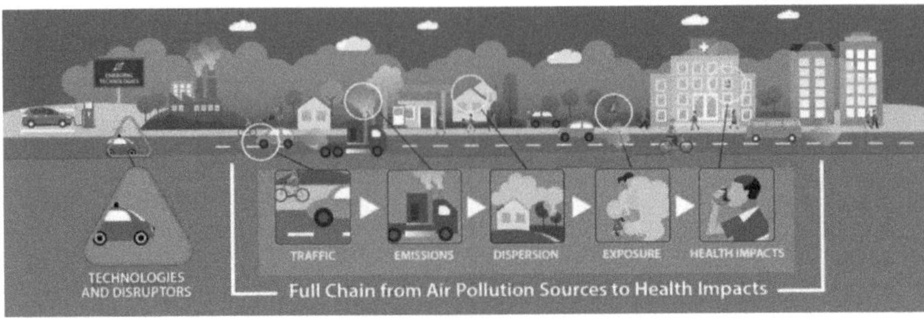

Fig. 23.1 The full-chain: Linking TRAP to health impacts.

combination thereof. Unfortunately, there is currently a lack of research involving the full-chain approach (Nieuwenhuijsen, Khreis, Verlinghieri, Mueller, & Rojas-Rueda, 2017).

This chapter aims to describe the development of a literature library that compiles 946 key studies addressing separate full-chain elements and the full chain as a whole in order to identify relevant research and knowledge gaps. There are three key goals of this chapter, each of which contribute to either the state of art or practice in this field.

Goals contributing to the state of art are

1. Describe the development of a literature library on traffic activity, traffic-related emissions, air pollution, human exposures, health, and technology.
2. Assess the collective body of literature that exists relating to TRAP and human health and highlight:
   a. Where the literature is clustered
   b. Where the literature is missing
   c. Key research and knowledge gaps
   d. Recommendations for the design of relevant future studies

   Goals contributing to the state of practice are
3. Improve the time and ease students, researchers, and practitioners need to identify and access relevant articles as they conduct or design their studies.

Literature library development

The literature library discussed in this chapter is unique in that it is one of the first to gather highly specialized content into a single location. It will provide users with efficient access to a hub of articles related to traffic activity, traffic-related emissions, air quality/dispersion, human exposure, health impacts, and new technologies, in addition to helping the users in identifying the full-chain element(s) assessed in each article. The literature library was further developed into an open access tool which can improve the time and ease students, researchers, and practitioners need to access relevant articles as they conduct or design their studies. Additionally, the library is beneficial to journals by improving their dissemination efforts. For example, there is a significant amount of research documented in Transportation Research Record (TRR) that is not readily available to researchers. As the library grows with the addition of new material, such as more articles published in TRR, users will be able to identify and access them more readily and frequently. Information about similar libraries is extremely limited at this point. To the authors' best knowledge, this chapter is at the forefront of reviewing the literature with a collection of 946 articles and continuous updates. The development of the literature library includes three main steps: collection, organization, and analysis of the literature, which are briefly described next.

Literature collection

Relevant articles were collected via experts' knowledge of relevant literature in this field, article recommendations from email list subscriptions and a nonsystematic literature search. Keywords used to search for studies related to traffic, emissions, air quality/dispersion, exposure, health impacts, and technologies and are shown in Table 23.2. Studies prior to the year 2000 were excluded from the library to maintain a collection with the most recent information. The search for and addition of studies is currently ongoing.

Literature organization

Each study was alphabetically entered into a Microsoft Excel sheet with respect to its Reference ID (Ref ID), which is the shortened version of a full citation. This Excel sheet contains article information including Ref ID, author(s), article title, article citation, URL link to the abstract and/or full paper, the publication year, the publication source (journal paper, book, thesis, report, or conference paper), literature topic(s) addressed, and study type(s). Two independent researchers (KS and HK) categorized studies in the library. To maintain consistency, each researcher cross-checked 10% of study categorizations completed by the other researcher. Discrepancies were very minimal and were resolved by consensus. The Excel sheet was then made

Table 23.2 Literature search keywords.

Topic	Keywords
Traffic	"Traffic monitoring", "traffic counting", "traffic modeling", "traffic measurement", "traffic flow", "motor way", "public transport", "motor vehicle", "motorized road transport", "freeway traffic", "traffic congestion", "urban transport", "trip", "travel modes", "cycling", "activity-based travel", "commute", "traffic volume", "traffic type", "traffic distance", "traffic density", "vehicle miles traveled", or equivalent.
Traffic emissions	"Emission modeling", "emission monitoring", "emission measurement", "emission factors", "emission testing", "traffic emissions", "traffic-related pollutant emissions", "exhaust emissions", "non-exhaust emissions", "vehicle emissions", "mobile source emissions", "acceleration/deceleration emissions", "cold start emissions", "fuel use and emissions", "greenhouse gas emissions", "tailpipe emissions", "(NO_x/NO_2/BC/CO_2/$PM_{2.5}$/PM_{10}) emissions", "(NO_x/NO_2/BC/CO_2/$PM_{2.5}$/PM_{10}) production", "fuel use, energy consumption", "low emission zones", "diesel", or equivalent.
Air quality/dispersion	"Air quality/pollution modeling", "air quality/pollution monitoring", "air quality/pollution measurement", "air pollution", "traffic-related air pollution", "outdoor/indoor air pollution", "ambient air pollution", "air pollutants", "(NO_x/NO_2/BC/CO_2/$PM_{2.5}$/PM_{10}) concentration", "ultra-fine particulate matter", "pollutant dispersion", "particulate distribution", "background pollutant levels", "green space impact", or equivalent.
Exposure	"Long-term exposure", "short-term exposure", "air pollution exposure", "emissions exposure", "traffic-related air pollution exposure", "human exposure", "early life exposure", "maternal exposure", "(preconceptional/prenatal/postnatal) exposure", "exposure assessment", or equivalent.
Health impacts	"Adverse health outcomes", "improved health outcomes", "morbidity", "mortality", "health impact assessment", "burden of disease assessment", "health risks", "health benefits", "public health impacts", "population health", "DALY", "hospitalization", "epidemiological study", or equivalent.

Table 23.2 Literature search keywords—cont'd

Topic	Keywords
Technologies	"New technology", "emerging technology", "autonomous vehicles", "electric vehicles", "electric bikes", "plug in vehicles", "new sensor technology", "wearable sensors", "mobile phone technology", (impact of) "vehicle emission standards", (impact of) "emerging fuels", (impact of) "alternative fuels", (impact of) "technologies", "exhaust after treatment", "hybrid vehicle", "advanced navigation systems", "investigation of old and new technology", or equivalent.

openly available at: https://www.carteeh.org/carteeh-literature-library/. More information on the categorization by literature topic(s) and study type(s) is described below.

Literature topics

The literature topics (subject areas) used to categorize studies refer to elements of the full chain as described in the introduction, in addition to a "technologies/disrupters" topic. The full chain includes traffic activity, emissions, dispersion, exposure, and health impacts (Khreis et al., 2018).

- *Traffic activity* refers to the amount and patterns of motorized vehicle usage at a certain time/place.
- *Emissions* are the amount of pollutants emitted from motorized vehicles, whether directly from the exhaust or nonexhaust emissions resulting from brake wear, tire wear, road surface abrasion, and resuspension in the wake of passing traffic (Thorpe & Harrison, 2008). These pollutants include NO_x, $PM_{2.5}$, PM_{10}, nitrogen dioxide (NO_2), and black carbon (BC), among others.
- *Air quality/dispersion* occurs when vehicle emissions are dispersed through the ambient air. Dispersion is influenced by factors such as wind speed, wind direction, time of day, traffic volume, and site layout (Baldwin et al., 2015). The resulting pollutant concentrations influence overall air quality.
- *Exposure* describes how severely/frequently a population is exposed to the different traffic-related air pollutants.
- *Health impacts* are the tail end of the full chain. This element refers to health-related effects or impacts due to the exposure to TRAP.
- *Technologies/disrupters* refers to emerging or new innovations in the transportation realm such as autonomous, connected and electric vehicles

alongside emerging fuels and new technologies in measurement methods and sensors. Although not part of the full chain, this area is included to keep users up-to-date with the latest innovations in the field and their impacts on the elements of the full chain.

Studies were categorized under a particular literature topic when the study specifically measured, modeled, or qualitatively discussed the topic as it is described above. For example, a systematic review focused on air pollution, how often a population is exposed to it and air pollution's impact on diabetes would be categorized under the air quality/dispersion, exposure, and health impacts topics, respectively. Another example is a study measuring the number of miles traveled by a vehicle and modeling the amount of emissions released; this study would be categorized under the traffic and emissions categories, respectively.

Literature study types

The literature study types used to categorize the included articles were measurement, modeling, practice/policy, and review. *Measurement* studies refer to studies that conduct direct and indirect forms of measurement and monitoring of traffic, emissions, air quality, exposure, health impacts, and/or technology. Measurements may include data obtained from surveys, monitors, or counting. *Modeling* studies refer to studies that use prediction models, simulation models, mathematical models, and empirical models of traffic, emissions, air quality, exposure, health impacts, and/or technology. Each model type is geared toward predicting real-world outcomes; however, each model uses a different approach to accomplish this. For example, prediction models utilize data and probability variables (Ishak & Al-Deek, 2002), simulation models use a digital version of a dynamic physical event (Antoniou et al., 2012), mathematical models utilize mathematical concepts to express real-world relationships in quantitative terms (Antoniou et al., 2012) and empirical models use observations and available data sets (Sommer, Eckhoff, German, & Dressler, 2011). Modeling is used to make predictions, or estimate traffic, emissions, dispersion, exposure, and health impacts, when measurements are infeasible or impossible. *Practice/policy* studies refer to studies with clear guidelines, guidance, policy implications, "how to" documents, policy analysis, policy scenario analysis, investigation of the impact of policies, policy measures, and strategies. These types of studies either serve as a useful guide in practice, have clear policy implications and/or recommendations, or both. *Review* includes studies involving meta-analyses, systematic reviews, literature reviews, or equivalent.

Categorization by "Literature Topic" and "Study Type" is not mutually exclusive; that is, a specific article may include one or more applicable topics or types of methods/analysis. Additionally, it should be noted that categorizations are determined in a semisubjective manner. Two independent authors, KS and HK, cross-checked categorizations for consistency. However, users are encouraged to highlight any inconsistencies that may remain using the library Suggestion Box described later.

Literature analysis

All 946 included studies were analyzed to determine the

- Number of studies addressing each literature topic
- Number of studies addressing all five elements of the full chain
- Number of studies addressing each literature study type
- Distribution of literature sources (journal, report, conference paper, or thesis)
- Distribution of studies by publication year and literature topic addressed
- Distribution of studies by literature topic and study types used

The results from this analysis are discussed later in this chapter.

In addition to these quantitative analyses, a qualitative analysis tool (NVIVO) was utilized to identify the most frequently occurring words across all text contained in the literature library articles. Fig. 23.2 displays a Word Cloud with the 70 most used terms (with a minimum length of 5 letters) across the library articles' entire text. This visual displays frequently used words, such as "exposure," "health," and "pollution" to name a few.

Literature library features

Additional features have been developed to accompany the literature library's Excel sheet, including an Online Library Search tool and a Suggestion Box. A snapshot of the Online Library Search tool is depicted in Fig. 23.3. Furthermore, Fig. 23.4 displays a snapshot of part of the literature library's Excel sheet organized alphabetically with respect to Ref ID. This Excel sheet may be downloaded in its entirety or exported in part, depending on the user's search via the online tool.

Online library search tool

The Online Library Search tool provides users with a method to filter studies by literature topic(s) and/or study type(s). Users can select checkboxes for listed study topics and study types and/or enter a specific search term

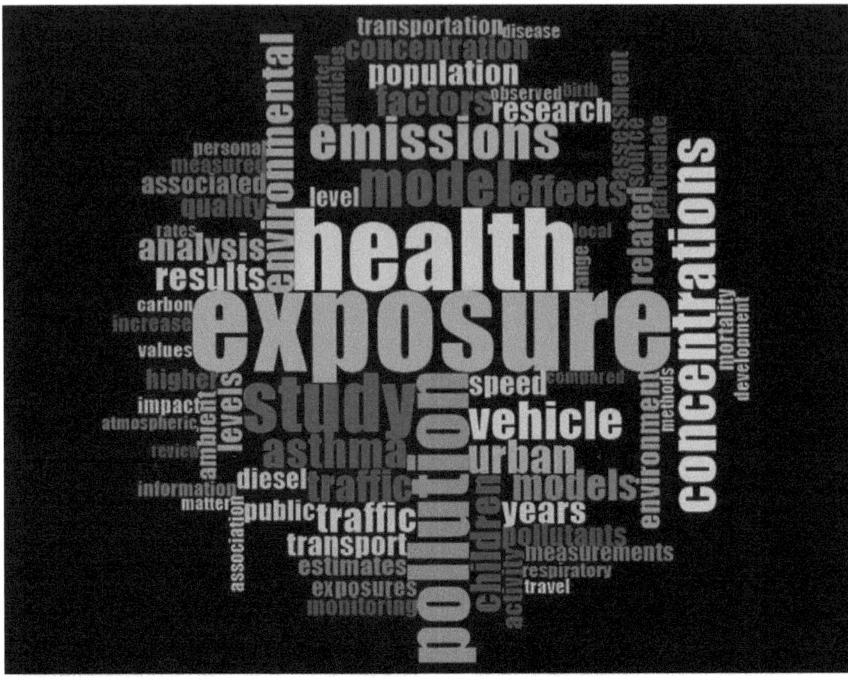

Fig. 23.2 Most frequently used terms across current literature library content.

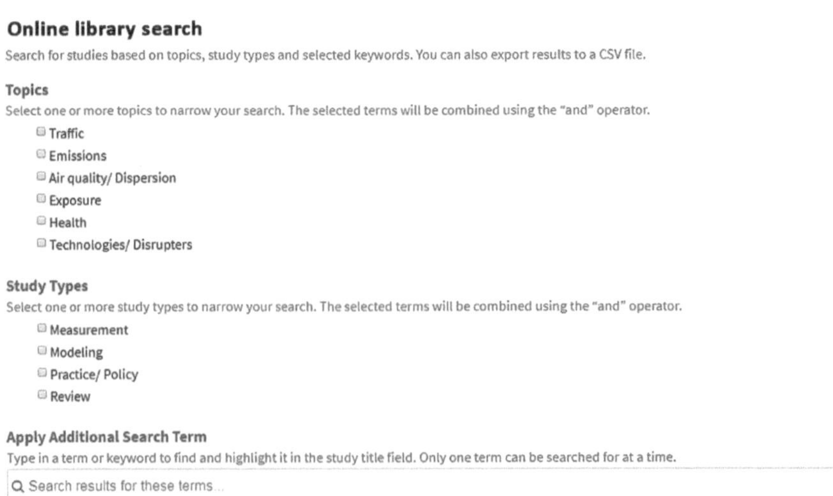

Fig. 23.3 Online library search tool.

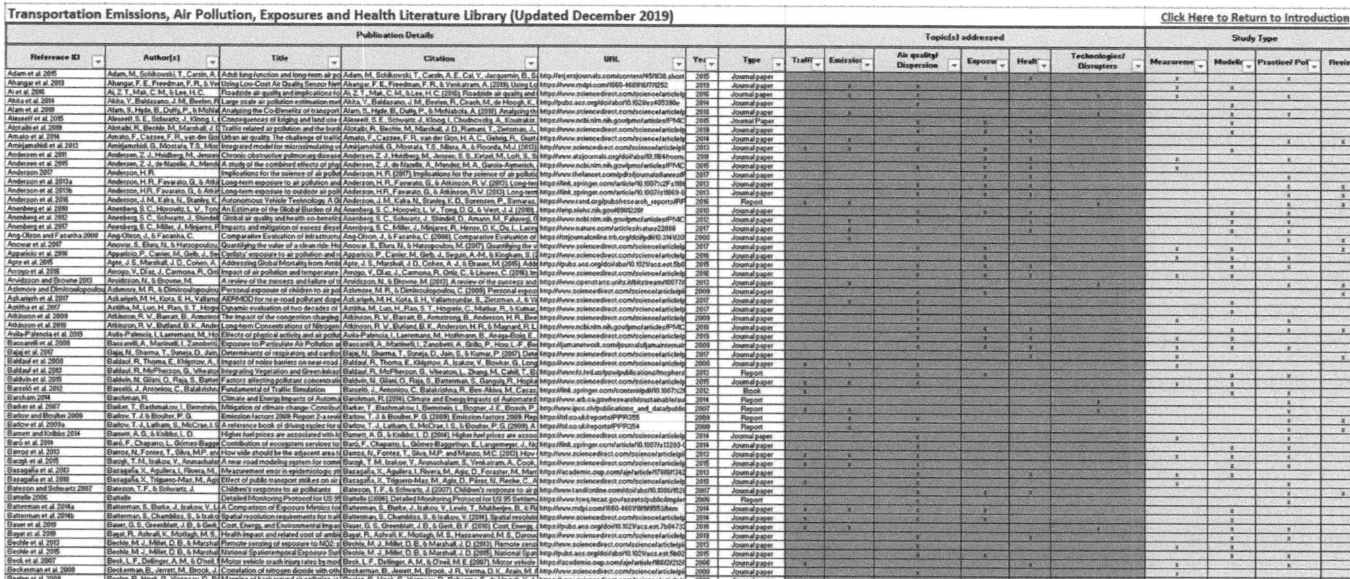

Fig. 23.4 Literature library excel sheet.

into a search bar to narrow their results. Library studies matching the users' search criteria will directly display on the webpage and can be exported as a CSV file. This makes it easy for users to access and save material that they are specifically searching for.

Suggestion box

The Library Suggestion Box provides users with a platform to send feedback to library managers. Feedback may include questions, request for information/access, content suggestions, general comments, or other miscellaneous comments. Users are encouraged to submit their input so library managers may implement feedback to enhance the library experience.

Literature library analysis

All 946 studies were analyzed and results from these analyses are described below. The data sheds light as to where information provided in the literature library is clustered as well as where it lacks.

As mentioned before, categorization by "Literature Topic" and "Study Type" is not mutually exclusive; that is, a specific article may include one or more applicable topics or types of methods/analysis. Air quality is the most addressed literature topic for studies in the library (62.2%), followed by exposure (53.8%), health impact (46.2%), emissions (30.3%), traffic (29.2%), and technologies (7.7%) (Fig. 23.5). A major highlight is only 26 studies address all elements in the full chain—merely 2.7% of the library. Further analysis reveals practice/policy (55.2%) is the most common study type addressed in the library, followed by modeling (50.3%), then measurement (47%), and review (22.9%) (Fig. 23.6). It is surprising that solely 55.2% of library studies align under the practice/policy category as research is conducted to gain knowledge and ultimately initiate policy change. Under this assumption, all or most studies should have some sort of policy implication; however, this is not the case. One theory for this gap is that policy implications may not always be explicit in articles, making it difficult to accurately categorize them as such. Further, researchers might also be reluctant to make clear policy recommendations based on their works. Analysis by literature source demonstrates that approximately 90.7% of literature library studies are journals, followed by reports at 6.6%, conference papers at 1.6%, theses at 1%, and books at 0.2%. This is because current literature collection methodology yields primarily journal articles rather than gray literature. The distribution of studies by publication year and literature topic was also

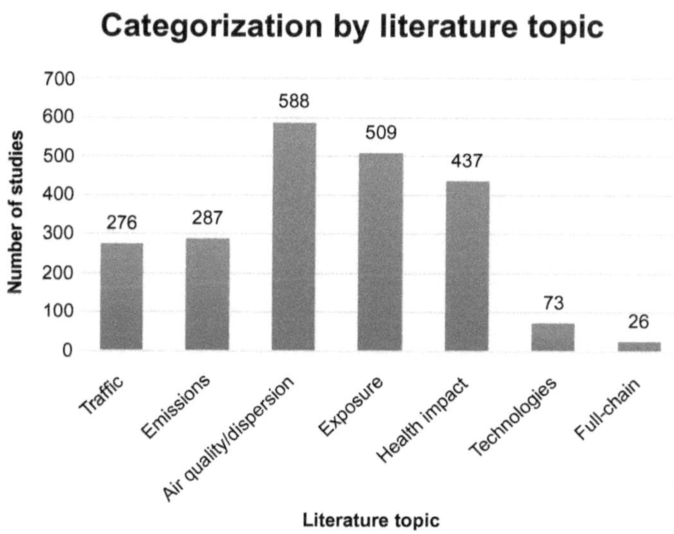

Fig. 23.5 Categorization by literature topic.

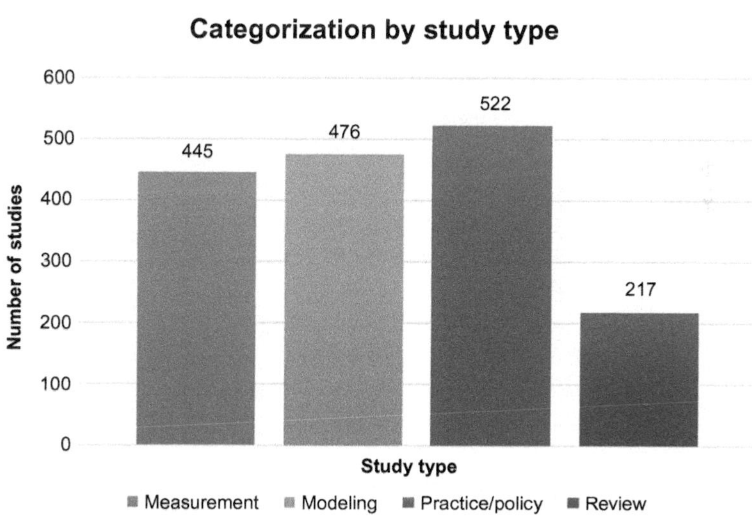

Fig. 23.6 Categorization by study type.

analyzed. Fig. 23.7 displays a visual of each literature topic dispersed across each year between 2000 and 2019. Studies prior to the year 2000 were not considered for addition to the library in order to maintain the most recent collection possible. Most studies in the library have been published between 2012 and 2018, as shown in Fig. 23.7. Finally, the distribution of studies

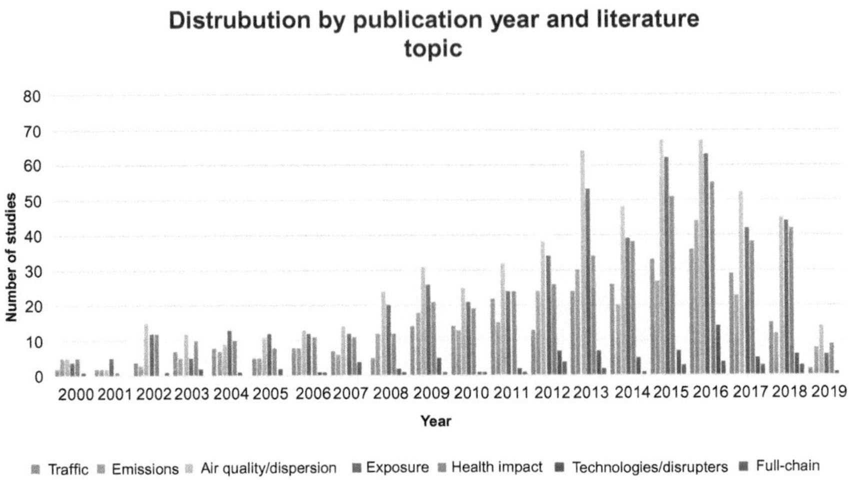

Fig. 23.7 Distribution by publication year and literature topic.

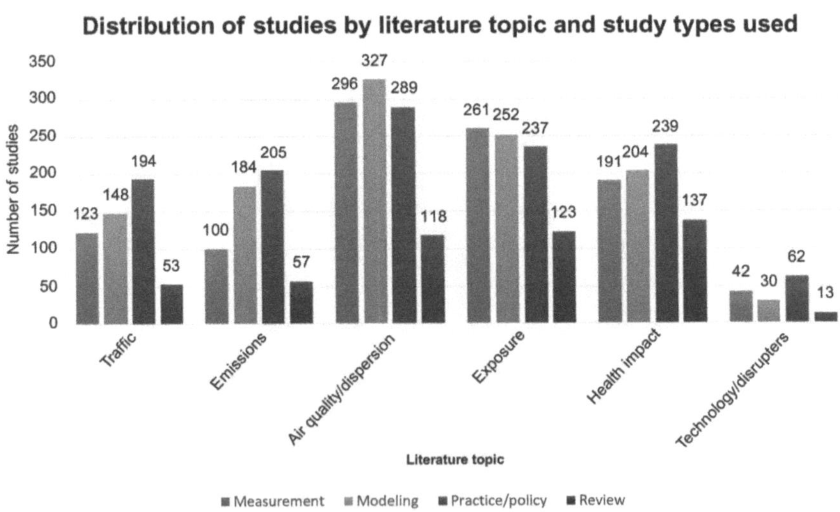

Fig. 23.8 Distribution by literature topic and study types used.

by literature topic and study types used was analyzed. Fig. 23.8 compares the number of studies incorporating measurement, modeling, practice/ policy or review with respect to each topic (traffic, emissions, air quality, exposure, health impact, and technologies). An interesting finding to highlight is that library studies involving traffic, emissions, air quality/dispersion, and health impact were more likely to incorporate modeling than measurement while exposure and technologies/disrupters were more likely to incorporate measurement than modeling. Perhaps this is an indication of practice differences in these fields. Another interesting point to mention is that practice/policy was implemented more than any other study type in traffic, emissions, health impact, and technology/disrupter-related studies. Traffic and health impact topics sit on each end of the full chain, and their high number of practice/policy studies emphasizes the call for change in these fields. However, practice/policy is not addressed as often for studies involving elements within the full chain, such as air quality/dispersion and exposure. This further demonstrates the gap lying within the full chain. Ideally, there would be consistency in practice/policy utilization across all elements of the full chain with the intent to promote policies mitigating emissions, air pollution, human exposures, and health effects of TRAP.

An important theme throughout this analysis is there is a lack of studies addressing all five elements of the full chain. There are studies that address one, two, three, or even up to four topics at a time but do not provide the full "head to tail" picture of TRAP's impact on human health. Understanding the relationship between all elements of the full chain is important because it paves way for discussion about practice/policy implications and making more concrete/specific policy recommendations. Research with a start-to-finish, evidence-based emphasis provides a fuller picture on air pollution sources and the human health effects/impacts of those particular sources. In practice, modeling and analyzing the full chain will create a more integrated knowledge base in this multidisciplinary field and will facilitate a more comprehensive, wholesome investigative approach. Typically, TRAP health research is segregated into separate stages and assessed by different disciplines (Khreis, 2018). There is difficulty in addressing all full-chain elements at once because most students, practitioners, and researchers have expertise in different areas. However, collaboration across these disciplines can aid in considering all full-chain elements. Perhaps a set of related, or companion studies could be conducted by a group of experts from the different areas of the full chain. Therefore, all parts of the full chain are still taken into consideration across the set of studies. Additionally, a multidisciplinary group could choose to collaborate on one project if all full-chain elements were to be considered in the same study. In order to progress toward policy change, researchers are highly encouraged to conduct studies involving or at least considering the full chain. Further, there is great importance in collaborating across fields to connect the beginning and end of the full chain.

Researchers can refer to the studies that we identified for more insight on how to conduct full-chain analyses. Likewise, this book covers the different elements on the full chain between transportation sources and human health, including the assessment of traffic sources, emissions, air pollution/dispersion, human exposures, and health impacts and as such is a useful reference on assessment methods for each element. This book's highly specialized content and detailed nature makes it a prime resource in this field as it can be used by students, researchers, and practitioners conducting relevant studies with the intent to influence policy. Thorough and specific knowledge can help policymakers in the decision-making process as they may be more confident in considering regulation changes (e.g., creating stricter emissions standards or phasing out a particular vehicle fleet) to improve the population's health.

Library limitations

There are three main limitations of the library laid out in Table 23.3. The first limitation is the literature selection thus far has not been systematic. A systematic search method may facilitate the discovery of more relevant papers that have not been included in the library. However, it should be noted that the effort in the literature search thus far came about organically. The purpose was to collect material for a library hub, rather than answering a specific research question. Limited time and resources did not allow for the option of conducting a systematic review.

The nonsystematic nature of literature collection leads us into to our second limitation: there is a limited amount of content available in the library thus far. This library was created in recent years and is in early stages of development, so some relevant content may be missing from our collection at this point in time. However, the addition of studies is ongoing and aims to include a wider variety of studies as more literature is identified and added. Library developers encourage users to submit material that they feel relevant to be considered for addition into the library via email or Library Suggestion Box.

Table 23.3 Limitations of the literature library.

Library limitations
Nonsystematic literature search
Limited content available in the library
Literature topics and study type categorization is conducted in a semisubjective manner

The third limitation is that the literature topics and study types were categorized and double checked by only two independent researchers. Although a portion of current library studies' categorizations were checked by both researchers for consistency, there is still an outstanding portion of studies only categorized by one person. Categorization followed keyword and theme guidelines; however, the researcher ultimately decides what topic(s) and study type(s) to mark in a semisubjective manner. For example, one researcher may view an article as having policy implications, whereas the other may not. This may skew the results obtained from library analysis in regard to the number of studies addressing each topic and study type. There would be potential benefits from including additional personnel in cross-checking categorization for greater consistency. Additionally, we encourage any students, researchers, or practitioners who use this resource to highlight any inconsistencies to library developers via the Library Suggestion Box.

Literature library benefits

Aside from the aforementioned library limitations, there are multiple benefits that stem from the literature library. As mentioned earlier in the chapter, the library reduces the time it takes for students, researchers, and practitioners to identify and access relevant material related to full-chain elements and technologies, highlights literature gaps, and emphasizes the need for more consideration of the full chain in research and practice. In addition, the literature library is beneficial in other aspects including its versatility and application.

Versatility

The literature library is an open-access tool available at https://www.carteeh.org/carteeh-literature-library/. There is potential for a diverse audience to utilize this resource. Of course, the primary audience for the library includes students, researchers, and practitioners; however, that does not mean others are limited in its use. The tool may be used by various professionals across many disciplines. The versatile nature of this tool, therefore, may facilitate interdisciplinary interactions as different professionals may conduct work across different elements of the full chain.

Application of concepts

The content compiled in the library includes literature addressing topics across the full chain in addition to technology. The literature library is a hub of materials that expand on and/or put into practice all concepts discussed in previous chapters, such as traffic, emissions, air pollution, exposures, health, measurement, modeling, policy, and reviews. Persons interested in this field and the applications of concepts discussed in this book can access specialized papers from this resource.

Future research recommendations

A recommendation for future research includes conducting more studies using all full-chain elements. This recommendation is geared toward students, researchers, and practitioners to encourage them to incorporate, or at least symmetrically consider, all full-chain elements in their studies. As described before, this start-to-finish approach highlights the different components of the full chain and links them together to pinpoint the impacts of particular air pollution sources. Implementing the full chain in more research will pave way for discussion about transportation emissions, air quality, exposure, and health policy and practice implications. Benefits of this include influencing policymakers in their decision-making process and creating a more integrated knowledge base highlights the areas responsible for the most uncertainties and errors. An example to demonstrate this point is given next.

Khreis et al. (2018), which is referenced in this chapter and included in the literature library, incorporated all five elements of full-chain analyses and highlighted its importance regarding policymaking. This work assessed the burden of disease, specifically childhood asthma, which was attributable to traffic-related air pollution in Bradford, UK (Khreis et al., 2018). Traffic, vehicle emission, and atmospheric dispersion modeling were used to assess children's exposures and estimate the number of childhood asthma cases attributable to TRAP in this area. Ultimately, the article's results stated that 18% of annual childhood asthma cases in Bradford are attributable to air pollution, specifically NO_2. A total of 3% of cases in Bradford are specifically attributed to traffic-NO_2 without minor roads and cold starts and 7% of cases are specifically attributed to traffic-NO_2 with minor roads and cold starts (Khreis et al., 2018). The main takeaway here is the study pinpointed the source (traffic) of this particular health outcome (childhood asthma), thus providing policymakers with clear, detailed information they can consider while making relevant policy decisions to improve children's health. For example, policymakers might create stricter emissions standards or phase out a particular vehicle fleet in response to the information found in this article. This study is one example of a real-life application of important concepts discussed throughout this book, including the full chain, traffic modeling, vehicle emissions modeling, air pollution modeling, exposure assessment, source apportionment, health impact assessment, burden of disease assessment, and TRAP's impact on health in policymaking. More detailed information on each of these concepts can be found in other chapters.

Summary

This chapter highlighted the development, analysis, and benefits of a literature library, emphasized the full-chain assessment of TRAP and health in research, practice, and policymaking and described the current state of literature regarding traffic activity, traffic-related emissions, air pollution, human exposures, and health. The literature library is unique in its highly specialized content and efficiency. A total of 946 articles related to full-chain elements, in addition to technologies, were collected and categorized in one open-access location, accompanied by a criteria-based online search tool. The compilation of these studies was the first step in assessing the collective body of existing literature relating to TRAP and human health. We provided a higher-level analysis of where the literature is clustered and where it is missing and identified research and knowledge gaps to guide the composition of similar repositories and the design of relevant studies in the future. This work identified a lack of material incorporating all full-chain elements (only 2.7%), therefore encouraging students, researchers, and practitioners to implement, or at least systematically consider, the full chain in their works. Utilization of the full-chain assessment approach will ultimately help pinpoint traffic sources of air pollution that impact health outcomes. Thus, policymakers will be aided in their decision-making process regarding TRAP and adverse health outcomes. Additionally, the full chain in practice will help facilitate a more interdisciplinary, comprehensive investigative approach, rather than a segregated discipline-based assessment. Finally, this chapter is only one part of an entire book that is a prime resource in this field due to its specialized content and valuable insight on the full-chain elements, policy, and technology.

Acknowledgements

This work was partially funded from the Texas A & M Transportation Institute's Center for Advancing Research in Transportation Emissions, Energy, and Health (CARTEEH), a U.S. Department of Transportation's University Transportation Center, College Station, TX. The grant number is 69A3551747128. More information about CARTEEH is available at: https://www.carteeh.org/.

References

Antoniou, C., Balakrishna, R., Barceló, J., Ben-Akiva, M., Casas, J., Choudhury, C., et al. (2012). Fundamental of traffic simulation. *International Series in Operations Research & Management Science, 145*. Available at: https://link.springer.com/content/pdf/10.1007%2F978-1-4419-6142-6.pdf.

Baldwin, N., Gilani, O., Raja, S., Batterman, S., Ganguly, R., Hopke, P., et al. (2015). Factors affecting pollutant concentrations in the near-road environment. *Atmospheric Environment, 115*, 223–235. Available at: https://www.sciencedirect.com/science/article/pii/S1352231015300972?via%3Dihub.

Barker, T., Bashmakov, I., Bernstein, L., Bogner, J. E., Bosch, P., Dave, R., et al. (2007). Mitigation of climate change: Contribution of working group III to the fourth assessment report of the intergovernmental panel on climate change. *Intergovernmental Panel on Climate Change (IPCC)*. Available at: http://www.ipcc.ch/publications_and_data/publications_ipcc_fourth_assessment_report_wg3_report_mitigation_of_climate_change.htm.

Beelen, R., Raaschou-Nielsen, O., Stafoggia, M., Andersen, Z. J., Weinmayr, G., Hoffmann, B., et al. (2014). Effects of long-term exposure to air pollution on natural-cause mortality: An analysis of 22 European cohorts within the multicentre ESCAPE project. *The Lancet, 383*(9919), 785–795. Available at: https://www.sciencedirect.com/science/article/pii/S0140673613621583?via%3Dihub.

Brook, R. D., Rajagopalan, S., Pope, C. A., Brook, J. R., Bhatnagar, A., Diez-Roux, A.V., et al. (2010). Particulate matter air pollution and cardiovascular disease. *Circulation, 121*(21), 2331–2378. Available at: http://circ.ahajournals.org/content/121/21/2331.

Cesaroni, G., Forastiere, F., Stafoggia, M., Andersen, Z. J., Badaloni, C., Beelen, R., et al. (2014). Long term exposure to ambient air pollution and incidence of acute coronary events: Prospective cohort study and meta-analysis in 11 European cohorts from the ESCAPE project. *BMJ, 348*, f7412. Available at: http://www.bmj.com/content/348/bmj.f7412.

Eze, I. C., Hemkens, L. G., Bucher, H. C., Hoffmann, B., Schindler, C., Künzli, N., et al. (2015). Association between ambient air pollution and diabetes mellitus in Europe and North America: Systematic review and meta-analysis. *Environmental Health Perspectives, 123*(5), 381. Available at: https://www.ncbi.nlm.nih.gov/pmc/articles/PMC4421762/.

Fleischer, N. L., Merialdi, M., van Donkelaar, A., Vadillo-Ortega, F., Martin, R. V., Betran, A. P., et al. (2014). Outdoor air pollution, preterm birth, and low birth weight: Analysis of the World Health Organization global survey on maternal and perinatal health. *Environmental Health Perspectives, 122*(4), 425–430. Available at: https://ehp.niehs.nih.gov/1306837/.

Gasana, J., Dillikar, D., Mendy, A., Forno, E., & Vieira, E. R. (2012). Motor vehicle air pollution and asthma in children: A meta-analysis. *Environmental Research, 117*, 36–45. Available at: http://www.sciencedirect.com/science/article/pii/S0013935112001442.

Health Effects Institute Panel on the Health Effects of Traffic-Related Air Pollution. (2010). *Traffic-related air pollution: a critical review of the literature on emissions, exposure, and health effects (Special report no. 17)*. Health Effects Institute. Available at: https://www.healtheffects.org/publication/traffic-related-air-pollution-critical-review-literature-emissions-exposure-and-health.

Ishak, S., & Al-Deek, H. (2002). Performance evaluation of short-term time-series traffic prediction model. *Journal of Transportation Engineering, 128*(6), 490–498. https://doi.org/10.1061/(ASCE)0733-947X(2002)128:6(490).

Jerrett, M., McConnell, R., Wolch, J., Chang, R., Lam, C., Dunton, G., et al. (2014). Traffic-related air pollution and obesity formation in children: A longitudinal, multilevel analysis. *Environmental Health, 13*(1), 49. Available at: https://doi.org/10.1186/1476-069X-13-49.

Khreis, H. (2018). *Early-Life Exposure to Traffic-Related Air Pollution and Risk of Development of Childhood Asthma: A Systematic Review and Meta-Analysis, a Novel Exposure Assessment Study and a Health Impact Assessment* [Doctor of philosophy thesis]. The University of Leeds, Faculty of Environment, Institute for Transport Studies. [Doctor of philosophy thesis] https://www.researchgate.net/publication/323586326_Early-Life_Exposure_to_Traffic-Related_Air_Pollution_and_Risk_of_Development_of_Childhood_Asthma_A_Systematic_Review_and_Meta-Analysis_a_Novel_Exposure_Assessment_Study_and_a_Health_Impact_Assessment.

Khreis, H., de Hoogh, K., & Nieuwenhuijsen, M.J. (2018). Full-chain health impact assessment of traffic-related air pollution and childhood asthma. *Environment International, 114*, 365–375. Available at: https://www.sciencedirect.com/science/article/pii/S0160412017320184.

Khreis, H., Kelly, C., Tate, J., Parslow, R., Lucas, K., & Nieuwenhuijsen, M. (2017). Exposure to traffic-related air pollution and risk of development of childhood asthma: A systematic review and meta-analysis. *Environment International, 100*, 1–31. Available at: http://www.sciencedirect.com/science/article/pii/S0160412016307838.

Lafuente, R., García-Blàquez, N., Jacquemin, B., & Checa, M.A. (2016). Outdoor air pollution and sperm quality. *Fertility and Sterility, 106*(4), 880–896. Available at: https://ac.els-cdn.com/S001502821662719X/1-s2.0-S001502821662719X-main.pdf?_tid=ff4e3b22-fc87-11e7-8479-00000aacb361&acdnat=1516304849_d301f64cc7be8b4226678f4791f6b3ca.

Lindgren, A., Stroh, E., Montnémery, P., Nihlén, U., Jakobsson, K., & Axmon, A. (2009). Traffic-related air pollution associated with prevalence of asthma and COPD/chronic bronchitis. A cross-sectional study in Southern Sweden. *International Journal of Health Geographics, 8*(2). Available at: https://doi.org/10.1186/1476-072X-8-2. .

Link, M. S., & Dockery, D.W. (2013). Air pollution and the triggering of cardiac arrhythmias. *Current Opinion in Cardiology, 25*(1), 16–22. Available at: https://www.ncbi.nlm.nih.gov/pmc/articles/PMC3750956/.

MacIntyre, E.A., Gehring, U., Mölter, A., Fuertes, E., Klümper, C., Krämer, U., et al. (2014). Air pollution and respiratory infections during early childhood: An analysis of 10 European birth cohorts within the ESCAPE project. *Environmental Health Perspectives, 122*(1), 107–113. Available at: https://www.ncbi.nlm.nih.gov/pmc/articles/PMC3888562/.

Mustafić, H., Jabre, P., Caussin, C., Murad, M. H., Escolano, S., Tafflet, M., et al. (2012). Main air pollutants and myocardial infarction: A systematic review and meta-analysis. *JAMA, 307*(7), 713–721. Available at: https://jamanetwork.com/journals/jama/fullarticle/1104975.

Nieuwenhuijsen, M. J., Khreis, H., Verlinghieri, E., Mueller, N., & Rojas-Rueda, D. (2017). Participatory quantitative health impact assessment of urban and transport planning in cities: A review and research needs. *Environment International, 103*, 61–72. Available at: http://www.sciencedirect.com/science/article/pii/S016041201730128927.

Pedersen, M., Giorgis-Allemand, L., Bernard, C., Aguilera, I., Andersen, A. M. N., Ballester, F., et al. (2013). Ambient air pollution and low birthweight: A European cohort study (ESCAPE). *The Lancet Respiratory Medicine, 1*(9), 695–704. Available at: http://www.sciencedirect.com/science/article/pii/S2213260013701929.

Pedersen, M., Stayner, L., Slama, R., Sørensen, M., Figueras, F., Nieuwenhuijsen, M.J., et al. (2014). Ambient air pollution and pregnancy-induced hypertensive disorders: A systematic review and meta-analysis. *Hypertension, 64*(3), 494–500. Available at: http://hyper.ahajournals.org/content/hypertensionaha/early/2014/06/16/HYPERTENSIONAHA.114.03545.full.pdf.

Power, M. C., Adar, S. D., Yanosky, J. D., & Weuve, J. (2016). Exposure to air pollution as a potential contributor to cognitive function, cognitive decline, brain imaging, and dementia: A systematic review of epidemiologic research. *Neurotoxicology, 56*, 235–253. Available at: https://www.sciencedirect.com/science/article/pii/S0161813X1630105X?via%3Dihub.

Raaschou-Nielsen, O., Andersen, Z. J., Beelen, R., Samoli, E., Stafoggia, M., Weinmayr, G., et al. (2013). Air pollution and lung cancer incidence in 17 European cohorts: Prospective analyses from the European study of cohorts for air pollution effects (ESCAPE). *The Lancet Oncology, 14*(9), 813–822. Available at: http://www.sciencedirect.com/science/article/pii/S1470204513702791.

Raz, R., Roberts, A. L., Lyall, K., Hart, J. E., Just, A. C., Laden, F., et al. (2015). Autism spectrum disorder and particulate matter air pollution before, during, and after pregnancy: A nested case–control analysis within the nurses' health study II cohort. *Environmental Health Perspectives, 123*(3), 264–270. Available at: https://www.ncbi.nlm.nih.gov/pmc/articles/PMC4348742/.

Sapkota, A., Chelikowsky, A. P., Nachman, K. E., Cohen, A. J., & Ritz, B. (2012). Exposure to particulate matter and adverse birth outcomes: A comprehensive review and meta-analysis. *Air Quality, Atmosphere & Health, 5*(4), 369–381. Available at: https://doi.org/10.1007/s11869-010-0106-3.

Sommer, C., Eckhoff, D., German, R., & Dressler, F. (2011). A computationally inexpensive empirical model of IEEE 802.11p radio shadowing in urban environments. In *8th international conference on wireless on-demand network systems and services, WONS 2011* (pp. 84–90). https://doi.org/10.1109/WONS.2011.5720204.

Stafoggia, M., Cesaroni, G., Peters, A., Andersen, Z. J., Badaloni, C., Beelen, R., et al. (2014). Long-term exposure to ambient air pollution and incidence of cerebrovascular events: Results from 11 European cohorts within the ESCAPE project. *Environmental Health Perspectives, 122*(9), 919–925. Available at: https://www.ncbi.nlm.nih.gov/pmc/articles/PMC4153743/.

Thorpe, A., & Harrison, R. M. (2008). Sources and properties of non-exhaust particulate matter from road traffic: A review. *Science of the Total Environment, 400*(1–3), 270–282. Available at: http://www.sciencedirect.com/science/article/pii/S004896970800658X.

United Nations Department of Economic and Social Affairs Population Division. (2017). *World population prospects: The 2017 revision key findings and advance tables* (pp. 1–46). United Nations. Available at: https://esa.un.org/unpd/wpp/Publications/Files/WPP2017_KeyFindings.pdf.

Vrijheid, M., Martinez, D., Manzanares, S., Dadvand, P., Schembari, A., Rankin, J., et al. (2011). Ambient air pollution and risk of congenital anomalies: A systematic review and meta-analysis. *Environmental Health Perspectives, 119*(5), 598. Available at: https://www.ncbi.nlm.nih.gov/pmc/articles/PMC3094408/.

CHAPTER 24

How emerging technology and its integrations is advancing our understanding of urban and traffic-related air pollution

John Medeiros and Haneen Khreis
Center for Advancing Research in Transportation Emissions, Energy and Health (CARTEEH), Texas A&M Transportation Institute (TTI), College Station, TX, United States

Abbreviations

3D	three dimensional
ADHD	attention deficit hyperactivity disorder
AQ-SPEC	Air Quality Sensor Performance Evaluation Center
BC	black carbon
CART	classification and regression tree
CO	carbon monoxide
COPD	chronic obstructive pulmonary disease
CVD	cardiovascular disease
EDF	environmental defense fund
GIS	geographical information systems
GPS	global positioning system
LR	linear regression
LUR	land use regression
MLR	multiple linear regression
MSE	mean squared error
NO	nitrogen oxide
NO_2	nitrogen dioxide
O_3	ozone
Pb	lead
PM	particulate matter
PNC	particle number concentration
R	correlation coefficient
R^2	coefficient of determination
RMSE	root mean squared error
SCK	smart citizens kit
SO_2	sulfur dioxide
TRAP	traffic-related air pollution

Traffic-Related Air Pollution
https://doi.org/10.1016/B978-0-12-818122-5.00024-7

UFP	ultrafine particles
US EPA	United States Environmental Protection Agency
US	United States
WHO	World Health Organization

Introduction

Air pollution is generally defined as air containing elevated concentrations of contaminants or toxic chemicals above natural levels (IARC, 2016). The United States Environmental Protection Agency (US EPA) has identified six pollutants as criteria air pollutants because they pose significant threats to environmental and human health. These include particulate matter (PM), ozone (O_3), carbon monoxide (CO), lead (Pb), sulfur dioxide (SO_2), and nitrogen dioxide (NO_2). Many of these pollutants, such as PM, CO, and NO_2, along with others such as hydrocarbons and Black Carbon (BC), result from traffic activity, especially when occurring in urban areas, which are often referred to as traffic-related air pollution (TRAP) (Brewer, 2019; HEI, 2010).

These pollutants have been associated with numerous negative health effects such as cardiovascular disease, respiratory disease, cancers, and premature death (HEI, 2010; IARC, 2016). While children, the elderly, and citizens with preexisting health conditions are most susceptible to the effects of air pollution, it is still a global issue that affects everyone. The relevance of this issue is also increasing as populations continue to grow at unprecedented rates and the projection is that nearly 70% of people will live in cities by 2050, where air pollution due to traffic emissions is mostly concentrated (United Nations, 2018). The World Health Organization (WHO) identified that 92% of the world's population is living in areas that do not meet their guidelines for healthy air quality and that ambient pollution contributes to 6%–58% of deaths and diseases from lung cancer, respiratory infections, ischemic heart disease, and chronic obstructive pulmonary disease (COPD) (WHO, 2018). The issue is so great that outdoor air pollution as a whole was responsible for an estimated 4.2 million deaths in 2016, and a recent study has attributed nearly 10,000 more deaths in the United States (US) to a 5.5% increase in $PM_{2.5}$ pollution from 2016 to 2018 (Clay & Muller, 2019). In addition, air pollution is not only being linked to adverse physical health effects but also many seemingly less apparent effects on various aspects of one's physiological health. A recent systematic review published in the journal of Current Opinion in Psychology examined many of these associations (Lu, 2019). The review found that air pollution had associations with numerous mental disorders such as anxiety, depression, autism, and even suicide and self-harm, as well as effects on cognitive functions including attention, memory, decision-making, dementia, and Attention Deficit Hyperactivity Disorder (ADHD).

Air pollution monitoring has primarily been accomplished with regulatory monitors at fixed sites owned by the official government or environmental organizations. These regulatory monitors are able to measure multiple pollutants with a high degree of accuracy; however, they are often only present in limited quantities due to high maintenance levels and cost (Castell et al., 2017). In addition, air pollution concentrations have demonstrated significant spatial and temporal variability, especially TRAP. Traffic-related emissions contribute to local and regional air pollution concentrations and are known to vary significantly in space and time in intra-urban and urban settings (HEI, 2010). Wu, Reis, Lin, Beverland, & Heal (2015) found the spatial variability of BC and ultrafine particles (UFP) can be as much as 3× greater than $PM_{0.5-2.5}$ and Apte et al. (2017) found the concentrations of primary pollutants (BC, NO, NO_2) can vary 5× within a single city block. Furthermore, the chemical and physical reactions of pollutants along with their complex interactions with various aspects of the urban environment such as meteorology and land characteristics lead to complex dispersion patterns which further complicate our understanding of the behavior of air pollution (Lateb et al., 2016). Reasons like these make air pollution very difficult to accurately assess; despite this, numerous epidemiological and health studies have had to rely on the insufficient data from these monitors over the years.

A common approach to address measurement limitation is the use of statistical methods and interpolation to create spatially averaged concentration maps, which are then applied over large areas and time durations to conduct exposure assessments of people residing within the defined area (Steinle, Reis, & Sabel, 2013). This is a rather simplified method to conduct exposure assessment because of the many factors not accounted for in these methods, such as ignoring intermediate sources of emissions between monitoring locations which may affect local populations differently. Studies have shown that actual individual exposure should be determined with higher spatiotemporal resolution maps while considering the mobility patterns of each subject as they travel through various environments in space and time (Park & Kwan, 2017). It is because factors like these, that sparse air pollution monitors and traditional modeling methods are not able to characterize the large amount of variability inherent to TRAP and personal exposure.

In an attempt to address the primary limitations of past monitoring networks and modeling methods, there has been a rise in new technology for air pollution measurement and modeling in recent years. Prominent categories of technology include low-cost sensors, mobile sensors, satellites sensing, and advanced modeling techniques such as machine learning and hybrid models. These new methods are able to capture air pollution at higher spatial and temporal resolution and are better suited to capture the variability of TRAP in urban settings. These new methods are also capable of extending air pollution monitoring to areas without networks and capture personal exposure of air pollution, which is what truly affects one's health. This rise in technology has also allowed studies to explore the integration of various combinations of these technologies together in a single campaign. The goal of this chapter is to demonstrate how the low-cost and mobile sensing component of these new technologies are improving results and advancing modeling, exposure assessment, and public health analyses. We will then explore examples of studies applying integrated approaches of these technologies as well as highlight the advantages and insights gained from them.

Methods

Searches were primarily conducted from May 29th to August 2nd, 2019 in Google Scholar and Science Direct. Combinations of key words, including "air pollution," "mobile," "low-cost," "model," "integration," and "stationary," were used. The selected studies were recently published (primarily 5 years or less) and came from peer-reviewed international journals such as Environmental Science and Technology, Building and Environment, Journal of Cleaner Production, and Environmental Health. We included studies of our targeted measurement categories or techniques (mobile, low-cost, modeling) and/or the integration of multiple methods and techniques together. Various literature reviews, articles, and websites were also utilized to compare our observations, provide quick access to general background information and explore current air quality monitoring projects. This paper is not a systematic review and does not cover every aspect of each measurement category, but rather intends to provide the reader with an updated perspective on the current efforts in the industry based on the selected reviewed literature. We chose studies that applied uses of new technology and were good examples of each method in order to demonstrate how these studies were conducted, and the advances made by each one.

Results: Emerging measurement methods

This results section consists of four subsections: Traditional Modeling, Low-Cost Sensors, Mobile Sensing, and Integration. "Traditional Modeling" is included to provide an overview of how modeling is utilized to overcome the limitations of traditional monitoring methods and how LUR models are incorporating mobile monitoring. Next, we included two methods of measurement in "Low-cost Sensors" and "Mobile Sensing" followed by an "Integration" section which features various combinations of mobile, low-cost, modeling, and/or regulatory stationary monitoring together. We chose to include these measurement methods because of their increased popularity and their ability to characterize the spatial and temporal variability of urban and traffic-related air pollution. We also believe these have the potential for more immediate impacts regarding new knowledge of air pollution, citizen awareness, exposure reduction, epidemiology, and policy interventions. We entertained the idea of including satellite remote sensing of air pollution; however, while these operations are increasingly capable of capturing higher spatial resolution, they are more likely to examine air pollution on large geographical scales and therefore are less likely to produce data leading to new insights at the city and street level, which ground-based mobile and low-cost monitors are better suited for. The results section is followed by a "discussion" section, which provides an overview of findings, comparisons of the studies, and recommendations for future research.

Traditional modeling

In order to gain a better understanding of how new measurement technology and its integration together with modeling is helping to advance exposure assessment in the next section; it is first helpful to have a basic understanding of the modeling applications often necessary to improve data from traditional monitoring. Common modeling approaches include statistical averaging or interpolation models, regression models, and dispersion models. There are four main interpolation approaches. Spatial averaging methods calculate the mean of concentrations at monitoring locations and assume an equal weight of each location to predict values at areas of interest despite differing distances to monitoring stations, which therefore disregards potential factors contributing to spatial variability such as geography and proximity. Nearest neighbor improves on this slightly by assigning concentrations to the desired area of interest from the nearest monitor. These two interpolation approaches are no longer as commonly used due to their inability to consider spatial variability and consider pollution from sources in between monitoring locations (Xie et al., 2017). More common interpolation approaches include Inverse Distance Weighting and Kriging. Unlike the past two methods, Inverse Distance Weighting considers closer measurements to contribute more to concentrations through distance-dependent weighted averages while Kriging assigns weights to each location based on spatial correlations among measurements and minimizing error variance (Bayraktar, 2005; Xie et al., 2017). These interpolation methods are able to provide good spatial resolution; however, they function best with distributed and dense monitoring networks, especially when characterizing highly variable pollutants like those associated with TRAP (HEI, 2010; Yu et al., 2018).

On the other hand, land use regression (LUR) models pollutant concentrations through multiple regression based on their relationship and interactions with land use features called predictor variables in a Geographical Information Systems (GIS) framework (Xie et al., 2017). Typical predictor variables included in such models are traffic intensity, road type, land cover, population density, and others (Xie et al., 2017). Pollutant concentrations are treated as the dependent variables while the predictor variables are treated as the independent variables, and pollutant levels can then be predicted in other locations based on the model containing the best-identified predictor variables. These models are often used in epidemiological studies because of their relative practicality and ability to better characterize finer-scale variation in air pollution, which are important for accurate exposure assessment and the minimization of exposure misclassifications (Ryan & Lemasters, 2007). Similar to the other methods, LUR can be limited from low input data, such as limited pollutant concentration measurements and land-use variables. Additionally, the developed models are often location-specific and are difficult to apply to other study areas. Another limitation is that LUR also assumes that observations are independent of each other; however, this is not the case as observations are usually similar in nearby areas (Ryan & Lemasters, 2007; Xie et al., 2017). The methods above have been described in more detail in Chapter 6 by Beevers and Williams.

Atmospheric dispersion models utilize mathematical equations and calculations on input data consisting of pollutant emissions, land use, and meteorological conditions in order to simulate the processes affecting the dispersion and predict concentrations in space and time (IARC, 2016; Xie et al., 2017). There are numerous types of dispersion models which operate using different mathematical formulas and assumption in order to be better suited for certain situations or environments. The most common is the steady-state Gaussian-based dispersion model which approximates pollutants in the downwind and vertical directions using a partial differential equation as a function of mean wind speed. This model has a fast-computational time but does not operate well in situations with low wind speeds and significant three-dimensional (3D) dispersion. Other types include Lagrangian and Eulerian models. Lagrangian models operate using ordinary differential equations and consider the effects of buoyancy and turbulence. Eulerian models operate using second-order partial differential equations with initial and boundary conditions to provide an evolution of pollutant concentrations (Leelőssy et al., 2014; Xie et al., 2017). In general, dispersion models do not require dense monitoring networks, which make them well suited for use with past and current measurement capabilities. Despite this, their accuracy depends on the quality of input data which can be expensive and time-consuming to obtain, their assumptions or equations used could be unrealistic and they often require specialized personnel to operate (HEI, 2010; IARC, 2016; Xie et al., 2017).

Low-cost sensors

An important recent addition to air pollution measurement technology is low-cost air quality sensors. Low-cost air sensors were identified as an integral aspect of a "changing paradigm of air pollution monitoring," consisting of a shift from reliance on government and regulatory monitoring to the use of low-cost, portable, and easy-to-use sensors made possible due to advances in micro-fabrication techniques, microelectromechanical systems, and energy-efficient radios and sensor circuits (Snyder et al., 2013). Today, numerous studies have been conducted which have demonstrated the capabilities of these sensors and there now exists a multitude of low-cost sensors capable of measuring a variety of PM and gaseous pollutants. PM sensors operate based on light scattering principles, in which aerosol sizes and numbers are derived based on the amount of light reflected by the aerosols and subsequently absorbed by a photodetector (Morawska et al., 2018). Gaseous sensors are primarily based on metal-oxide semiconductors and electrochemical sensors, in which gaseous concentrations are derived from changes measured in chemical and electrical properties such as conductivity, capacitance, and resistance or sensors which detect the absorption of light by chemiluminescence (Snyder et al., 2013).

The largest downsides of low-cost sensors are their accuracy when compared to regulatory monitors and susceptibility to environmental factors such as temperature, relative humidity, pressure, drift, and cross-sensitivities to other pollutants (Clements et al., 2017). R^2 values for low-cost sensors evaluated in field conditions against reference monitors typically range from 0.4 to 0.8 (Clements et al., 2017). Despite low-cost sensors not being able to provide accuracy levels of reference monitors, their small size, relatively low-cost, and relative ease of use measuring pollutants have allowed flexibility in their intended application, making them suitable for a variety of different deployments. A 2018 literature review by Morawska et al. (2018) conducted on the state of application of low-cost sensors attempted to answer whether the technology was fit for purpose and how far its application has progressed. The authors concluded that low-cost sensors were successful in expanding community awareness and supplementing routine networks, while work was needed on enhancing source compliance and wide-scale monitoring of personal exposure (Morawska et al., 2018). One of the most common applications of low-cost sensors is the creation of a network by a scientific organization or research team, which consists of numerous interconnected sensors transmitting data to a central location or database. Deployment in this manner can provide a more complete assessment of spatiotemporal variability of air pollution when compared to traditional monitoring through the identification of hotspots, temporal variability, and concentrations affecting personal exposures (Morawska et al., 2018).

Caubel, Cados, Preble, and Kirchstetter (2019) conducted such a study in Oakland, California in 2017 containing the largest and most dense ever network of BC monitors by using 100 low-cost PM sensors for 100 days, allowing a spatial density of 6.7 monitors/km^2, which was approximately 100× greater than traditional regulatory networks. The sensors were developed specifically by Caubel et al. (2019) and featured increased precision and accuracy through an algorithm that diminished sensitivity to temperature, making them more appropriate for outdoor use. Prior to deployment, they were zero calibrated in a lab for 1–14 days and then colocated with reference monitors for 2–7 days to derive calibration factors and validate measurements. Sensors were then placed at 1.5 m height on fences, trees and porches within a variety of locations categorized as industrial, residential, truck route, port, and upwind in West Oakland. Each sensor wirelessly transmitted data to an online database while also simultaneously recording 0.5 Hz measurements to an internal SD card which was periodically collected and uploaded to the database. The data were then combined to generate 240,000 hourly average concentrations for BC, of which 16% were excluded due to power loss, pump failure, and technical errors. The results of the study revealed that BC concentrations varied significantly based on land use, traffic patterns, wind direction, and time (Caubel et al., 2019). The highest concentrations were obtained near heavy-duty diesel trucking routes, during both nighttime and daytime shifts, while the lowest was at the upwind destinations near the San Francisco Bay, where concentrations were some 200% lower (0.2–0.4 $\mu g/m^{-3}$) than the trucking routes. Residential sites in the West Oakland community recorded values in between the upwind locations and trucking routes, some 67% higher and 29.1% lower respectively. In addition, the resulting data revealed the sharp decay in concentrations between various sensor locations and a 75%–120% fluctuation over diurnal cycles and 50%–90% over weekly cycles.

Low-cost sensors are also playing an important role in what some have called, the "democratization" of air quality monitoring, allowing ordinary citizens to take matters into their own hands when it comes to measuring air pollution and their personal exposure. This is made possible by several consumer level and easy to use low-cost sensors on the market today. In addition, the Air Quality Sensor Performance Evaluation Center (AQ-SPEC) and the US Environmental Protection Agency (US EPA) have provided online resources for citizens and scientists to help purchase, use and evaluate the performance of many popular sensors on the market (http://www.aqmd.gov/; https://www.epa.gov/air-sensor-toolbox). Many of these low-cost sensors are designed to be deployed either for stationary use indoor or outdoor and in some event to be worn on one's person, often having a dedicated application where the user can view data gathered by the sensors. Such personal sensors can be critical in evaluating personal exposure to air pollution for people with known susceptibility and increased sensitivity to air pollution.

Mobile sensing

Mobile monitoring presents benefits over traditional stationary monitors because of its ability to measure air pollution with much higher spatial resolution over larger areas at lower costs. Because of air pollution's high spatial variability, having a method to better capture this is important for epidemiological studies, which typically relied on estimates from stationary monitors from sparse networks or distinct spatial modeling techniques like statistical interpolation or LUR models. In order to conduct mobile monitoring campaigns, these studies have often utilized numerous transportation forms such as cars and public transit vehicles including tramcars and busses to characterize air pollution over city-scale areas. Mobile monitoring campaigns can be structured, in which specific routes set for travel to target specific areas or pass by a reference monitor for calibration purposes, or unstructured in which no specific routes or requirements for sampling are set. Structured campaigns require more labor, planning, and cost, but are usually more reliable because of the nature of repeated sampling in each location, which results in more data and may provide better estimates of long-term pollutant concentrations and their spatial patterns.

An excellent example of a structured mobile monitoring on cars is the monitoring campaign conducted by Apte et al. (2017), which used Google Street View cars mounted with laboratory-grade Aclima Environmental Intelligence monitors measuring BC, NO, and NO_2 at high temporal resolution (1-Hz) for 6–8 h over 1 year in residential, commercial, and industrial areas of Oakland, California. The objective of the project was to better characterize the spatial variability of various pollutants while addressing how past studies consisting of mobile sensing units only operating for short durations lacked sufficient data to accurately represent long-term spatial patterns. Therefore, this study was unique because of the high coverage density and repeatedly sampling due to its structured routes, which resulted in massive amounts of data generated (3×10^6 1-Hz measurement) over 750 road kilometers. In order to manage this data, a data reduction approach called "snapping" was applied. This assigned the many 1-Hz measurements to a common central location of each 30-m roadway segments based on GPS coordinates, which created high resolution (30 m) maps over the 30 km^2 area. Median concentration values for each roadway segment were then chosen as a measure for central tendency to minimize bias estimations from rare concentration peaks and capture more typical concentrations.

It was found that the primary pollutant concentrations could vary much more than anticipated in such fine-scale environments. Values varied anywhere from 2 to 8× within a city block or individual neighborhood and consistent daytime hotspots were identified throughout the area near vehicle repair facilities, trucking routes, and industrial sites. In addition, pollutant concentrations were found to differ from the reference monitor by as much as 61%, which demonstrates that traditional sparse monitoring networks can provide imprecise approximations for air pollution concentrations contributing to population exposures (Apte et al., 2017). The high resolution data and fine-scale variations in intra-urban environments obtained from the Google Street View cars allowed the first believed cohort study to examine the link between cardiovascular disease (CVD) outcomes within individual neighborhoods using such fine-scale NO, NO_2, and BC data to be conducted (Alexeeff et al., 2018). The overall results (hazard ratios) between pollutants and health outcomes were found to be consistent with other studies utilizing less dense monitoring networks and over larger areas; moreover, it was found that elderly populations contained a statistically significant increased risk to CVD events. Thus, it characterized the effects of TRAP also found in traditional studies which assume the same concentration for larger areas; however, in the context of this study, the use of high resolution data obtained from the mobile monitoring campaign allowed these findings to also be observed in a much smaller geographic region with a highly localized population. The success of the Oakland study led to the expansion of measurement locations in various areas of California and cities in the US and Europe such as Houston, Salt Lake City, Copenhagen, and Amsterdam. In early 2018, Google announced an additional expansion of the project, which included the goal of developing 50 cars by the end of 2019, allowing further expansion of mobile monitoring campaigns around the world to cities in Asia, Africa, and South America. In June 2019, Google made its data obtained in areas of California accessible to scientists and organizations through an online request.

While mobile monitoring on vehicles has been used to better characterize spatial distributions of air pollution, an increasingly common application of mobile monitoring now made possible by the miniaturization of sensors technology and the creation of more compact low-cost sensors is to more directly characterize personal exposure through wearable sensors. Sensors can be worn around one's neck, placed in pouches or backpacks and mounted onto bicycles. This is believed to provide the best estimates of actual individual exposure to air pollution due to the sensor's constant colocation with the carrier as he or she travels throughout different environments, which inevitably consist of varying types of pollutants and concentration levels. The importance of accounting for travel behavior in air pollution exposure assessment is very evident in a recent study conducted in Edinburgh, Scotland in the summer of 2018. In this study, the authors utilized Particle Number Concentration (PNC) sensors to examine UFP exposure while commuting on a bicycle (Luengo-Oroz & Reis, 2019). The goal was to examine the effects of three routes consisting of varied land and traffic characteristics and identify sources of PNC peaks. All of the routes were to and from the same location and rather close in proximity. A Testo DiSCmini handheld counter was placed into a backpack with a flexible Tygon polymer tube protruding out near the bikers breathing zone along with a smartwatch for GPS coordinates and a GoPro Hero 2014 camera on the biker to help identify events contributing to peak PNC.

While PNC values showed large variations depending on route and day, overall trends and their causes were observed among the routes. Important factors contributing to variations in PNC concentrations were identified as shared bus/bike lanes, intersections, the act of overtaking and being overtaken by the busses, construction sites, and green spaces. Higher PNC values were measured near the heavier traffic routes with large amount of bus and truck travel adjacent to the bike lane. Lower PNC values were measured on the residential route with more green space and less traffic; however, there was no bike lane on this route. It was then concluded that cyclists should avoid routes with large amounts of truck and bus traffic (diesel) and use green spaces and residential streets (off-road paths) more often. This study represents a successful use of mobile monitoring to measure personal exposure to UFP because of its identification of higher polluted routes, areas, and their contributing factors. In addition, it helped corroborate knowledge regarding the high spatial variability of UFP and the importance of localized sources in UFP generation. Furthermore, while route choice can reduce exposure, tackling air pollution with policy and regulations is necessary to reduce UFP emissions. It was hoped that the results of the study and the identification of factors influencing UFP exposure could help facilitate the establishment of more bike-friendly roads, reduced emissions, and inform cyclist about safer routes to choose (Luengo-Oroz & Reis, 2019).

Integration

Increased integration of new technology to advance air pollution exposure has been recognized by governmental bodies like the US EPA and in academic articles. Applications of integration of multiple methods together have been happening for some time, but it is the concurrent rise of these many technological measurement advancements previously discussed and the studies incorporating combinations of them together in a single operation that make this time exciting. Larkin and Hystad (2017) explored many of the technological advances made in low-cost, wearable, mobile and remote sensing, smartphones, and models to aid in personal exposure and health research. They also identified the current state of air pollution exposure research and the broad division between methods examining large populations and small populations, which differ in their designs, methodology, strengths, and limitations. From our review, this also applies to studies utilizing a more integrated approach of these technologies, which appear to be split between smaller scale operations on the ground utilizing the integration of various types of mobile and stationary sensors together, and large geographical scale operations based on networks of reference monitoring and satellite remote sensing. Similar observations have been made in academic studies and governmental organization reports. In an EPA 2014–2018 Progress Report (Williams et al., 2018) one section entitled "Advanced Applications and Integrated Monitoring for Air Quality" explored the "harmonization of existing air quality monitoring networks and satellite measurements for improved quantitative use of satellite data" and a paper by Singla, Bansal, Misra, and Raheja (2018) entitled "Towards an Integrated Framework for Air Quality Monitoring and Exposure Estimation—A Review," focused on integration as it pertains to wearable sensors and mobility information to aid in personal exposure estimation. Many of the measurement methods discussed in the previous sections are then incorporated into various modeling applications, including both traditional modeling methods and those based on artificial intelligence, machine learning, and hybrid models. This integrated approach is now contributing to significant improvements and new capabilities in many studies, which often utilize remote sensing and/or local ground-based operations. While Larkin and Hystad (2017) provided a more broad overview on newer technology for advancing personal exposure, this section aims to provide examples of studies incorporating mobile and stationary sensors together in a single sampling campaign while exploring the methodology, performance, and insights of each ground-based integrated approach.

Ground-based integration

The benefits of integration are present in many recent studies utilizing both mobile and stationary measurement for LUR modeling. One such study carried out the use of tramcar-mounted monitors along the northern Hong Kong Island in the winter and summer of 2014–2015 (Li, Fung, & Lau, 2018). Similar to the study conducted in Oakland using Google Street View cars, this study incorporated a median concentration snapping procedure (50 m) of 1-Hz measurements. It used an optical instrument mounted inside a tramcar which sampled ambient $PM_{2.5}$ fed through conductive tubing, as well as two fixed-site regulatory monitors. One was collocated along the tramcar route in a busy town area mounted at a similar height (3.5 m) to the tramcar in order to measure $PM_{2.5}$ for calibration purposes. The other was on a remote northeastern island away from the city to measure background concentrations in order to normalize daily urban variations of $PM_{2.5}$ measured in the tramcar. Despite tramcar measurements overestimating concentrations by 2.02× and 1.40× in the winter and summer respectively, spatial and seasonal patterns of $PM_{2.5}$ were still determined after calibration was made possible by the two reference monitors. This resulted in high correlations (R=0.84–0.93) between tramcar and fixed monitors. It was found that the winter tramcar $PM_{2.5}$ was approximately twice as high as in the summer due to regional air mass transport determined through the background concentration monitors, which was consistent with other studies. Daily median $PM_{2.5}$ varied greatly in both seasons, but winter $PM_{2.5}$ was found to be less spatially variable. Even though high correlations were obtained between tramcar and the fixed monitors, daily ratios between the two still differed enough for the authors to warn against using traditional monitoring networks to characterize personal exposure in microenvironments and declare that mobile monitoring can provide better data for epidemiology (Li et al., 2018). Spatial variation along the tramcar route was also easily determined due to the high resolution of 50 m, which helped corroborate knowledge of poor air quality in two areas and led to the identification of a new hotspot.

The use of a tramcar mounted PM monitor in Hong Kong allowed one of the first studies examining how building morphology affects the dispersion of street-level air pollution on such a large scale to be conducted (Shi, Xie, Fung, & Ng, 2018). The authors stated that studies examining pollution dispersion in urban environments are usually based on time-consuming and complex numerical simulations, making it difficult for urban planners to optimize planning schemes for better pollutant dispersion. In order to effectively examine the dispersion of pollutants and building morphology's effect, small-scale spatial variability of air pollution must be known, but this was not possible with the limited number of fixed monitors in Hong Kong (Shi et al., 2018). Thus, the utilization of a mobile monitor on public transport along fixed routes provided sufficient data to investigate the building morphology factors using GIS and multivariate statistical analysis. It was found that building morphology explains as much as 37% of the spatial variability of street-level $PM_{2.5}$. Four of the most important model predictor variables identified through stepwise multiple linear regression (MLR) were building volume density (3D spatial distribution of building density, %), building coverage ratio (ground coverage intensity, %), podium layer (0–15 m), frontal area index (horizontal building morphological permeability at street level and dimensionless area ratio) and the variability in building heights (standard deviation of building heights within a specific area, m). In addition, the authors were able to use simple linear regression (LR) to identify the critical buffer width of each morphology factor, which is a defined radius in which the factor strongly affects the street-level air quality. Because this study identified critical building design factors, their critical buffers and correlations with urban air quality, it is hopeful that the information will better inform decisions considering pollution dispersion in urban planning and policy (Shi et al., 2018).

We are also now seeing more novel ways to combine measurement methods together. Simon et al. (2018) conducted a study which was believed to be the first one to evaluate the benefits of combining stationary and mobile measurements for developing regression models of UFP. Their study was conducted in an area of Boston and nearby town Chelsea, with each area featuring a central stationary monitor and mobile monitoring on structured routes. In addition, both areas featured stationary monitors (9 in Chelsea and 11 in Boston) at selected residential locations to evaluate the performance of the generated regression models in predicting UFP concentrations. Two separate multivariable regression models were created for each study area: one based on mobile monitoring only and a hybrid model based on the combination of mobile and stationary monitoring. The hybrid models were based on PNC measurements and featured calculated spatial factors: the ratio of the mobile laboratory to the central reference site measurements. The spatial factor values were determined from a spatial factor model, which used natural log-transformed values of the spatial factors and the covariates in a regression model. Factors were then multiplied by the hourly fifth percentile of the reference site monitor to determine PNC. Model performance was evaluated with adjusted R^2 and root mean squared error (RMSE) on predictions of 50% of the data not used in the model and both models were used to predict hourly ln(PNC) values for the residential locations. Results were then compared to the ambient values measured at the locations with various statistical tests.

The results of the study found the hybrid LUR models had significantly better performance over the mobile-only models. Not only were the models more accurate and precise, but they demonstrated better ability to capture the spatial and temporal trends. Correlations (R) were improved from 0.47 to 0.74 in Boston and from 0.51 to 0.73 for Chelsea, and RMSE was reduced in both models as well. Predictions were overall consistent with measured values for both models, but the hybrid models performed much better in estimating overnight concentrations. Hybrid models also showed greater ability to predict high and low values of PNC at the residence locations compared to the mobile-only models. Despite this, hybrid models for both areas over predicted concentrations, which was actually lowered after including the reference monitor in Chelsea but increased in Boston. This was believed to be related to the location of the reference monitor, which was placed at ground level in a heavy traffic area near a bus station in Boston, while the Chelsea monitor was on a third story away from potential local sources. Due to the overprediction still present in the hybrid compared to the mobile models, systematic error was not reduced and future studies were recommended to consider the location of stationary reference monitors and its impacts on model predictions. In addition, additional reference monitors were speculated to help the systematic error and therefore recommended in future studies. While systematic error was still high, the hybrid models all saw increased precision and therefore classical error, which could reduce potential bias and be beneficial for short term health studies. For epidemiology studies, the authors stated that the systematic error was likely to have limited influence, but it is important to address when interpreting the model outputs as it relates to exposure assessment. This study is important because it one of the first ones to implement a novel hybrid modeling method for LUR demonstrating the improvements in predictive capabilities using both mobile and stationary monitoring (Simon et al., 2018).

The addition of larger low-cost networks and mobile measurement operations results in larger amounts of data which require processing and interpretation. This makes advanced modeling techniques such as machine learning and data-driven not only good candidates to be applied on large-scale areas utilizing remote sensing as commonly seen but also further extended to new ground monitoring methods. While these models have provided encouraging results, the majority of studies using them have only utilized fixed monitoring stations with sparse networks which are often not representative of personal exposure due to their low spatial resolution and high installation height (Mihăiţă, Dupont, Chery, Camargo, & Cai, 2019). In addition, because these methods offer improved resolution and hotspot identification, they are important for quickly interpreting data and will play a crucial role in real-time alert systems. These ideas inspired one study to incorporate machine learning approaches on both stationary and mobile monitoring targeting personal exposure levels in the Grand Nancy Metropolis in France for 2 weeks of Spring 2015. Seven fixed passive tubes measuring NO_2 on street pillars at 3-m height were first placed around two of the most congested intersections and two portable sensors. One sensor was a crowd-funded low-cost unit called the Smart Citizen Kit (SCK) developed at the Institute for Advanced Architecture of Catalonia, capable of measuring NO_2, noise, temperature, humidity, and light. The other sensor was the Azimut mobile electrochemical station capable of measuring NO_2, noise, temperature, and humidity. This sensor which was carried by volunteers at normal human height along routes designed to pass by and wait near each of passive tubes. The monitor also had the ability to transmit its data in real time over GPRS to the MyGreenServices open data portal allowing evaluation and visualization of the data collected, an important factor for data-driven air quality predictions and emergency alert generation: the hopeful end goal for the authors. The passive tubes demonstrated highly accurate average concentrations when compared to a local reference monitor with only a 3.9% error, which paved the way for comparison and analysis of the performance of the other two sensors.

Results of the study showed that the Azimut mobile station was capable of accurately identifying hotspots and concentrations between the mobile monitoring at the human level and the higher passive tubes, which could differ up to 5×. Weather conditions also played an important role in influencing NO_2 concentrations. High humidity, low temperature, and high precipitation appeared to reduce NO_2 levels and accumulation. Comparisons with the SCK sensor demonstrated how different sensors can record varied results despite measuring the same pollutant at the same moment. While the two sensors were not deployed in the exact same manner (SCK was deployed on balcony near closest passive tube), this finding still indicated the need for more long-term data collection and analysis to better determine the reasons behind the variability and understand more about each sensor (Mihăiţă et al., 2019). The next step in the study was to evaluate the factors that determine NO_2 concentrations through various data-driven models. Input variables stored in a matrix included traffic noise, location, temperature, and humidity while the output NO_2 values were stored in a vector. A basic regression model was used on 75% of the data to establish a baseline performance for comparison of additional models through Mean Squared Error (MSE). The first model was a decision tree following a Classification and Regression Tree (CART) algorithm. It identified humidity and location as the most important features and reduced MSE by 50.6% from the base model. The second model was a neural network containing 10 neurons and trained with a Levenberg–Marquardt algorithm. This model generated a 25% reduction in MSE over the decision tree and resulted in R^2 of 0.71. Finally, applying a Bayesian regularization approach saw significant improvements: reducing MSE by another 4.09% and bringing R^2 up to 0.81 (Mihăiţă et al., 2019).

The clear stepwise improvements in R^2 through the data-driven modeling iterations provide a good demonstration of the potential to predict NO_2 concentrations. In addition, the observed discrepancies in concentrations from the mobile monitoring and the passive monitors (3–5 × higher) helped to validate the potential risk in estimating personal exposure from monitors located at higher levels and suggests that mobile monitoring may be better at quantifying actual human exposure to air pollution (Mihăiţă et al., 2019). This study is also unique because it is one of the first studies applying these modeling methods to data captured from personal mobile sensing units. Even with important benefits and insights gained, the authors recognized their study was on a limited area and used only two types of sensors and that scaling to a larger area presents additional challenges such as "higher cost, human resources, material aquistion, data processing, and interpretation." However, it is unique due to its application of integrating stationary and mobile sensing along with advanced modeling techniques and helped provide a proof-of-concept and the conclusion that machine learning methods can be efficient for air quality prediction and real-time alert systems by detecting pollution anomalies in certain areas (Mihăiţă et al., 2019).

Perhaps the most exciting and extensive application of an integrated approach to air pollution monitoring is Breathe London, a large-scale project to better characterize the poor air quality affecting the city of London and its inhabitants, made possible by the partnership between the mayor of London, Environmental Defense Fund (EDF), Google and a consortium of numerous companies. This operation is unique because it utilizes many of the categories of air pollution measurement and modeling previously discussed and more importantly provides an excellent demonstration of the benefits of integrating these technologies together in a single operation. The project consists of a network of 100 AQMesh sensors mounted on a lamppost for stationary monitoring around the city. Sensors were placed near various traffic levels and distances from roadways, areas lacking monitors from the existing network and locations suspected to have sensitive populations such as schools and medical facilities. The project also consists of a mobile monitoring application in the form of Google Street View cars and personal wearable sensors inside the backpacks of 250 school children during their commute to sample diverse areas throughout London. Lastly and perhaps most important, is the Google Cloud-based platform used to manage all of the data obtained and allowing for an interactive map of London containing the locations of mobile monitoring zones and stationary pods while providing relatively real-time air pollution measurements throughout the city. The map is entirely public and can be accessed on the Breathe London website (https://www.breathelondon.org/), where users can view each pod's location and mobile sampling routes, and have immediate access to information on the concentration of pollutants across the city as well as examine one's personal exposure. This is aided with a "find my location" feature that displays recent concentrations from the nearest monitor to the user. This project aims to support policy making, increase citizen engagement, and provide public data visualization services, which can all hopefully go a long way in educating the public and creating policy initiatives to reduce air pollution. Such policy implications are already taking form and found as areas on the map designated as Low and UltraLow Emission Zones. These are areas designated for lower emission vehicles advocated by the city mayor and enforced with the use of vehicle restrictions and violation fines while the evaluation and compliance are carried out with the low-cost and mobile sensing techniques. Initial data released has concluded that over 40% of the sensors are being exposed to NO_2 Level exceeding WHO guidelines (Barret, 2019). In addition, commuting children who traveled by automobile or walking/cycling near main roadways were exposed to significantly higher levels of $PM_{2.5}$ than children who traveled on backroads (Diana Varaden & Barratt, 2019).

Table

The next pages feature a table summarizing important information and takeaways of each of the studies listed in the results section above. Categories include their location/duration, sampling medium, monitoring equipment/pollutants, exposure assessment/methodology, and results (Table 24.1).

Discussion and conclusion

We have now seen a variety of new technological methods for air pollution measurement and aspects of integration between them. The addition of low-cost sensors and mobile monitoring is now paving the way to measure air quality in a variety of ways not possible just years ago. Mobile monitoring campaigns such as the one conducted in Oakland using Google Street View cars and in Hong Kong using a tramcar demonstrated the advantages of such an approach by characterizing urban air pollution with much higher spatial resolution, which led to improved source and hotspot identification. Advantages of mobile monitoring were

Table 24.1 Summary of studies and sampling operations.

Author/title	Location/duration	Sampling medium	Monitoring equipment/ pollutants	Exposure assessment/ methodology	Results and key insights
Caubel et al. (2019) /A Distributed Network of 100 Black Carbon Sensors for 100 Days of Air Quality Monitoring in West Oakland, California	– West Oakland, California – May 19 to August 27, 2017	– Low-cost sensor network	– Aerosol Black Carbon Detector (ABCD) developed by Caubel et al. (2019) – BC	– Stationary sampling-based campaign with 100 sensors placed at 1.5 m height on fences, porches and trees across a 15 km^2 area consisting of industrial, port, residential and various traffic locations for 24 h/day and 7 days/week – Sensors featured a software algorithm to diminish sensitivity to temperature to enhance accuracy and precision	– BC concentrations displayed significant spatiotemporal variation over the area depending on time, location, land use, traffic patterns, and wind direction. Lowest concentrations were obtained at upwind locations whereas concentrations obtained near Port terminal roads frequently visited by heavy-duty diesel trucks were 200% higher. – Zero-calibration of the sensors after the project revealed their performance did not degrade over the entire duration – Subsequent analysis revealed that 2–4 weeks of measurements were sufficient to calculate similar averages as the 14 weeks

Apte et al. (2017)/ High-Resolution Air Pollution Mapping with Google Street View Cars: Exploiting Big Data	– Oakland, California – May 282,015 to May 14, 2016	– Mobile monitoring,	– Two Laboratory grade Aclima Environmental Intelligence fast-response monitors on Google Street View Cars, GPS – BC, NO, NO_2	– Mobile monitoring campaign using two Google Street View cars in 30 km^2 of mixed residential, commercial and industrial areas on weekdays creating median concentration values for 30-m roadway segments – Conducted additional analysis of spatial patterns of pollution-based on distance-decay relationships and the fraction of pollution due to local sources above a baseline value	– Concentrations varied from a single regulatory monitor by 32% (BC), 61% (NO), and 35% (NO_2) – Concentrations varied 2–8× within individual blocks and neighborhoods which suggest that spatial variability is much finer than detected with conventional methods – Consistent daytime hotspots were observed in trucking route intersections, vehicle repair facilities and industrial sites – 2–3× fewer sampling days were sufficient to obtain similar results with good precision and low bias

Continued

Table 24.1 Summary of studies and sampling operations—cont'd

Author/title	Location/duration	Sampling medium	Monitoring equipment/ pollutants	Exposure assessment/ methodology	Results and key insights
Alexeeff et al. (2018)/ High-resolution mapping of traffic related air pollution with Google street view cars and incidence of cardiovascular events	– Oakland, California	– Mobile monitoring	– Utilized mobile monitoring approach and data obtained from above study by Apte et al. (2017)	– Cohort epidemiology study examining the incidence of cardiovascular events including stroke, coronary heart disease and myocardial infarction with covariates such as sex, race, (BMI), smoking status, COPD, hypertension and socioeconomic status in the 30 km^2 area of Oakland sampled by Apte et al. (2017)	– Magnitude of effects among general population were overall consistent with other studies using lower resolution data over larger areas – Corroborated evidence that elderly populations were more susceptible to CVD events than nonelderly – Found no differences in susceptibility to air pollution among diabetics and nondiabetics – Advanced epidemiology studies by utilizing fine-scale data obtained from mobile monitoring and demonstrated how even intraurban variations in TRAP can disproportionately affect local populations in a small study area

Luengo-Oroz and Reis (2019)/Assessment of cyclists' exposure to ultrafine particles along alternative commuting routes in Edinburgh	– Edinburgh, Scotland – June 18 to July 1, 2018	– Mobile monitoring	– Testo DiSCmini handheld nanoparticle counter (Testo Se & Co. KGaA, Germany) with Tygon flexible polymer sampling tube – POLAR M200 (Polar Eletro, Oy, Finland) smartwatch for GPS coordinates – GoPro Hero 2014 for source identification – PNC (UFP)	– Sampling-based method with sensor placed in bike riders backpack sampling air from tubing near rider breathing level along three separate routes for commuting university students characterized by varying traffic levels and infrastructure between 8:00 and 9:00AM on weekdays	– Revealed that route choice had a substantial influence on exposure to UFP for the cyclists – Use of a camera and GPS allowed excellent source identification capabilities which identified shared bus/bike lanes, intersections, overtaking, bicycle boxes, construction sites, and green spaced as important contributing factors – Provided insights for future urban planning such as avoiding shared bus/bike lanes, providing shielding bicycle boxes at intersections to reduce pollution, additional off-road paths and providing information to cyclist about various route choice

Continued

Table 24.1 Summary of studies and sampling operations—cont'd

Author/title	Location/duration	Sampling medium	Monitoring equipment/ pollutants	Exposure assessment/ methodology	Results and key insights
Li et al. (2018)/High spatiotemporal characterization of on-road $PM_{2.5}$ concentrations in high-density urban areas using mobile monitoring	– Hong Kong, China – Dec./Jan. 2014 (winter) and June/July 2015 (summer)	– Mobile and stationary regulatory monitoring	– DustTrak II 8530, GPS (Mobile Tramcar Monitor) – Tapered Element Oscillating Microbalance Model (TEOM) 1405-DF monitor (ThermoFisher Scientific, Waltham, MA, USA) (Stationary Regulatory Monitors) – $PM_{2.5}$	– Mobile and stationary-based campaign with monitor mounted in a tramcar sampling air through ambient tubing on a single route passing a colocated regulatory monitor for calibration as well as a included a fixed regulatory monitor on a remote island measuring background concentrations – Mobile monitoring assigned median concentrations to the nearest 50-m segment along route	– The use of the two regulatory monitors allowed over-estimation ratios and background concertation corrections to be applied to the mobile measurements – Calibration ratios and factors derived from above comparisons resulted in high r=0.84–0.93 correlations of mobile measurements to regulatory monitors – Winter mobile concentrations were found to be 2× summer and less spatially variable, whereas summer was found to be lower and more spatially variably with a larger influence from local on-road concentrations than background – Localized pollution hotspots were identified throughout the route from the mobile monitor which could result in significant errors when estimating personal exposure for epidemiology studies using the single regulatory monitor

Shi et al. (2018)/ Identifying critical building morphological design factors of street-level air pollution dispersion in high-density built environment using mobile monitoring	– Hong Kong, China	– Mobile monitoring	– Utilized mobile monitoring approach and data obtained in above study by Li et al. (2018) – $PM_{2.5}$	– Evaluated 17 critical building morphology factors and their correlations with $PM_{2.5}$ using ArcGIS and multivariate statistical analysis – Identified critical buffer widths (radius for each morphology factor that explains $PM_{2.5}$ to the greatest extent) through simple LR and morphological factors contribution to $PM_{2.5}$ variation with MLR	– Building morphology explained 37% of $PM_{2.5}$ variation in the winter and 31% in the winter – Important contributing factors were identified as building volume density, building coverage ratio, podium layer frontal area index and variability in building heights – The quantitative characterization of building morphology and its affects on TRAP exposure can have important implications for urban planning, policy making, and dispersion analysis

Continued

Table 24.1 Summary of studies and sampling operations—cont'd

Author/title	Location/duration	Sampling medium	Monitoring equipment/ pollutants	Exposure assessment/ methodology	Results and key insights
Simon et al. (2018)/ Combining Measurements from Mobile Monitoring and a Reference Site to Develop Models of Ambient Ultrafine Particle Number Concentration at Residences	– Boston and Chelsea, Massachusetts – Dec 2011 to May 2015	– Mobile, stationary reference and residential monitoring	– TSI butanol-based condensation particle counter, Model 3775 (mobile monitoring) – TSI water-based condensation particle counter, Model 3783 (stationary reference and residential) – PNC (UFP)	– Each of the two study areas featured a central stationary reference monitor, mobile monitor vehicle and stationary monitors at residence sites – Two separate multivariable regression models were created, one based on mobile only measurements and a hybrid model considering both mobile and central stationary measurements – Each model's performance was evaluated by comparing predicted natural log concentrations to those obtained at the residence locations	– Hybrid models performed significantly better than mobile models regarding accuracy and precision (correlations improved from 0.47 to 0.74 in Boston and from 0.51 to 0.73 in Chelsea) – Hybrid models better captured spatiotemporal trends including overnight concentrations as well as extreme high and low values at residence locations – All models overpredicted concentrations, but overprediction was lowered in the hybrid Chelsea model and the location of the reference monitor was suspected to have an important influence – Demonstrated the benefits of combining mobile and stationary monitoring in a single model, but the extent to which the improvements affected epidemiology was recommended for future research

Mihăiţă et al. (2019)/ Evaluating air quality by combining stationary, smart mobile pollution monitoring and data-driven modeling	– Grand Nancy Metropolis, France – April 29 to May 13, 2015	– Mobile and fixed Low-Cost Sensor Monitoring	– Stationary Passive tubes – Azimut Motoring Station (AMS) – Smart Citizen Kit (SCK) – Central regulatory monitor – NO_2	– 7 Stationary passive tubes were placed at 3-m height near congested and busy streets – AMS carried by volunteers at 1.5-m height along route passing by passive tubes – SCK deployed near a passive tube for comparison with AMS – Applied various machine learning models including decision trees, neural network and Bayesian regularization	– The carried mobile Azimut monitor was capable of identifying hotspots and revealed significantly higher NO_2 concentrations (3–5×) compared to passive tubes at higher elevation – The mobile units had higher sensitivity to weather conditions such as temperature, humidity, wind and precipitation, which also influenced NO_2 concentrations – Machine learning was found to be an efficient tool for predicting NO_2 concentrations (R^2 = 0.81) and provided a proof of concept of real time alert systems which could influence citizen travel behavior

Continued

Table 24.1 Summary of studies and sampling operations—cont'd

Author/title	Location/duration	Sampling medium	Monitoring equipment/ pollutants	Exposure assessment/ methodology	Results and key insights
– Breathe London – https://www.breathelondon.org/about/	– London, England – Late 2018–mid-2020	– Mobile, fixed low-cost sensor and personal low-cost sensor monitoring	– AQMesh sensors (stationary) – Two Laboratory grade Aclima Environmental Intelligence fast-response monitors on Google Street View Cars, GPS – Custom wearable sensors inside backpacks manufactured by Dyson – NO_2, $PM_{2.5}$, PM_{10}, PM_1, NO, NO_2, CO_2, O_3	– A network of 100 AQMesh sensors on lamp posts and buildings across mixed traffic areas and "sensitive" locations including schools and medical facilities all across London – Two Google Street Views cars sampling air pollution on various routes and targeted areas from morning to evening on weekdays – 250 school children sampled ambient air during their commute with custom wearable sensors in backpacks to analyze how different modes of travel affect exposure to air pollution	– Monitoring helps evaluate and enforce the enactment of low emission zones and provide "hyperlocal" insights about air pollution across the city – Public interactive map of London containing the locations of mobile monitoring zones and stationary pods that provides relatively real-time air pollution measurements throughout the city – Wearable application demonstrated school children which traveled by cycling/walking or automobile near main streets were exposed to significantly higher $PM_{2.5}$ levels that those who took back roads – The visualization of air pollution and documentation of policy interventions can hopefully lead to healthier air in London

also evident in the study utilizing a bicycle to measure UFP in Edinburgh, in which three separate routes in close proximity linking two destinations demonstrated drastically different concentrations of UFP, stressing the importance of route choice when considering one's exposure to TRAP. In addition, the use of a camera and GPS in this study proved beneficial when examining the causes of elevated levels of UFP across the routes. Data obtained from mobile monitoring and its enhanced source identification and spatial capabilities can, therefore, be essential when enacting policy efforts to mitigate human exposures to air pollution. This can be seen in studies using mobile monitoring and networks of stationary low-cost sensors such as Breathe London, where these two sensor types are crucial for the evaluation of the city's air quality and the areas designated as low emission zones.

Integration of stationary and mobile monitoring also brought about additional benefits not possible when utilized separately. These methods make a good combination because of the mobile monitors' ability to better capture air pollution in space and the stationary monitors' ability to capture temporal variations. The flexibility of the two methods is also beneficial because of the ability to deploy mobile monitoring on specific routes designed to be colocated with stationary monitors. Stationary monitoring is also important for determining background concentrations because it allows for adjustments to observe concentration levels and this helps to quantify pollution due to local emissions sources. In addition, stationary monitors or reference monitors can be used to adjust for temporal correlations not captured in mobile monitoring campaigns. This was evident with the use of both vehicle mobile monitoring and a central fixed monitor to create a single hybrid LUR model, which brought about more accurate and precise data as well as a better ability to capture spatial and temporal trends when compared to models generated from only mobile monitoring. In addition, the use of both mobile and stationary monitoring can allow better identification of pollutant concentrations affecting personal exposure and the potential large discrepancies obtained from lower elevated mobile or low-cost sensors to those measured from often higher elevated reference monitors. This was evident in the small-scale operation exploring various data-driven modeling approaches to personal carried mobile sensors and stationary passive monitors for the prediction of NO_2. One of the most notable findings of this study was the significantly higher concentrations recorded by the street-level mobile monitors when compared to passive monitors just 1.5 m higher (3 m height).

While performance was improved with many studies, there are still some complications present among them. The study which utilized a network of 100 low-cost sensors conducted in Oakland speculated that mobile monitoring and LUR which use a fixed-site monitor to adjust for temporal variation may represent a "flawed assumption." This speculation was due to the high variability of BC concentrations observed in the study and relatively weak correlations found between the network of sensors and one low-cost monitor at a central regulatory location. Because both the 100×100 network and Google Street View study were conducted in the same area of Oakland, initial comparisons between the two methods could be conducted. While both methods were able to identify high spatial variations of BC, the limitation of mobile monitoring to daytime weekday operation meant that it was unable to adequately characterize the diurnal and weekly variation in concentrations obtained from the network of sensors. From the studies examined, it appears that fixed-site monitoring still provides adequate data for adjusting concentrations obtained from mobile monitoring as seen in the Hong Kong study and Boston hybrid model study examining UFP. In the Boston study, temporal adjustments from the fixed monitors helped to lower the overpredictions of concentration obtained from the mobile monitoring in one area. Additionally, the authors noted that overprediction could be further decreased with the inclusion of additional stationary reference monitors and that careful consideration of elevation and location should be taken into account when using them for comparisons and adjustments of values obtained from mobile or low-cost networks. The use of multiple reference monitors can also provide additional data when assessing the performance of mobile monitors and low-cost networks, and allow the nearest reference monitor to be utilized when adjusting or analyzing data among the other monitors. The speculation that stationary reference monitors may not be adequate for such adjustments in the Oakland study was only based on comparisons with one monitor at a central location. Therefore, the ability to obtain improved spatial and temporal resolution can be enhanced by integrating both a network of sensors or mobile monitoring with stationary reference monitor(s); however, the relationships among the monitors play an important role.

Common reservations and conclusions reached from each study include stressing the importance of pre-calibration and validation of sensors in appropriate meteorological conditions before deployment as well as for lengths longer than the duration of the desired sampling period to ensure adequate data throughout the study. Filtering protocols should be developed to get rid of faulty data and an additional data capture backup method for obtaining data is recommended to prevent data loss due to various factors such as poor connections or sensor failure. Additionally, cited by studies utilizing LUR was the limited flexibility of the models to be applied to other areas and perhaps future efforts could explore the creation of more versatile models applicable to other study areas. However, the use of low-cost networks and mobile monitoring could potentially make the transferability of LUR models less of an issue in the future due to their relative ease of use and versatile deployment capabilities which can help to quickly gather data to create new models, or explore correlations with targeted predictor variables, such as the operation in Hong Kong by Shi et al. (2018).

While results of these integration studies have been encouraging with regards to characterizing air pollution at higher spatial and temporal resolutions, which can have outcomes in air pollution control and policy, additional studies should be conducted to further examine the extent of which these results affect epidemiological findings through comparisons to those relying on traditional exposure assessments. A cross-comparison of epidemiological studies utilizing multiple exposure assessments generated from methods highlighted above could be one way to achieve this. Many of the studies reviewed also entertained the idea of expanding to larger scale operations and identified that this will present additional complications due to the need for increased data management, material acquisition, and a rapidly increasing cost; however, many of the studies observed a relatively quick conversion of values and diminished return of sampling durations necessary to obtain acceptable accuracy relative to the complete campaign duration. For example, Caubel et al. (2019) found that 2–4 weeks of measurements were sufficient to calculate values similar to those from the complete 14 weeks and Apte et al. (2017) found that 2–3× less driving days were capable of reproducing spatial patterns with good precision and low bias. This suggests that the use of less intensive sampling campaigns and durations can yield sufficient results, which provides an encouraging outlook for future efforts incorporating mobile, low-cost monitors, integrated campaigns, and larger scale operations such as Breathe London, to characterize air pollution in additional cities, while potentially requiring reduced labor and cost.

In all, the use of new mobile and low-cost monitoring, modeling and their integration together appear to be making large strides in how urban air pollution and TRAP is measured and understood. This has been observed in the studies examined, which have demonstrated a number of improvements and new capabilities. Some advances include source attribution and hotspot identification, higher spatial and temporal resolution, personal exposure characterization, and enhanced predictive capabilities. It is now hoped that these advances can lead to individual citizen awareness, government regulations, policy initiatives, and expanded characterization of air pollution. These outcomes have the potential to produce a collective improvement in public health through improved epidemiology studies, consideration of travel behavior influencing personal exposure and air pollution control.

References

Alexeeff, S. E., Roy, A., Shan, J., Liu, X., Messier, K., Apte, J. S., et al. (2018). High-resolution mapping of traffic related air pollution with Google street view cars and incidence of cardiovascular events within Neighborhoods in Oakland, CA. *Environmental Health: A Global Access Science Source*, *17*(1), 1–13. https://doi.org/10.1186/s12940-018-0382-1.

Apte, J. S., Messier, K. P., Gani, S., Brauer, M., Kirchstetter, T. W., Lunden, M. M., et al. (2017). High-resolution air pollution mapping with Google street view cars: Exploiting big data. *Environmental Science and Technology*, *51*(12), 6999–7008. https://doi.org/10.1021/acs.est.7b00891.

Barret, T. (2019). *40% of breathe London sensors detecting illegal levels of NO_2*. Retrieved from airqualitynews.com website. https://airqualitynews.com/2019/07/24/40-of-breathe-london-sensors-detecting-illegal-levels-of-no2/.

Bayraktar, H. (2005). A kriging-based approach for locating a sampling site—In the assessment of air quality. *Stochastic Environmental Research and Risk Assessment*, *19*, 301–305. https://doi.org/10.1007/s00477-005-0234-8.

Brewer, T. L. (2019). Black carbon emissions and regulatory policies in transportation. *Energy Policy*, *129*(March), 1047–1055. https://doi.org/10.1016/j.enpol.2019.02.073.

Castell, N., Dauge, F. R., Schneider, P., Vogt, M., Lerner, U., Fishbain, B., et al. (2017). Can commercial low-cost sensor platforms contribute to air quality monitoring and exposure estimates? *Environment International*, *99*, 293–302. https://doi.org/10.1016/j.envint.2016.12.007.

Caubel, J. J., Cados, T. E., Preble, C. V., & Kirchstetter, T. W. (2019). A distributed network of 100 black carbon sensors for 100 days of air quality monitoring in West Oakland, California [research-article]. *Environmental Science & Technology*, *53*, 7564–7573. https://doi.org/10.1021/acs.est.9b00282.

Clay, K., & Muller, N. (2019). *Recent increases in air pollution: Evidence and implications for mortality*. National Bureau of Economic Research. https://doi.org/10.3386/w26381.

Clements, A. L., Griswold, W. G., Rs, A., Johnston, J. E., Herting, M. M., Thorson, J., et al. (2017). Low-cost air quality monitoring tools: From research to practice (a workshop summary). *Sensors (Basel)*, *17*, 1–20. https://doi.org/10.3390/s17112478.

Diana Varaden, E. L., & Barratt, B. (2019). *The breathe london wearables study engaging primary school children to monitor air pollution in London*. (pp. 1–12). Environmental Research Group.

HEI. (2010). *Traffic-related air pollution: A critical review of the literature on emissions, exposure, and health effects a special report of the HEI panel on the health effects of traffic-related air pollution-executive summary*. Health Effects Institute.

IARC. (2016). *Outdoor air pollution. IARC monograph evaluation carcinogenic risks human. Vol. 109* (pp. 1–454).

Larkin, A., & Hystad, P. (2017). Towards personal exposures: How technology is changing air pollution and Health Research. *Current Environmental Health Reports*, *4*(4), 463–471. https://doi.org/10.1007/s40572-017-0163-y.

Lateb, M., Meroney, R. N., Yataghene, M., Fellouah, H., Saleh, F., & Boufadel, M. C. (2016). On the use of numerical modelling for near-field pollutant dispersion in urban environments—A review. *Environmental Pollution, 208*, 271–283. https://doi.org/10.1016/j.envpol.2015.07.039.
Leelőssy, Á., Molnár, F., Izsák, F., Havasi, Á., Lagzi, I., & Mészáros, R. (2014). Dispersion modeling of air pollutants in the atmosphere: A review. *Central European Journal of Geosciences, 6*(3), 257–278. https://doi.org/10.2478/s13533-012-0188-6.
Li, Z., Fung, J. C. H., & Lau, A. K. H. (2018). High spatiotemporal characterization of on-road PM2.5 concentrations in high-density urban areas using mobile monitoring. *Building and Environment, 143*(May), 196–205. https://doi.org/10.1016/j.buildenv.2018.07.014.
Lu, J. G. (2019). Air pollution: A systematic review of its psychological, economic, and social effects. *Current Opinion in Psychology*, https://doi.org/10.1016/j.copsyc.2019.06.024.
Luengo-Oroz, J., & Reis, S. (2019). Assessment of cyclists' exposure to ultrafine particles along alternative commuting routes in Edinburgh. *Atmospheric Pollution Research, 10*(4), 1148–1158. https://doi.org/10.1016/j.apr.2019.01.020.
Mihăiţă, A. S., Dupont, L., Chery, O., Camargo, M., & Cai, C. (2019). Evaluating air quality by combining stationary, smart mobile pollution monitoring and data-driven modelling. *Journal of Cleaner Production, 221*, 398–418. https://doi.org/10.1016/j.jclepro.2019.02.179.
Morawska, L., Thai, P. K., Liu, X., Asumadu-Sakyi, A., Ayoko, G., Bartonova, A., et al. (2018). Applications of low-cost sensing technologies for air quality monitoring and exposure assessment: How far have they gone? *Environment International, 116*(April), 286–299. https://doi.org/10.1016/j.envint.2018.04.018.
Park, Y. M., & Kwan, M. P. (2017). Individual exposure estimates may be erroneous when spatiotemporal variability of air pollution and human mobility are ignored. *Health and Place, 43*(February 2016), 85–94. https://doi.org/10.1016/j.healthplace.2016.10.002.
Ryan, P. H., & Lemasters, G. K. (2007). A review of land-use regression models for characterizing intraurban air pollution exposure. *Inhalation Toxicology, 19*(Suppl. 1), 127–133. https://doi.org/10.1080/08958370701495998.
Shi, Y., Xie, X., Fung, J. C. H., & Ng, E. (2018). Identifying critical building morphological design factors of street-level air pollution dispersion in high-density built environment using mobile monitoring. *Building and Environment, 128*(September 2017), 248–259. https://doi.org/10.1016/j.buildenv.2017.11.043.
Simon, M. C., Patton, A. P., Naumova, E. N., Levy, J. I., Kumar, P., Brugge, D., et al. (2018). Combining measurements from Mobile monitoring and a reference site to develop models of ambient ultrafine particle number concentration at residences [research-article]. *Environmental Science and Technology, 52*(12), 6985–6995. https://doi.org/10.1021/acs.est.8b00292.
Singla, S., Bansal, D., Misra, A., & Raheja, G. (2018). Towards an integrated framework for air quality monitoring and exposure estimation—A review. *Environmental Monitoring and Assessment, 190*(9). https://doi.org/10.1007/s10661-018-6940-8.
Snyder, E. G., Watkins, T. H., Solomon, P. A., Thoma, E. D., Williams, R. W., Hagler, G. S. W., et al. (2013). The changing paradigm of air pollution monitoring. *Environmental Science and Technology, 47*(20), 11369–11377. https://doi.org/10.1021/es4022602.
Steinle, S., Reis, S., & Sabel, C. E. (2013). Quantifying human exposure to air pollution-moving from static monitoring to spatio-temporally resolved personal exposure assessment. *Science of the Total Environment, 443*, 184–193. https://doi.org/10.1016/j.scitotenv.2012.10.098.
United Nations. (2018). *World urbanization prospects 2018*. In Webpage. Retrieved from https://population.un.org/wup/.
WHO. (2018). *Ambient air pollution fact sheet*. Retrieved from World Health Organization website. https://www.who.int/en/news-room/fact-sheets/detail/ambient-(outdoor)-air-quality-and-health.

Williams, R., Vasu Kilaru, Conner, T., Clements, A., Colon, M., Breen, M., et al. (2018). New paradigm for air pollution monitoring: Emerging sensor technologies 2014-2018 progress report (181 pp). US EPA. Retrieved from https://cfpub.epa.gov/si/si_public_record_report.cfm?dirEntryId=339746&Lab=NERL.

Wu, H., Reis, S., Lin, C., Beverland, I. J., & Heal, M. R. (2015). Identifying drivers for the intra-urban spatial variability of airborne particulate matter components and their interrelationships. *Atmospheric Environment, 112*, 306–316. https://doi.org/10.1016/j.atmosenv.2015.04.059.

Xie, X., Semanjski, I., Gautama, S., Tsiligianni, E., Deligiannis, N., Rajan, R., et al. (2017). A review of urban air pollution monitoring and exposure assessment methods. *ISPRS International Journal of Geo-Information, 6*(12), 389. https://doi.org/10.3390/ijgi6120389.

Yu, H., Russell, A., Mulholland, J., Odman, T., Hu, Y., Chang, H. H., et al. (2018). Cross-comparison and evaluation of air pollution field estimation methods. *Atmospheric Environment, 179*(January), 49–60. https://doi.org/10.1016/j.atmosenv.2018.01.045.

CHAPTER 25

Traffic-related air pollution: Emissions, human exposures, and health—The way forward

Haneen Khreis[a,b,c,d], Mark J. Nieuwenhuijsen[b,c,d], Josias Zietsman[a], and Tara Ramani[a]
[a]Center for Advancing Research in Transportation Emissions, Energy, and Health (CARTEEH), Texas A&M Transportation Institute (TTI), College Station, TX, United States
[b]Barcelona Institute for Global Health (ISGlobal), Centre for Research in Environmental Epidemiology (CREAL), Barcelona, Spain
[c]Universitat Pompeu Fabra (UPF), Barcelona, Spain
[d]CIBER Epidemiología y Salud Pública (CIBERESP), Madrid, Spain

Abbreviations

BC	black carbon
PM	particulate matter
TRAP	traffic-related air pollution
UFP	ultrafine particles

Introduction

In this book, we offered the reader a comprehensive overview of the state-of-the-art knowledge on traffic-related air pollution (TRAP) and its ultimate impacts on public health. TRAP is a public health threat with increasing relevance, especially in cities and urban areas, where people congregate and are in close proximity to traffic activity (Khreis et al., 2020). This book reflects on TRAP as an adverse but modifiable environmental exposure with a significant public health burden. The book consists of 25 chapters, including this one, from a diverse set of international experts coming from a wide variety of disciplines including transportation engineering, planning and policy, urban planning, environmental sciences, climate change sciences, economics, epidemiology, public health, toxicology, medicine, and more. This diversity of perspectives is necessary to understand such a complex topic; it also speaks to the need for systems thinking and cross-disciplinary approaches to understand current and future issues and proactively identify solutions. In this final chapter, we reflect on the entirety of the book,

Traffic-Related Air Pollution
https://doi.org/10.1016/B978-0-12-818122-5.00025-9

consolidating the key messages from each chapter and highlighting the recent advances and remaining gaps in current knowledge and practice.

Where are we now?

The full chain between traffic activity and human health impacts

As discussed in Chapter 1 (Khreis et al., 2020), the relationship between traffic activity and human health impacts is a complicated one, which can be considered through the full-chain framework depicted in Fig. 25.1. Policies, technologies, and market solutions can modify this relationship by modifying one or more of the elements of the full chain. For example, transportation demand management policies to reduce travel demand can impact traffic activity and, therefore, impact vehicle emissions, resulting TRAP concentrations, human exposures and health impacts. Technologies such as the introduction of electric vehicles in urban areas can impact vehicle emissions, resulting TRAP concentrations, human exposures and health impacts. Market solutions, such as placing urban air purifiers along major roads and the use of face masks can reduce human exposures and adverse health impacts. Changes in any element of the full chain will result in changes in the subsequent elements, which lead to worse or improved public health outcomes. The full-chain approach offers a useful framework to understand and explore the individual elements and their interactions. Implementing the full chain in more research will pave way for discussion about transportation emissions, air quality, exposure, and policy and practice implications. Benefits of this include influencing policymakers in their decision-making process and creating a more integrated knowledge base highlighting the areas responsible for the most uncertainties and errors (Sanchez et al., 2020).

Recent advances

To date, the research on TRAP and human health has witnessed important advances, which led to important improvements in practice and policy decision-making. These advances have improved our understanding of TRAP and its distribution, its associated health effects and large-scale health impacts and the wide variety of policy options available to mitigate this ubiquitous exposure, and how to appraise these options in terms of impacts and cost effectiveness (Khreis et al., 2020). Countries across the world have taken regulatory action on air quality, and technological advancements and engine emission controls which have significantly reduced individual

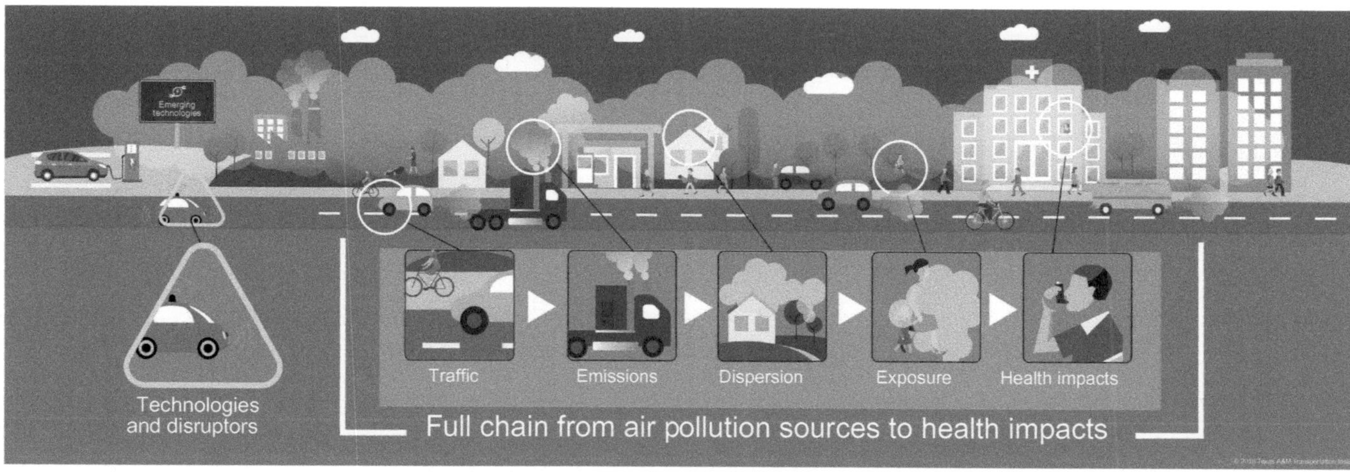

Fig. 25.1 The full chain: linking TRAP to health impacts. *Source: Center for Advancing Research in Transportation Emissions, Energy, and Health (CARTEEH), available from: https://www.carteeh.org/.*

vehicles' emissions. This has resulted in improved air quality and reduced traffic emissions, especially in developed nations. On the other hand, especially in developing nations, these technological advances are lagging and are outpaced by the sheer increase in traffic activity due to economic growth, and the increase in the activity of heavy-duty vehicles for freight and goods movement. In recent decades, ambient air pollution levels have fallen across most of Europe, North, and South America, whilst they increased across most countries in Asia and Africa (Ritchie & Roser, 2020). Furthermore, while air quality has improved in many developed nations (Clark, Millet, & Marshall, 2017; Laden et al., 2006; U.S. Environmental Protection Agency, 2019), there is yet evidence that the racial disparity in air pollution exposure persisted in the same time period (Clark et al., 2017), indicating that improvements in air quality were not accompanied by improvements in environmental justice.

Guidelines, standards, and regulations

Key advances in the field have been discussed in more depth across this book and are briefly summarized next. Over the years, the development and tightening of vehicle emissions and air quality guidelines and regulations have resulted in overall reductions in emissions and measurable improvements in air quality across a range of pollutants (Li, 2020). In terms of air quality guidelines to curb anthropogenic air pollution, the United States is viewed as a world leader, having initiated the process with the Air Pollution Control Act in 1955, with the landmark Clean Air Act of 1977, and the Clean Air Act Amendments of 1990, which established much of the regulatory framework as applied today (Rodgers & Rodgers, 2020). Other nations have followed similar paths in the establishment of air quality regulations, with differing timelines and approaches, all resulting in marked improvements in air quality. From the vehicle emission standards perspective, improvements in engine technologies and aftertreatment devices have allowed vehicle manufacturers to respond to increasingly stringent emission standards, with new vehicles today often emitting 1% of the pollutant emission levels compared to a typical vehicle from the 1970s (Rodgers & Rodgers, 2020).

Assessment of the elements of the full chain

The science and relevant methods and tools have also witnessed great leaps forward including advancements in the assessment of traffic activity, enabling the collection of fine-grained vehicle speed, location and routing data (Xu, 2020), and vehicle emissions, including the ability to more accurately measure and model vehicle emissions (Ropkins, Ibarra-Espinosa, & Bernard, 2020). These, in turn have translated to the ability to better characterize air pollution (Askariyeh, Khreis, & Vallamsundar, 2020) and human exposures (Beevers & Williams, 2020), both through modeling and monitoring approaches, and the combination thereof. These assessments have been revolutionized by new data sources and availability, in addition to parallel technological advances in computation, sensors that allow for broader deployment at lower costs, and communications (Medeiros & Khreis, 2020; Xu, 2020). On the other side of the spectrum, health research has also taken great leaps forward, capitalizing on some of the advances mentioned above which enabled, for example, the better characterization of the high spatial and temporal variability of TRAP (Beevers & Williams, 2020).

Studies and syntheses of air pollution epidemiology, toxicology, and mechanisms

Air pollution epidemiology started to focus more on the local intra-urban variations of air pollution, driven by traffic activity across many cities (Andersen, 2020; Khreis et al., 2020). Both short- and long-term chronic exposures to TRAP have been studied and have been associated with a wide spectrum of global diseases (Andersen, 2020; Fox, Koehler, & Johnson, 2020). These studies continue to evolve at a rapid pace and the associations between TRAP exposures and a variety of emerging and concerning health effects are being shown (Andersen, 2020; Fox et al., 2020). We no longer think of TRAP as a factor leading to premature mortality and the development of respiratory and cardiovascular diseases only, but we now know that it also has adverse effects on cognitive, immune and metabolic function, and reproductive outcomes (Andersen, 2020; Fox et al., 2020; Khreis, 2020). The body of epidemiological studies in this area has grown substantially, requiring new syntheses in the form of systematic reviews and meta-analyses to elucidate on conflicting findings, assess the strength of the overall body of evidence and come up with clear and concise bottom-line conclusions, which help make this emerging science actionable in policy decision-making (Lam, Vesterinen, & Woodruff, 2020). These forms of syntheses have also increased in the recent years, signaling the need for overall critical assessments, which can inform policy decision-making. Syntheses methods have greatly evolved and there is now a wealth of guidance on conducting high-quality systematic reviews and meta-analyses, appraising the individual underlying studies and assessing the overall body of evidence (Lam et al., 2020).

This information is critical to consult when conducting such syntheses as there is also evidence that many published systematic reviews are not of high quality, do not adhere to best practices and are, therefore, of limited capacity to influence the policy decision-making process (Lam et al., 2020).

To further strengthen the case for action, epidemiological findings, and their syntheses, have been grounded in mechanistic and toxicological research. Mechanistic and toxicological evidence lends biological plausibility to observations made in select and potentially biased samples, with a high potential for exposure misclassification, especially relevant for the exposure to TRAP, which is hallmarked by its high spatial and temporal variability (Beevers & Williams, 2020). Indeed, a large body of mechanistic research now supports epidemiological associations showing that TRAP is associated with wide-ranging detrimental effects throughout the body (Miller & Raftis, 2020), including airway remodeling, inflammation, oxidative stress, and a shift in immune function (Barthelemy et al., 2020; Miller & Raftis, 2020; Thurston et al., 2020). Plausible biological mechanisms support the case for causality in many of these associations. This body of work, combined with findings from air pollution epidemiology, systematic reviews, and meta-analyses, now emphasizes the need to place vehicle emissions high on the agenda of policies to reduce air pollution (Miller & Raftis, 2020). The realization and effective implementation of such policies will further reduce the detrimental effects of TRAP and will likely be accompanied by improvements in health in many societies, especially if realized at a global level (Miller & Raftis, 2020).

Closely related to mechanistic and toxicological studies is the new science of -omics (genomics, transcriptomics, proteomics, or metabolomics) as applied to TRAP and health research. The concept of the exposome and omics approaches was recently introduced to address the difficulties associated with establishing the health effects of a complex, diverse, and relatively low-concentrated mixture such as the TRAP mixture, which may have long latency periods, and largely unknown modes of action and interactions with underlying genetic and other factors that modulate susceptibility (Demetriou & Vineis, 2020). The exposome has allowed for an empowerment of environmental research, by improving measurements of external stressors and of internal biological changes, taking advantage of advancements in high-throughput technologies of omics. Studying the biomarkers associated with biological changes in response to environmental stressors enhanced research on the health effects of TRAP by improving the exposure assessment, increasing the understanding of mechanisms, and enabling the investigation of individual susceptibility (Demetriou & Vineis, 2020).

While the importance of epidemiological and toxicological observations, and their integration, cannot be stressed enough, the impacts of TRAP at the population level are the next step in translating these observations into large-scale implications useful for policy decision-making. In the recent years, the use of qualitative and quantitative health impact and burden of disease assessments filled this gap and provided practitioners and policy makers with tools to assess the baseline and projected health impacts of current line conditions and policy changes (Mindell & Birley, 2020; Mueller, Nieuwenhuijsen, & Rojas-Rueda, 2020). There is now a range of methods and tools, which can be used in health impact and burden of disease assessments, and the assessment of inequalities has a central role in these exercises (Mindell & Birley, 2020; Mueller et al., 2020). Quantitative health impact and burden of disease assessments have been particularly useful in this context as these exercises can inform numerically on the health risks associated with the exposure of interest and provide a magnitude of the health impacts (Mueller et al., 2020). This is especially important for influencing policy proposals, because decision-makers may give more weight to outcomes that are measurable (i.e., quantifiable) (Joffe & Mindell, 2005); informing the analysis of health benefit/risk trade-off of public policies (Mueller, 2017). Quantifying health outcomes also aids in the inclusion of these outcomes in transportation investment appraisal tools such as cost-benefit analysis (Burris, 2020; Mueller, 2017; World Health Organization, W.H.O, 2015), which facilitates and adds defensibility to the inclusion of health in public policy. Examples of such exercises have been overviewed briefly in Chapter 1 (Khreis et al., 2020) and Chapter 13 (Mueller et al., 2020).

Policy decision-making and best practices

Like many other societal issues, policies and decision-making regarding TRAP, however, is driven by broader societal issues, well beyond public health (Rodgers & Rodgers, 2020). When understanding the context of policy and decision-making regarding TRAP and health, it is important to acknowledge the social context of the time. For example, when motor vehicles were first introduced in large cities, they were viewed as providing a net environmental health benefit, by reducing the number of horse-drawn carriages in cities that caused health problems due to the large quantities of horse manure (Rodgers & Rodgers, 2020). While motor vehicles solved one environmental and social problem at the time, they have resulted in several others, including TRAP. Similarly, when considering policies and decisions to address TRAP today, there is a need to take a broader view of potential benefits and disbenefits beyond pollutant or exposure reductions, and also acknowledge that the reduction of TRAP and its adverse health effects is only one objective in the policy decision-making realm. Policy makers may view this objective as a barrier in achieving other important objectives with more immediate returns such as economic growth and efficiency, or as an enabler to achieving more long-term environmental and social welfare objectives such as the mitigation of climate change and improving equity (Khreis et al., 2020; May, 2020; Sudmant et al., 2020). Luckily, practitioners and decision-makers now have access to a wide range of measures, which can achieve a multitude of objectives and tackle the multifaceted issues cities face nowadays, beyond poor air quality and its adverse health effects. These measures may be categorized under the six umbrella categories of land use (e.g., development density and mix), infrastructure (e.g., cycling networks), management and services (e.g., low emission zones), attitudinal and behavioral (e.g., ride sharing), information provision (e.g., trip planning systems), and pricing (e.g., fuel taxes) (May, 2020). A new seventh category can be technologies, with automates, connected and electric vehicles being key examples (Glazener & Khreis, 2020; Tanvir, Hao, & Boriboonsomsin, 2020). Furthermore, policy option and package generation tools and guidance on the selection of the most appropriate policies for local contexts are becoming more available, although the formal process of policy option generation is still not utilized in many cities (May, 2020).

Likewise, cities across the world have been implementing a wide range of policies, which are now being documented in various literature reviews (Glazener & Khreis, 2020; Sudmant et al., 2020) and relevant databases of case studies (May, 2020). These policies seem to act best when integrated in packages, which combine interventions that address the health burdens associated with the various exposures and lifestyles related to urban and transportation planning, in addition to contributing to other objectives such as efficiency and climate change mitigation (Glazener & Khreis, 2020; Khreis et al., 2019; May, 2020). These interventions also include solutions offered by nature, the so-called Nature-based Solutions, such green space and vegetation barriers (Baldauf, 2020), which have other benefits in relation to the mitigation of noise, climate change, urban heat and improvements in physical activity, social contacts, and mental health (Nieuwenhuijsen et al., 2017). However, evidence now shows that not all green space is equal or effective, and there are many design considerations and local factors such as physical and species-specific characteristics and the thickness and porosity of the vegetation, which need to be carefully considered when using green space to achieve air quality and health benefits (Baldauf, 2020).

The cost effectiveness or the feasibility of policy options is a critical factor in the discussion on how to best mitigate TRAP and reduce its public health burden. The monetary evaluation of the costs and benefits of transportation policies offers a method to take the health impacts of TRAP into account in transportation investment appraisal and these methods have been well described in the literature (Burris, 2020). The ideal exercise of monetization, however, would consider the full range of impacts associated with a policy change, including the health impacts of changes in air quality and other adverse exposures associated with transportation (Khreis, Glazener, et al., 2019), but also other impacts on the society, the environment, and the economy. These exercises are very few but are beginning to emerge, especially in the climate change literature where the belief that climate action would require reducing attention toward other environmental and societal issues, stand in the way of action (Sudmant et al., 2020). Synthesis of the literature now shows that low-carbon urban transportation policies including the three core categories of land use changes, modal shift and public transportation improvements, and fleet improvement and transportation electrification may have substantial benefits beyond climate change and air pollution mitigation, though the quantification and monetization of these impacts are still scarce (Sudmant et al., 2020). These benefits extend to improving road safety, promoting economically efficient urban development, increasing social inclusivity, and reducing congestion (Sudmant et al., 2020).

Environmental injustice and health inequalities

Current TRAP levels and their potential changes due to new policies do not have the same effect on all segments of the population. The bulk of available evidence points toward traffic on highways and major roadways disproportionately affecting lower socioeconomic status and racial/ethnic minority communities (Fuller & Brugge, 2020). The increased air pollution exposure faced by these communities results in disparities in health outcomes. Prevalence of asthma, diabetes, and hypertension is higher in low-income and minority communities when compared to overall national averages. Other analyses have shown an increased risk for cancer in environmental justice communities that is linked to air pollution exposures, especially those from local TRAP (Fuller & Brugge, 2020). This environmental justice concern should be considered when devising policies to mitigate adverse impacts of TRAP with an emphasis on disadvantaged neighborhoods (Fuller & Brugge, 2020). The most effective direction of scarce resources may be made by better considering these susceptible populations in the planning and policy decision-making process and addressing the range of intrinsic and extrinsic factors which makes them more exposed, but also more susceptible to their exposures (Khreis et al., 2016, 2019). These factors include their living locations and conditions, their occupations, malnutrition and the lack of antioxidant intakes, exposure to stress, exposure to violence, genetics, and others. These factors can modify and often amplify the disproportionate adverse health effects of TRAP and other traffic-related exposures, contributing further to the gross inequalities in health (Marmot, 2005).

Emerging technologies and market solutions

Emerging transportation technologies are also significantly changing the field and are introducing new questions and uncertainties. The issue of electric vehicles and transportation decarbonization is particularly relevant to understanding the future implications for TRAP as a health issue. While studies have started to investigate and project the impacts of such technologies, including the impacts of connected and autonomous vehicles and on-demand mobility services, this evidence base is still in its infancy. There are expected benefits of some of these technologies on air quality and health due to an overall reduction in vehicle emissions, such as in the case of electric vehicles or due to reduced car ownership when shared mobility options become widely available. However, there is also the potential for unintended consequences, which need to be addressed proactively rather than reactively after the adaptation of these technologies. For example, rebound effects are a concern where vehicle miles traveled, and associated TRAP, can increase due to the adaptation of on-demand mobility services (Tanvir et al., 2020). Energy savings due to new technologies also do not always translate to a net reduction in emissions and TRAP (Tanvir et al., 2020). Specifically, in the case of electric vehicles, the generation of electricity from nonrenewable sources will move the source of pollutants away from vehicle tailpipes to power plants, potentially affecting nearby communities, which are likely to be ethnic minorities of lower economic status. This is an important consideration, especially in regions where electricity continues to be primarily generated from nonrenewable sources such as coal. Beyond implications for vehicle emissions and TRAP, there are also concerns with broader environmental justice issues surrounding the proliferation of emerging technologies, including access and affordability for disadvantaged sections of the society.

Apart from controlling vehicle emissions through preventative measures (which avoid the source of emissions entirely, such as reduction in vehicle trips) or primary measures (which reduce emissions at the source, through more efficient engines or exhaust aftertreatment devices), there are secondary options aimed at mitigating TRAP and its health effects by treating the air near pollutant sources, or by reducing exposure to pollutants at the corridor level through green barriers, or at the individual level through face masks or other technologies. In developing nations where the benefits of preventative and primary measures are outpaced by the sheer volume of vehicles, these secondary measures are being increasingly adopted, including commercially available solutions. National laboratories and research institutes in India, for example, have worked on these types of mitigation solutions, though their overall benefits and efficacy are not yet conclusively determined (Nagendra et al., 2020).

We are also seeing the adaptation of new technologies for the monitoring and modeling of TRAP such as low-cost air quality sensors and mobile monitoring, which are becoming increasingly popular (Medeiros & Khreis, 2020). Low-cost and/or high-end sensors are being mounted on various mobile platforms such as bicycles, cars, and buses. These methods are being combined together with modeling and/or other monitoring sources in a single operation to take advantage of their respective strengths and improve the exposure assessment exercise (Medeiros & Khreis, 2020). The use of these new methods and their integration are resulting in new air quality maps at higher spatial and temporal resolution with improved source attribution and hotspot identification and more accurate estimates of TRAP at the street level (Medeiros & Khreis, 2020). These methods have also improved the understanding of meteorology effects and land use on urban pollution (Medeiros & Khreis, 2020). Finally, there are new efforts to synthesize and organize the literature on the full-chain covering studies of transportation, emissions, air quality, human exposure, and health, and providing a full "head to tail" picture of TRAP's impact on human health (Sanchez et al., 2020). Understanding the relationship between full-chain elements paves a way for discussion about policy and practice implications and for making more concrete and specific policy recommendations. Since air pollution regulations mostly target pollution at the source, this approach allows for understanding the ultimate effect of policies on public health and is critical to understanding the TRAP-health nexus (Sanchez et al., 2020).

Where do we want to go and how will we get there?

Synthesis and mainstreaming of current knowledge

We have come a long way, but there are as yet critical knowledge gaps which need to be filled offering exciting research and practice opportunities and a pathway to push and track progress toward the goal of clean air and protection of the public's health. While many advances have been made in TRAP and health research and practice, the above emerging knowledge remains sporadic. The wide range of existing and new methods and tools to quantify TRAP, its health impacts, and the impacts of policy and emerging technology options are as yet not mainstreamed in teaching, research, practice, or policy decision-making. The real-world impact of scientific and methodological advances is, therefore, limited. To a large extent, this book fills this gap by synthesizing information from a wide variety of relevant disciplines, offering a resource for educators, undergraduate and postgraduate students, researchers, practitioners, advocacy groups, and policy decision-makers interested in the area of TRAP and health.

Transcending beyond silos and narrow focus on isolated issues

However, a critical issue that remains ingrained in current education, practice, and policy systems is the operation in silos and sometimes the development and use of specialized methods and tools remain constrained to the silo where it originated from. This issue is exacerbated by the different terminologies that the different fields, such as transportation and health, use. The spoken languages are different and also differ by geography, as shown throughout this book. While this book is a good representation of languages spoken across the different professions and countries, it is not a full account for these differences. To truly understand the nature of TRAP and its health effects, and positively influence policy decision-making, a cross-disciplinary approach which prompts the sharing of knowledge, languages, and methods is needed.

Furthermore, while we focused on TRAP as one transportation-related exposure that negatively impacts public health, there is a real need to consider this issue in the broader context of transportation and health. Emerging knowledge suggests that transportation can affect health through 14 intertwined pathways namely: (1) green space and esthetics, (2) physical activity, (3) access, (4) mobility independence, (5) contamination, (6) social exclusion, (7) noise, (8) urban heat islands, (9) road crashes, (10) air pollution, (11) community severance, (12) electromagnetic fields, (13) stress, and

(14) greenhouse gas emissions (Khreis, Glazener, et al., 2019). This multitude of exposures is what occurs in the real world and what ultimately drives the health impacts we observe. Studies on the interactions and effect modifications of air pollution effects through other exposures such as green space, heat, noise, and physical activity are yet scarce, but are needed as they reflect real-world conditions and may further advance our understanding of cumulative risk and impact of policies. Such research can also advance the estimation of exposure-response functions and, therefore, health impact and burden of disease assessments, and can result in findings that are more relevant for policy decision-making, especially at the local level.

Identifying mitigation strategies for TRAP should involve consideration of the potential co-benefits or disbenefits along those other transportation and health pathways, which often have burdens of disease comparable to that of TRAP. The most effective and prudent mitigation efforts would consider this wider range of exposures and health effects (Khreis, Glazener, et al., 2019). The time has come to adopt systemic and holistic approaches as the one exposure-one outcome approach is no longer valid and will only lead to a narrow focus and negative, "unintended," consequences (Khreis, 2020).

Revising guidelines, standards, and regulations

Other advances in the field point to the inadequacy of air quality guidelines and regulations. While the tightening of vehicle emission standards and air quality guidelines and regulations resulted in measurable improvements in air quality across a range of pollutants (Li, 2020), there is now ample evidence that adverse health effects still occur at air pollution levels below these thresholds (Andersen, 2020; Beelen et al., 2014; Belanger et al., 2006; Castro et al., 2009; Chen & Omaye, 2001; Loxham, Davies, & Holgate, 2019; MacIntyre et al., 2014; Nishimura et al., 2013; Pedersen et al., 2013; Scoggins et al., 2004; Wei et al., 2019; World Health Organization, W.H.O, 2013). The effects of low-level air pollution are a relatively new focus area in air pollution epidemiology and insights from this area will be instrumental in refining the exposure-response functions, the burden of disease estimates from TRAP, and the assessment of impacts of policy changes.

There is also a need to study the whole plethora of traffic-related air pollutants, and to ground new epidemiological observations in toxicological and mechanistic research. This is particularly true for pollutants such as particulate matter (PM), which contain a wide range of particles with different compositions/chemistry, oxidative potential, sizes, surface areas, and other characteristics, which are closely related to toxicity and health effects. Furthermore, certain traffic-related air pollutants, such as ammonia (a by-product of selective catalytic reduction technology and an important contributor to the formation of secondary particles), black carbon (BC), and ultrafine particles (UFPs), which have been shown to elicit adverse health effects (Cape et al., 2004; Dennekamp et al., 2002; Durbin et al., 2001; Khreis et al., 2017; Luben et al., 2017; Onat & Stakeeva, 2013; Perrino et al., 2002; Perrino, Catrambone, & Di Menno Di Bucchianico, 2003; Skjøth & Hertel, 2013; Tomlin, Sutton, & Tate, 2010), are yet unregulated and not routinely measured or studied (Khreis et al., 2020).

These pollutants may be particularly challenging to study. Studies examining local intra-urban air pollution have shown BC and UFPs' spatial variations by as much as 5 times within a single city block and by as much as 3 times compared to fine particulate matter (Medeiros & Khreis, 2020). Therefore, an important question that remains open is which are the putative agent or agents in the TRAP mixture? Many epidemiological studies lack data on and control for multiple pollutants, making the distinction of pollutant-specific effects, if any, not possible (Khreis et al., 2019). This is a knowledge gap that would benefit from thoughtful future epidemiological research, which should be planned in light of toxicological and epidemiological findings. As the evidence accumulates, it may prompt the revision of air quality guidelines and regulations and perhaps the addition of new pollutants.

In the same context, another contemporary issue which warrants further attention is the relative importance of non-tailpipe emissions in the health effects and the burden of disease assessments of TRAP (Andersen, 2020). These emissions may increase, both in the developed and developing world, with the expected introduction of electric vehicles and in many regions across the world, these emissions are unregulated (Timmers & Achten, 2016). Despite this expected future trend, very few studies address the health effects of non-tailpipe emissions, which again relates to the need for better PM speciation and analysis, through, for example, full-chain or source apportionment methods, and not treating all PM as one pollutant with homogenous health effects.

Utilizing data, methods, and tools from the different discipline to inform policy decision-making

Another avenue for further research, which might improve future practices and policy decision-making, is the more accurate estimation of the burden of disease related to TRAP. This area will benefit from utilizing advances in research across each of the elements along the full chain, which by default speaks to the need of integrating new information and knowledge from the different disciplines. For example, advances made in the monitoring and modeling of traffic activity (Xu, 2020) are yet to be utilized to better model TRAP and assign human exposures. Traffic modeling has many inherent uncertainties, including in the estimation of traffic speeds and flows, the fleet mix, the exclusion and agglomeration of minor roads in traffic models, the treatment of diurnal and seasonal effects, and the accurate geocoding of road links in relation to human populations' locations and their mobility patterns.

There are now a variety of new data sources that can be used to overcome some of these shortcomings including remote sensing (Holt et al., 2009; Toth, Grejner-Brzezinska, & Merry, 2003), satellite imaging and air photos (Larsen, Koren, & Solberg, 2009; McCord et al., 1855; Wang et al., 2016), and telematics (Pellecuer, Tate, & Chapman, 2016), which can provide observed traffic activity data at a finer spatial and/or temporal scale when compared to traffic models. This data are yet to be used for health impact and burden of disease assessments, and in theory can provide a larger and more detailed traffic network and fleet data, potentially at the street and the minute level, which might improve the subsequent modeling of vehicle emissions, air quality, human exposures, and health impacts.

The same holds true for vehicle emissions, air pollution, and human exposure modeling (Askariyeh et al., 2020; Beevers & Williams, 2020; Medeiros & Khreis, 2020; Ropkins et al., 2020). For example, vehicle emissions monitoring evolved away from dynamometer testing and fixed duty cycles to in-use real-world testing and remote sensing, facilitated by new technologies and small low-cost sensors, which offer possibilities for expanded low-cost data collection and the evolution of vehicle emission models (Ropkins et al., 2020). The consultation of new monitoring data sources should not only be done for the validation of models across the full chain, but also should be considered as a complementary element in the development and calibration of the different models. On their own, monitoring and modeling approaches are limited. The integration of these approaches has shown great promise in capturing the high spatial and temporal variability of TRAP in urban settings (Medeiros & Khreis, 2020), a critical prerequisite to assessing the health effects of these exposures. These new methods are also capable of extending air pollution monitoring to areas without networks, an issue of particular importance in the developing world, and to capture personal exposures of air pollution, which is what truly affects one's health (Medeiros & Khreis, 2020).

On the other end of the full-chain spectrum lies many limitations related to the analysis of health effects and the estimation of population-level health impacts. Individual susceptibility to TRAP exposures is an under-researched area in the literature, but one which can improve the estimation of exposure-response functions and subsequent burden of disease assessment, if appropriately developed. TRAP exposures seem to have differential effects on different populations, potentially modified by factors like sex, age, genetics, ethnicity, diet, occupation, socioeconomic status, comorbidities, and co-exposures to other environmental stressors such as violence, stress, and road traffic noise (Andersen, 2020; Barthelemy et al., 2020; Demetriou & Vineis, 2020; Fox et al., 2020; Miller & Raftis, 2020).

Shedding light on susceptible subgroups can strengthen the causal reasoning by addressing mechanisms (Demetriou & Vineis, 2020), improve the burden of disease estimates (Mueller et al., 2020), and also pave the way for adequately allocating mitigation resources and reducing environmental injustice (Fuller & Brugge, 2020). Burden of disease and health impact assessments would also benefit from including the wider range of health effects, which have been recently associated with TRAP (Fox et al., 2020), especially those which have been established by high-quality systematic reviews and meta-analyses (Lam et al., 2020). In addition, these assessments can increase their policy utility by monetizing quantitative estimates of the burden of disease (Burris, 2020), which is not commonly practiced (Sudmant et al., 2020). Health impact and burden of disease assessment studies have also generally been limited to research and academic purposes. Recent studies have emphasized the need for health impact assessment in the decision-making processes for federal, state, and local urban planning, land use, and environmental regulation (Rodgers & Rodgers, 2020). More user-friendly tools and models are needed to allow planners, practitioners, and policy decision-makers to routinely integrate quantitative health impact and burden of disease assessment methodologies in their transportation appraisal Schemes (Mueller et al., 2020). There is also a need for more stakeholder involvement (Mueller et al., 2020; Sudmant et al., 2020), and full-chain modeling approaches, integrating the different fields of research and practice from the exposure source to the health outcome (Mueller et al., 2020).

Learning lessons from the past

When it comes to policy decision-making, there is a value in the analysis of historical trends of air pollution and how various cultures have developed public responses to respond to this issue overtime (Rodgers & Rodgers, 2020). Vehicles were once considered as a solution to an environmental problem caused by horses—and it is ironic that they are now a cause of global issues related to health and the environment (Rodgers & Rodgers, 2020). Even if we move toward a decarbonized future, current policy and decision-making operate within the context of current technologies and the vehicle fleet and until now, air quality policy and decision-making have been reactive (i.e., addressing things after they have become a problem); as technological advances proliferate worldwide, we cannot afford to do this in the future, and must take more proactive action (Rodgers & Rodgers, 2020).

Mainstreaming policy options generation, knowledge transfer and more research on the effectiveness, transferability, and interactions of policy options

There is also a value of utilizing policy option generation tools and guidance on policy selection, which are now more readily available, although not mainstreamed in the policy decision-making process (May, 2020). Despite the recent availability of these tools and guidance documents, there remain some key gaps which future research can fill. For example, more information on the performance of select policy options is needed to evaluate their impacts and effectiveness against specific objectives and in specific contexts. Indeed, there is a wealth of disparate information out there on the performance of policy options and these can be synthesized under umbrellas of specific policy objectives, for example, reducing air pollution from traffic and improving public health.

More information on the transferability of policy options is also needed to better understand contextual factors, barriers, and facilitators to good practice (May, 2020; Sudmant et al., 2020). Better understanding the interaction of different policy measures in packages and best practices in the design of packages can further aid the policy decision-making process. There is little knowledge available on how integrated policy packages can address more than one of these issues and how effective these could be in achieving a multitude of goals such as increasing physical activity to promote public health and mitigating climate change (Glazener & Khreis, 2019; Gouldson et al., 2015; May, Khreis, & Mullen, 2018).

Finally, the explicit inclusion of public health as an objective in transportation planning and policy tools is yet to be done and can help prompt practitioners and policy makers to consider the health impacts of their practices (Khreis, May, & Nieuwenhuijsen, 2017). Adding public health as an explicit objective can be a good framing device from a transportation planning perspective—more palatable than reducing car use, for example, in terms of getting public and political support. Exploring best practices to reduce TRAP and improve physical activity from around the world also highlighted several knowledge gaps (Glazener & Khreis, 2020). For example, the literature suggests that there are still uncertainties about the net impact of green spaces on air pollution (Glazener & Khreis, 2020), which might be a result of the various types and designs of green spaces which affect air pollution levels differently (Baldauf, 2020).

Research needs remain on developing methods to quantify the benefits of green space on air quality, particularly at the local scale. For modeling vegetation effects on local air quality, research is also needed on characterizing the key factors and characteristics of vegetation that contribute to air pollution reductions for both increased pollutant transportation and dispersion as well as deposition of particles onto the vegetation surfaces (Baldauf, 2020).

Furthermore, while the integration of vehicle technologies may be beneficial for reducing TRAP, this integration will first occur in the developed nations (Glazener & Khreis, 2020). About 90% of the population growth in the next 30 years, however, will occur in the developing nations (United Nations, 2018), and the adoption of modern technologies will be delayed in these nations, creating a period where increasing populations will put a stress on transportation systems, potentially increasing emissions and exposure hotspots (Glazener & Khreis, 2020). There is the potential that the proven best practices displayed in the developed urban areas can be replicated in the developing countries but determining these best practices will rely on the development of studies that can trace air pollution back to transportation sources, and these studies are still very few (Glazener & Khreis, 2020).

The importance of considering the whole range of social, health, environmental, and economic impacts associated with policies has been stressed throughout the book, in addition to the need to carefully consider negative impacts, and the potential for negative unintended consequences. Where assessment that includes these elements provides a case for action, there remain many barriers and facilitators to best practices.

Of importance, municipal actors rarely have the opportunity or capacity to unilaterally implement interventions at a meaningful scale. To the extent they are able to, however, it would be wise to develop and implement actions with the engagement and support of a wider array of actions. Collaboration and partnership between transportation policymakers vertically (to regional, national and international policymakers, as well as local and community organizations), and horizontally (between cities and departments within cities), can facilitate learning, provide much needed capacity, and help coordinate the transportation planning and policy process. Urban policymakers can also work to address the barriers affecting private and community actors as they look to support policy changes. This can include helping to address financial barriers to action, but it can also relate to the socially and territorially contextualized practices that are much a part of any transportation system as the cars, buses, trains, and other infrastructures. In this respect, policymakers need to think about the diverse needs of citizens and the ways that they can engage with the nuanced and sometimes place-specific barriers and enablers of behavioral and social change (Sudmant et al., 2020).

Projecting the impacts of technologies and innovations

Many uncertainties arise due to the rapidly changing nature of the transportation sector, where new technologies and innovations such as autonomous and electric vehicles are being adopted across the world. While electric vehicles have the potential to reduce emissions from the tailpipe, the net emissions impact is highly dependent on emissions associated with the electric grid (Tanvir et al., 2020). Similarly, the overall impacts of shared mobility services and automated and connected vehicles on travel patterns as a whole, is unclear, and there are also potential challenges such as induced demand and increased vehicle miles of travel, which can have a negative impact on TRAP (Glazener & Khreis, 2020; Tanvir et al., 2020). Further, these issues need to be placed in the broader social and environmental context, such as environmental justice (Tanvir et al., 2020). Real-world emissions measurement and testing is needed in the context of the developing nations to better model their conditions and assess impact of mitigation scenarios, including market-based solutions (Nagendra et al., 2020).

Finally, analyzing a vast literature on elements of the full chain between traffic air pollution sources and associated health outcomes: traffic, emissions, air quality, exposures, and health impacts, in addition to technology, demonstrated that there is a lack of studies addressing all full-chain elements: only 2.7% of articles indexed in a newly developed literature library covered all elements of the full chain (Sanchez et al., 2020). There is a greater need to implement such analyses, especially when aiming to influence policy to mitigate TRAP and its adverse health impacts.

Summary and conclusions

Air pollution is an established health concern and one of the principal causes of premature mortality globally, with vehicle traffic as a leading contributor to this, especially in urban areas. A recent visible reminder of the link between traffic and air pollution was during the lockdowns due to the COVD-19 pandemic, where cities across the world experienced marked improvements in air quality due to less vehicle activity. However, the problem of air quality, and specifically TRAP is complicated, as shown in this book. On considering the elements of the full chain between traffic and health, we showed that there are significant research advances in several areas that collectively support more rigorous practice and policy decision-making to mitigate TRAP and its adverse health impacts. However, the transportation sector is facing rapid changes on many fronts, which require a proactive, rather than a reactive approach, to mitigating TRAP and its adverse health impacts and addressing environmental injustice and health inequalities. Also, importantly, TRAP and health issues cannot be viewed in isolation and should be placed in the broader landscape of urban and transportation planning and policy that can support public health through modifying a wide variety of relevant pathways, beyond air quality. We have come a long way, but there are as yet critical knowledge gaps which need to be filled offering exciting research and practice opportunities and a pathway toward the goal of clean air and protection of the public's health. Key recent advances and remaining gaps are outlined in this chapter, paving the way forward to more informed research, better understandings, and improved practice and policy.

Acknowledgments

Fig. 25.1 was created with funding from the Texas A & M Transportation Institute's Center for Advancing Research in Transportation Emissions, Energy, and Health (CARTEEH), a US Department of Transportation's University Transportation Center, College Station, TX. The grant number is 69A3551747128. More information about the Center for Advancing Research in Transportation Emissions, Energy, and Health is available at: https://www.carteeh.org/.

References

Andersen, Z. J. (2020). Air pollution epidemiology. In H. Khreis, et al. (Eds.), *Traffic-related air pollution*: Elseveir.

Askariyeh, M. H., Khreis, H., & Vallamsundar, S. (2020). Air pollution monitoring and modeling. In H. Khreis, et al. (Eds.), *Traffic-related air pollution*: Elseveir.

Baldauf, R. (2020). Air pollution mitigation through vegetation barriers and green space. In H. Khreis, et al. (Eds.), *Traffic-related air pollution*: Elseveir.

Barthelemy, J., et al. (2020). New opportunities to mitigate the burden of disease caused by traffic related air pollution: Antioxidant-rich diets and supplements. *International Journal of Environmental Research and Public Health*, *17*(2), 630.

Beelen, R., et al. (2014). Effects of long-term exposure to air pollution on natural-cause mortality: An analysis of 22 European cohorts within the multicentre ESCAPE project. *The Lancet, 383*(9919), 785–795.
Beevers, S. D., & Williams, M. L. (2020). Traffic-related air pollution and exposure assessment. In H. Khreis, et al. (Eds.), *Traffic-related air pollution*: Elseveir.
Belanger, K., et al. (2006). Association of indoor nitrogen dioxide exposure with respiratory symptoms in children with asthma. *American Journal of Respiratory and Critical Care Medicine, 173*(3), 297–303.
Burris, M. (2020). Cost-effectiveness of projects and policies. In H. Khreis, et al. (Eds.), *Traffic-related air pollution*: Elseveir.
Cape, J., et al. (2004). Concentrations of ammonia and nitrogen dioxide at roadside verges, and their contribution to nitrogen deposition. *Environmental Pollution, 132*(3), 469–478.
Castro, H. A., et al. (2009). Effect of air pollution on lung function in schoolchildren in Rio de Janeiro, Brazil. *Revista de Saúde Pública, 43*(1), 26–34.
Chen, L., & Omaye, S.T. (2001). Air pollution and health effects in northern Nevada. *Reviews on Environmental Health, 16*(2), 133–149.
Clark, L. P., Millet, D. B., & Marshall, J. D. (2017). Changes in transportation-related air pollution exposures by race-ethnicity and socioeconomic status: Outdoor nitrogen dioxide in the United States in 2000 and 2010. *Environmental Health Perspectives, 97012*, 1.
Demetriou, C., & Vineis, P. (2020). Biomarkers and omics of health effects associated with traffic-related air pollution. In H. Khreis, et al. (Eds.), *Traffic-related air pollution*: Elseveir.
Dennekamp, M., et al. (2002). Exposure to ultrafine particles and PM2. 5 in different micro-environments. *Annals of Occupational Hygiene, 46*(suppl 1), 412–414.
Durbin, T. D., et al. (2001). Emissions of ammonia from light duty vehicles. In *10th International Emission Inventory Conference*. Denver: United States Environmental Protection Agency.
Fox, M., Koehler, K., & Johnson, N. (2020). Established and emerging effects of traffic-related air pollution. In H. Khreis, et al. (Eds.), *Traffic-related air pollution*: Elseveir.
Fuller, C. H., & Brugge, D. (2020). In H. Khreis, et al. (Eds.), *Environmental justice in traffic-related air pollution*: Elseveir.
Glazener, A., & Khreis, H. (2019). Transforming our cities: Best practices towards clean air and active transportation. *Current Environmental Health Reports, 6*(1), 22–37.
Glazener, A., & Khreis, H. (2020). Best practices for air quality and active transportation. In H. Khreis, et al. (Eds.), *Traffic-related air pollution*: Elseveir.
Gouldson, A., et al. (2015). Accelerating low-carbon development in the world's cities. In *Contributing paper for seizing the global opportunity: Partnerships for better growth and a better climate*. London and Washington, DC: New Climate Economy. Available at: http://newclimateeconomy.report/misc/working-papers.
Holt, A. C., et al. (2009). Object-based detection and classification of vehicles from high-resolution aerial photography. *Photogrammetric Engineering & Remote Sensing, 75*(7), 871–880.
Joffe, M., & Mindell, J. (2005). Health impact assessment. *Occupational and Environmental Medicine, 62*(12), 907–912.
Khreis, H. (2020). Traffic, air pollution, and health. In *Advances in transportation and health*. (pp. 59–104): Elsevier.
Khreis, H., Glazener, A., Ramani, T., Zietsman, J., Nieuwenhuijsen, M. J., Mindell, J. S., et al. (2019). *Transportation and Health: A Conceptual Model and Literature Review*. College Station, Texas: Center for Advancing Research in Transportation Emissions, Energy, and Health. Available at: http://www.carteeh.org/wp-content/uploads/2019/04/14-Pathways-Project-Brief_Final-version_24April2019.pdf.
Khreis, H., May, A., & Nieuwenhuijsen, M. (2017). Health impacts of urban transport policy measures: A guidance note for practice. *Journal of Transport & Health, 6*, 209–227.

Khreis, H., et al. (2016). The health impacts of traffic-related exposures in urban areas: Understanding real effects, underlying driving forces and co-producing future directions. *Journal of Transport & Health*, *3*, 249–267.
Khreis, H., et al. (2017). Exposure to traffic-related air pollution and risk of development of childhood asthma: A systematic review and meta-analysis. *Environment International*, *100*, 1–31.
Khreis, H., et al. (2019). Traffic-related air pollution and the local chronic burden of childhood asthma in Bradford, UK. *International Journal of Transportation Science and Technology*, *8*(2), 116–128.
Khreis, H., et al. (2020). In H. Khreis, et al. (Eds.), *Traffic-related air pollution: Emissions, human exposures, and health—An introduction in traffic-related air pollution*: Elseveir.
Laden, F., et al. (2006). Reduction in fine particulate air pollution and mortality: Extended follow-up of the Harvard six cities study. *American Journal of Respiratory and Critical Care Medicine*, *173*(6), 667–672.
Lam, J., Vesterinen, H., & Woodruff, T. (2020). Systematic reviews and meta-analyses of air pollution epidemiological studies. In H. Khreis, et al. (Eds.), *Traffic-related air pollution*: Elseveir.
Larsen, S.Ø., Koren, H., & Solberg, R. (2009). Traffic monitoring using very high resolution satellite imagery. *Photogrammetric Engineering & Remote Sensing*, *75*(7), 859–869.
Li, W.-W. (2020). In H. Khreis, et al. (Eds.), *Air pollution, air quality, vehicle emissions, and environmental regulations, in traffic-related air pollution*: Elseveir.
Loxham, M., Davies, D. E., & Holgate, S. T. (2019). *The health effects of fine particulate air pollution*. British Medical Journal Publishing Group.
Luben, T. J., et al. (2017). A systematic review of cardiovascular emergency department visits, hospital admissions and mortality associated with ambient black carbon. *Environment International*, *107*, 154–162.
MacIntyre, E. A., et al. (2014). Air pollution and respiratory infections during early childhood: An analysis of 10 European birth cohorts within the ESCAPE project. *Environmental Health Perspectives*, *122*(1), 107.
Marmot, M. (2005). Social determinants of health inequalities. *The Lancet*, *365*(9464), 1099–1104.
May, A. D. (2020). Policy option generation and selection. In H. Khreis, et al. (Eds.), *Traffic-related air pollution*: Elseveir.
May, A. D., Khreis, H., & Mullen, C. (2018). Option generation for policy measures and packages: An assessment of the KonSULT knowledgebase. *Case Studies on Transport Policy*, *6*(3), 311–318.
McCord, M., et al. (1855). Estimating annual average daily traffic from satellite imagery and air photos: Empirical results. *Journal of the Transportation Research Board*, *2003*, 136–142.
Medeiros, J., & Khreis, H. (2020). How emerging technology and its integrations is advancing our understanding of urban and traffic related air pollution. In H. Khreis, et al. (Eds.), *Traffic-related air pollution*: Elseveir.
Miller, M. R., & Raftis, J. B. (2020). Evidence from toxicological and mechanistic studies. In H. Khreis, et al. (Eds.), *Traffic-related air pollution*: Elseveir.
Mindell, J. S., & Birley, M. H. (2020). Qualitative health impact assessment. In H. Khreis, et al. (Eds.), *Traffic-related air pollution*. Elseveir.
Mueller, N. (2017). *Health impact assessment of urban and transport planning policies*. Barcelona, Spain: Department of Experimental and Health Sciences, Universitat Pompeu Fabra.
Mueller, N., Nieuwenhuijsen, M. J., & Rojas-Rueda, D. (2020). Quantitative health impact and burden of disease assessment of traffic-related air pollution. In H. Khreis, et al. (Eds.), *Traffic-related air pollution*. Elseveir.
Nagendra, S. S., et al. (2020). Human exposure and commercially available market solutions: Perspectives from the developing nation context. In H. Khreis, et al. (Eds.), *Traffic-related air pollution*: Elseveir.
Nieuwenhuijsen, M. J., et al. (2017). Fifty shades of green: Pathway to healthy urban living. *Epidemiology*, *28*, 63–71.

Nishimura, K. K., et al. (2013). Early-life air pollution and asthma risk in minority children. The GALA II and SAGE II studies. *American Journal of Respiratory and Critical Care Medicine, 188*(3), 309–318.

Onat, B., & Stakeeva, B. (2013). Personal exposure of commuters in public transport to PM2.5 and fine particle counts. *Atmospheric Pollution Research, 4*, 329–335.

Pedersen, M., et al. (2013). Ambient air pollution and low birthweight: A European cohort study (ESCAPE). *The Lancet Respiratory Medicine, 1*(9), 695–704.

Pellecuer, L., Tate, J., & Chapman, S. (2016). *How do traffic flow and the emissions they produce vary through the day, week, season and year: Evidence from big telematics data*. In *21st international transport and air pollution conference, Lyon, France*.

Perrino, C., Catrambone, M., & Di Menno Di Bucchianico, A. (2003). Gaseous ammonia from traffic emissions in the urban area of Rome. *Advances in air pollution series*. (pp. 601–609).

Perrino, C., et al. (2002). Gaseous ammonia in the urban area of Rome, Italy and its relationship with traffic emissions. *Atmospheric Environment, 36*(34), 5385–5394.

Ritchie, H., & Roser, M. (2020). *Our world in data: Outdoor air pollution*. [cited 8th May 2020]; Available from: https://ourworldindata.org/outdoor-air-pollution#citation.

Rodgers, M. O., & Rodgers, A. W. (2020). Impacts of traffic-related air pollution on policy and decision making. In H. Khreis, et al. (Eds.), *Traffic-related air pollution*: Elseveir.

Ropkins, K., Ibarra-Espinosa, S., & Bernard, Y. (2020). Vehicle emissions measurement and modeling. In H. Khreis, et al. (Eds.), *Traffic-related air pollution*: Elseveir.

Sanchez, K. A., et al. (2020). In H. Khreis, et al. (Eds.), *The state of the literature on traffic-related emissions, air pollution, human exposures, and health, in traffic-related air pollution*: Elseveir.

Scoggins, A., et al. (2004). Spatial analysis of annual air pollution exposure and mortality. *Science of the Total Environment, 321*(1), 71–85.

Skjøth, C. A., & Hertel, O. (2013). *Ammonia emissions in Europe*. In *Urban air quality in Europe*. (pp. 141–163): Springer.

Sudmant, A., et al. (2020). The social, environmental, health and economic impacts of low carbon transport policy: A review of the evidence. In H. Khreis, et al. (Eds.), *Traffic-related air pollution*: Elseveir.

Tanvir, S., Hao, P., & Boriboonsomsin, K. (2020). Emerging transportation technologies and implications for traffic-related emissions, air pollution exposure, and health. In H. Khreis, et al. (Eds.), *Traffic-related air pollution*: Elseveir.

Thurston, G. D., et al. (2020). Outdoor air pollution and new-onset airway disease. An official American Thoracic Society workshop report. *Annals of the American Thoracic Society, 17*(4), 387–398.

Timmers, V. R., & Achten, P. A. (2016). Non-exhaust PM emissions from electric vehicles. *Atmospheric Environment, 134*, 10–17.

Tomlin, A., Sutton, A., & Tate, J. (2010). The effect of commuter route choice on particulate exposure of an urban cyclist. In *The 2010 annual UK review meeting on outdoor and indoor air pollution research meeting*: Department of Health: Cranfield University.

Toth, C., Grejner-Brzezinska, D., & Merry, C. (2003). Supporting traffic flow management with high-definition imagery. In *Proceedings of the Joint ISPRS Workshop on High Resolution Mapping from Space*.

U.S. Environmental Protection Agency. (2019). *E.P.A. our nation's air*. [cited 8th May 2020]; Available from: https://gispub.epa.gov/air/trendsreport/2019/#home.

United Nations. (2018). *2018 Revision of World Urbanization Prospects*. Available from: https://www.un.org/development/desa/publications/2018-revision-of-world-urbanization-prospects.html.

Wang, G., et al. (2016). Vehicles detection using GF-2 imagery based on watershed image segmentation. In *Geoscience and remote sensing symposium (IGARSS), 2016 IEEE international*: IEEE.

Wei, Y., et al. (2019). Short term exposure to fine particulate matter and hospital admission risks and costs in the medicare population: Time stratified, case crossover study. *BMJ, 367*.

World Health Organization, W.H.O. (2013). *Health effects of particulate matter. Policy implications for countries in Eastern Europe, Caucasus and Central Asia.* World Health Organization.
World Health Organization, W.H.O. (2015). *Quantitative assessment of environmental health impacts at population level.* [cited 2017 12 January]; Available from: http://www.who.int/heli/tools/quantassess/en/.
Xu, Y. A. (2020). Traffic monitoring and modeling for energy, air quality, and health. In H. Khreis, et al. (Eds.), *Traffic-related air pollution*: Elseveir.

Index

Note: Page numbers followed by *f* indicate figures, *t* indicate tables, and *b* indicate boxes.

Q

R

S

U

V

FSC
www.fsc.org
MIX
Papier | Fördert gute Waldnutzung
FSC® C083411

Kolibri 360 GmbH
Ferdinand-Jühlke-Straße 7
99095 Erfurt, Deutschland
produktsicherheit@kolibri360.de